NAPL Removal:
Surfactants, Foams, and Microemulsions

Edited by

Stephanie Fiorenza
Rice University, Houston, TX

Clarence A. Miller
Rice University, Houston, TX

Carroll L. Oubre
Rice University, Houston, TX

C. Herb Ward
Rice University, Houston, TX

Authors

Michael D. Annable

George J. Hirasaki

Richard E. Jackson

James W. Jawitz

Minquan Jin

James B. Lawson

Jon Londergan

Hans Meinardus

Clarence A. Miller

Gary A. Pope

P. Suresh C. Rao

R. Dean Rhue

Thomas J. Simpkin

Robert Szafranski

Dicksen Tanzil

CRC Press
Taylor & Francis Group
Boca Raton London New York

CRC Press is an imprint of the
Taylor & Francis Group, an **informa** business

Library of Congress Cataloging-in-Publication Data

NAPL removal: surfactants, foams, and microemulsions / edited by Stephanie Fiorenza
... [et al.].
    p. cm. -- (AATDF monographs)
Includes bibliographical references and index.
ISBN 1-56670-467-7
1. Dense nonaqueous phase liquids--Environmental aspects. 2. Hazardous waste site
remediation. 3. Soil remediation 4. Groundwater--Purification 5. Surface active agents.
6. Emulsions. I. Fiorenza, Stephanie. II.Series.

TD1066.D45 N37 2000
628.5'2—dc21
                                         99-086278
                                              CIP

© 2000 by CRC Press LLC
Lewis Publishers is an imprint of CRC Press LLC

No claim to original U.S. Government works
International Standard Book Number 1-56670-467-7
Library of Congress Card Number 99-086278

Foreword

The field of *in situ* surfactant and cosolvent flushing for remediation of nonaqueous phase liquids (NAPL) has grown rapidly since the Advanced Applied Technology Demonstration Facility (AATDF) Program decided to investigate these technologies. The AATDF was interested in surfactant/cosolvent flushing because it constitutes an *in situ* technique with the potential to reduce the mass of NAPL that can serve as a continuing source of groundwater contamination. Although the flushing approach will not remove all traces of contamination, it might be possible to remove enough mass so that less intensive remediation technologies, such as natural attenuation, can be used in a treatment train approach to meet long-term cleanup goals.

The surfactant/foam and single-phase microemulsion (SPME) projects were selected by the AATDF because they had the potential to develop the field of cosolvent/surfactant flushing by addressing needed breakthroughs. The field tests that had been conducted at the time that the AATDF funded this work were primarily solubilization of chlorinated solvent dense nonaqueous phase liquids (DNAPLs), although one field demonstration of DNAPL mobilization/solubilization with polymer as a mobility control agent had also occurred. One of the major technical problems confronting the use of a flushing technology is that of inefficient sweep of heterogeneous, contaminated media. The surfactant/foam process first uses surfactant to flush the more permeable zones (the usual approach in a surfactant flood). The difference is the subsequent pulsing of air, which also enters the permeable flow paths, creating an *in situ* foam. The foam then blocks the more permeable zones of an aquifer with foam, allowing surfactant to perfuse the less permeable zones. The surfactant/foam process therefore has the potential for efficient sweep and removal of NAPL mass. Another problem facing the surfactant/cosolvent flushing technology is that of cost: the cost of the chemical and the cost of treating the effluent. The single-phase microemulsion method is an efficient solubilization-only technique for removing NAPLs in situations where mobilization is undesirable. By using fewer pore volumes of surfactant or cosolvent, costs can be reduced. Ultimately, the chemical amendment may need to be separated and reused to reduce cost. As part of the AATDF overall strategy for the development of surfactant/cosolvent flushing technology, two other projects — demonstration of surfactant recovery and reuse, and the *Technologies Practices Manual* — were funded. Their results are included as separate books in this monograph series.

Researchers from Rice University, the University of Texas at Austin, INTERA, Inc. (now Duke Engineering and Services), and the University of Florida conducted the surfactant/foam and single-phase microemulsion tests, respectively, at Hill Air Force Base in Utah, although they were at different sites on the base. The surfactant/foam process was tested at the DNAPL-contaminated Operable Unit 2, and SPME was tested at the LNAPL-contaminated Operable Unit 1. The single-phase microemulsion project is a good example of leveraging funding — the AATDF; the Strategic Environmental Research & Development Program (SERDP), which is a collaborative effort involving the U.S. EPA, U.S. DOE, and U.S. DOD; and the Environmental Restoration & Management Division of Hill Air Force Base all supported this project.

This monograph records the results of the field demonstrations, which were the first field applications of both technologies. The reader will find it to be complete and quantitative. The monograph also includes a section on the potential costs of a full-scale implementation of the surfactant/foam and single-phase microemulsion demonstrations at their respective sites. This design and evaluation was conducted by CH2M Hill.

AATDF Monographs

This monograph is one of a 10-volume series that records the results of the AATDF Program:

- Surfactants and Cosolvents for NAPL Remediation: A Technology Practices Manual
- Sequenced Reactive Barriers for Groundwater Remediation
- Modular Remediation Testing System
- Phytoremediation of Hydrocarbon-Contaminated Soil
- Steam and Electroheating Remediation of Tight Soils
- Soil Vapor Extraction Using Radio Frequency Heating: Resource Manual and Technology Demonstration
- Subsurface Contamination Monitoring Using Laser Fluorescence
- Remediation of Firing-Range Impact Berms
- Reuse of Surfactants and Cosolvents for NAPL Remediation
- NAPL Removal: Surfactants, Foams, and Microemulsions

Advanced Applied Technology Demonstration Facility (AATDF)
Energy and Environmental Systems Institute MS-316
Rice University
6100 Main Street
Houston, TX 77005-1892

U.S. Army Corps of Engineers
Waterways Experiment Station
3909 Halls Ferry Road
Vicksburg, MS 39180-6199

Rice University
6100 Main Street
Houston, TX 77005-1892

Acknowledgments

The contributions and cooperation of the following organizations and individuals enabled the surfactant/foam and single-phase microemulsions field demonstrations to be conducted successfully and in a timely manner: (1) Hill Air Force Base, especially Steve Hicken, Dr. Jon Ginn, Dr. Dan Stone, and Robert Elliott of the Environmental Management Directorate, which agreed to host these field tests (a very important factor) and provided utilities and countless other measures of general support; (2) Rob Stites of EPA Region VIII and Mo Slam and Duane Mortenson at the Utah Department of Environmental Quality, all of whom acted most expeditiously in their review of work plans and on other requests.

The authors worked closely with the AATDF staff and project advisor throughout both of the field demonstrations. The advisor who reviewed and commented extensively on these projects was Dr. William Rixey, University of Houston.

The authors of Section I, the Surfactant/Foam Process for Aquifer Remediation, would also like to acknowledge the contributions of the following individuals to this document: Jacqueline Avvakoumides, Varadarajan Dwarakanath, Laureen Kennedy, Paul Mariner, INTERA, Inc.; Dr. Tim Oolman, Radian International. Dr. William Wade of The University of Texas at Austin developed the surfactant used in the field demonstration of surfactant/foam; without his initial contribution, the demonstration would not have been possible. Dr. Mojdeh Delshad provided tremendous technical support in the utilization and modification of UTCHEM for the foam process. Dr. Bruce Rouse and Konstantinos Kostarelos designed and tested the large-scale aquifer model at UT. Curt Himle, Steve Godard, and Grant Cooper of Radian International, LLC, operated the source recovery system that treated the effluent from the foam field test. The following people are also acknowledged for their technical contributions or other involvement in the project: Anthony Holder, Naoko Akiya, Stephanie King, and Matt Tenny, Rice University; Sonny Casaus, Curtis Chester, Mike Fort, Grace Hunter, Lisa Rottinghaus, Jeff Scheerhorn, Anthony West, and Carl Young, INTERA, Inc.; Esther Barrientes, Sharon A. Bernard, Joanna Castillo, Jeff Edgar, Egidio Leito, Meng Lim, Taimur Mailk, Rehan Queshi, Kanaka Prasad Saripalli, Greg Wachel, Jonathan Witham, and Ding Zhu, the University of Texas at Austin; and Gregg Hadlock, Craig Holloway, Richard Lewis, Trevor Lindley, and Cannon Silver, Radian International.

The authors of Section II, Single-Phase Microemulsions, would like to acknowledge Meifong Zhou and William Reve for their assistance in collecting the laboratory data. Professor William Wade, University of Texas at Austin, provided technical advice and consultations during many visits to Gainesville throughout this project. The authors appreciate the assistance of the following individuals who helped conduct the field experiments, collect samples, and perform laboratory analyses: Randall Sillan, Clayton Clark II, A. Dale Helms, Andrew James, Dongping Dai, William Reve, Gloria Sillan, Heonki Kim, Kirk Hatfield, K. Prasad Saripalli, Stewart Finkel, Balaji Ramachandran, Meifong Zhou, and Yan Zhang, all from the University of Florida; Mark Bastasch, Rice University; and A. Lynn Wood, Susan Mravik, John Hoggit, Brian Newell, Dan Coulson, Bob Lien, Ray Adcock, and Troy Lackey, U.S. EPA National Risk Management Research Lab at Ada, Oklahoma.

The Technology Advisory Board and other AATDF advisory committees provided valuable assistance at the start and during the project in such areas as development of the surfactant strategy, directional changes in the projects, and technology transfer.

The AATDF Program was funded by the United States Department of Defense under Grant No. DACA39-93-1-0001 to Rice University. The Program was given oversight by the U.S. Army Engineer Waterways Experiment Station in Vicksburg, Mississippi. Although AATDF, SERDP, and Air Force funds were used to support the single-phase microemulsion study, this document has not been subjected to peer review within these agencies, and the conclusions stated herein do not necessarily reflect the official views of these agencies, nor does this document constitute an official endorsement by these agencies.

Preface

Following a national competition, the Department of Defense (DOD) awarded a $19.3 million grant to a university consortium of environmental research centers led by Rice University and directed by Dr. C. Herb Ward, Foyt Family Chair of Engineering. The DOD Advanced Applied Technology Demonstration Facility (AATDF) Program for Environmental Remediation Technologies was established on May 1, 1993 to enhance the development of innovative remediation technologies for the DOD by facilitating the process from academic research to full-scale utilization. The AATDF focus is to select, test, and document performance of innovative environmental technologies for the remediation of DOD sites.

Participating universities include Stanford University, the University of Texas at Austin, Rice University, Lamar University, University of Waterloo, and Louisiana State University. The directors of the environmental research centers at these universities serve as the Technology Advisory Board (TAB). The U.S. Army Engineer Waterways Experiment Station manages the AATDF Grant for the DOD. Dr. John Keeley is the Technical Grant Officer. The DOD/AATDF is supported by five leading consulting engineering firms: Remediation Technologies, Inc., Battelle Memorial Institute, GeoTrans, Inc., Arcadis Geraghty and Miller, Inc., and Groundwater Services, Inc., along with advisory groups from the DOD, industry, and commercialization interests.

Starting with 170 preproposals that were submitted in response to a broadly disseminated announcement, 12 projects were chosen by a peer-review process for field demonstrations. The technologies chosen were targeted at the DOD's most serious problems of soil and groundwater contamination. The primary objective was to provide more cost-effective solutions, preferably using *in situ* treatment. Eight projects were led by university researchers, two projects were managed by government agencies, and two others were conducted by engineering companies. Engineering partners were paired with the academic teams to provide field demonstration experience. Technology experts helped guide each project.

DOD sites were evaluated for their potential to support quantitative technology demonstrations. More than 75 sites were evaluated in order to match test sites to technologies. Following the development of detailed work plans, carefully monitored field tests were conducted and the performance and economics of each technology were evaluated.

One AATDF project designed and developed two portable Experimental Controlled Release Systems (ECRS) for testing and field simulations of emerging remediation concepts and technologies. The ECRS is modular and portable and allows researchers, at their sites, to safely simulate contaminant releases and study remediation techniques without contaminant loss to the environment. The completely contained system allows for accurate material and energy balances.

The results of the DOD/AATDF Program provide the DOD and others with detailed performance and cost data for a number of emerging, field-tested technologies. The program also provides information on the niches and limitations of the technologies to allow for more informed selection of remedial solutions for environmental cleanup.

The AATDF Program can be contacted at: Energy and Environmental Systems Institute, MS-316, Rice University, 6100 Main, Houston, TX, 77005, phone 713-527-4700; fax 713-285-5948; or e-mail <eesi@rice.edu>.

The DOD/AATDF Program staff include:

Director:
  Dr. C. Herb Ward
Program Manager:
  Dr. Carroll L. Oubre
Assistant Program Manager:
  Dr. Kathy Balshaw-Biddle
Assistant Program Manager:
  Dr. Stephanie Fiorenza

Assistant Program Manager:
  Dr. Donald F. Lowe
Financial/Data Manager:
  Mr. Robert M. Dawson
Publications Coordinator/Graphic Designer:
  Ms. Mary Cormier
Meeting Coordinator:
  Ms. Susie Spicer

This volume, *NAPL Removal: Surfactants, Foams, and Microemulsions,* is one of a 10-monograph series that records the results of the DOD/AATDF environmental technology demonstrations. Many have contributed to the success of the AATDF program and to the knowledge gained. We trust that our efforts to fully disclose and record our findings will contribute valuable lessons learned and help further innovative technology development for environmental cleanup.

Stephanie Fiorenza
Carroll L. Oubre
C. Herb Ward

About the Editors

Stephanie Fiorenza

Stephanie Fiorenza is an assistant program manager with AATDF at Rice University, where she manages five projects involving the field demonstration of innovative remediation technologies. Dr. Fiorenza has a Ph.D. in environmental science and engineering from Rice University and a B.A. in environmental studies from Brown University. In her role as an assistant program manager for AATDF, Dr. Fiorenza provides the managerial guidance and technical expertise required for the successful field demonstration of each technology. She has also been an active participant in the preparation of all the projects' reports. Prior to joining AATDF, Dr. Fiorenza worked as an environmental specialist for Amoco Corporation in their Groundwater Management section. Her areas of interest are the biodegradation of organic chemicals, microbial ecology of the subsurface, and development of remediation technologies.

Carroll L. Oubre

Carroll L. Oubre is the program manager for the DOD/AATDF Program. As such, he is responsible for the day-to-day management of the $19.3 million DOD/AATDF Program. This includes guidance of the AATDF staff, overview of the 12 demonstration projects, assuring project milestones are met within budget, and that complete reporting of the results is timely.

Dr. Oubre has a B.S. in chemical engineering from the University of Southwestern Louisiana, an M.S. in chemical engineering from Ohio State University, and a Ph.D. in chemical engineering from Rice University. He worked for Shell Oil Company for 28 years, with his last job as manager of environmental research and development for Royal Dutch Shell in England. Prior to that, he was Director of Environmental Research and Development at Shell Development Company in Houston, Texas.

C. H. (Herb) Ward

C. H. Ward is the Foyt Family Chair of Engineering in the George R. Brown School of Engineering at Rice University. He is also Professor of Environmental Science and Engineering and Ecology and Evolutionary Biology.

Dr. Ward has undergraduate (B.S.) and graduate (M.S. and Ph.D.) degrees from New Mexico State University and Cornell University, respectively. He also earned the M.P.H. in environmental health from the University of Texas.

Following 22 years as chair of the Department of Environmental Science and Engineering at Rice University, Dr. Ward is now Director of the Energy and Environmental Systems Institute (EESI), a university-wide program designed to mobilize industry, government, and academia to focus on problems related to energy production and environmental protection.

Dr. Ward is also Director of the Department of Defense Advanced Applied Technology Demonstration Facility (AATDF) Program, a distinguished consortium of university-based environmental research centers supported by consulting environmental engineering firms to guide selection, development, demonstration, and commercialization of advanced applied environmental restoration technologies for the DOD. For the past 18 years, he has directed the activities of the National Center for Ground Water Research (NCGWR), a consortium of universities charged with conducting long-range exploratory research to help anticipate and solve the nation's emerging groundwater problems. He is also co-director of the EPA-sponsored Hazardous Substances Research Center/South and Southwest (HSRC/S and SW), which focuses its research on contaminated sediments and dredged materials.

Dr. Ward has served as president of both the American Institute of Biological Sciences and the Society for Industrial Microbiology. He is the founding and current editor-in-chief of the international journal *Environmental Toxicology and Chemistry*.

About the Authors

Part I

George J. Hirasaki

Professor Hirasaki received his B.S. from Lamar University and his Ph.D. in chemical engineering from Rice University. He recently joined the Rice University faculty after a 26-year career with Shell Development Company. His research in fluid transport through porous media ranges from the microscopic scale, studying intermolecular forces governing wettability, to the megascopic scale, using numerical reservoir simulators for fieldwide modeling. A recurring theme throughout this research is the dominance of interfaces in the determination of fluid transport processes. In addition to applying the foam process for aquifer remediation, he has ongoing projects investigating the phenomena governing the wetting and spreading of petroleum systems and the underlying mechanisms of NMR well-logging measurements to improve methods of interpretation. Professor Hirasaki is a member of the National Academy of Engineering.

Richard E. Jackson

Richard Jackson manages the Geosystems & Geochemistry section of Duke E&S Austin, which has pioneered the use of partitioning interwell tracer tests across the United States. He has degrees in hydrology, civil engineering, and hydrogeology (Ph.D., Waterloo, 1979). He was formerly chief of the Groundwater Contamination Project at the National Water Research Institute of Canada.

Minquan Jin

Minquan Jin, DE&S Austin, TX, has M.S. and Ph.D. degrees in petroleum engineering from the University of Texas at Austin. He has been responsible for PITT and surfactant-flood design at DE&S since 1996.

James B. Lawson

James Lawson received a Ph.D. in physical chemistry from Rice University and conducted research at Shell Oil Company for 31 years, primarily in the areas of tertiary recovery and production. Dr. Lawson joined the Chemical Engineering Department at Rice and developed the laboratory operations for the DOD/AATDF surfactant/foam project. He has since joined Baroid Drilling Fluids and serves as consultant on operations.

John Londergan

John Londergan, DE&S Austin, TX, is senior hydrogeologist with DE&S. He has a B.S. in geology from the University of Massachusetts–Amherst and an M.S. in geophysics from Texas A&M University. He has been responsible for the management of partitioning tracer tests and surfactant floods at Hill AFB in Utah; Camp Lejeune, North Carolina; and the Savage Well Superfund site in New Hampshire.

Hans Meinardus

Hans Meinardus, DE&S Austin, TX, is senior hydrogeologist with DE&S. He has a B.S. in geology and an M.S. in hydrogeology from Texas A&M University. He has been responsible for the management of four DE&S projects at Hill AFB in Utah and has completed nine partitioning interwell tracer tests and two surfactant floods.

Clarence A. Miller

Clarence Miller, Associate Director of the Energy and Enviromental Systems Institute at Rice University, is Louis Calder Professor of Chemical Engineering at Rice and a former chairman of the department. He received B.A. and B.S. degrees from Rice University and a Ph.D. from the University of Minnesota, all in chemical engineering. After serving with the Division of Naval Reactors in Washington, D.C., Professor Miller was a member of the chemical engineering faculty at Carnegie-Mellon University. His research centers on interfacial phenomena, including equilibrium and dynamic phenomena in foams, emulsions, and microemulsions. Previously, his research involved enhanced oil recovery processes; currently, his research includes investigation of processes employing surfactants for removal of liquid contaminants from groundwater aquifers and of the mechanism of detergency. Professor Miller is a Fellow of the American Institute of Chemical Engineers and has been a Visiting Scholar at Cambridge University, Delft University of Technology, and the University of Bayreuth (Germany).

Gary A. Pope

Gary Pope is a Professor of Petroleum Engineering at the University of Texas at Austin. He has 25 years of experience with the experimental investigation of surfactants, cosolvents, and water-soluble polymers with applications to both oil recovery and cleanup of organics from aquifers. He and his colleagues and students developed a three-dimensional multicomponent, multiphase number simulator (UTCHEM) that is widely used to solve problems of flow in permeable media. Dr. Pope was the first to propose the use of partitioning tracers to locate and estimate the amount and saturation distribution of NAPL in aquifers. Recently, he has been leading an interdisciplinary research team in the development and design of characterization and remediation methods at six DNAPL sites across the United States. Professor Pope is a member of the National Academy of Engineering.

Robert Szafranski

Robert Szafranski received a B.A. degree in mathematics and a Ph.D. degree in chemical engineering from Rice University. His Ph.D. thesis dealt with laboratory studies and field testing of the surfactant/foam process. He is currently employed by Exxon Production Research Company.

Dicksen Tanzil

Dicksen Tanzil has a B.S. degree in chemical engineering from Purdue University. He is currently a Ph.D. student in chemical engineering at Rice University. His research emphasizes numerical simulation of the surfactant/foam process.

Part II

Michael D. Annable

Dr. Annable has been a faculty member in the department of Environmental Engineering Sciences at the University of Florida since 1992. He received his Ph.D. from Michigan State University, working on soil vapor extraction of multicomponent nonaqueous phase liquids. His current interests are in physical-chemical processes related to field-scale application of innovative technologies for subsurface remediation. He is currently involved in a number of interdisciplinary research and education efforts in hydrologic sciences at the University of Florida.

James W. Jawitz

James W. Jawitz is a graduate research assistant in the Department of Environmental Engineering Sciences at the University of Florida. He holds B.S. and M.E. degrees in Environmental Engineering from the University of Florida. His research to date has focused on the characterization and remediation of soil and groundwater contaminants.

P.S.C. Rao

Dr. Suresh Rao has been on the faculty at the University of Florida since 1975, where he is now a Graduate Research Professor of Soil and Water Science, and also serves as the Director of the Center for Natural Resources. He earned his degrees from A.P. Agricultural University (India), Colorado State University, and the University of Hawaii. His research interests include theoretical and experimental aspects of the coupling between hydrologic and biogeochemical processes that influence the behavior of agricultural chemicals and hazardous waste constituents in soils, sediments, and aquifers. His current research focuses on the development and testing of innovative technologies for remediation of contaminated soils and groundwater. Dr. Rao was elected (1989) Fellow of Soil Science Society of America (SSSA) and the American Society of Agronomy (ASA). He received the 1991 Environmental Quality Research Award (American Society of Agronomy) and the 1998 Soil Science Research Award (Soil Science Society of America). He has served a 3-year term as an associate editor for the *Journal of Environmental Quality* (ASA), and was elected the Chair of ASA Division A-5 (Environmental Quality). He was one of the four founding Editors-in-Chief for the *Journal of Contaminant Hydrology* (Elsevier), and was an associate editor for *Environmental Toxicology & Chemistry* (Society of Environmental Toxicology & Chemistry) and also for *Water Resources Research* (American Geophysical Union). Dr. Rao served on several committees of the National Research Council (NRC), including a 3-year term as a member of the Water Science and Technology Board. He recently chaired the NRC Committee on Commercialization of Innovative Remediation Technologies.

R. Dean Rhue

Dr. Rhue is a Professor of Soil and Water Science at the University of Florida, where he has been employed for the last 21 years. His research has focused on the colloidal and physical chemistry of soils, with emphasis on environmental problems associated with organic and inorganic pollutants in soil and innovative methods of remediating sites containing certain pollutants. Most recently, his research has concentrated on three areas: (1) evaluating a microemulsion technology for soil remediation, (2) relating residual levels of arsenic contamination at cattle-dipping vats to soil taxonomic and morphologic properties, and (3) developing a phosphorus retention index for animal-waste disposal on sandy soil. He teaches soil chemistry and methods of soil chemical analysis, and his courses have emphasized the application of basic principles of chemistry and physics to the solution of problems associated with pollutant behavior and groundwater systems.

Part III

Thomas J. Simpkin

Thomas Simpkin is a senior environmental engineer with CH2M Hill, where he serves as a process and design engineer on soil and groundwater remediation projects. He has Ph.D. and M.S. degrees in sanitary engineering and a B.S. in civil engineering from the University of Wisconsin–Madison. Dr. Simpkin has worked on a variety of projects ranging from municipal and industrial wastewater treatment to remediation of hazardous wastes with a focus on biological treatment of wastewater and soils, and the application of other remediation technologies. He served as a design and operations engineer for large-scale pilot tests of several *in situ* remediation technologies at the UPPR–Laramie Tie Plants and as the project manager for full-scale remediation of the Broderick Wood Products site in Denver, Colorado.

AATDF Advisors

UNIVERSITY ENVIRONMENTAL RESEARCH CENTERS

National Center for Ground Water Research
Dr. C. H. Ward
Rice University, Houston, TX

Hazardous Substances Research Centers
South and Southwest
Dr. Danny Reible and Dr. Louis Thibodeaux
Louisiana State University, Baton Rouge, LA

Waterloo Centre for Groundwater Research
Dr. John Cherry and Mr. David Smyth
University of Waterloo, Ontario, Canada

Western Region Hazardous Substances Research Center
Dr. Perry McCarty
Stanford University, Stanford, CA

Gulf Coast Hazardous Substances Research Center
Dr. Jack Hopper and Dr. Alan Ford
Lamar University, Beaumont, TX

Environmental Solutions Program
Dr. Raymond C. Loehr
University of Texas, Austin, TX

DOD/ADVISORY COMMITTEE

Dr. John Keeley, Co-Chair
Assistant Director, Environmental Laboratory
U.S. Army Corps of Engineers
Waterways Experiment Station, Vicksburg, MS

Mr. James I. Arnold, Co-Chair
Acting Division Chief, Technical Support
U.S. Army Environmental Center
Aberdeen, MD

Dr. John M. Cullinane
Program Manager, Installation Restoration
Waterways Experiment Station
U.S. Army Corps of Engineers, Vicksburg, MS

Mr. Scott Markert and Dr. Shun Ling
Naval Facilities Engineering Center
Alexandria, VA

Dr. Jimmy Cornette, Dr. Michael Katona and Major Mark Smith
Environics Directorate
Armstrong Laboratory
Tyndall AFB, FL

COMMERCIALIZATION AND TECHNOLOGY TRANSFER ADVISORY COMMITTEE

Mr. Benjamin Bailar, Chair
Dean, Jones Graduate School of Administration
Rice University, Houston, TX

Dr. James H. Johnson, Jr., Associate Chair
Dean of Engineering
Howard University, Washington, D.C.

Dr. Corale L. Brierley
Consultant
VistaTech Partnership, Ltd.
Salt Lake City, UT

Dr. Walter Kovalick
Director, Technology Innovation Office
Office of Solid Wastes & Emergency Response
EPA, Washington, D.C.

Mr. M. R. (Dick) Scalf
EPA Robert S. Kerr
Environmental Research Laboratory (retired)
Ada, OK

Mr. Terry A. Young
Executive Director
Technology Licensing Office
Texas A&M University, College Station, TX

Mr. Stephen J. Banks
President
BCM Technologies, Inc., Houston, TX

List of Acronyms and Abbreviations

AATDF	Advanced Applied Technology Demonstration Facility
AFB	Air Force Base
AFCEE	Air Force Center for Environmental Excellence
AMSL	Above mean sea level
ANOVA	Analysis of variance
ANSI	American National Standards Institute
ASTM	American Society for Testing and Materials
bgs	Below ground surface
BOD	Biological oxygen demand
Br$^-$	Bromide ion
BTC	Breakthrough curve
C	Centigrade
Ca^{+2}	Calcium ion
CaCl$_2$	Calcium dichloride
CC	Continuing Calibration Check Sample
CERCLA	Comprehensive Environmental Response, Compensation, and Liability Act
cfm	Cubic feet per minute
CFR	Code of Federal Regulations
cfs	Cubic feet per second
Cl$^-$	Chloride ion
CLP	Controlled laboratory procedures
cm/s	Centimeter per second
cm^3	Cubic centimeter
CMC	Critical micelle concentration
COC	Chain of custody
COD	Chemical oxygen demand
cp	Centipoise
DAS	Data Acquisition System
d	Days
DCB	1,2-Dichlorobenzene
DCE	Dichoroethene
dL	Deciliter
DMP	2,2-Dimethyl-3-pentanol
DNAPL	Dense nonaqueous phase liquid
DOD	U.S. Department of Defense
DODEC	Dodecane
DOT	U.S. Department of Transportation
DQO	Data quality objective
EOR	Enhanced oil recovery
EPA	United States Environmental Protection Agency
eq	Equivalent
EW	Extraction well
FDA	U.S. Food and Drug Administration
FID	Flame ionization detector
FRTR	Federal Remediation Technologies Roundtable
ft	Feet
g	Gram
GAC	Granular activated carbon
gal	Gallon
GC	Gas chromatography
GC/MS	Gas chromatography/mass spectrometry
gpm	Gallons per minute
HC	Hydraulic control (well)

HDPE	High-density polyethylene
HEX	Hexanol
HLB	Hydrophilic-lipophilic balance
HPLC	High-pressure liquid chromatography
h	Hour
H&S	Health and safety
ID	Inner diameter
IFT	Interfacial Tension
in.	Inch
IPA	Isopropanol
IW	Injection well
IWTP	Industrial wastewater treatment plant
K	Hydraulic conductivity
kg	Kilogram
kL	Kiloliter
K_{ow}	Octanol/water partition coefficient
L	Liter
lb	Pound
L/m	Liter per minute
LC	Liquid chromatography
LTTD	Low-temperature thermal desorption
m	Meter
M	Molar
m^3	Cubic meter
MDL	Method detection limit
M.E.	Microemulsion
MEHEP	6-Methyl-2-heptanol
meq	Milliequivalent
METH	Methanol
mg	Milligram
mg/L	Milligram per liter
mL	Milliliter
mm	Millimeter
mN/m	MilliNewton per meter
MLS	Multilevel sampler
mol	Mole
MS/MSD	Matrix Spike/Matrix Spike Duplicate
MTU	Michigan Technological University
MW	Monitoring well
NaCl	Sodium chloride
NAPTH	Naphthalene
NAPL	Nonaqueous phase liquid
OD	Outer diameter
O&M	Operation and maintenance
OSHA	Occupational Safety and Health Administration
OU1	Operable Unit 1
OU2	Operable Unit 2
OV	Organic Vapor
P&T	Pump-and-treat
PCB	Polychlorinated biphenyl
PCE	Tetrachloroethene or perchloroethylene
pH	Negative log of hydrogen ion concentration
PI	Principal Investigator
PID	Photoionization Detector
PITT	Partitioning Interwell Tracer Test

POTW	Publicly Owned Treatment Works
ppm	Parts per million
PQL	Practical Quantitation Limit
psi(g)	Pounds per square inch (gauge)
PV	Pore Volume
PVC	Polyvinyl Chloride
P-XYL	*para*-Xylene
QAPP	Quality Assurance Project Plan
QA/QC	Quality Assurance/Quality Control
%R	Percent Recovery
R^2	Correlation coefficient
RICE	Rice University
RI/FS	Remedial Investigation/Feasibility Study
s	Standard deviation
SB	Soil boring
scfpm	Standard cubic feet per minute
SDBS	Sodium dodecylbenzene sulfonate
SERDP	Strategic Environmental Research and Development Program
SOP	Standard operating procedure
SPME	Single-phase microemulsion
SRS	Source recovery system
TCA	1,1,1-Trichloroethane
TCB	Trichlorobenzene
TCE	Trichloroethene, trichloroethylene
TMB	1,3,5-Trimethylbenzene
TOC	Top of casing
TRIDEC	Tridecane
UF	University of Florida
μg	Microgram
μL	Microliter
μm	Micrometer
UNDEC	Undecane
UT	University of Texas at Austin
UTCHEM	A 3-dimensional simulator developed at the University of Texas
VOA	Volatile organic analysis
VOC	Volatile organic compound
V_o	Volume of oil or NAPL
V_s	Volume of surfactant
WBS	Work breakdown structure
wt%	Weight percent
WWTP	Wastewater treatment plant

List of Figures

List of Tables

Contents

PART I

Field Demonstration of the Surfactant/Foam Process for Remediation of a Heterogeneous Aquifer Contaminated with DNAPL

George J. Hirasaki
Rice University

Richard E. Jackson
DE&S

Minquan Jin
DE&S

James B. Lawson
Rice University

John Londergan
DE&S

Hans Meinardus
DE&S

Clarence A. Miller
Rice University

Gary A. Pope
University of Texas

Robert Szafranski
Exxon Production Research Company

Dicksen Tanzil
Rice University

EXECUTIVE SUMMARY

Numerous military and industrial sites are contaminated with trichloroethene (TCE) and other chlorinated solvents. Pump-and-treat processes have proved unsuccessful in remediating these sites because they are unable to remove much of the liquid contaminant that acts as a long-term source by slowly dissolving in groundwater that flows past it. The dissolved plume is regenerated over a period of years or even decades.

No satisfactory method currently exists for removal of all or nearly all of the liquid chlorinated solvent from a contaminated groundwater aquifer. Surfactants offer the possibility of such removal by solubilization and/or mobilization of the liquid source, as demonstrated by laboratory and previous field work. However, a significant limitation of surfactant processes and, indeed, of all processes involving injection of fluids to effect remediation is that the injected fluids flow preferentially in zones of high hydraulic conductivity in a heterogeneous aquifer. In fact, most of the injected fluids continue to flow through these high-conductivity zones even after they have been cleaned and thus make a minimal contribution to remediation. Only a small portion of the injected fluids flow through the zones of low hydraulic conductivity that remain contaminated. As a result, the time and cost of remediation are much higher than in a homogeneous aquifer and the total quantity of surfactants or other materials introduced into the subsurface is greater.

The main objective of the surfactant/foam process is to improve process performance in a heterogeneous aquifer by providing a more uniform sweep of the formation. After an initial slug of an aqueous surfactant solution is introduced into the aquifer, entering mainly the high conductivity zones, some air is injected, which forms a "foam" in these zones and significantly increases the resistance to flow of liquid there. Hence, when additional surfactant solution is injected, a larger portion of it enters the low-conductivity zones, leading to more rapid remediation. The validity of this concept was confirmed during the present project by experiments conducted in a laboratory model containing two layers of sands with different hydraulic conductivities. For example, in one case where conductivities differed by a factor of 20, nearly all TCE was removed by the surfactant/foam process after 1 pore volume (PV) of surfactant solution had been injected. In contrast, about a third of the TCE originally present remained even after continuous injection of 26 PV of the same surfactant solution without foam.

A field demonstration, the first for the surfactant/foam process, was conducted during the spring of 1997 by Rice University and INTERA (now Duke Engineering and Services) at Operable Unit 2 (OU2) of Hill Air Force Base, Logan, Utah. Funding was provided by the Advanced Applied Technology Demonstration Facility (AATDF) of DOD. The University of Texas at Austin (UT) helped with some of the laboratory and simulation work and provided the basic surfactant formulation although it was optimized by Rice for use with foam. Hill AFB provided extensive logistical support, and Radian operated the effluent treatment facility.

3

OU2 is underlain by an alluvial sand aquifer that forms a channel confined on its sides and below by thick clay deposits constituting a capillary barrier to contaminant migration. The deepest part of the sand in the test area is about 45 feet (ft) below the surface and 18 ft below the water table. The contaminant was waste from past degreasing operations at the base and contained about 70% TCE, with smaller amounts of tetrachloroethene (PCE) and 1,1,1-trichloroethane (TCA).

The test site spanned the width of the channel and was about 20 ft long. It did not include a pool where extensive contaminant accumulated, but instead was a portion of the migration path of contaminant between the disposal trenches and two pools where tests of other remediation technologies had been carried out. Borings taken about a year before the test and samples from the wells themselves revealed that the contaminant in the pattern was confined to the bottom 3 or 4 ft of the channel and was present in relatively small quantities, viz., local saturations of 2% to 14% pore volume. A partitioning interwell tracer test (PITT) conducted before the test indicated that about 21 gal of liquid contaminant was present, which is generally consistent with estimates based on data from the borings and wells.

About 105 samples from the borings were subjected to sieve analysis to obtain information on variation of the hydraulic conductivity with lateral position and especially elevation. The contaminant was located in sands with conductivities of $1-3 \times 10^{-4}$ m/s (permeabilities of 10 to 30 darcy). However, more-permeable sands containing no contaminant and having conductivities on the order of 10^{-3} m/s (100 darcy) were seen at elevations 4 to 5 ft above the base of the sand. Since injection was planned over a 5-ft screened interval, these coarse sands, if continuous, would provide a preferential flow path or "thief zone" for the injected surfactant solution in the absence of foam.

Three injection and three extraction wells were completed in a 20-ft line-drive pattern spanning the channel (the outermost wells were about 12 ft apart). Also, two monitoring wells were located near the center of the channel at positions about one third and two thirds of the distance between injection and extraction wells. Each monitoring well was screened so that it could be sampled in three locations: the bottom foot above the clay, 4 to 6 ft above the clay (corresponding to the thief zone mentioned above), and 10 to 13 ft above the clay. Two hydraulic control wells were located outside the pattern along the channel about 10 ft from the central injector and central extractor respectively.

Following the initial PITT, a surfactant-free solution containing about 1 wt% NaCl, the optimum value for mobilizing contaminant with the surfactant used, was injected for 1 d. The total volume of this solution was approximately equal to the swept volume of the pattern as determined by the PITT. Next, injection of a solution containing about 3.5 wt% of the anionic surfactant sodium dihexyl sulfosuccinate at the same salinity commenced and was continued at the same rate for slightly over 3 d. After some 8 hours (h) of surfactant injection, air injection began with each well in turn receiving air for approximately 2 h. Air pressure was controlled to allow air to enter the upper part of the screened interval while surfactant solution continued to flow into the lower part. The total amount of surfactant solution with the above concentration was about 3.2 times the swept volume of the pattern. Afterward, a more dilute NaCl solution (0.8 wt%) was injected for about 12 h, followed by a water flood to break the foam and remove most of the surfactant and, finally, a second PITT to determine the amount of contaminant remaining.

No significant problems were encountered with air injection, and pressure rose at the injection wells, indicating an increase in the resistance to flow. Moreover, foam was observed in samples from the two upper screened intervals of both monitoring wells throughout much of the test. Even in the bottom screened intervals of the monitoring wells, which were located just above the clay aquitard, foam was seen at various times during the test.

The contaminant produced at the extraction wells during the test (beyond that which would have been present in dissolved form if no surfactant had been used) was approximately 34 gal, based on analysis of the effluent from each well as a function of time. An additional 3 gal of contaminant was recovered in a dense microemulsion phase, which was pumped to the surface from the bottom foot of one of the monitoring wells, MW-2. This well, located about 7 ft from the central extractor, was situated in a small local depression and apparently continued to receive mobilized contaminant intermittently during the process. The total production of 37 gal exceeded the initial estimate of 21 gal given above. It thus appears that contaminant from outside was able to enter the pattern during the test, probably from the region beyond the injection wells that was known to be contaminated. Data from both monitoring wells and the results of numerical simulation support this conclusion. Owing to this effect, and to changes in phase behavior resulting from ion exchange between the surfactant solution and clays, contaminant in the form of a free phase or a dense middle phase microemulsion continued to be present at the base of MW-2 for about 3 d after surfactant injection had ended. Also, contaminant continued to be produced at a relatively low concentration — about 500 parts per million (ppm) compared with a background of about 100 ppm — in the central extraction well for about 4 d beyond the period of surfactant injection. However, 21 gal, the amount believed to be in the pattern initially, had been produced by about the time that surfactant and air injection ended.

Despite the apparent influx of contaminant, both the final PITT and data from five borings taken at the end of the test showed that very little liquid contaminant remained at the end of the test. According to the PITT, only about 2.6 gal were present in the entire swept volume of about 8300 gal. Two of the final borings showed no contamination. The others showed slight contamination over intervals of 2, 5, and 7 inches, respectively, the last being for the deepest boring. Based on these results and the fact that only a small portion of the test volume was located at the depths where contaminant was found, total contamination was estimated to be 2.8 gal. In view of the uncertainty of both estimates, their agreement is satisfactory. The average their final contaminant saturation using the PITT results was 0.03%. This value is equal within experimental error to the lowest value achieved previously with surfactant remediation, which was during the INTERA/UT test in a nearby portion of OU2 in the summer of 1996. Although both that test, which used no mobility control, and the present one were conservatively designed, it is worth noting that the surfactant/foam process used only about 60% as much surfactant per unit of swept volume.

The PITT has been highly successful at several sites as a tool for estimating liquid contaminant content of an aquifer. When the aquifer is nearly clean, however, as at the end of the present test and the INTERA/UT test, the average saturation obtained depends on extrapolating asymptotic behavior that occurs near the end of the PITT when tracer concentrations are low. Under these conditions, it is desirable to ensure that analytical methods for the tracer are highly accurate and that the test is continued long enough to minimize uncertainty involved in the extrapolation.

As mentioned above, one of the monitoring wells was located at a local depression in the aquitard. This situation occurred by chance, as it was not practical to determine the fine details of the aquitard contour prior to drilling the wells. Mobilized contaminant initially within the pattern and that entering from outside tended to migrate to this low spot. A surfactant or surfactant/foam remediation process should be designed to ensure that liquid contaminant or dense middle phase microemulsions can be recovered from small depressions even if their existence and locations are unknown and flow from them to the extraction wells has an upward component that is opposed by gravity. It may be desirable, for example, to include a period during the last part of surfactant injection where salinity is low enough that contaminant remaining in such depressions can be solubilized into the aqueous surfactant solution.

While the test was successful in reducing contaminant content to very low levels, it revealed that better characterization is desirable to better define aquitard contours before injection and extraction wells are drilled. Had more information been available in this case, the wells could have been located to eliminate or minimize contaminant inflow from outside the pattern and to extract from the local depression.

Chapter 1

Description of Surfactant/Foam Process and Surfactant-Enhanced Aquifer Remediation

A field trial of the surfactant/foam process for aquifer remediation was conducted in the spring of 1997 at Hill Air Force Base in Logan, Utah. The principal investigators of the project were Drs. George Hirasaki and Clarence Miller of Rice University; their co-principal investigators were Drs. Gary Pope and Bill Wade of The University of Texas at Austin. Their engineering partner was INTERA Inc. of Austin, Texas (now Duke Engineering and Services), led by Dr. Richard E. Jackson. INTERA conducted all the field operations, including the design and analysis of the partitioning interwell tracer tests. Radian International, LLC (Radian) treated the effluent from the surfactant/foam flood in the Source (i.e., dense nonaqueous phase liquid DNAPL) Recovery System (SRS) that they operate at OU2.

The objectives of this project were as follows:

1. Reduction of NAPL in the test region of an aquifer at the demonstration site
2. Reduction in the amount of surfactant required to remove the NAPL
3. Assessment of the performance of the surfactant/foam process through the use of partitioning tracers and subsurface cores
4. Assessment of the performance of partitioning tracers in measuring the amount of NAPL before and after a remediation process
5. Development of simulation model to scale-up the process for field application

The first two objectives of the surfactant/foam project are a result of the interest and desire to use the least amount of surfactant to minimize costs and time. It is believed that the cost of chemicals and the time required for remediation may limit the economic attractiveness of surfactant-based clean-up technologies. The surfactant/foam project promised to do both because of the improved sweep efficiency of the process.

DNAPL removal by surfactant-induced solubilization or mobilization can occur only where the formation is contacted or "swept" by the injected surfactant solution. In heterogeneous formations, the injected fluids tend to flow through the higher conductivity regions and bypass the lower conductivity regions. All aquifers exhibit some heterogeneity in the distribution of hydraulic conductivity (e.g., stratification caused by variations over time in the conditions for deposition). As a result, the flow within them is not uniform. Indeed, nonuniform flow occurs even in an aquifer with uniform hydraulic conductivity if it contains an organic liquid contaminant that is not uniformly distributed and hence occupies different fractions of the pore space in different locations. Heterogeneity constitutes one of the greatest challenges to remediation because success requires removal of contaminant throughout the aquifer, not just along the preferential flow paths. No matter how effective the injected fluid is in removing the contaminant it contacts, no remediation occurs in bypassed portions of the aquifer.

Therefore, by improving the sweep efficiency of a surfactant solution, a smaller volume of the solution will be required to remove DNAPL from a heterogeneous formation. The surfactant/foam process improves sweep efficiency. First, surfactant solution is injected and flows along the higher conductivity pathways. Then, air or other inert gas is injected, which also flows preferentially along the same pathways and generates foam there. The high gas content of the foam increases the resistance to liquid flow with the result that additional injected surfactant solution is diverted to lower conductivity pathways.

The third and fourth objectives came about because the partitioning interwell tracer test (PITT) was used at Hill AFB as one of the performance evaluation methods for the various innovative remedial technologies demonstrated at OU1 and OU2. The aquifer remediation process needs some measure of its effectiveness in removing DNAPL from the aquifer. In the surfactant/foam field test, both cores and PITTs before and after the surfactant/foam flood were used as performance measures. In addition, the production of DNAPL and solubilized components in the extraction wells was monitored. Since the area chosen for the Advanced Applied Technology Demonstration Facility (AATDF) demonstration had less DNAPL than other areas of OU2, it was recognized that the volume of DNAPL recovered, or the percent of the initial DNAPL recovered, would not be a meaningful measure of the success of the project. Instead, the amount of DNAPL remaining after the remediation process would be a better measure of the effectiveness of the process in removal of DNAPL.

The last objective arose from the need to design the foam flood quantitatively. Such a design should be "robust" in the sense that minor variations in the hydraulic conductivity or DNAPL distribution from that which had been simulated would not cause a failure of the surfactant/foam flood to achieve its primary objective. The advantages of a numerically designed surfactant flood are that the operating parameters can be optimized by multiple simulations and the performance of the process at another site can be simulated by changing the geosystem model.

Additional objectives developed after the laboratory and field work commenced. In particular, it was necessary to demonstrate that foam could be injected into a shallow aquifer and that it could be propagated through the formation with the micellar surfactant solution. This objective arose quite simply because foam had not been previously used for aquifer remediation, and there were two concerns in its use. First, if the mobility of the foam were too low, the injection pressure would become too high and damage to the formation such as fracturing or fluidization of the sand could occur. Second, the hydrostatic gradient due to the density difference between water and air is greater than the pressure gradient due to horizontal flow in high-permeability aquifers. Thus, there was a concern that the injected air would migrate upward into the vadose zone rather than flow laterally with the surfactant solution. Another objective was to acquire cost and performance data so that this technology could be applied at other sites. Such information was used by CH2M Hill in the development of the Technology Practices Manual (Simpkin et al., 1999) and Section III of this volume.

A brief review of surfactant-enhanced subsurface remediation, which follows, will place the surfactant/foam approach in perspective and demonstrate its importance in the development of mass removal technologies. Surfactant-enhanced aquifer remediation (SEAR) is an innovative remedial technology that has been developed and refined over the last 10 years. There are several approaches to surfactant-enhanced aquifer remediation; these generally vary by the degree to which the interfacial tension between the injected surfactant solution and the nonaqueous phase contaminant is reduced. There are also various constraints on the application of surfactant-enhanced aquifer remediation, one of the most important of which is the effect of aquifer heterogeneities on the performance of the surfactant flood. Traditionally, in petroleum reservoir engineering, polymer has been used to compensate for heterogeneities because it increases the viscosity of the injected surfactant solution. A more recent approach to this problem of "mobility control" is the use of injected air to produce a foam in the more permeable sedimentary units of the aquifer. The foam forms a reduced mobility, gas/liquid dispersion that diverts any subsequently injected surfactant solution into lower permeability layers, thus ensuring an efficient sweep. The surfactant/foam process is, therefore, an advanced form of surfactant-enhanced aquifer remediation especially appropriate for heterogeneous alluvial aquifers.

Surfactant-enhanced aquifer remediation has been acknowledged as a promising innovative technology for the removal of DNAPLs primarily because of the history of the use of surfactant floods in the petroleum industry (Pope and Wade, 1995) and its subsequent application to the environmental restoration of a shallow, creosote-contaminated aquifer at Laramie, WY in 1988–1989 (Pitts et al., 1993). The surfactant solution that was injected into the alluvial aquifer at Laramie removed 84% of the creosote. This promising application led to a controlled experiment in 1991 and 1992 in which perchloroethylene (PCE) was spilled into a test cell installed in an alluvial aquifer at Borden, Ontario, by the University of Waterloo (Kueper et al., 1993). A surfactant solution injected into the test cell by the State University of New York (SUNY) at Buffalo (Fountain et al., 1996) removed 75% of the PCE by enhancing the effective solubility of this organic solvent in the groundwater. Both of these tests were conducted within sheet pile walls that hydraulically isolated the contaminated parts of the alluvial aquifers. It is noteworthy that the spatial distribution and total volume of the DNAPL were well understood in both the Laramie and Borden field tests and that both sites contained some free-phase DNAPL that resulted in relatively high initial DNAPL saturations (i.e., volume of DNAPL per unit pore volume).

Although these two tests were of very different sizes (the Laramie test cell had a pore volume of about 500 m³ [140,000 gal] while the Borden cell was only about 10 m³ [2400 gal]), they indicated that a well-characterized site could be substantially cleaned up. Table 1.1 shows the results of these and other, more recent tests, and includes as performance criteria the percent reduction in DNAPL mass and the average DNAPL saturation following the surfactant flood. The relative importance of these two criteria is site dependent, although the latter criterion is considered dominant in the present project.

The Borden demonstration was a micellar flood, indicating that an aqueous micellar solution of surfactant, alcohol, and electrolyte was injected into the aquifer. A micelle is a colloidal-sized aggregate of surfactant monomers. The solubilization of the DNAPL within the micelle produces a microemulsion, which is thermodynamically stable as opposed to a normal emulsion, often known as a macroemulsion. When polymer is added to the micellar solution, as with the Laramie demonstration, the resulting flood is termed a micellar-polymer flood. Polymer is employed as a mobility-controlling agent to ensure that both low- and high-permeability zones of the contaminated aquifer are swept by the surfactant. That sweep is, in fact, nonuniform in the absence of mobility control was shown by the results of a surfactant field test in a formation more heterogeneous than the Borden site (Fountain et al., 1996).

During 1996, INTERA, the University of Texas (UT), and Radian conducted a demonstration of surfactant-enhanced aquifer remediation at OU2, Hill AFB. This site is especially suitable for demonstrations of DNAPL remediation in that there is a large amount of trichloroethene (TCE)-rich DNAPL present in a permeable, shallow sand aquifer that has been fairly well characterized. In addition, there is an efficient treatment system on site, which includes a steam-stripper that can reduce high levels of TCE in an aqueous microemulsion to less than 1 mg TCE/L in the effluent. The main surfactant flood conducted by the INTERA/UT/Radian team at OU2 used a surfactant formulation with a sufficiently low interfacial tension to mobilize the contaminant and not simply solubilize it as in the Borden test. When mobilization occurs, much less surfactant is required. It was initially planned to include polymer at OU2. However, insufficient funding was available to include polymer. Nevertheless, the initial plans to undertake a micellar-polymer flood at OU2 made it especially attractive to the Rice/UT/INTERA team to use the OU2 site for a surfactant/foam demonstration because it would allow a comparative test of polymer vs. foam flooding for mobility control to improve sweep efficiency in heterogeneous aquifers. Even without polymer, the comparison is useful.

Table 1.1 Recent Surfacant/Cosolvent Floods for Removal of NAPL from Alluvial Aquifers

Team, Site (Ref.)	Flooding Agent	NAPL Composition	Swept Pore Volume (m³)	PVs Surfactant Injected	NAPL Mass Reduction (%)	Post-flood NAPL Saturation (%)
CH2M Hill Surtek, Laramie, WY (Pitts, 1993)	Alkali-surfactant-polymer	Creosote	530	3	84	2.7
SUNY/Univ.Waterloo, Borden, ON (Fountain, 1996)	2% Surfactant	PCE	9.1	14	77	0.2
UFlorida Hill AFB OU1 (Jawitz, 1998)	3% Brij 97, 2.5% *n*-pentanol	Multicomponent LNAPL	4.5	9.5	86	0.8
Intera/UT, Hill AFB OU2 (Brown, 1997)	8% MA-80l, 4% isopropanol	Multicomponent DNAPL	57	2.4	99	0.04
Intera/SUNY UT/DOE, Portsmouth, OH, 1996	4% surfactant, 4% alcohol, 0.2% NaCl/CaCl$_2$	Multicomponent DNAPL	5.6	3.0	50	0.1

CHAPTER 2

Surfactant/Foam Process Development

2.1 INTRODUCTION TO SURFACTANT PROPERTIES AND PHASE BEHAVIOR

The selection of the surfactant system is one of the most critical steps in designing a successful surfactant flood. The surfactant chosen must solubilize and lower the interfacial tension between the contaminant and aqueous phases so that the DNAPL can be freed from the pore structure and displaced. Moreover, the surfactant must exhibit the appropriate phase behavior under the conditions expected in the target zone of the aquifer. The tendency to form gels, liquid crystals, and emulsions should be minimal, because surfactants prone to this type of behavior are likely to cause a loss of hydraulic conductivity due to plugging problems, present effluent treatment problems, and create phases that are difficult to recover hydraulically. In addition, the selected surfactant should also have a high contaminant solubilization capacity to maximize DNAPL recovery, and fast equilibration/coalescence times to minimize mass transfer and kinetic effects. Other desirable characteristics include no or very low toxicity, and the ability to biodegrade in a reasonable time frame (neither too quickly nor too slowly).

Once a candidate has been identified, the phase behavior of the surfactant solution with and without contaminant present must be determined so that the most efficient process can be designed. Phase behavior studies are performed to characterize the interaction of surfactant solutions with DNAPL. Important information — such as contaminant solubilization by surfactant, equilibration/coalescence times, and any liquid crystal/gel/emulsion-forming tendencies — is measured in these experiments, which are carried out not only to select the best surfactant for the site, but also to design the optimum surfactant solution (including surfactant and additive concentrations).

The term "microemulsion" is frequently confused with the term "emulsion" (also called macroemulsion), which is a dispersion of one liquid phase in an immiscible liquid (e.g., oil and vinegar). Microemulsions, however, are thermodynamically stable solutions composed of submicroscopic structures, e.g., TCE dissolved in a surfactant micelle (Bourrel and Schechter, 1988). When designing a surfactant flood for aquifer remediation, surfactants that readily form microemulsions are desirable, while surfactants that exhibit stable macroemulsions should be avoided.

In a phase behavior study, known volumes of DNAPL and surfactant solutions are placed in graduated tubes, which are then sealed. Then, the contents are mixed and allowed to equilibrate for at least several hours. The resulting phases of the water/surfactant/DNAPL system appearing in the graduated tube and the positions of their interfaces are then noted. The difference in the interface readings before and after equilibration are used to calculate the solubilization of DNAPL components by the surfactant. The solubilization capacity of the surfactant solution is also verified analytically using gas chromatography. All phase behavior experiments were conducted at surfactant concentrations greater than the critical micelle concentration (CMC) of the surfactant under investigation.

2.1.1 Selection of Sodium Dihexyl Sulfosuccinate

The surfactant used during this project was one of three suggested by researchers at the University of Texas at Austin: sodium dihexyl sulfosuccinate, sodium diamyl sulfosuccinate, or sodium dioctyl sulfosuccinate. All three of the related surfactants generated Winsor Type III (three-phase) behavior with a variety of chlorinated solvents (Baran et al., 1994). In a Type III system, a middle-phase microemulsion exists along with oil and water phases. In such a system, there is ultra-low interfacial tension between the contaminant and the surfactant solution. Additional information on surfactant flooding can be found in Reed and Healy (1977).

During preliminary tests, it was noticed that aqueous solutions of both the diamyl and dioctyl surfactants separated into two phases, with the surfactant concentrating in one of the two phases. This behavior could create problems in field application, so both of these surfactants were dropped from consideration.

The remaining surfactant, dihexyl sulfosuccinate, is properly called sodium di(1,3-dimethylbutyl)sulfosuccinate, but the dihexyl nomenclature is more commonly used. The chemical structure of the surfactant is shown in Figure 2.1. It is commercially available from CYTEC Inc. under the trade name Aerosol MA-80I. The product of commerce is 80% MA-80I, with the balance made up of water and isopropanol. In the laboratory and in the field, the surfactant was used as received from the manufacturer, but the concentrations referred to in the text are the concentrations of the active material.

Dihexyl sulfosuccinate is an example of an anionic surfactant. It is referred to as such because, in aqueous solution, the sodium ion dissociates from the rest of the molecule, leaving a negative charge. A major benefit of selecting an anionic surfactant is that the phase behavior can be controlled by changing the salinity of the surfactant solution, which is relatively easy to accomplish. Another advantage is that anionic surfactants generally do not adsorb to soils as much as nonionic or cationic surfactants because soil surfaces often have a (repellent) anionic charge. This lessened adsorption potential is a benefit for application of a given surfactant.

FDA approval of dihexyl sulfosuccinate is another boon for its use. The U.S. Hazard Category describes MA-80I as relatively harmless. The 96-h LC-50 for bluegill sunfish is greater than 1000 mg/L and the 96-h LC50 for rainbow trout is 1200 mg/L, suggesting that the surfactant has low toxicity. These points may ease a regulator's fears about the addition of a new chemical into the subsurface.

Aerosol MA-80I was selected as the surfactant for this project primarily because it could generate ultra-low interfacial tensions between chlorinated solvents and water. However, the observation that the surfactant did not generate multiple phases in aqueous solution and its commercial

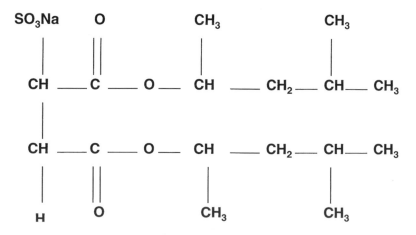

Figure 2.1 Chemical structure of sodium dihexyl sulfosuccinate (Aerosol MA-80I).

availability were also important factors. The anionic nature of the surfactant imparted a control mechanism and lessened the potential for surfactant loss to adsorption. Finally, the FDA food-grade designation was an added bonus for this particular surfactant.

2.1.2 Critical Micelle Concentration

The CMC is the concentration at which a surfactant begins to form micelles as well as monomers in water. A micelle is a more ordered formation of surfactant molecules, often spherical, in which the hydrophilic heads of the surfactant molecules are on the exterior of the micelle in contact with water while the hydrophobic tails are together in the core of the micelle, not in contact with the surrounding water. A micellar solution is capable of solubilizing much more contaminant than a monomeric solution because the nonaqueous molecules can partition into the hydrophobic cores of the micelles.

The CMC was determined by measuring the surface tension of a series of surfactant solutions of varying concentrations. The concentration at which the surface tension ceases to change is the CMC. Since the CMC can be a function of salinity for ionic surfactants, it was determined at several different salinities. The results are tabulated in Table 2.1. The optimal salinity for MA-80I is in the range of the 11,000 ppm NaCl measurement, so these data show that the surfactant/foam flood should have a surfactant concentration greater than 0.81%. A surfactant concentration of 4% was used throughout most of this project, although a few unsuccessful experiments were run with a surfactant concentration of 1%.

Table 2.1 Critical Micelle Concentration for MA-80I at Various Salinities

NaCl Concentration, (ppm)	Critical Micelle Concentration, (wt%)
0	2.03
1000	1.33
4000	1.02
8000	0.88
11,000	0.81

2.1.3 Adverse Phase Formation

Surfactants can form stable lamellar liquid crystal phases under certain conditions. During the course of the two-dimensional sand pack studies (see Chapter 2.3), the effluent from one sand pack experiment was found to contain a viscous, lamellar liquid crystal phase. The viscosity of one sample containing the liquid crystal phase was found to be 60 cp at a shear rate of 1.5 s^{-1}. Because a highly viscous TCE-containing phase could become effectively trapped by viscous forces within a porous medium, the range and extent of the liquid crystal phase were investigated further.

The two phase diagrams shown in Figures 2.2 and 2.3 were developed at room temperature and at 15°C, near the aquifer temperature, using a 4% MA-80I solution with varying amounts of TCE and NaCl. These phase diagrams aided in the interpretation of experimental results. The phase behavior of the effluent samples was used to track the contaminant concentration even before the samples could be analyzed by gas chromatography.

The MA-80I/NaCl solution at room temperature was isotropic, but with the addition of approximately 0.5% TCE to the samples, a lamellar liquid crystal and its dispersions formed. Above a TCE concentration of 2% to 4%, the system became an isotropic phase once again. Above 15% TCE, classic Winsor phase behavior was observed (see Figure 2.2).

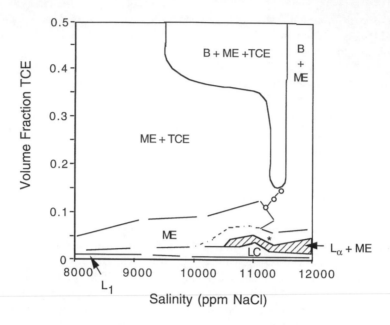

Figure 2.2 Partial phase diagram at 22°C for 4% MA-80I at various salinities and TCE concentrations.

At 15°C, the phase behavior of the surfactant/TCE system became more adverse. Addition of even a small quantity of TCE (about 0.2%) to the surfactant solution generated a liquid crystal phase; no isotropic phase was observed for even the lowest non-zero TCE concentration samples. Depending on the salinity, the lamellar liquid crystal persisted until the TCE concentration reached 3% to 5%, after which the system again became isotropic.

The formation of the viscous liquid crystal phases was recognized as a potential problem in the application of MA-80I but, unfortunately, no way to prevent the formation of liquid crystals could be found that did not impair the effectiveness of the foam (e.g., addition of alcohol). If the surfactant/foam process operated well, there was only a small mixing zone where surfactant solution and TCE could come into contact. The surfactant solution should displace the TCE, which would accumulate in a bank that would not mix much with the surfactant. The effluent contaminant concentration would then drop from excess to TCE levels too low to form liquid crystals. It was decided that the best approach to controlling the formation of the adverse phase was to operate the surfactant/foam flood optimally to minimize the potential for liquid crystals.

2.1.4 Salinity Scans

The technique for determining the optimal salinity for a given surfactant solution and contaminant is referred to as a *salinity scan*. The surfactant concentration of interest (4% MA-80I) was mixed at various salinities. The different solutions were then placed into test tubes with an equal volume of the contaminant. The tubes were shaken to mix and allowed to equilibrate. The transition

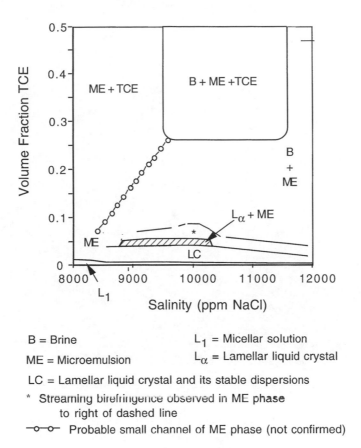

Figure 2.3 Partial phase diagram at 15°C for 4% MA-80I at various salinities and TCE concentrations.

from Winsor Type I (under-optimum) to Type III (three-phase) to Type II (over-optimum) phase behavior was then noted, and the salinity at which equal volumes of water and contaminant were solubilized into the middle-phase microemulsion was taken to be the optimal salinity for the tested conditions. An example of a salinity scan is shown in Figure 2.4, where the optimal salinity for the system is seen to be between 11,000 and 12,000 ppm (mg/L) NaCl.

Salinity scans were run for a variety of systems. Surfactant concentrations were 4% and 1% MA-80I; contaminants tested were TCE and Hill AFB DNAPL; and temperatures tested were room temperature (22°C), 15°C, and 12°C. These combinations were chosen so that the surfactant/foam process could be tested and developed for optimum performance on the benchtop and also in the field. The results are tabulated in Table 2.2.

2.1.5 The Effect of Ca²⁺ and Alcohol on Phase Behavior and Foaming

As part of the development of the surfactant/foam process, the substitution of $CaCl_2$ for NaCl as an electrolyte and the addition of alcohols to the system were tested. $CaCl_2$ was tested as a potential method for controlling foam strength, while isopropanol (IPA) was added in an effort to control liquid crystal formation.

Substituting $CaCl_2$ for NaCl weakened the foam by a factor of two. In addition, the three-phase region of the phase diagram was significantly narrower when $CaCl_2$ was used instead of NaCl, which indicated that dilution of the surfactant solution *in situ* could lead to loss of surface activity more easily with $CaCl_2$ than with NaCl. The use of divalent calcium ion made the optimal salinity more dependent on the surfactant concentration, which made maintaining the optimal solution

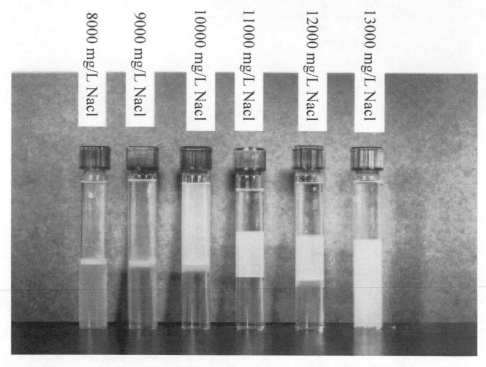

Figure 2.4 Phase behavior of the 4.8% MA-80I/TCE system with increasing salinity.

chemistry more difficult. Loss of effectiveness due to dilution was later observed in a layered 2-D model experiment. Since the formation of too-strong foam had been found to be less of a problem in the layered systems, CaCl$_2$ was dropped in favor of NaCl as the electrolyte.

Isopropanol was also investigated as an additive to the surfactant solution. It was found that an IPA concentration of 4.5% could prevent the formation of liquid crystals at room temperature while still allowing the generation of ultra-low interfacial tensions in the MA-80I/TCE system (Szafranski, 1997). However, the surfactant/foam floods tested with the added IPA did not perform well due to the inability to form a sufficiently strong foam. The IPA destabilized the foam to such an extent that most of the mobility control was lost and the contaminant recovery was slow. Because generation of a foam was critical to the success of the process, the addition of IPA as a method for controlling liquid crystal formation was rejected.

2.1.6 Effect of DNAPL on Foam

Another area of consideration in designing a surfactant/foam flood is the effect of oleic phase contaminant on the foam lamellae. If the DNAPL were to break the foam, this could actually enhance the operation of the surfactant/foam flood. Because the foam would be stable in clean

Table 2.2 Optimum Salinities for Various MA-80I/Contaminant Systems

MA-80I Concentration, (%)	Contaminant	Temperature, (°C)	Optimal Salinity, (ppm NaCl)
4	TCE	22	11,500
4	TCE	15	10,500
4	Hill DNAPL	15	10,750
4	Hill DNAPL	12	10,250
1	TCE	22	10,250

areas and unstable in contaminated regions, the foam would tend to divert surfactant solution through the regions that still contained DNAPL.

Laboratory tests found that pure TCE did not break the MA-80I foam, although a mixture of 20% n-decane with 80% TCE did. The Hill DNAPL was found to break the foam lamellae, probably due to the nonchlorinated components. During the foam flooding in the field, foam produced from the clean segments of the monitoring wells was quite stable, while foam produced from the contaminated intervals was fairly unstable and broke quickly.

2.1.7 Biodegradation Studies

Although sulfosuccinate surfactants are widely used in personal care products, household cleaning products, and a variety of industrial products, there is little information currently available on the biodegradability of these surfactants. A "Ready Biodegradability: Closed Bottle Test" conducted in 1991 indicated a removal of approximately 17% MA-80I after 28 days. Under industry classification guidelines, this surfactant was not readily biodegradable. Before conducting the surfactant flood field test, however, it was felt that more information on the biodegradability of this surfactant was needed.

Subsequently, an "Aerobic Aquatic Biodegradation" test was conducted on MA-80I and two similar surfactants (Britton, 1996). Based on the results of this experiment, it was concluded that MA-80I was inherently biodegradable, although none of the surfactants tested would have been classified as readily biodegradable. Thus, in addition to the superior qualities as a remediation agent for DNAPLs, MA-80I may also display biodegradation qualities that make it suitable for environmental use, without the tendency to biodegrade so fast that large material losses or biochemical fouling could occur.

2.2 COLUMN TESTS

A number of experiments were run in sand columns that approximated field conditions to determine what parameters were most important in controlling the strength of a generated foam. Some of the column experiments used contaminant while others displaced only water from the column. These experiments provided information on potential adverse interactions between the surfactant, the DNAPL, and the aquifer sediments, such as surfactant adsorption, clay mobilization, or the formation of liquid crystal, gel, or emulsion phases. Mean residence times required for both partitioning tracers and surfactants in the subsurface to avoid nonequilibrium effects are also obtained from column experiments.

The column used for most of the experiments was a 1-ft long, 1-in. diameter glass column. The column was packed with sand with a permeability of approximately 150 darcy (hydraulic conductivity = 0.144 cm/s). The pore volume (PV) of the column was approximately 60 mL. Two pressure transducers were connected across the length of the pack to measure the pressure drop across the column. Surfactant solutions and air were alternately injected by a pair of pumps controlled by a computer.

The base case surfactant/foam flood was a 4% surfactant solution at its optimal salinity of 11,500 ppm NaCl. The surfactant solution was injected in 0.10 PV slugs alternating with 0.10 PV air slugs. Both surfactant and air were injected at 20 ft/d interstitial velocity. All of the column experiments were run at room temperature.

2.2.1 Foam Propagation

The first significant result found during column testing was that a strong foam could be propagated under different conditions from those used during its creation. A short pulse of air injected at 10 times the nominal flow rate could generate a foam near the inflow end of the column and then the foam could be pushed through the column at a slower rate (20 ft/d). The foam remained stable throughout the movement down the length of the column. This result was important because a high flow rate is sometimes needed to create a foam in a high-permeability sand. The high rates could probably not be sustained in field operations, but perhaps could be used only initially to generate the foam.

2.2.2 Slug Size

The next series of laboratory experiments studied the effect of slug size on the generation of foam. These tests demonstrated that smaller slugs of surfactant solution and air generated foam more quickly than larger additions. In fact, the largest slugs tested, 1 PV, never generated any significant foam pressure. Figure 2.5 illustrates the effects of slug size on foam generation. In the figure, the significant finding is seen at the point that the graphs rise sharply rather than when the final pressure is reached. Note that the pressure in the 0.10 PV slug flood begins to rise quickly, and the sharpest rise comes earlier than the 0.33 PV slug flood. The results of this set of experiments indicated that the air and surfactant slugs should be kept as small as feasible to generate strong foam as quickly as possible.

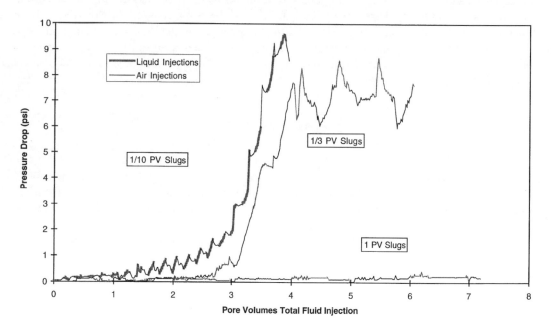

Figure 2.5 Effects of slug size on foam strength.

2.2.3 Flow Rate

The effect of flow rate on foam generation was also investigated. The experimental flow rates were 40, 60, 80, and 120 ft/d as well as 20 ft/d. All the experiments were run with 0.10 PV slugs of surfactant and air. As Figure 2.6 illustrates, the faster flow rates generated foam more quickly than the slower rates. Obviously, the injection rates would be more limited in a field application than in the laboratory, but these results indicate that increasing the flow rates helps to generate as strong a foam as possible. In Figure 2.6, the point of interest is again the point at which the individual plots begin to rise sharply. The data are plotted as the apparent viscosity of the fluid in order to normalize the pressure drop to the flow rate. Another important observation is that the

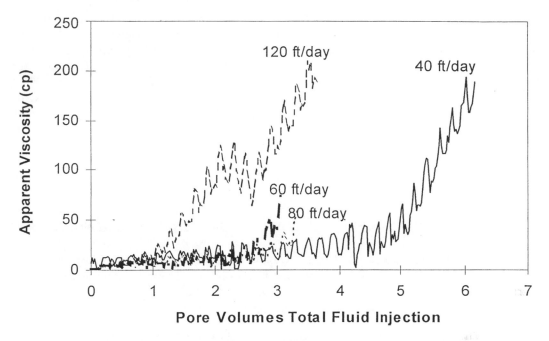

Figure 2.6 Effects of flow rate on foam strength.

foam generation at 120 ft/d is noticeably different from that at the lower flow rates. Apparently, a transition point is passed for the particular sand in use. The same foam behavior is noted at 20 ft/d in a 10-darcy sand packed in the same column, so it seems that some of the foam generation depends on the sand within which it is taking place.

2.2.4 Additional Parameters

Additional parameters tested for their effect on foam generation and strength included the effect of extended throughput, use of a surfactant preflood, and salinity. It was found that the only effective method for lowering an excessive pressure drop was to wash the surfactant solution from the column with water or brine and restart the foam generation.

One of the most important lessons learned during the column experiments described above was that the foam parameters did not scale directly to 2-D situations. In both Figures 2.5 and 2.6, the final pressure drops reached were far in excess of the acceptable pressure gradients for field application. However, in the layered 2-D model experiments that were run before the column tests, the same degree of blockage was never noted, not even in the highest pressure drop 2-D experiments. It was decided to use the information from the columns as general trends in foam generation and behavior, but not to expect the same pressure gradients, even in the same sands. The reason for the higher pressure drop in column tests is speculated to be the absence of alternate flow paths as in heterogeneous 2-D and 3-D systems.

2.2.5 Vertical and Heterogeneous Column Experiments

A series of experiments was also run in vertical and heterogeneous columns. The tests were designed to determine the potential for vertical migration of the foam generated in an aquifer. Because the foam would be much lighter than the surrounding water, buoyancy would tend to drive it up and away from the target areas.

The columns were packed with either a 150-darcy sand, a 300+-darcy sand, or a three-segment pack with a 50-darcy sand between two layers of 300-darcy sand (when viewed vertically). The injection of air vertically upward into columns that contained a slug of surfactant solution generated a strong foam, increasing the apparent viscosity of the air by 2 orders of magnitude. The vertically traveling foam displaced nearly all the liquid from the column before breaking through at the outlet. The low-permeability layer did not have any additional effect on the strength of the foam when injected upward.

These results indicated that a lower permeability layer of soil at a field site would not necessarily generate a confining foam barrier that was stronger than the strong foam that was generated by flowing air vertically upward through surfactant solution in a homogeneous soil. However, under the same conditions, vertically moving foam was stronger than horizontally moving foam, which indicated that foam created in the field would not simply rise up around the well boring and contribute nothing to the remediation process.

2.3 SMALL 2-D MODEL CONFIGURATION AND CHARACTERIZATION

The 2-D model was 20 in. long, 3 in. high, and 0.75 inch thick. A plate-glass window served as the front face of the model so that photos and visual observations could be taken. Fluids were injected through three screens on one end and produced at an autocollector at the other. A syringe pump was used to inject the surfactant and a pressure regulator was used to control air injection. Several transducers were plumbed to measure pressure across various intervals. A schematic diagram of the set-up of the 2-D model is shown in Figure 2.7.

The sand pack used in the experiments consisted of a high-permeability layer overlying a lower permeability layer. The permeability ratio between the two layers was about 20:1. This ratio was higher than expected in the field demonstration, but the greater ratio served to illustrate more dramatically the effects of differing permeability. The surfactant solution injected was 4% MA-80I

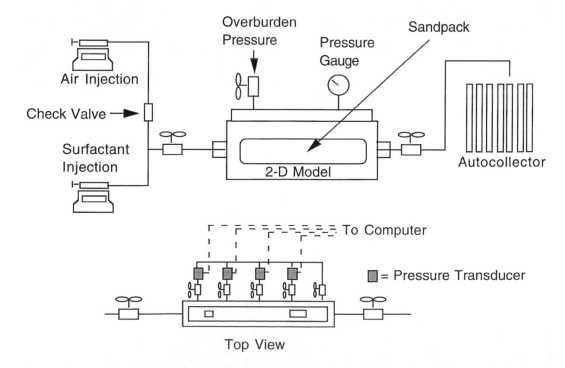

Figure 2.7 Schematic diagram of 2-D model experiments.

with 11,500 ppm NaCl. The surfactant solution was injected at 20 ft/d interstitial velocity in the high-permeability layer. TCE that had been dyed red was used as a representative contaminant.

2.3.1 Pressure-Regulated Foam Generation

The first observation in the 2-D model was that an effective foam could not be generated using alternating air and surfactant slugs injected at 20 ft/d. Because the maximum fluid flow rate that could be reasonably sustained under field conditions was about 20 ft/d, some other method for injecting air was needed. The solution to this problem was to pressure-constrain the air injection rather than rate-constrain it. Initially, the flow rate would be quite high, causing rapid generation of the foam (see column experiments, Chapter 2.2.5). As foam was created, the air injection rate would slow and the strong foam would then propagate at a slower rate, as indicated in the column experiments. If the resistance to air flow became great enough, air flow could even cease. In this manner, the overpressurization of the aquifer formation could be avoided. The air injection was designed to continue until it was no longer improving foam generation, after which time a small portion (0.17 PV) of surfactant solution would be injected before air injection resumed.

The first test of the pressure-regulated foam in the 2-D model worked quite well. Setting the air injection pressure at the maximum desired pressure resulted in a strong foam being generated without overpressuring the model. The initial air injection entered the model at approximately 250 ft/d, which led to the rapid formation of foam in the high-permeability layer. The foam swept through the high-permeability layer and diverted much of the subsequent surfactant solution into the low-permeability layer. The pressure history during the experiment is illustrated in Figure 2.8. Note the consistent pressure drop throughout the run.

One-third PV of surfactant was injected initially, which displaced much of the TCE from the high-permeability layer. The first air injection pushed surfactant through the high-permeability layer, completing the TCE removal from that layer. During the next four surfactant additions (0.17

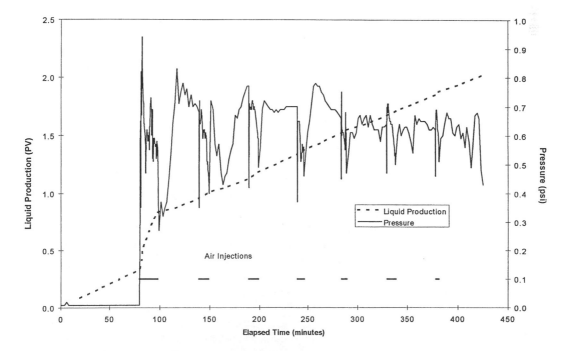

Figure 2.8 Pressure data during pressure-regulated foam flood.

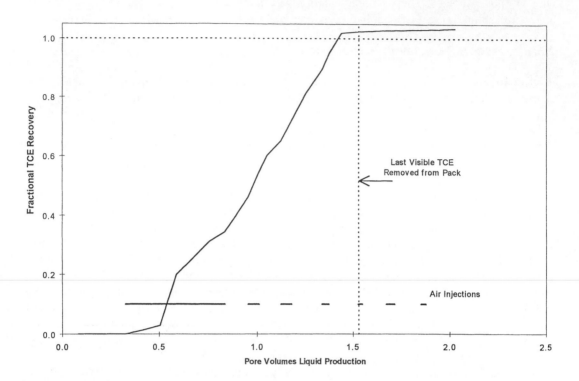

Figure 2.9 TCE recovery during pressure-regulated foam flood.

PV each), approximately half of each portion was diverted into the low-permeability layer, indicating that the foam had impeded liquid flow by a factor of 20, so that the apparent permeabilities of the two layers were about equal. The removal of TCE from the low-permeability layer continued at approximately the same rate as the removal of the TCE from the high-permeability layer. In the TCE recovery curve (Figure 2.9), the slope of the curve did not change much until all of the TCE had been recovered, showing that the removal of TCE from the model occurred at a uniform rate in both layers. Only 1 PV total surfactant was needed to remove all of the visible TCE from the model. The mass balance of the TCE indicated a 104% recovery, which was within the measurement error.

2.3.2   Surfactant/Foam Flood vs. Surfactant Flood

For comparative purposes, a surfactant flood was run in the same pack as the pressure-regulated surfactant/foam flood described above. The surfactant solution was the same and the flow rate was initially 20 ft/d. Later in the experiment, the flow rate was increased in order to complete the test. The surfactant flood worked quite well in cleaning out the high-permeability layer of the model, and the TCE was rapidly displaced from that layer. However, after the TCE was removed from the high-permeability layer, surfactant solution continued to preferentially flow through the high-permeability sand and contributed nothing to the removal of the remaining TCE. TCE slowly flowed out of the low-permeability sand and the experiment was ended after 26 PV surfactant had removed only 65% of the TCE from the model. The TCE recovery curve is shown in Figure 2.10, along with that for the surfactant/foam flood. Note that the initial recovery is similar; it is in the later stages of cleanup that the foam flood shows its strength. Similar results were found in other layered sand packs with different permeability ratios and packing configurations.

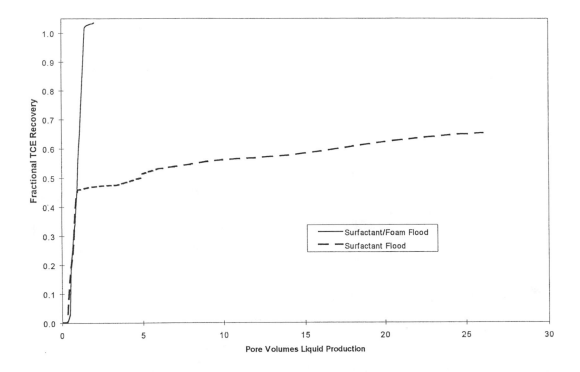

Figure 2.10 Comparison of TCE recovery with and without foam as a mobility control process.

2.4 DESIGN AND MODIFICATION OF UT LAYERED 2-D SAND TANK

The experiments conducted at the University of Texas at Austin (UT) used a glass-walled tank measuring 8 × 5 ft with stainless steel frame and supports. The glass walls were made of tempered, laminated glass of 1.5-in. thickness to provide sufficient strength to support the anticipated pressures that were to be applied. In addition, several cross members were included in the design to provide additional support. Provision was made in both end-pieces of the tank for multiple inlet/extraction ports, as well as a series of ports for pressure monitoring throughout the length of the tank during experimental runs. The glass plate was sealed against leakage using Viton O-rings around the entire perimeter of the plate, and all metal-to-metal junctions were sealed.

Before constructing the tank, calculations were made to determine the glass thickness required to avoid breakage as well as to minimize bowing of the glass during the pressure loading. However, during the course of the experiments, some bowing of the glass wall occurred. The result was that the sand slumped in the central region of the pack and additional sand was needed to fill the space.

2.4.1 Framing Details

The outer frame consisted of three separate pieces: a central piece (2 in. × 2 in.) that formed the 2-in. space for the actual sand pack, two spacers (1 in. × 1.5 in.) for the glass plate, and two outer struts (2 in. × 2 in.) to provide structural support for the tank. The total width of the frame for the model was 9 in. The central piece had an O-ring groove machined into the surfaces that were in contact with the glass plate to provide a sealing surface. Bolts were inserted at 6-in. intervals along the perimeter through the outer struts and spacers and were secured to the central piece.

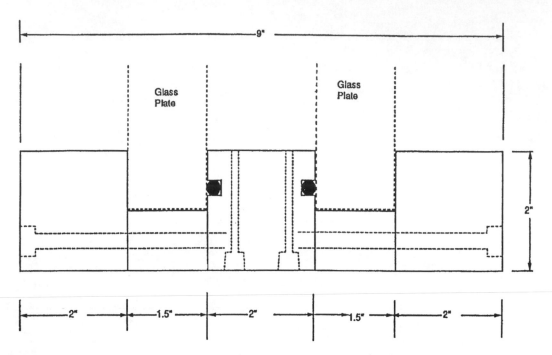

Figure 2.11 Cross-sectional view of UT layered sand tank: end-piece and strut assembly.

These were tightened uniformly to provide the necessary compression of the O-ring against the glass surface. A cross-sectional view of the assembled structure is shown in Figure 2.11.

Two vertical cross struts (1 ft × 2 in.) were bolted to the outside struts to divide the horizontal distance into three equal sections. A horizontal cross support (1 in. × 2 in.) was placed 29 in. from the base. This divided the tank vertically into an asymmetric configuration since the majority of the pressure would be exerted on the lower portion of the tank. The cross-supports were notched at the intersections to allow them to fit together and were bolted together during assembly as well as being bolted to the corresponding strut of the frame. A surface view of the assembled tank is shown in Figure 2.12.

2.4.2 End-piece Details

The end-pieces were designed to allow for several sections for injection and intervening sections for pressure monitoring and placement of DNAPL spill wells. To accomplish this, two ports with a 2-in. horizontal spacing between them were drilled and tapped for 1/8-in. fittings at 2-in. vertical intervals along the entire length of the end-piece. The fittings were bored out to allow 1/8-in. tubing to be inserted through the fittings. For the injection/recovery ports, the inner surface of the end-piece was milled to a depth of 1/16 in. to allow the fitting of screens. The screens served to prevent migration of sand grains into the ports. The unscreened ports were bored through to a diameter slightly greater than 1/8 in. to allow 1/8-in. stainless steel tubing to slide through the fittings and end-piece to the desired location within the tank. The stainless steel tubing was plugged with a small glass wool plug to prevent migration of sand into the tubing while maintaining fluid continuity for pressure measurements.

Nylon ferrules were used on the Swagelok end of the fittings. This provided a secure pressure fitting while enabling the reuse of the tubing during reassembly and packing. Stainless steel shut-off valves, either two- or three-way, were attached to each of the lines to enable individual fittings to be opened or closed as dictated by the experimental procedure being used.

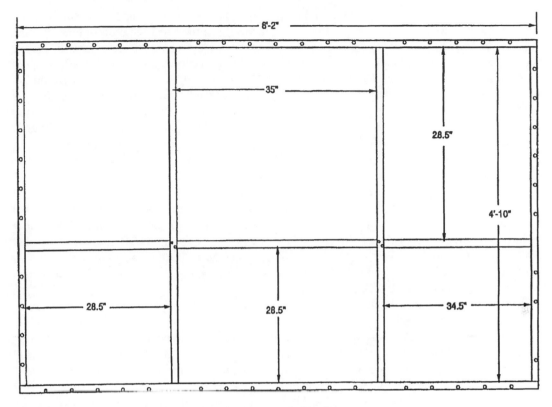

Figure 2.12 Surface view of UT layered sand tank: overall side view.

2.4.3 Assembly and Leak-Testing of the Tank

To assemble the tank, it was necessary to first assemble one side lying flat on the floor, without the glass plate, to provide the necessary support. The glass plate was carefully lifted into place using a floor crane. After properly positioning the glass, the spacer and struts with the cross-members were positioned and bolted down. The entire assembly was then turned over and the same procedure was used to position the other glass plate. All metal-to-glass contact areas were isolated with a 1/32-in. Neoprene gasket to provide a cushion between the glass and metal to minimize the possibility of creating excess stress on the glass that might cause it to fail. Once the entire structure was assembled, the tank was then lifted back into its final vertical stance. Because the assembled tank weighed in excess of 1 ton without any packing, the top bar of the tank was fitted with two heavy-duty swivel bolts for lifting.

Once the tank had been assembled and positioned appropriately, it was water-tested. This consisted of completely filling the tank with water and looking for any leaks. Some small leaks were detected and were corrected by retightening the bolts. After water-testing, the tank was drained and dried completely before being packed with sand.

2.4.4 Packing and Initial Testing of the Tank

A series of five coarse mesh screens were used in a wooden frame, which provided a 2-in. vertical spacing between the individual screens to disperse the flow of sand across the entire 2-in. width of the pack. After placement of the screen assembly, the sand was slowly drained through the screens into the tank. This was accomplished using a large container filled with sand and suspended from an I-beam with a roller. A small hole with tubing allowed the sand to drain slowly from the reservoir onto the center of the screen assembly. During filling, the reservoir was moved along the I-beam from end to end of the tank to provide a uniform filling along the length of the tank. The walls of the tank were subjected to periodic vibration by pounding the glass walls with a rubber mallet to assist in settling the sand. The sand flow was adjusted to a rate such that filling occurred at a rate of 0.5 in. per hour.

A thin layer of silica flour was placed in the bottom of the tank to provide a tight permeability barrier along the bottom of the tank. Two primary layers of sand were added to the tank after the silica flour layer. The lower layer was an Ottawa F-96 sand (grain size range of 106 to 212 μm), while the higher permeability layer was an Ottawa F-35 sand (grain size range of 300 to 600 μm). The lower layer was 24 in. thick, while the upper layer was 26 in. thick. Another thin (4 in.) layer was added using the Ottawa F-95 sand to provide an upper permeability barrier to the flow. The remainder of the space was then filled with the Ottawa F-35 sand.

Initial design of the tank was based on discussions that indicated that the pressures involved would not necessitate the use of over-burden pressures for the experiments. To add an over-burden required for the final design of the experiments, it was necessary to construct a fluid-filled Viton bladder to place on top of the sand pack. This was constructed using a 1/32-in. Viton sheet that was cut and glued to the proper size and form to fit into the tank. A 1/8-in. stainless steel tube was mounted in the bladder wall and fed through a fitting in the top of the tank to enable over-burden pressure to be applied to the tank contents.

After filling the tank with sand, the top was replaced. The pack was flooded with carbon dioxide overnight and then saturated with water at a slow rate from the bottom ports on the end-piece. Once the tank was filled, it was water flooded to determine the water permeability and then an initial tracer flood was conducted. During these tests, a considerable amount of settling of the sand occurred due to a slight bowing of the glass plates. This resulted in overextension and tearing of the Viton bladder. Attempts to resolve the problem resulted in an uncorrectable situation that necessitated removal of the sand and repacking of the model. During this time, the leaks in the bladder were repaired as well. Repacking of the tank was accompanied by some vibration of the tank using a heavy-duty vibrator unit to achieve a tight pack.

The tank was then again saturated with carbon dioxide, followed by water saturation and initial evaluation for permeability by the use of a tracer test. The permeability was measured at 60 darcy as compared with the predicted permeability of 42 darcy based on grain sizes used. The tracers included a dye for visualization as well as ethanol, 3-methyl-3-pentanol, 2-ethyl-1-butanol, and 2,3-dimethyl-3-pentanol tracers. The alcohol tracers were measured using gas chromatography and showed the effective pore volume for the tank to be 70.9 L. Some slight settling occurred again during these tests, which was corrected by removing the top and refilling the space with sand.

2.4.5 Experimental Problems and Corrections Applied

During the first foam flood experiment, it was observed that the foam did not propagate completely through the length of the tank. Careful observation of the injection end of the tank indicated that the gas being injected had formed a channel vertically along the injection end-piece and was moving over the top surface of the model. During the course of the flood, some additional sand settling occurred because the pressure of 4 psi was greater than that used in the initial tests. Some slight additional bowing of the glass plates also occurred. It was decided to make additional modifications and repeat the experiment.

The foam was broken and the injected surfactant was flushed out of the model. The cover of the tank was removed and sand was added to level out the upper surface. A slurry of silica flour was poured over the surface, and the bladder and cover were replaced. The second foam test was conducted and was proceeding well until the generated pressure reached 6.5 psi, at which time the glass walls of the tank bowed and immediate settling of the sand occurred. The resultant space allowed gas to be diverted immediately through the top of the tank, which terminated the foam propagation through the sand pack.

Again, the foam was broken and the injected surfactant was flushed out of the model. The cover of the tank was removed and sand at the injection end of the model was excavated to a depth just above the injection ports, a depth of approximately 6 in., and angled upward to about 6 in. from the injection end-piece. This was filled with a slurry of silica flour to provide a permeability barrier to the movement of the injected materials. Several iterations of the procedure were required before an adequate seal was achieved. In addition to the slurry of silica flour over the top layer of the sand, steel blocks (2 in. × 2 in. × 2 in.) were cemented into the corners before setting the bladder in place. After situating the bladder, another layer of silica flour slurry was placed on top of the bladder and the cover replaced on the tank. Once the bladder was filled and inflated with water, the pressure on the bladder was maintained slightly higher than the fluid injection pressure to ensure that the bladder would not collapse from hydrostatic pressure during the foam flood.

A vacuum pump was used in conjunction with a manifold at the production end of the tank to enable generation of a differential pressure sufficient to cause propagation of the foam through the sand while maintaining a lower internal pressure within the tank. Using this technique, a successful foam flood was accomplished.

2.4.6 Performance in the UT Tank

The surfactant/foam process was tested in the 2-D (8 ft × 5 ft × 2 in.) model at the University of Texas at Austin. The surfactant solution used was 4% MA-80I at 11,500 ppm NaCl. The interstitial flow rate was 20 ft/d in the high-permeability layer. The primary difference in operating the larger model was that surfactant was continuously injected as in the planned field operations. Air was injected at a set pressure. The air pressure depressed the liquid injection level, allowing the air to enter the model above that point.

After an initial surfactant injection, air was injected into the UT model. The foam was quickly generated and began to propagate across the model. It was noted that vertical migration of the foam appeared to be hindered by the low-permeability confining layer at the top of the model, in contrast to the results of the vertical column tests (see Chapter 2.2.5). As the foam propagated across the model, the air injection rate slowed as the pressure gradient rose. In order to continue injecting air, the injection pressure was increased. The increase was possible because the air injection had not been set to the maximum possible pressure due to the nature of the apparatus.

Foam was propagated along about half the length of the model in the high-permeability sand before the experiment was interrupted for several days. Upon resuming the run, it was found that the foam had become more intractable while shut in and could not be propagated at the desired pressure gradient. Finally, the foam was broken by the injection of fresh water, and a fresh slug of surfactant solution was injected to regenerate the foam. When air was injected into the unblocked model, it generated a foam that moved through the cleared paths and crossed the length of the model.

This experiment was important because it illustrated that foam could be propagated over a longer distance than the 20-in. 2-D model used at Rice University. In the case of this project, the 8-ft distance was about 40% of the distance needed for the field test, and thus served well as an interim step. A second point of interest was the apparent strengthening of the foam after it had been shut in for several days. It was not clear that this would be a problem in a 3-D setting, but it was information that was helpful to have in planning for the field test.

2.5 SUMMARY OF FINDINGS FROM SURFACTANT/FOAM DEVELOPMENT EXPERIMENTS

Sodium dihexyl sulfosuccinate (commercial name Aerosol MA-80I) was identified as the candidate surfactant due to its high contaminant solubilization, minimal tendency to form gels/liquid crystals/emulsion, quick coalescence, low toxicity, biodegradability, and, most importantly, the ability to generate and sustain foam.

- Foam stability was reduced by the use of calcium chloride as the electrolyte, by the addition of an alcohol such as isopropanol, and by the presence of Hill DNAPL.
- Small alternating slugs of surfactant and air or high air velocities were required to generate foam in horizontal columns. Once formed, foam could be propagated at lower velocities. However, foam characteristics were dependent on both flow rates and soil permeability.
- Foam could be generated and propagated through 2-D sand packs. Pressure-regulated air injection was found to be the best procedure because it provided high initial air velocities to generate the foam, but lower velocities during propagation.
- Sandpack experiments in two-layered soil with a 20:1 permeability contrast showed that only 1 PV of surfactant was needed to recover 100% of the DNAPL in the presence of foam, whereas 26 PV of the same surfactant in the absence of foam removed only 65% DNAPL.
- Foam could be propagated in a large 2-D model (8 ft × 5 ft × 2 in.) and showed strengthening after a few days. Injection of water was observed to break the foam.

CHAPTER 3

Surfactant/Foam Field Site Description

Hill AFB is located in northern Utah, approximately 25 miles north of Salt Lake City and about 5 miles south of Ogden (Figure 3.1). The base covers approximately 6700 acres in Davis and Weber Counties, with the majority of the base located within Davis County. The base is situated just west of the Wasatch Mountains, on a plateau approximately 300 ft above the valley floor. Surface elevations at Hill AFB range from about 4600 ft above mean sea level (amsl) along the western boundary of the base to over 5000 ft amsl near the eastern boundary. Operable Unit 2 (OU2) is located on the northeastern boundary of Hill AFB along Perimeter Road on a bench situated about one third the distance down the hillslope of the western valley wall of the Weber Valley.

OU2 is located on one of the upper terraces (locally known as "benches") of the Weber Valley, near the upper edge of a landslide complex. The head of a large rotational slump block has been identified between the site and the Davis-Weber Canal. The slide extends down past the canal onto the floodplain of the Weber River (Radian, 1992). At the site, a surficial sequence of unconsolidated alluvial deposits rests unconformably on the Alpine Formation, a silty clay deposit at least 150 ft thick. The shallow alluvium, a heterogeneous mixture of predominantly sand and gravel deposits ranging from 10 to over 50 ft in thickness, is classified as belonging to the Provo Formation of the Pleistocene Age. The heterogeneous alluvial sediments draped across the hillslope and deposited in the buried channel on top of the deltaic sediments range in grain size from silts and clays to large cobbles, an indication of the wide fluctuations that occurred in the hydraulic energy that controlled their deposition. The alluvial sediments are predominantly silicate in nature, with a clay fraction of less than 2% by weight. The clays present in the alluvium are mostly illites.

3.1 CONTAMINATION AT SITE

Base records indicate that from 1967 to 1975, the OU2 site, known then as "Chemical Disposal Pit #3," was used to dispose of unknown quantities of chlorinated organic compounds, primarily trichloroethene (TCE) from a solvent recovery unit, and sludge from vapor degreasers. There are also reports that this site received an unknown volume of plating tank bottoms in the early 1940s (Engineering-Science, 1982). The disposal area consisted of at least two unlined disposal trenches trending north-northwest, which are estimated to have been approximately 6 ft deep. Size estimates based on geophysical surveys, aerial-photo interpretations, and soil-boring data indicate that the two trenches were about 10 ft wide and had lengths of approximately 50 and 100 ft (Radian, 1993). The approximate location and footprint of each known disposal trench is shown in Figure 3.2.

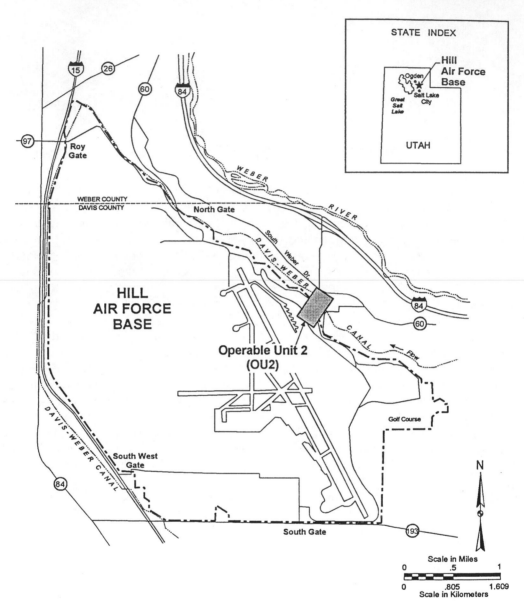

Figure 3.1 Location map of Hill Air Force Base.

The area impacted at OU2 by these past disposal practices can be loosely divided into two zones. The first, known as the DNAPL zone or source area, is the portion of the site affected by free-phase and residual-phase DNAPL in the subsurface. This includes both the saturated and unsaturated portions of the aquifer beneath the trenches (DNAPL entry location) and that portion of the aquifer through which DNAPL has migrated horizontally as an immiscible phase. At OU2, the source area is located in sediments deposited in a buried channel in the area west of Perimeter Road (see Figure 3.2). The second zone of contamination is the dissolved phase plume, which consists of organic constituents dissolved into the groundwater. The plume emanates from the source area and extends downgradient in a northeasterly direction to the valley floor.

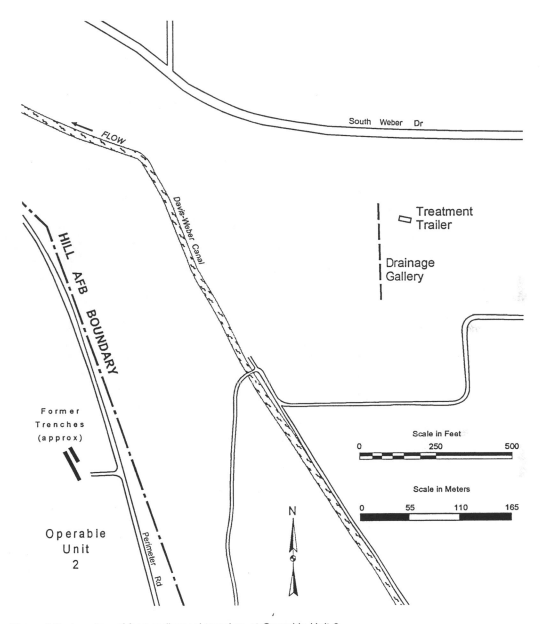

Figure 3.2   Location of former disposal trenches at Operable Unit 2.

A source recovery system (SRS) was constructed at OU2 in 1992. The SRS was designed to pump free-phase DNAPL from the contaminated shallow aquifer in the source area and treat the associated contaminated groundwater removed by the extraction wells. A plan view of the source zone at OU2 showing site features and the SRS facilities is presented in Figure 3.3. The SRS system, operated by Radian International LLC, consists of a well field of five extraction wells connected to an on-site treatment plant via an above-ground pipeline. The fluids extracted from the source area are piped into a series of phase separators to recover DNAPL, and the remaining groundwater is then sent through a steam stripper. Recovered DNAPL is transported to an off-site, permitted treatment facility, while the treated groundwater is transferred via a pipeline to the Industrial Wastewater Treatment Plant (IWTP) at Hill AFB. The SRS has been operational since 1993, recovering over

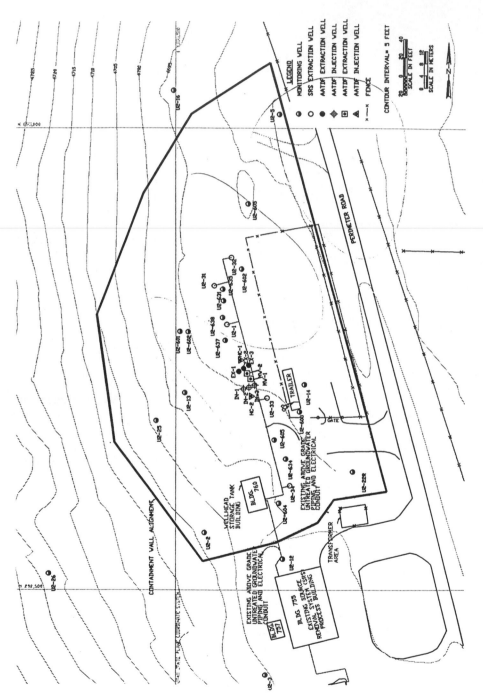

Figure 3.3 Location of wells used during surfactant/foam process demonstration at Hill AFB.

23,000 gal DNAPL and treating over 1,000,000 gal contaminated groundwater (Radian, 1994). Although the SRS extraction wells no longer continuously produce free-phase DNAPL, it has been conservatively estimated that 30,000 to 40,000 gal DNAPL remain in the aquifer.

Two pump-and-treat systems are currently operating in the plume area. One of these is a conventional air-stripping system located at the downgradient extent of the plume. The other system utilizes granular activated carbon to treat a low volume of flow from a drainage gallery situated in the central portion of the plume (see Figure 3.2). As part of the Record of Decision (ROD) for the site, a slurry wall was constructed around the source area in the fall of 1996 prior to the surfactant/foam demonstration described in this book. The slurry wall is protected on the upgradient side by a shallow groundwater interceptor trench. Current work at OU2 under the ROD includes the construction of two interceptor trenches in the plume area and the installation of two tray-type air strippers in the SRS groundwater storage building to treat the trench effluent.

3.2 HYDROGEOLOGY OF SITE

Hydrogeological investigations conducted in the past have identified two shallow aquifers beneath OU2. Groundwater in the upper system is found in the Provo alluvium, primarily under unconfined (water table) conditions in the channel sediments beneath the former disposal trenches. The saturated zone in this area is typically about 19 ft thick, with a depth to water of 20 to 28 ft below ground surface (bgs). The water levels fluctuate greatly from season to season, due to the limited extent of this aquifer. The aquifer itself is of very limited extent in the DNAPL source area because its saturated zone is entirely contained in the narrow channel incised into the underlying clay of the Alpine Formation. Downslope, the stratigraphy and hydrogeology of the upper zone flow system have been altered significantly by landslide processes. The hill slope below the test site contains numerous springs and seeps during the wet season, some contaminated with organic chemicals from the OU2 area.

The deeper of the shallow water-bearing systems is located in a zone of thin beds and laminae of silt and fine sand present as discontinuous lenses within the dominantly clay Alpine Formation. This water-bearing zone is under semiconfined and confined conditions approximately 70 to 90 ft below ground surface. The extent of the hydraulic connection between individual lenses, if any, is not known, but the lenses are relatively discontinuous. There is no known drinking-water usage from either this hydrostratigraphic unit or the upper shallow aquifer.

There are two deeper regional confined aquifers known as the Delta and Sunset aquifers that underlie Hill AFB. Based on logs from nearby deep wells on the base, the Delta aquifer is probably located 500 to 600 ft below ground surface at OU2. The shallower Sunset aquifer has been encountered to the northeast of OU2, but the logs for deep wells completed in the Delta aquifer close to the site do not show any hydrogeologic units that would represent the Sunset aquifer (Radian, 1992). If the Sunset aquifer does exist in this area of the base, it should be encountered at about 300 ft below ground surface at OU2. These regional aquifers are separated from the shallow local systems at OU2 by a thick aquiclude composed of the lacustrine clays and fine-grained deposits of the Alpine Formation.

Results of pumping tests conducted by Radian at OU2 in the Provo Formation indicate that hydraulic conductivity (K) values range from 42.6 ft/d to 116 ft/d for the channel alluvium. The storage coefficient is approximately 0.16 (Radian, 1994). The moisture content of the vadose zone sediments of the Provo Formation generally ranges from 2% to 8% by weight. Sand cores from the Provo Formation are determined to have an average porosity of 27%, and an average density of 2.64 g/cm^3 (Radian, 1994). Hydraulic conductivity of the Alpine Formation from core sample measurements indicate conductivities of 2.7×10^{-5} ft/d (9.6×10^{-6} darcy) for the clay aquiclude and a porosity of 31% (Core Labs, 1996).

A map of the shallow aquifer's natural (precontainment wall) water table surface at OU2 is depicted in Figure 3.4. The direction of groundwater flow in the saturated sediments of the Provo Formation is controlled by the topography of the aquiclude and the structure resulting from landslide processes that occurred at the site in the past. The regional flow direction is northeast, perpendicular to the hill slope. Locally, however, at the site of the former disposal trenches, the flow direction follows the channel axis both to the north and the south, and is constrained by the aquiclude from flowing to the northeast. The presence of intermittent seeps and the migration of dissolved contaminants downslope and to the east of the ends of the channel

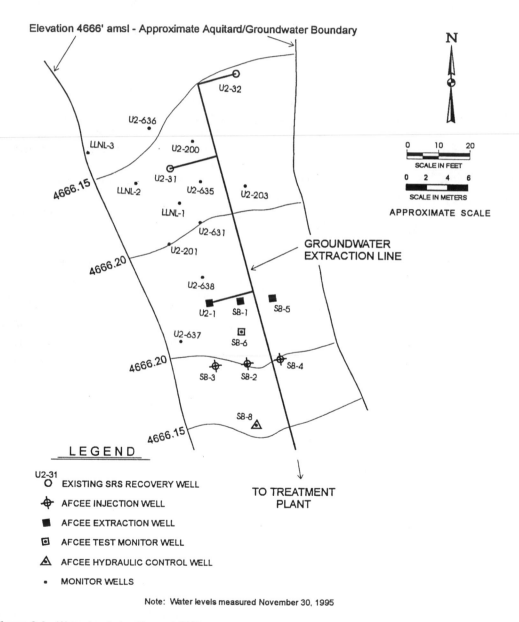

Figure 3.4   Water level elevations at OU2.

indicate that groundwater flows out of the channel and to the northeast when the water table elevation rises above the natural spillways formed by the rise in the channel bottom to the north and south. Groundwater velocities in the disposal trench area are estimated to range from 0.28 to 0.85 ft/d. Groundwater in the OU2 area is characterized as Ca-Mg-bicarbonate type water, as plotted on a trilinear diagram (Radian, 1992). The total dissolved solids range from 200 to 600 mg/L.

3.3 AQUICLUDE PROPERTIES AND STRUCTURE

One of the most important features of the geosystem at OU2 is the Alpine clay aquiclude. The channel structure eroded into it constrains the extent of the alluvial aquifer, and has a large influence on the aquifer's response to hydraulic perturbations. The Alpine Formation sediments are typically composed of 50:50 silt and clay, and are classified as lean clays with low to medium plasticity (Radian, 1992). Silty partings are common, but no fractures were observed in any of the cores recovered during the drilling program. The Alpine Formation is a deltaic deposit that contains occasional thin, interbedded, fine-grained sands and silts. These units do not appear to be continuous, and are not able to be correlated from borehole to borehole (Radian, 1992). At OU2, the Alpine Formation is separated from the Cambrian and Precambrian bedrock by a thick unnamed sequence of unconsolidated basin-fill deposits.

In addition, the clay acts as a competent capillary barrier. A mercury-injection porosimetry test conducted on a clay sample acquired from the aquiclude during the AFCEE project demonstrated that a DNAPL pool depth of at least 11 ft would be required to drive the contaminant into the clay (Core Labs, 1996). The aquiclude, therefore, controls the distribution of mobile DNAPL at the bottom of the alluvial aquifer, trapping it in local depressions in the bottom of the channel.

Over the course of the remedial investigations at OU2, the existence of a large topographic low in the clay surface located in the central portion of OU2 was established (Radian, 1992; Oolman et al., 1995). Known as the "northern pool," this depression contains a large amount of DNAPL. During the drilling program for the AFCEE surfactant field test, a second, smaller depression in the channel bottom, also containing a DNAPL pool, was discovered. The DNAPL pools at the OU2 site are trapped stratigraphically by a large, northwest-to-southeast trending depression or trough in the Alpine clay. In plan view, it is a curvilinear feature with the concave side roughly facing east, as seen on the contour map of the buried clay surface shown in Figure 3.5. The depression, filled with Provo alluvium, is interpreted to be a paleo-channel utilized by high-energy flow events in the past, making it an irregular erosional surface, as illustrated on the three-dimensional view of the clay surface depicted in Figure 3.6. The coarse-grained sediments found in this channel generally fine upward, supporting the interpretation that these deposits were left behind by the lateral migration of a stream (Radian, 1992). The area in which the AATDF surfactant/foam demonstration was conducted is located to the south of these pools. Therefore, one of the objectives of the initial drilling program was to investigate the geometry of the channel to the south of these pools, with an emphasis on the topography at the bottom of the channel.

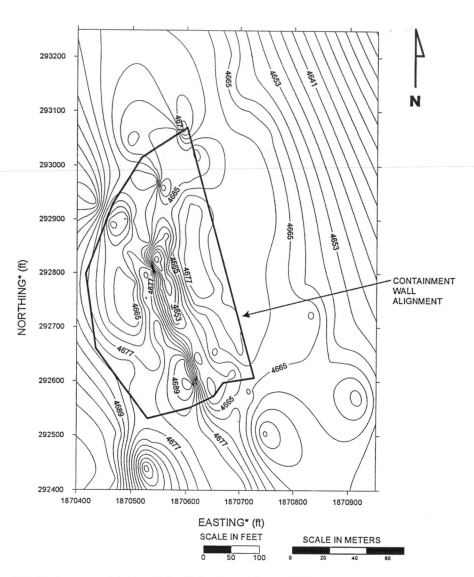

Figure 3.5   Contour map of the top of the Alpine clay surface at OU2.

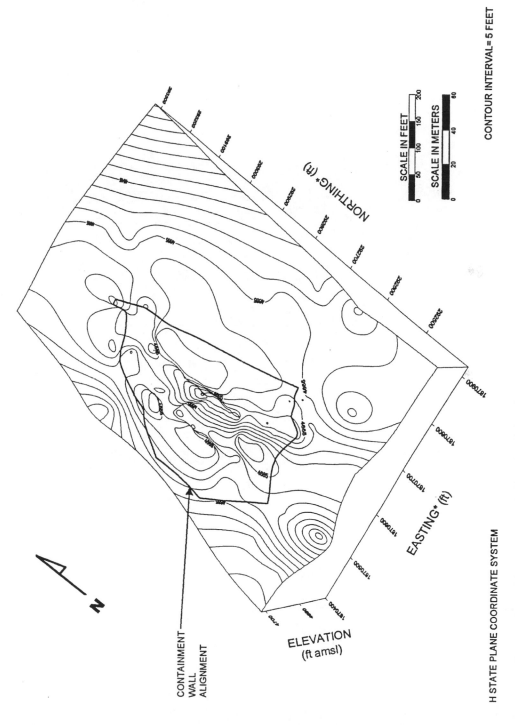

Figure 3.6 Block diagram of the channel eroded into the Alpine clay.

CHAPTER 4

Field Demonstration Design

4.1 SITE CHARACTERIZATION

Five borings, SB-9 through SB-13, were drilled to determine whether there was adequate DNAPL in the proposed test zone to conduct a surfactant/foam flood (see Figure 4.1). Additionally, the typical permeability distribution of selected alluvial samples from the test area was assessed by measurement of their grain-size distribution. The vertical contrast in permeability is important because it determines the vertical migration of surfactant and foam during a surfactant/foam flood as well as variation with depth of the horizontal velocity of injected fluids. A total of 105 samples from the five borings drilled in March 1996 were analyzed to determine their grain-size distributions for particles less than 4 mm. A final objective was to refine the coring, sampling, and analytical procedures to be used for subsequent borings drilled at OU2 for the surfactant/foam demonstration. The information obtained from the borings was used to pack the UT 8 × 5-ft transparent tank and was also incorporated into design simulations of the field test.

Previous drilling experience at OU2 had shown that sand flow into the auger annulus could create problems when drilling and sampling at the bottom of the aquifer. During the reconnaissance drilling program, sand flow problems were mitigated or minimized by pumping nearby well SB-8 in conjunction with the SRS extraction wells to reduce the head at the base of the aquifer. As a result, the water table was encountered at about 40 ft below the ground surface during drilling, almost 15 ft lower than is typical for the shallow aquifer at OU2.

The first boring drilled, Soil Boring 9 (SB-9), was located approximately 10 ft south of well SB-8. Three of the borings (SB-10, SB-12, and SB-13) were located approximately 20 ft south of AFCEE well SB-8 in a row perpendicular to the axis of the buried channel in order to define its cross-sectional profile. Another boring (SB-11) was drilled about 30 ft south of SB-8 on the anticipated trend of the channel axis. To determine the stratigraphy of the upper section of the alluvium, borings SB-9 and SB-10 were cored continuously from the surface because it was suspected they were in the vicinity of the chemical disposal pits. For the three remaining borings, 2-ft cores were obtained at 5-ft intervals to approximately 25 ft bgs. Below 25 ft, the boreholes were sampled continuously to the completion depth, approximately 2 ft into the clay aquiclude. Sampling and analytical methods are discussed in Chapter 6.1 and Appendix 1, respectively. The hydrogeologic information obtained from the boring logs and the DNAPL distribution in the test volume determined by analysis of the soil samples are discussed in Chapters 4.1.1 and 4.1.2, respectively. In this initial drilling program, 40 to 80 core samples were collected from each of five borings, for a total of 286 samples (including QA/QC samples). Of these, a subset of 105 samples was selected for grain-size determinations. The sampling frequency was roughly one sample per foot in the vadose zone, and one sample every 6 in. in the upper portions of the saturated zone. In the DNAPL zone at the base of the aquifer, samples were collected every 2 to 4 in. This sampling frequency was designed to provide an adequate representation

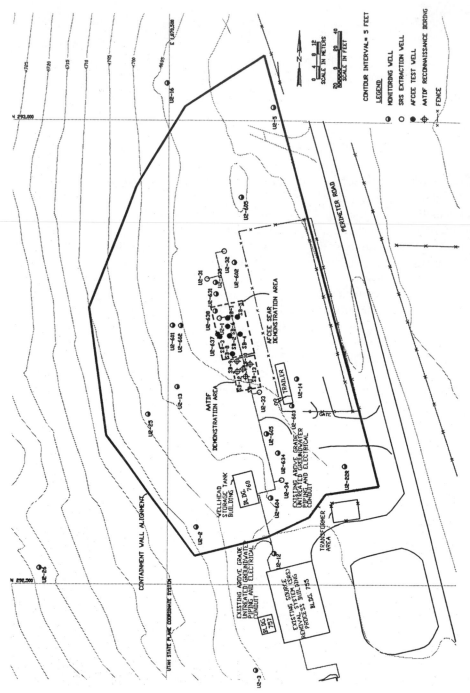

Figure 4.1 AATDF surfactant/foam and AFCEE surfactant-enhanced remediation demonstrations at Hill AFB, OU2.

of the variation in grain size and DNAPL content observed in the zone of interest. If an interface between two stratigraphic units was encountered in the core, a sample was collected from each side of the interface. In addition to the alluvial samples, at least three samples were collected from the clay aquiclude in each borehole to measure the penetration of DNAPL into the clay aquiclude. The samples were analyzed to determine DNAPL saturation.

The contour map in Figure 4.2 shows an interpretation of the clay surface generated after an analysis of these initial AATDF borings. Note that borings SB-9 through SB-13 indicated that the alignment of the channel bottom bends to the southeast, and that the channel seems to rise relatively rapidly to the south from the AFCEE test area depression (shown on the northern half of the map in Figure 4.2). The AATDF well field array was designed using this interpretation, which was subsequently changed as more information became available.

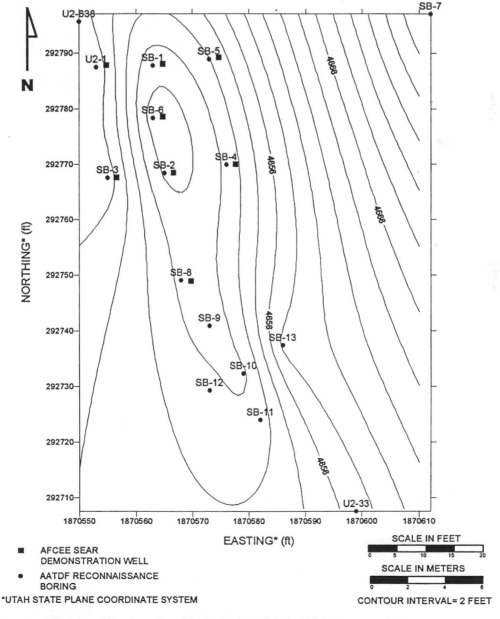

Figure 4.2 Aquiclude surface based on the initial AATDF field investigation.

The purpose of the next field effort was to install the AATDF well array to be used for the surfactant/foam flood. A second objective was to obtain additional information on the hydrogeology of the test area, and to confirm the results of the first soil borings. Two surfactant floods were conducted just north of the AATDF test area prior to well installation. Great care was taken to maintain hydraulic control and prevent surfactant incursion into the AATDF area during those floods. However, it was important to reestablish the baseline contaminant concentrations for the surfactant/foam flood. Soil sampling conducted during the installation of the test wells confirmed the results obtained from the initial borings, and allowed the team to determine if any changes had occurred in the test zone. The nine borings drilled during the installation of the AATDF well field, along with existing wells and borings drilled previously in the vicinity, are shown in Figure 4.3. These borings included EX-1, EX-2, and EX-3 for completion as

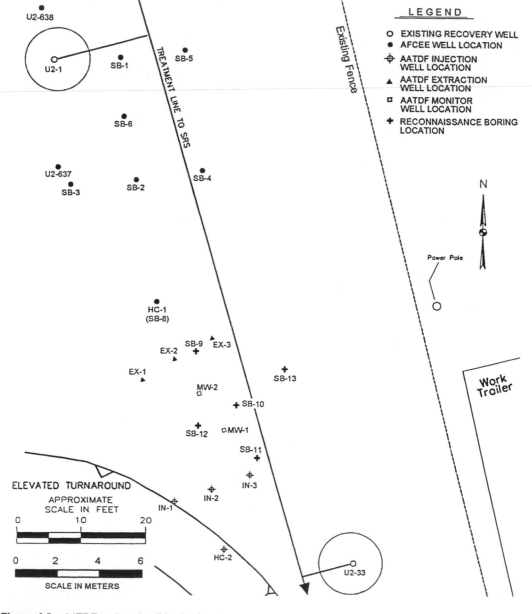

Figure 4.3 AATDF well and soil boring locations.

extraction wells, IN-1, IN-2, and IN-3 for completion as injection wells, MW-1 and MW-2 for completion as multilevel monitoring wells, and HC-2, to be used as the southern hydraulic control well. The borehole locations were selected so as to locate the boreholes in such a manner that the extraction and injection wells could be installed across the deepest portion of the DNAPL zone in the buried channel.

The clay contact in the borehole that was meant to become the westernmost injection well (the original IN-1 location) was encountered at a depth much greater than anticipated, forcing a reevaluation of the interpreted position of the channel centerline. Subsequent drilling along a line to the southwest of this location confirmed that the channel centerline actually continued on the trend discovered during the drilling of the AFCEE wells to the north. Subsequently, the positions of the injection wells were adjusted so that the central well could be completed close to the centerline of the channel. The resulting realignment of these wells is apparent in Figure 4.4. Therefore, as the drilling proceeded, the remaining undrilled boring locations were adjusted based on the elevation of the Alpine clay encountered in previous boreholes. The resulting well pattern is discussed in detail in Chapter 4.2.

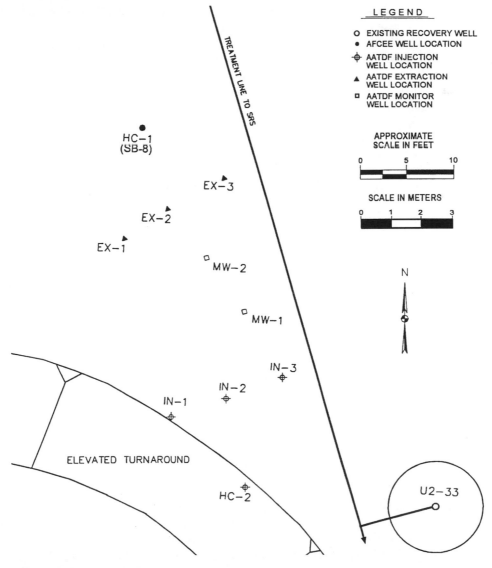

Figure 4.4 AATDF well field array.

To alleviate pressure in the aquifer and thereby minimize sand flow, a pneumatic pump in well SB-8 was used to extract groundwater during the drilling. The pump was operated intermittently due to maintenance on the SRS treatment plant. Each borehole was cored continuously from the ground surface to the completion depth, approximately 2 ft into the clay aquiclude. Twenty to twenty-six samples were collected from each of the nine borings, for a total of 227 samples (including QA/QC samples). The sampling frequency used was roughly one sample per foot in the upper sections of the saturated alluvium for each 2-ft interval of recovered core. In the DNAPL zone at the base of the aquifer, samples were collected every 2 to 6 in., depending on the core recovery obtained with the sampling device. If a sedimentary interface between two samples was encountered, a sample was collected from each side of the interface. Four samples were taken from each clay aquiclude core. The stratigraphy of the AATDF demonstration area is discussed below.

4.1.1   Stratigraphy Observed During Drilling Programs

The AATDF demonstration area straddled two small local depressions in the bottom of the channel, with a small rise or "saddle" in the middle of the array, as shown on the contour map of the aquiclude surface generated with the information obtained after drilling and logging the AATDF well field (Figure 4.5). Figure 4.6 is a block diagram view of the clay surface in this area, illustrating the relationship between these local depressions and the topography of the channel bottom to the north of the AATDF well field.

The stratigraphic data acquired from the boreholes drilled in the AATDF demonstration area were incorporated into three geological cross-sections: two perpendicular to the channel axis and one along the channel axis, as shown on the index map depicted in Figure 4.7. The cross sections, presented as Figures 4.8, 4.9, and 4.10, respectively, show the heterogeneous nature of the alluvium contained in the buried channel, as well as the geometry of the channel itself. The width of the buried channel in the AATDF well field was much narrower than originally anticipated, and although the wells were only 6 ft apart, the screened intervals differed by as much as 8 ft (between EX-1 and EX-2). In the demonstration area, the channel bottom was only 1 to 3 ft wide.

In general, the sediments found in the buried channel increased in grain size with increasing depth. The top 14 ft was composed of silty clays and clayey silts, with occasional occurrences of entrained well-rounded gravel. This is typical of terrace (ancient floodplain) deposits left behind by slow-moving backwaters during flood events. Some of this material, however, was composed of fill used to grade the site and fill in the old chemical disposal pits. Below the upper silty section of the alluvium, very fine sand graded into medium and then coarse sand. The depth at which a significant amount of gravel was encountered varied somewhat, but averaged about 25 ft bgs. The diameter of gravel particles ranged from 0.25 to 3 in., with occasional cobbles (over 3 in.) present. There was also some variability in the degree of rounding, but most of the gravel was subrounded or rounded. Moreover, the percentage of gravel present was also variable; there were occurrences of well-sorted fine gravel as well as poorly sorted sequences (i.e., those containing a continuous distribution of grain sizes).

The geometry of these heterogeneous deposits was predominately lenticular rather than layered, and it is impossible to correlate most individual units between boreholes. These sediments were interpreted to be stacked alluvial channel deposits lain down by a fast-flowing stream that intermittently carried a large bed load. However, very fine sand with some silt was found at the bottom of the deeper depressions in the aquiclude (e.g., at MW-2) and occasionally higher up in the section. These sediments probably represent lower energy environments that existed between erosional events and the subsequent depositional events that filled the channel with the coarse-grained alluvium. A subtle lateral trend in sediment grain size also exists in the channel. The aquifer material generally becomes finer-grained in character both to the north and south of the AATDF demonstration area.

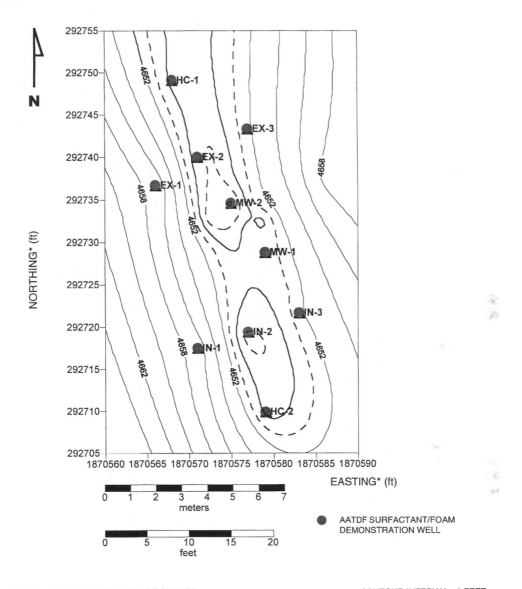

Figure 4.5 Aquiclude surface in the AATDF demonstration area determined after installation of the well field array.

4.1.2 Contaminant Distribution Found in Drilling Programs

DNAPL saturations (volume percent of DNAPL in the sample's pore space) for each of the borings are shown in Figures 4.11 and 4.12. The maximum local contaminant saturations found within the pattern were 22% near the bottom of hydraulic control well HC-2 and 19% in monitoring well MW-2. Generally, much lower values of approximately 1% to 6% saturation were seen over a depth interval of 3 to 4 ft at the base of MW-1, EX-2, and IN-2. The other borings within the pattern showed both lower maxima and smaller vertical extents of contamination.

The volume of DNAPL initially present in the well pattern can be estimated from the sum of products of the saturation values assumed at each elevation, based on cores and the pore volume at each elevation taken from an aquifer model. Using the aquifer model discussed in Chapter 4.4, an estimate of 18 gal initial trapped DNAPL (below its residual saturation) was obtained.

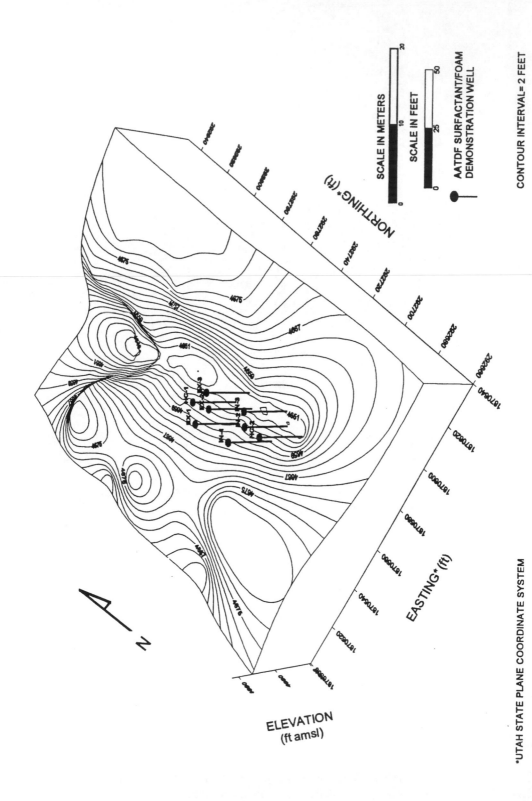

Figure 4.6 Block diagram of aquiclude surface in the AATDF demonstration area.

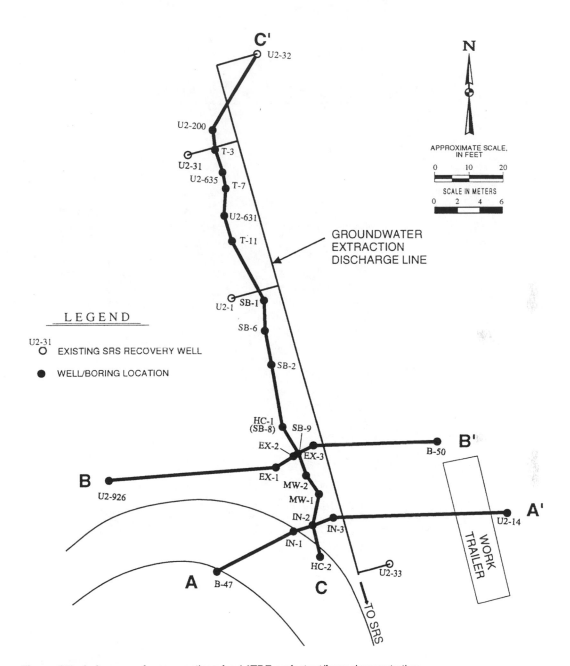

Figure 4.7 Index map of cross sections for AATDF surfactant/foam demonstration.

Additionally, the model assumes that there is 6 gal mobile DNAPL present at the very bottom of the aquifer near IN-2, making a total estimate of 24 gal initial DNAPL between the lines of injectors and producers.

A cross section through the deepest part of the channel is depicted in Figure 4.13; the initial DNAPL contaminated region is shown as the lightly shaded zone in this figure. As seen in the figure, DNAPL was encountered in the deepest portions of the aquifer along a centerline of the channel. The slope of the aquitard shown is that used in the numerical simulations (see Chapter 4.2.5).

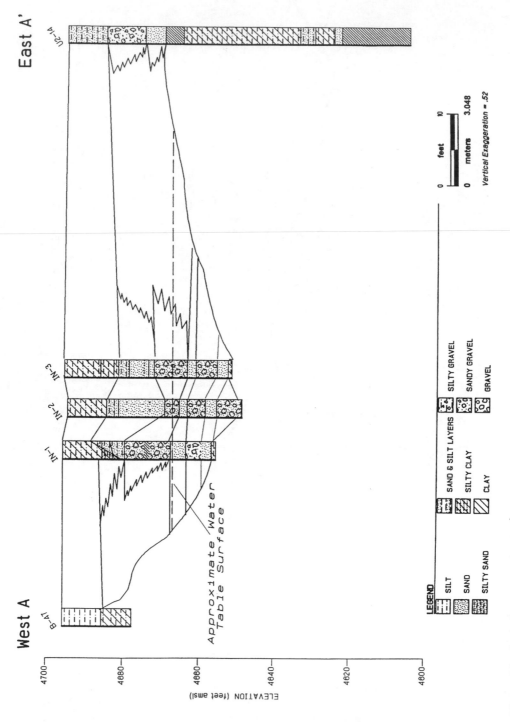

Figure 4.8   Geologic cross section A-A'.

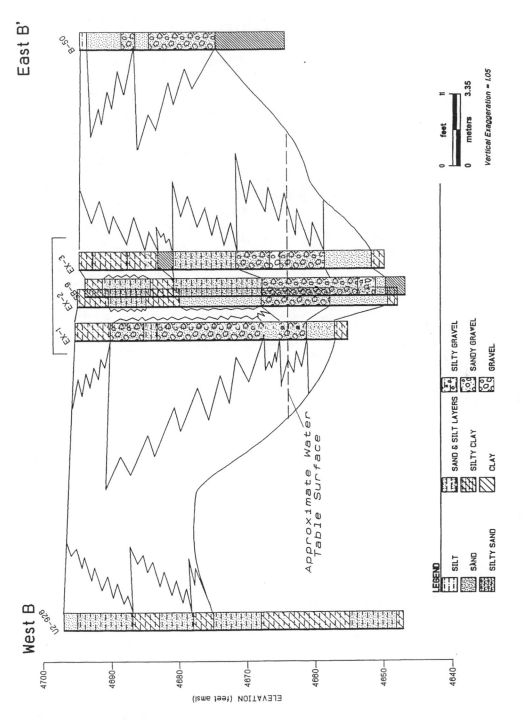

Figure 4.9 Geologic cross section B-B'.

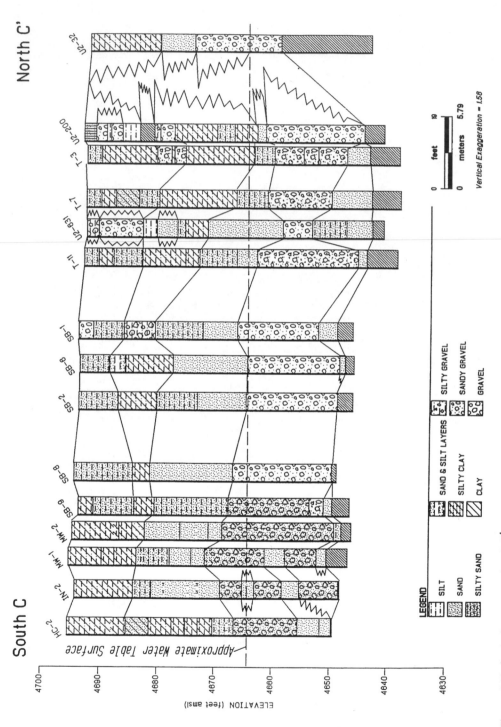

Figure 4.10 Geologic cross-section C–C'.

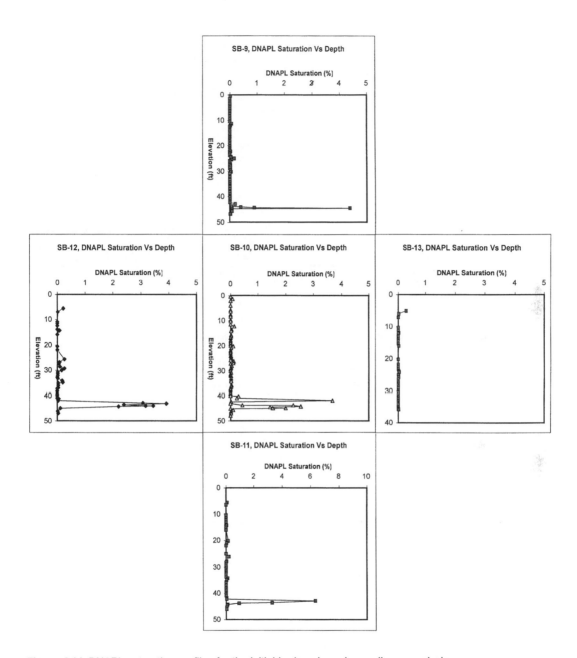

Figure 4.11 DNAPL saturation profiles for the initial borings based on soil core analysis.

4.1.3 Permeability Results

Each of the resulting grain size analyses was fitted to a log normal distribution (see Figure 4.14 for an example), and the permeability and porosity of the sample were estimated. To evaluate the accuracy of the permeability values obtained from the grain-size correlations, the permeabilities of five samples were measured in the laboratory and compared with those predicted from correlations. The alluvial samples chosen for permeability determinations did not contain a gravel fraction and were therefore easily dry-packed into a 1-cm (ID) column. A packed column length of 0.45 m was

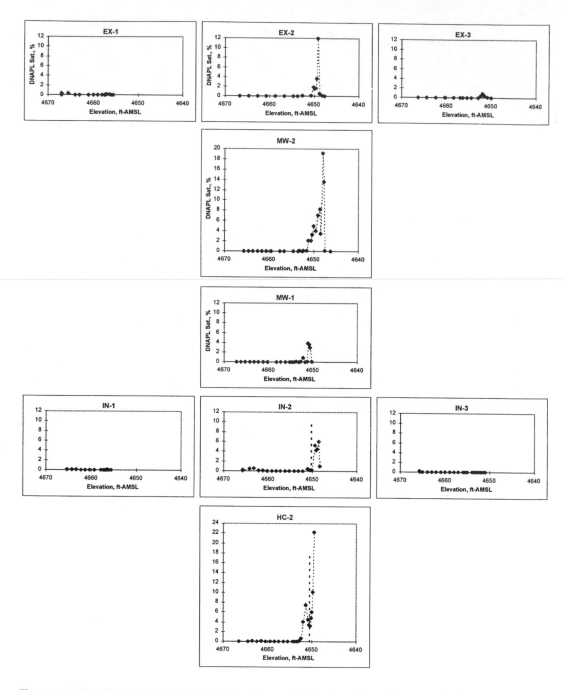

Figure 4.12 DNAPL saturation profiles for the well array borings based on soil core analysis.

used, and the porosity of the resulting packed column was approximately 40%. The results of the column permeability measurements are presented in Table 4.1.

Figure 4.15 shows the comparison between the predicted and measured permeabilities. The two are in very close agreement in the high-permeability region, i.e., above about 50 darcy (5×10^{-4} m/s). For permeabilities below 20 darcy (2×10^{-4} m/s), however, the predicted values are higher than those measured. One possible reason for this discrepancy is that the measured permeabilities

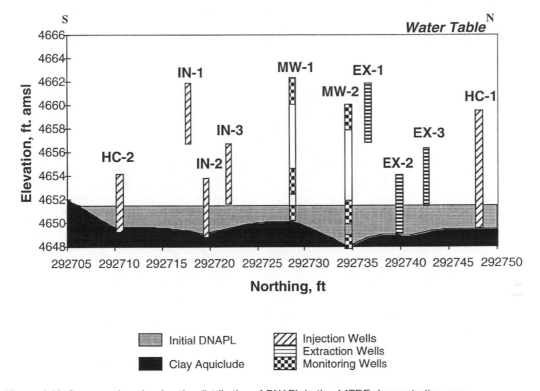

Figure 4.13 Cross section showing the distribution of DNAPL in the AATDF demonstration area.

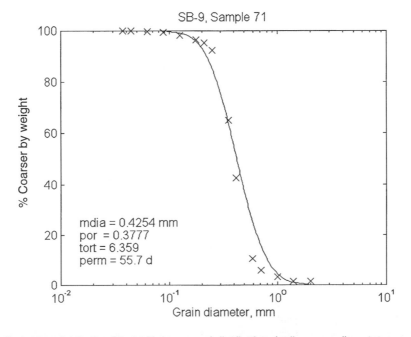

Figure 4.14 Grain size distribution fitted with log normal distribution. (mdia, mean diameter; por, porosity; tort, tortuosity; perm, permeability.)

Table 4.1   Results of Column Permeability Measurements on OU2 Samples

Sample #	Measured			Predicted		
	Porosity	k (darcy)	K (cm/s)	Porosity	k (darcy)	K (cm/s)
SB9-69	0.374	82.1	0.079	0.36	81	0.078
SB9-71	0.395	48.3	0.046	0.37	46	0.044
SB10-102	0.423	1.5	0.001	0.35	5	0.005
SB10-143	0.386	22.5	0.022	0.37	28	0.027
SB12-205	0.441	4.3	0.004	0.37	12	0.012

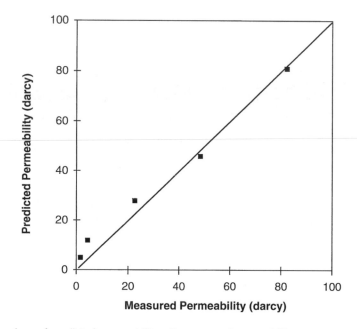

Figure 4.15 Comparison of predicted permeability with measured permeability.

of the small-particle-size (low-permeability) sands may have been overestimated due to nonuniform packing in the column. It is also possible that layers of very fine particles in the column may have contributed to higher pressure drop, hence, lower permeability measurements. Finally, the average particle diameter used for the prediction may have been overestimated due to the presence of clay-bound particle aggregates in the lower permeability samples from OU2.

The permeability of the sand fraction (<4 mm) below 30 ft did not vary as much as was expected, given the lithology recorded on the boring logs. Although the presence or absence of gravel in a hydrostratigraphic unit can give the appearance of large differences in the permeability, the permeability of sand samples with some gravel fraction will likely be dominated by the permeability of the sand fraction. Because the sieve analysis included only particles less than 4 mm, the effect of the gravel was neglected.

The permeabilities of the Hill AFB soil samples estimated from the grain size correlations are plotted against their corresponding gravel fractions in Figure 4.16. The permeability of each sample was estimated from the size distribution of grains less than 4 mm in size; whereas gravel is defined here as particles larger than 4 mm. A linear regression study on the resulting data suggests a weak but positive relationship between the estimated permeability and gravel fraction (see the lines in Figure 4.16). However, the "goodness-of-fit" is very poor, as shown by the R^2 values given in the figure. Furthermore, an analysis of variance (ANOVA) gave very small F values, less than the critical F value for a 95% confidence interval. Thus, the variation in permeability with respect to

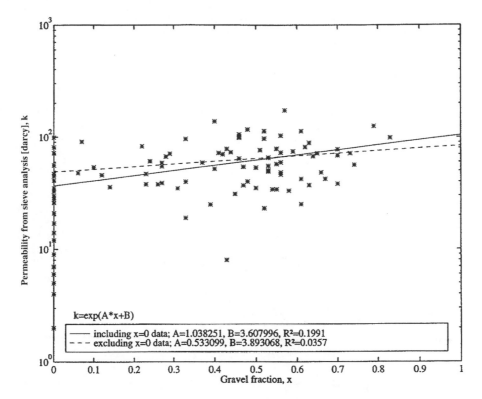

Figure 4.16 Correlation of the permeability of the sand fraction with the gravel fraction.

the gravel fraction in Figure 4.16 is likely to be due to random effects. This conclusion is further substantiated when the data points where no gravel is present in the samples (x = 0) are excluded from the ANOVA analysis. Therefore, no significant relationship is found between the permeabilities estimated from the sand and gravel fractions.

4.1.4 DNAPL Saturation and Permeability Distribution

In the surfactant/foam flood process, removal of DNAPL by a surfactant can only occur where the formation is contacted or "swept" by the injected surfactant solution. In heterogeneous formations, the injected fluids tend to flow through the higher permeability regions and bypass lower permeability regions. However, the sweep efficiency can be improved if the injected fluid is more viscous than the resident fluid. One of the objectives for the AATDF demonstration was to show the effectiveness of using foam for mobility control to improve the sweep efficiency of the surfactant solution.

As discussed in section 4.1.2, the DNAPL at the site is located at the base of the aquifer. To determine if low-permeability zones containing significant amounts of DNAPL existed in the surfactant/foam demonstration area, the permeability distribution estimated from the initial, or reconnaissance, borings was compared with the DNAPL distribution determined at the same time. The resulting comparison is shown on the plots in Figure 4.17, where the permeability is given in darcy and the DNAPL content is given in percent saturation. SB-13 is not shown because it had no significant DNAPL, and only two samples from this borehole were selected for grain-size and permeability determinations. The permeabilities greater than 100 darcy (conductivities greater than 0.001 m/s) were off-scale and are not shown so that there would be sufficient graphic resolution in the lower permeability regions. Notice that DNAPL was not detected in the samples from these high-permeability zones. Most of the DNAPL was trapped in the lower permeability sediments

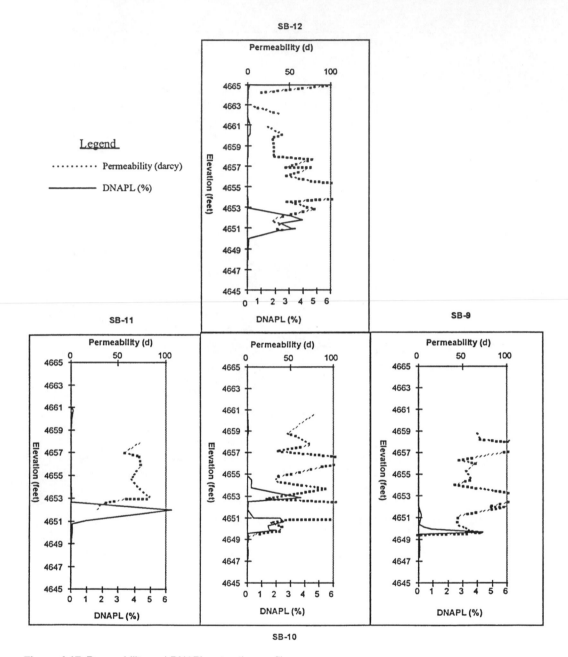

Figure 4.17 Permeability and DNAPL saturation profiles.

located at the base of the aquifer, making the OU2 AATDF demonstration a viable test of mobility control by foam in the surfactant/foam process.

4.2 TEST SITE PATTERN AND INSTRUMENTATION

The AATDF well array installed for the surfactant/foam demonstration was illustrated in Figure 4.4. Reference to this figure shows that the surfactant/foam injection and extraction wells were placed in a "line-drive" pattern to induce movement of the injected fluids along the axis of the buried alluvial channel. The AATDF well array consisted of a line of three extractors (EX-1, EX-2, and EX-3) spaced approximately 6 ft apart at the northern end of the test area, and a line of three injectors (IN-1, IN-2, and IN-3) spaced approximately 6 ft apart at the southern end. The injection and extraction well lines were 20 ft apart. In addition to the injection and extraction wells, two monitoring wells (MW-1 and MW-2) were installed near the center of the pattern. Two hydraulic control wells, HC-1 (SB-8) and HC-2, injected potable water during the flood and thereby created a groundwater barrier to minimize the loss of fluids from the test zone. During the flood, existing SRS monitoring wells U2-33 and U2-631 were used to monitor the aquifer pressure surrounding the AATDF well field.

4.2.1 Design of Injection and Extraction Wells

A schematic of the important features of the injection and extraction well construction is shown in Figure 4.18. The wells each had an inside diameter of 4 in. and were constructed using a Schedule 40 PVC casing connected to a 10-ft length of stainless steel casing atop a 5-ft-long stainless steel wire-wrapped screen with 0.02-in. slots. Solid screw-on sumps approximately 5 in. long were attached to the bottom of the screens, and two centralizers were used to ensure that the well was centered in the borehole during the placement of the filter pack. Special care was taken in constructing each well to ensure that a proper seal was achieved above the filter pack, preventing injected gas from migrating up the borehole annulus. To construct the filter pack, a 16–30 Colorado silica sand was placed with water by tremie tube around the well screens. The sand was brought a minimum of 2 ft above the well screens, as determined by measuring to the top of the filter packs with a weighted tape. A thin (approximately 6 in.) layer of fine (100-mesh) sand was placed on top of the coarser sand to complete each filter pack. Two feet of 1/4-in. bentonite pellets were then placed on top of each filter pack and hydrated with a minimum of 15 gal potable water. The bentonite seal was allowed to hydrate for a minimum of 4 h before well construction continued. To prevent the possibility of pressure loss along the well annulus, a concrete grout was then pumped upward into the annulus from the bentonite seal using a tremie tube. This cement mixture was formulated so that grout loss to the formation would be minimized. The volumes and quantities of materials required to construct the well were determined and recorded before placement of the material. Careful inventory and record were then kept of each material during its placement into the annulus to ensure that no void spaces were created by material "bridging off" in the annulus.

A 4-in. Schedule 40 PVC adapter with male NPT threads was glued to the top of the PVC casing of each injection and extraction well. These threaded collars were designed to provide an air-tight connection for the wellhead-injection manifold used during the foam flood. Surface completions, including the unreinforced concrete pads surrounding each group of wells, were constructed flush with the existing ground surface because the AATDF test area fell within an active construction zone. Flush-mounted steel vaults, which were lockable and provided protection for each well, covered the well risers. These vaults were installed with a raised lip above the concrete pad to facilitate the installation of a high-density polyethylene (HDPE) secondary containment liner for the surfactant/foam flood. Well completions were recorded and documented in the field on well completion diagrams. The screens for all the extraction and injection wells were 5 ft long, with the exception of well HC-1, which had been previously constructed with a 10-ft screen for a different project.

4.2.2 Monitoring Well Design

Two SOLINST multilevel samplers were used in the central portion of the test area between the line of extraction wells and the line of injection wells to monitor the less-permeable zone located at the base of the aquifer where the aquiclude was contacted and to monitor two high-permeability "thief zones" higher up in the formation (see Figure 4.4). The multilevel samplers were installed in special 4-in.-diameter monitoring wells to facilitate preventive and operational maintenance on

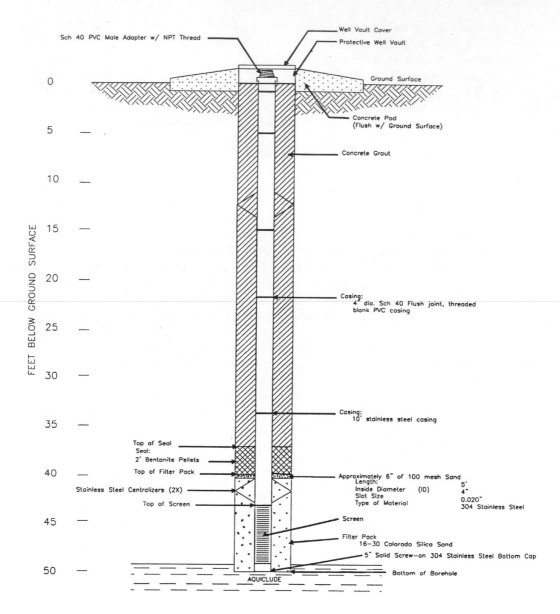

Figure 4.18 Injection/extraction well construction detail.

these instruments. Each monitoring well was outfitted with three short stainless steel screens spanning the stratigraphic intervals of interest.

Each monitoring well for the multilevel samplers had three screens: a 1-ft screen at the bottom (the A sampling point), a 2-ft screen located 2 ft above the top of the first (the B sampling point), and a 2-ft screen located 4 ft above the top of the second (the C sampling point). The stratigraphic zones monitored by the multilevel samplers were based on the analysis of the detailed grain-size distributions of borings SB-9 through SB-13. A dual-packer system seated in a stainless steel blank casing was used to isolate each sample port in the access well. After the multilevel samplers were inserted into each of the monitoring wells, the screened intervals were isolated by inflating packers to seat them in the blank casing sections. The multilevel samplers were fitted with sampling ports, pneumatic pumps, and pressure transducers to monitor each of the isolated screened intervals.

The monitoring wells were constructed using 4-in. ID Schedule 40 PVC casing connected to a section of blank stainless steel casing that contained the three lengths of stainless steel 0.01-in. slot wire-wrapped screens. Figure 4.19 shows the monitoring well construction. The well completions, including the flush-mounted steel vaults and unreinforced concrete pads, were constructed in the same fashion as the injection and extraction wells. The filter packs consisted of a 16–30 Colorado silica sand placed with potable water by tremie tube around the well screens. The sand was brought to a minimum of 2 ft above the highest well screen in each monitoring well, and then approximately 2 ft of bentonite pellets were placed on top of each filter pack and hydrated with a minimum of 15 gal potable water. The bentonite seals were allowed to hydrate for at least half an hour before the remaining well annulus was filled to the surface with a concrete grout. The grout was formulated with Portland cement and approximately 5% powdered bentonite to minimize grout loss to the formation. Centralizers were used to

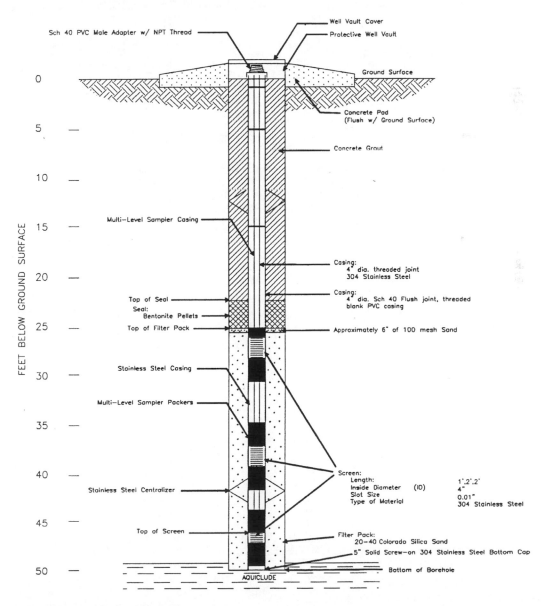

Figure 4.19 Monitoring well construction detail.

ensure that the monitoring well was properly centered in the borehole during placement of the filter packs, bentonite seals, and grout. Monitoring well completions were recorded and documented in the field on construction forms.

4.2.3 Surveying and Well Elevations

The last task completed in the well installation effort was surveying the location and elevation of each well in the pattern. An engineering tape was used to measure the distances between the wells and a known point (an existing monitoring well). The angles between these points were measured using a transit. The transit was also used to determine the elevation of both the ground surface at each boring location and the reference or measuring point at the top of each test well casing (TOC). The resulting data were then translated to Utah State Plane coordinates and to elevations above mean sea level (amsl). The survey data for each well in the AATDF test area are listed in Table 4.2.

Table 4.2   Well Field Survey Data

Well	Northing (ft amsl)	Easting (ft amsl)	TOC Elevation (ft amsl)	Ground Elevation (ft amsl)	Clay Elevation (ft amsl)
EX-1	292736.70	1870566.35	4694.63	4695.04	4657.04
EX-2	292739.98	1870571.30	4694.46	4694.66	4649.16
EX-3	292743.34	1870576.69	4694.36	4694.36	4651.56
MW-1	292728.83	1870579.16	4694.66	4695.19	4650.49
MW-2	292734.60	1870575.15	4694.39	4694.55	4647.85
IN-1	292717.54	1870571.32	4695.38	4695.36	4656.56
IN-2	292719.43	1870577.14	4694.90	4694.16	4649.06
IN-3	292721.67	1870583.31	4694.97	4694.97	4651.97
HC-1	292749.07	1870568.15	4696.74	4694.29	4649.60
HC-2	292709.85	1870579.24	4695.40	4695.40	4649.60

The elevations of the top and bottom of each AATDF well screen in the demonstration area are listed in Table 4.3. The screened interval of each well reflects the structure of the buried channel since the elevation of the bottom of the wells was dictated by the aquifer/aquiclude contact encountered at each well location. The width of the buried channel in the AATDF well field was much narrower than originally anticipated; and although the wells were only 6 ft apart, the elevations of the screened intervals differed by as much as 8 ft (between EX-1 and EX-2).

Table 4.3   Elevation of Screened Intervals of Wells

Well	Bottom of Screen (ft amsl)	Top of Screen (ft amsl)
EX-1	4656.53	4661.13
EX-2	4648.76	4653.36
EX-3	4651.36	4655.96
IN-1	4656.47	4661.07
IN-2	4648.96	4653.56
IN-3	4651.86	4656.46
MW-1A	4650.28	4651.28
MW-1B	4654.76	4656.76
MW-1C	4660.24	4662.24
MW-2A	4647.59	4648.60
MW-2B	4642.09	4654.08
MW-2C	4657.56	4659.57
HC-1	4648.46	4658.06
HC-2	4649.50	4654.10

4.2.4 Well Development

After the wells were installed, a minimum of 24 h was allowed to pass before each well was developed. To develop each well, a surge block was used to gently pull and push water across consecutive 1.5-ft sections of the well screen and filter pack. A bailer was used to periodically evacuate the wellbore of sediment-laden groundwater. The progress of the development effort was monitored by observing the amount of sediment in the purge water and measuring the pH, conductivity, and temperature of water samples collected from the bailer after each borehole volume of water was removed from the well.

A well was considered to be developed when at least 10 borehole volumes had been removed from the well, the purge water was relatively free of sediment, and the pH, conductivity, and temperature of the purge water had stabilized to within 10% of the previous set of readings. The water produced at each well during development was collected in 55-gal drums, and then transported and pumped into the SRS for treatment.

4.2.5 Slug Tests

Slug tests were used to determine the hydraulic conductivity of the area immediately surrounding a well. Generally, the greater the instantaneous head change induced during the slug test, the greater the stress placed on the aquifer and the greater the volume of aquifer tested. By testing each well in the line-drive pattern, a general indication of the spatial variation in hydraulic conductivity was obtained.

The slug tests conducted in the AATDF demonstration wells were both pneumatic, using air pressure to depress the water level in the well, and mechanical, using an object of known volume to raise or lower the water level in the test well. During the pneumatic testing, an air-tight wellhead manifold was used to seal off the annulus of the well being tested while allowing air to be delivered into the wellbore from a compressor. Once the well was pressurized to depress the water level the desired distance, a valve was closed (typically about 30 s) to maintain this pressure in the well while the aquifer regained equilibrium. Then, a valve on the wellhead manifold was opened, allowing the pressure in the wellbore to equalize instantaneously with atmospheric pressure.

The water level changes induced in the test well during the pneumatic and mechanical slug tests were measured automatically with electronic pressure transducers connected to a data acquisition system (DAS). The mechanical slug tests were monitored with a pressure transducer placed at the bottom of the well. For the pneumatic slug tests, a pressure transducer installed in the wellhead manifold was used to measure air pressure, and a transducer at the bottom of the well was used to record total pressure. During each pneumatic test, the water level in the well was monitored in real time by having the DAS subtract the air pressure measurement from the total pressure reading and displaying the resulting curve on a computer screen.

An example of a water level displacement plot from a pneumatic slug test conducted in well HC-2 is shown in Figure 4.20. Note that the initial displacement, or head change, in the wellbore obtained from a pneumatic slug is much greater than typically achieved with a mechanical slug, which relies on the volume of a cylinder dropped into the well to induce the initial head change. Moreover, the crucial early data acquired from a pneumatic slug are free of the water-level oscillations that commonly plague mechanical slug tests. In situations such as this, in which the aquifer is as transmissive as the shallow sand and gravel aquifer at OU2, pneumatic slug tests provide the distinct advantage of being able to measure induced responses before they dissipate. The slug test curves for each well were analyzed for conductivity using a method developed for unconfined aquifers by Bouwer and Rice (1976). The fitted line obtained from the Bouwer and Rice solution is also shown in Figure 4.20.

Table 4.4 lists the values of hydraulic conductivity determined from the pneumatic and mechanical slug tests for each well in the line-drive pattern. The hydraulic conductivities for the pneumatic

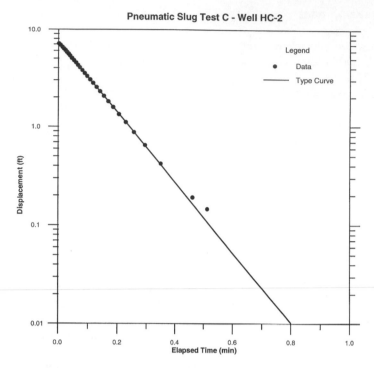

Figure 4.20 Example of a slug test curve and the solution fit.

and mechanical testing are similar, but in six out of the nine test wells, the mechanically derived conductivity is slightly higher than the pneumatic conductivity. This result can probably be attributed to the fact that, in general, the conductivities of the filter packs are slightly higher than the surrounding formation. Because the smaller displacements achieved by the mechanical slugs result in smaller radii of influence, the mechanical slug test results are biased by the filter pack to a greater degree than the pneumatic slugs. Hydraulic conductivity in the demonstration area generally decreased to both the north and to the south. The slug tests appeared to give hydraulic conductivity values very similar to those obtained previously (INTERA, 1997b).

Table 4.4   Hydraulic Conductivities from the Slug Tests

Well	Pneumatic m/s	Pneumatic ft/d	Mechanical m/s	Mechanical ft/d
EX-1	1.50E-04	42.5	1.61E-04	45.6
EX-2	2.64E-04	74.8	2.81E-04	79.6
EX-3	2.97E-04	84.1	3.31E-04	93.8
IN-1	2.88E-04	81.6	2.71E-04	76.8
IN-2	4.69E-04	132.9	5.23E-04	148.2
IN-3	2.19E-04	62.0	2.40E-04	68.0
MW-1	6.81E-04	192.9	5.26E-04	149.0
MW-2	4.39E-04	124.4	4.02E-04	113.9
HC-1	1.34E-04	38.0	1.47E-04	41.6
HC-2	3.13E-04	88.7	3.33E-04	94.3

4.2.6   Extraction/Injection Test

The injection, extraction, and the water-level monitoring systems used during this test were prototypes of the systems to be used during the actual demonstration. A 5.2-d extraction/injection test was conducted in the newly installed AATDF well field as a dry run of the field demonstration.

Well EX-2 was used as the producing (extraction) well for the test, and wells IN-2 and HC-2 were used as injection wells. The extraction/injection test was composed of several test segments, as shown in Figure 4.21, which is a compilation of the water levels recorded at the injection and extraction well during each test segment. These segments included:

- Background monitoring: a 10.5-h period during which the pre-test water levels in the AATDF demonstration area were monitored to establish the natural gradient and any transitions in the aquifer under ambient conditions.
- Extraction/injection test: 4.5 d of source (potable) water injection into well IN-2 at an average rate of 2.1 gpm while extracting from well EX-2 at an average rate of 4.3 gpm. Potable water was also injected into well HC-2 at an average rate of 4.0 gpm to provide hydraulic control over the test.
- Recovery test: post-test monitoring of the water levels in the well field for 4.5 h.

The injection system consisted of three polyurethane tanks with a recirculation system for injectate batch mixing and storage and a fluid delivery system. The fluid delivery system had a centrifugal injection pump, a flow control system with needle valves and digital flowmeters, an injectate sampling port, and drop tubing to deliver the fluid to the bottom of the well screen. The extraction system was a pneumatic pump installed in well EX-2 that was controlled with a needle valve and a digital flowmeter on the fluid discharge line. The discharge line from the pump was connected to the treatment line at existing SRS well U2-1, allowing the effluent from the test to be piped directly to the SRS system for treatment. The pneumatic pumps used for this project were chosen for their low shear characteristics. They are operated by air pressure cycles, creating a pulsed discharge flow and water level fluctuations in the well that are evident in the water level record for EX-2 in Figure 4.21.

During the aquifer test, water levels were monitored electronically with pressure transducers and the DAS. Pressure transducers were installed in the extraction well, injection well, hydraulic control well, and two additional wells (IN-3 and MW-2) in the AATDF array. For all of the test stages, water levels were automatically monitored every 20 min, and were augmented and verified by taking periodic manual readings. During the first day of the test, the discharge line from the extraction well was replaced with a larger-diameter line, which allowed the extraction rate to increase from 3 gpm to 4.3 gpm. Four days into the extraction/injection test, extremely cold weather created fluid flow problems, and the injection rates temporarily dropped by about 25%. The resulting rate fluctuation is reflected in the water level record seen in Figure 4.21.

4.2.7 Conservative Tracer Test

The application of the surfactant/foam process required the establishment of a forced-gradient flow field and the use of groundwater to transport the surfactant. A conservative (nonreactive) tracer test using sodium chloride was conducted during the injection/extraction test to understand how an induced flow system behaved in the AATDF demonstration area and to provide calibration data for the numerical model (UTCHEM) used in design. The results of the nonreactive tracer test were then analyzed to characterize dispersivity and travel time.

First, the flow field created by the simultaneous injection and extraction operations was allowed to reach steady state. Next, 1348 gal of a 2750 mg/L (6179 µS) NaCl solution was injected for 10.5 h into well IN-2 at a rate of 2.1 gpm while pumping from EX-2 at rate of 4.3 gpm. During the tracer test, the effluent of the extraction well was monitored for several water quality parameters, including conductivity, pH, temperature, and dissolved oxygen.

The injection of the electrolyte was followed by 3.5 d of water flooding to transport the tracer across the zone of interest. Monitoring of the tracer test effluent was terminated at the end of 2.5 d after extremely cold weather created fluid flow problems. The resulting tracer curve obtained at EX-2 is shown in Figure 4.22. The line depicted in the figure is an extrapolation of the tail of the

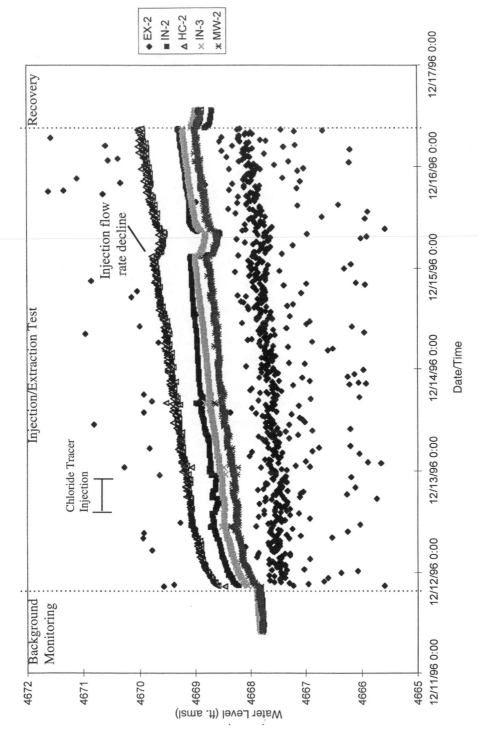

Figure 4.21 Extraction/injection test water levels.

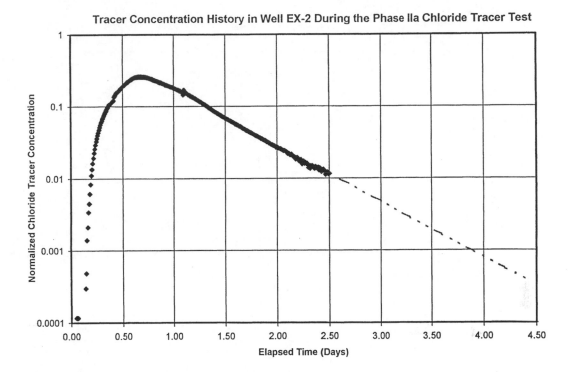

Figure 4.22 Conservative tracer test response.

tracer curve beyond 2.5 d. The volume of the aquifer between IN-2 and EX-2 swept by the chloride tracer during the test was calculated to be approximately 2760 gal. The calculated tracer recovery was above 100%, a reflection of the fact that the concentration of the tracer slug actually varied somewhat over the injection period.

The dispersivity measured by the tracer in the induced flow field was 5 ft over a transport distance of 20 ft. The Peclet number (a ratio expressing advective to diffusive transport) of the tracer response was determined to be 3.96. The tracer peak arrived in 15.8 h under a forced gradient of 0.023. The results of the tracer test were used to calibrate the UTCHEM model. This calibrated model was then used to design the PITTs used during the surfactant/foam flood, design the sampling program for the PITTs, and predict the behavior of the surfactant solution during the flood.

4.2.8   Single Well Foam Generation Test

A single well foam generation test was conducted to determine whether foam could be generated *in situ*. This field experiment took place in well SB-6, which was immediately adjacent to and north of the AATDF test area.

Air was supplied for the foam generation test with an air compressor. The air pressure was regulated by the operator, and flow rates were manually measured with a pair of rotometers. Pressures were measured with transducers at the bottom of the well and in the air flow line. Pressures (including differential pressure) were continually recorded and displayed on a simulated strip chart. The differential pressure is an indicator of the liquid level in the well.

The foam generation test well, SB-6, was screened over a 10-ft interval. The liquid was injected through tubing that extended close to the bottom of the well to avoid foam. A distributor was used at the end of the tubing to deflect the liquid stream to avoid dredging sediment from the bottom of the well. The air was injected directly into the well annulus. The wellhead was secured with a

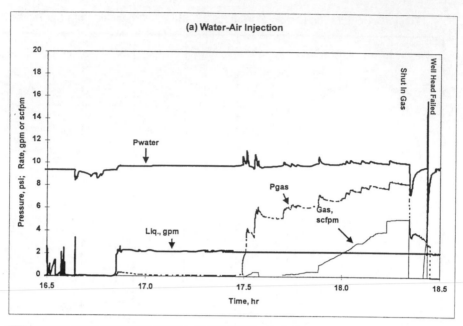

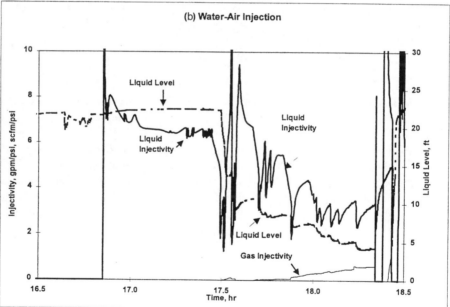

Figure 4.23 Water–air injection.

section of rubber sleeve and hose clamps. The wellhead failed twice during the test, accenting the need for a competent system during the field demonstration.

The results of the test are shown in Figure 4.23 for the water–air injection and in Figure 4.24 for the surfactant–air injection. The liquid level was determined by dividing the differential pressure by 0.433 psi/ft. The liquid injectivity was calculated by dividing the liquid injection rate, which is in gpm, by the difference of the water pressure at the bottom of the well and the hydrostatic pressure of 9.34 psi. The gas injectivity was calculated by dividing the gas injection rate (in standard cubic feet per minute [scfpm]) by the gas pressure. The gas injection rate was zero (except for a small amount of leakage from the wellhead) until the liquid level dropped below the top of the screened interval of 10 ft.

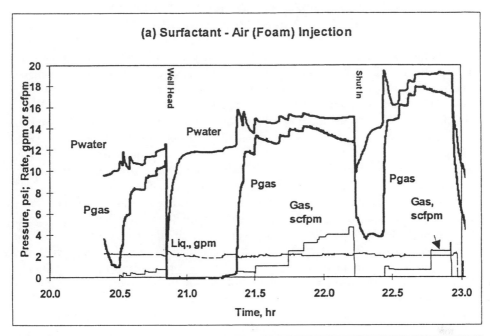

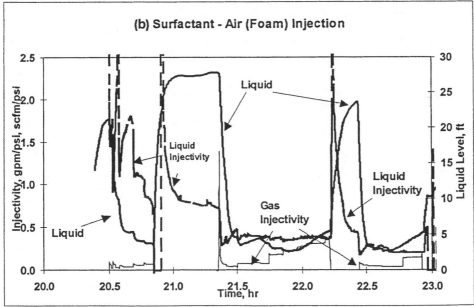

Figure 4.24 Surfactant–air injection.

As seen in Figure 4.23, the reduction of liquid injectivity and increase in gas injectivity corresponded to the change in liquid level in the screened interval. The gas injectivity increased to 0.62 scfpm/psi. Note that the injectivity scale in Figure 4.24 differs from that in Figure 4.23. Both the liquid and gas injectivity were reduced in the surfactant–air injection as compared with the water–air injection. The gas injectivity increased with time while gas was injected. After the gas injection was momentarily stopped and the liquid was allowed to enter over the entire screened interval, the subsequent cycle of gas injection had reduced gas injectivity, confirming that some foam had formed.

4.3 CHARACTERIZATION OF NAPL VOLUME USING
PARTITIONING INTERWELL TRACER TESTS

The partitioning interwell tracer test (PITT) was developed at the University of Texas by Professor G. A. Pope of the Department of Petroleum and Geosystems Engineering during the early 1990s. The PITT was developed from a concept that had first been used by the oil industry to measure the residual oil saturation that is homogeneously distributed in water-flooded well fields. Pope and colleagues modified the method so that not only the residual oil saturation, but also the swept pore volume and, therefore, the total volume of heterogeneously distributed DNAPL, could be determined. Therefore, the PITT allows the detection and volume estimation of DNAPL. A particularly important application of the test is remediation performance assessment (i.e., determination of the volume of DNAPL before and after a remediation effort).

The PITT involves the injection of a suite of tracers in one or more wells and subsequent extraction from other wells in a well field. Conservative (i.e., nonpartitioning) tracers pass unretarded through the DNAPL zone, whereas partitioning tracers are retarded due to their partitioning into and out of the DNAPL. In the unsaturated zone of an aquifer, the tracers employed are gases, whereas liquid tracers (e.g., alcohols) are used in the saturated zone. The chromatographic separation of the tracers due to this partitioning is used to measure the volume of DNAPL in the interwell zone.

Partitioning tracer column studies conducted with DNAPL and soil from the site provided important site-specific information on the behavior of the tracers. Simulation modeling prior to the field test with UTCHEM was used for this tracer test design.

4.3.1 Partitioning Tracer Column Studies

The objective of the partitioning tracer column experiments was to test the ability of partitioning tracers to determine residual DNAPL saturation in the alluvium. First, several partitioning tracer experiments were performed in columns packed with uncontaminated Hill alluvium to determine if there was any retardation of alcohol tracers by the organic material in the aquifer sediment. In these column studies, IPA was used as the nonreactive tracer and 3-methyl-3-pentanol, 1-hexanol, and 2,2-dimethyl-3-pentanol were the partitioning tracers. The results of these retardation experiments are presented in Table 4.5. Retardation factor measurements for selected partitioning tracers listed in this table range between 0.999 and 1.028, well within the ±0.035 experimental accuracy for measurement of tracer retardation. There was negligible retardation or adsorption of these particular partitioning alcohol tracers by the Hill aquifer material.

Table 4.5 Adsorption of Partitioning Tracers by OU2 Alluvium

Alcohol	Retardation	Kd	Adsorption (mg/kg)
3-Methyl-3-pentanol	0.999	−0.00011	−112.0
1-Hexanol	1.017	0.00295	294.7
2,2-Dimethyl-3-pentanol	1.028	0.00485	485.2

Next, partitioning tracers were perfused through columns packed with contaminated sediment from Hill AFB, OU2, to determine residual Hill DNAPL saturation. A typical example of the response of a partitioning tracer in a DNAPL-contaminated column is shown in Figure 4.25. Note that the position of the alcohol peak concentration is retarded relative to the peak of the tritium curve, which represents the nonpartitioning tracer response in this figure. Because of a lack of interaction with the material in the column, the tritium peak moves through the column at the velocity of the fluid flowing through it. The alcohol, however, partitions into and out of

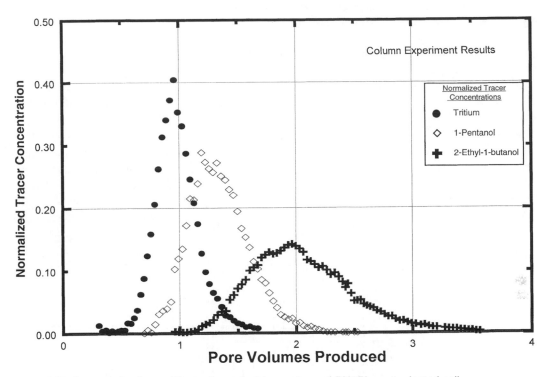

Figure 4.25 An example of a partitioning tracer test in a column of DNAPL-contaminated soil.

the DNAPL trapped in the pore spaces of the sediment in the column, so the velocity of its peak is slower than that of the fluid carrying it. Since the partition coefficient of the alcohol with DNAPL is known, the amount of DNAPL present in the column pore space can be determined by measuring the retardation of the alcohol relative to a conservative tracer. To verify this concept, an independent measurement of the residual DNAPL in the column must be compared with the estimate provided by the partitioning tracer. For these column studies, estimates of the amount of residual DNAPL present were calculated using both a mass balance (weighing the column before and after the addition of the DNAPL) and a volume balance (comparison between the known pore volume of the column and the volume of the DNAPL retained). The data analysis methods used to calculate the residual saturation of DNAPL from tracer curves are presented in Chapter 6.7.

A summary of residual DNAPL saturation estimates based on partitioning tracers, mass balance, and volume balance for selected partitioning tracer experiments conducted with DNAPL-contaminated columns is given in Table 4.6. The partitioning tracer estimates of residual DNAPL saturation agreed well with the volume and mass balance estimates, with a standard deviation of under 5% between the partitioning tracer estimates and the mass balance estimate. These results indicate that partitioning tracers can accurately determine residual DNAPL saturation in column experiments. However, the determination of residual DNAPL saturation using partitioning tracers requires

Table 4.6 Comparison of Residual DNAPL Estimates Based on Partitioning Tracers and Mass and Volume Balances

Soil Sample	Residual Saturation Mass Balance	Residual Saturation Volume Balance	Residual Saturation Partitioning Tracers
Ottawa	0.245	0.240	0.229
Hill AFB	0.254	0.225	0.256
Hill AFB	0.261	0.254	0.255

adequate residence time. The alcohol tracer must remain in contact with the DNAPL zone long enough for equilibrium partitioning to be achieved.

4.3.2 Simulations for Partitioning Tracer Test Design

A three-dimensional (3-D) model of the saturated subsurface using an appropriate geometry and grid and the best available estimates of hydraulic conductivity, porosity, dispersivity, fluid densities and viscosities, location and quantity of DNAPL, and other geochemical properties of the alluvium was constructed. The selection of the model area was based on the AATDF well pattern installed at the site. Based on the injection and extraction pattern, the simulation domain was divided into $28 \times 19 \times 15$ grid blocks. A plan view of the simulation grid is shown in Figure 4.26. The saturated thickness of 16 ft was divided into 15 vertical layers with 1 ft for each layer, except for the top layer, which was 2 ft thick. Smaller grid sizes were used for the grid blocks around the injection and extraction wells. These model dimensions and the number

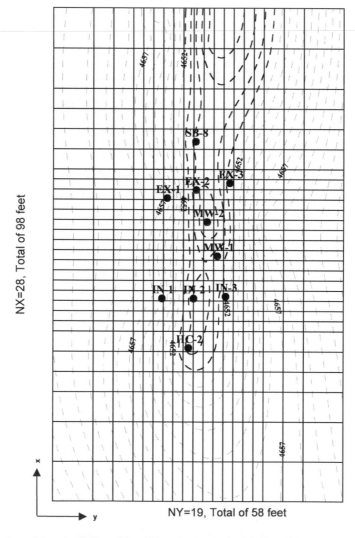

Figure 4.26 Plan view of the simulation grid, well location and top elevation of the aquiclude.

of grid blocks were chosen to minimize boundary effects on the flow field in the test area. The hydraulic heads at two outer boundaries of the simulation domain were kept constant at atmospheric pressure. The model incorporated the 3-D structure of the buried channel formed by the aquiclude and a permeability field based on analysis of the grain-size distributions of soil samples collected from the test area. The distribution of DNAPL in the model was based on data from soil samples. Most of DNAPL in the AATDF test zone was localized in a low-permeability sand layer lying on top of the aquiclude at the bottom of the channel, with residual saturations as high as 16% found in some samples. More residual DNAPL was also apparently trapped in a thin layer of finer-grained sediments found a few feet above the lower layer in some parts of the pattern, with a high-permeability layer containing relatively little DNAPL sandwiched between the two.

Next, the PITT model was calibrated to the nonpartitioning tracer test conducted after the installation of the well array. The model calibration was achieved by adjusting model parameters (such as porosity, dispersivity, and relative permeability) within their ranges of uncertainty until the model predictions of tracer concentrations matched those measured at extraction well EX-2 in the field. Figure 4.27 compares the UTCHEM simulation results and the field data. The model predictions and field data are in very good agreement, indicating that the simulation model was a reasonable representation of the aquifer in the AATDF test area.

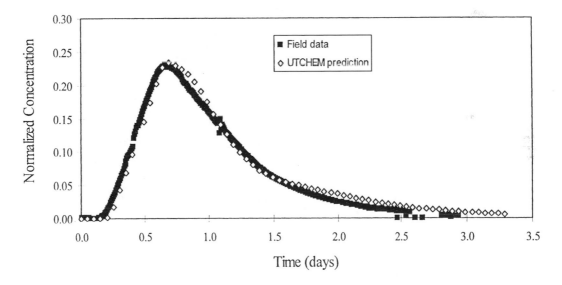

Figure 4.27 Simulation results of the calibrated model and the normalized tracer data.

Following calibration of the model, a number of sensitivity simulations studied the behavior of different partitioning tracers. These sensitivity studies varied the injection and extraction rates and the amount and distribution of DNAPL. The results were then used to select the partitioning tracers and determine the duration of the tracer test, the mass of each tracer needed, the injection and extraction rates, the extraction well effluent tracer concentrations, and the amount of tracer recoverable at the end of the tracer test. An optimum operation design was chosen, based on the results of these sensitivity studies. The design variables based on these sensitivity studies and model predictions are shown in Tables 4.7 and 4.8. The predicted tracer recovery at the end of the tracer test (6 d) was expected to be approximately 83%. The predicted swept aquifer pore volume was approximately 9930 gal.

Table 4.7 Design Summary of the Injection/Extraction Operations

Task	Injectate Composition in Well IN-1,-2,-3 @ 2 gpm	Injectate Composition in HC-1 @ 3 gpm HC-2@ 1 gpm	Duration (days)	Cumulative Time (days)	Extraction Rate at EX-1, 2, 3 (gpm)
Pre-flood PITT	H_2O	H_2O	1.0	1.0	3.3
	H_2O + tracers	H_2O	0.4	1.4	
	H_2O	H_2O	4.6	6.0	
	H_2O + NaCl	H_2O	1.0	7.0	
Surfactant/ foam flood	H_2O + NaCl + surfactant	H_2O	0.5	7.5	3.3
	H_2O + NaCl + surfactant + air	H_2O	1.5	9.0	
	H_2O + NaBr + surfactant + air	H_2O	1 h	9.0	
	H_2O+ NaCl + surfactant + air	H_2O	1.0	10.0	
	H_2O + NaCl	H_2O	0.5	10.5	
	H_2O	H_2O	11.0	21.5	
Post-flood PITT	H_2O	H_2O	1.0	22.5	3.3
	H_2O + tracers	H_2O	7.0	29.5	
Post-flood soil cores	H_2O	H_2O	3.5	33.0	3.3

Table 4.8 Summary of PITT Simulation Predictions

Well(s)	Injection Rate (gpm)	Extraction Rate (gpm)	Mean Residence Time (days)	Swept Volume (gal)	Tracer Recovery
IN-1 + EX-1	2	3.3	2.1	5270	72%
IN-2 + EX-2	2	3.3	0.9	2010	98%
IN-3 + EX-3	2	3.3	1.2	2650	80%
HC-1	3	—	—	—	—
HC-2	1	—	—	—	—
Total	10	10	—	9930	83%

4.3.3 Tracer Selection

A suite of tracers with varying partition coefficients was studied because there was a large range of uncertainty in the quantity and distribution of the DNAPL in the field test's swept zone. If the residual saturation had been known to be relatively high, selecting tracers with smaller partition coefficients would have been sufficient, because it is not mandatory to continue the test to obtain the response curves for the tracers with larger partition coefficients. However, because the residual saturation in the surfactant/foam test area was somewhat low, tracers with larger partition coefficients were included to ensure good separation of the tracer response curves, thereby giving a better estimate of DNAPL saturation. Many alcohols are available for use as partioning tracers. A number of alcohol tracers, including isopropanol, 1-propanol, 1-pentanol, 2-ethyl-1-butanol, n-hexanol, and 1-heptanol, were shown in experiments conducted at UT-Austin to have no sorption to the Hill alluvium and were used successfully in three recent PITTs at Hill AFB, OU2 (Brown et al., 1999).

It was desired that the partition coefficients of the tracers result in a retardation factor in the range of 1.2 to 4 for good separation of the nonpartitioning and partitioning tracers, given a reasonable test duration. Previous site investigation in the AATDF test area indicated that DNAPL saturation increased with depth. Therefore, tracers with larger partition coefficients such as heptanol were required for well pairs completed higher up in the aquifer (e.g., IN-1 and EX-1), while tracers with smaller partition coefficients such as 1-pentanol were needed for well pairs completed in the lowest section of the buried channel (e.g., IN-2 and EX-2). Tracers with middle-range partition coefficients, such as n-hexanol, were indicated for well pairs completed in between, such as IN-3 and EX-3. Finally, bromide, a nonpartitioning ion tracer, was added to the tracer suite as a backup for the nonpartitioning alcohol. Isopropanol (IPA or 2-propanol) was the conservative alcohol tracer of choice. Because the surfactant mixture supplied by the manufacturer contained some of this alcohol, 1-propanol was used instead. For the suite of tracers screened for use at Hill AFB, a residence time of approximately 18 h was sufficient to determine residual DNAPL saturation in the aquifer. The tracers chosen for the pre- and post-surfactant/foam flood PITTs are listed in Table 4.9, along with their respective static partition coefficients.

Table 4.9   Tracers Used Before and After the Surfactant/Foam Flood

Pre-flood Tracer	Concentration (mg/L)	Static Partition Coefficient
Bromide	1500	Nonpartitioning
Isopropanol	1500	Nonpartitioning
1-Pentanol	1500	3.9
1-Hexanol	1500	30.2
1-Heptanol	750	140.5
Post-flood tracer		
Bromide	750	Nonpartitioning
1-Propanol	750	Nonpartitioning
1-Hexanol	750	30.2
1-Heptanol	750	140.5
1-Octanol	200	420.0

4.3.4 Summary of Partitioning Tracer Test Design

- The static partition coefficients of several alcohol tracers were measured to identify tracers for the pre- and post-flood PITTs.
- Laboratory experiments in uncontaminated alluvium showed negligible retardation by alluvium.

- A close match between residual DNAPL saturation estimates based on mass balance and partitioning tracers in a soil column contaminated with a known volume of DNAPL indicated the use of partitioning tracers to determine residual DNAPL saturation.
- Initial conservative chloride tracer tests were used to calibrate the PITT design model.
- Sensitivity studies were made on variations in injection and extraction rates and the amount and distribution of DNAPL. Simulations were made to determine the duration of the tracer test, as well as the mass of each tracer needed, the injection and extraction rates, the extraction well effluent tracer concentrations and the amount of tracer recoverable at the end of tracer test.
- A residence time of 18 h in the subsurface will be required for equilibrium partitioning between the partitioning tracers and DNAPL.
- Pre-flood tracers were isopropanol, 1-pentanol, 1-hexanol, and 1-heptanol.
- Post-flood tracers were bromide, 1-propanol, 1-hexanol, 1-heptanol, and 1-octanol.

4.4 NUMERICAL MODELING OF THE SURFACTANT/FOAM PROCESS USING UTCHEM

UTCHEM is a multiphase, multicomponent, three-dimensional chemical compositional finite-difference simulator. It was originally developed to model surfactant-enhanced oil recovery, but was later modified for applications involving the use of surfactant for enhanced remediation of aquifers contaminated by NAPLs (Delshad et al., 1996).

The balance equations consist of mass conservation equations, an overall balance that determines the pressure for up to four fluid phases, and an energy balance equation to determine temperature. The assumptions imposed when developing the flow equations are local thermodynamic equilibrium except for tracers and dissolution of organic components, immobile solid phases, slightly compressible soil and fluids, Fickian dispersion, ideal mixing, and Darcy's law. The flow equations allow for chemical reactions and phase behavior and are complemented by constitutive relations. The boundary conditions are no-flow and no-dispersive flux across impermeable boundaries unless the option of open lateral boundaries is specified. The aqueous pressure equation is developed by summing the mass balance equations over all volume-occupying components. The volume-occupying components are water, surfactant, oil, alcohol, and gas. Darcy's law is substituted for the phase flux terms, using the definition of capillary pressure to relate pressures in the various phases.

The resulting flow equations are solved using a block-centered, finite-difference scheme. The solution method is implicit in pressure and explicit in concentration (IMPES type). One- and two-point upstream and third-order spatial discretization are available as options in the code. To increase the stability and robustness of the second- and third-order methods, a flux limiter that is total-variation-diminishing (TVD) has been added (Liu et al., 1994). The third-order method gives the most accurate solution.

The phases consist of a single-component gas phase and up to three liquid phases — aqueous, oleic, and microemulsion — depending on the relative amounts of the various components and effective electrolyte concentration (salinity) of the phase environment. The number of components is variable, depending on the application, but would include at least surfactant, oil, and water for modeling of surfactant-enhanced remediation. When gas, electrolytes, tracers, cosolvents, polymer, and other commonly needed components are included, the number of components may be on the order of 10 or more. When the geochemical option is used, a large number of additional aqueous components and solid phases may be used. Any number of water, oil, or gas tracers can be modeled. The tracers can partition, adsorb, and decay if they are radioactive. UTCHEM can model partitioning interwell tracer tests for the detection and estimation of contaminants and for remediation performance assessment in both saturated and vadose zones (Jin et al., 1995).

A significant portion of the research effort on chemical flooding simulation at the University of Texas at Austin has been directed at the development and implementation of accurate physical and chemical property models in UTCHEM. Heterogeneity and variation in relative permeability and capillary pressure are allowed throughout the porous medium, since for example, each grid block can have a different permeability and porosity.

Surfactant phase behavior modeling is based in part on the Hand representation of the ternary phase diagram (Hand, 1939). A pseudophase theory (Prouvost et al., 1984 ; Prouvost et al., 1985) reduces the water, oil, surfactant, and cosurfactant fluid mixtures to a pseudoternary composition space. The major physical phenomena modeled are density, viscosity, velocity-dependent dispersion, molecular diffusion, adsorption, interfacial tension, relative permeability, capillary pressure, capillary trapping, cation exchange, and polymer and gel properties such as permeability reduction, inaccessible pore volume, and non-Newtonian rheology. The phase mobilization is modeled through entrapped phase saturation and relative permeability dependence on trapping number.

The reaction chemistry includes aqueous electrolyte chemistry, precipitation/dissolution of minerals, ion exchange reactions with the matrix (the geochemical option), reactions of acidic components of oil with the bases in the aqueous solution (Bhuyan et al., 1991), and polymer reactions with cross-linking agents to form gel (Kim, 1995).

Nonequilibrium mass transfer of an organic component from the oleic phase to the surfactant-rich microemulsion phase is modeled using a linear mass transfer model similar to that given by Powers et al. (1991). Nonequilibrium mass transfer of tracer components is modeled by a generalized Coats-Smith model (Smith et al., 1988). The model includes options for multiple wells completed either horizontally or vertically. Aquifer boundaries are modeled as constant potential surfaces or as closed surfaces.

UTCHEM can model water tracer transport in fractured media using a dual-porosity formulation (Liang et al., 1995). Recently, a biodegradation model was incorporated in UTCHEM. Multiple organic compounds can be degraded by multiple microbial species using multiple electron acceptors (de Blanc et al., 1996).

The UTCHEM simulator used is based on UTCHEM version 5.3. Several modifications of the code were made to create the working version for this project.

4.4.1 Foam Modification

A simplified foam model was incorporated into the working version of UTCHEM to account for foam flow in porous media. When surfactant concentration in the liquid phases exceeds a certain threshold, C_s^*, gas is assumed to behave as foam. The foamed-gas mobility was reduced from that of ordinary gas by dividing gas relative permeability by a certain mobility reduction factor (MRF). The residual saturation of gas is increased when it behaves as foam to account for the additional gas trapping observed in other foam studies (e.g., Bernard et al., 1965). Below the threshold surfactant concentration C_s^*, ordinary gas-liquid multiphase flow occurs.

Shown schematically for each grid block:

IF

$$C_s > C_s^*$$

THEN

$$k_{rg} = \frac{k_{rg}^0}{MRF}$$

$$S_{gr} = S_{gr}^0 + S_{gr\,add}$$

IF NOT, THEN

$$k_{rg} = k_{rg}^0$$

$$S_{gr} = S_{gr}^0$$

where k_{rg} is the relative permeability of the gas phase and k_{rg}^0 is the relative permeability of the gas phase in the absence of foam. S_{gr}^0 is the ordinary gas residual saturation, while $S_{gr\,add}$ is the additional residual saturation when the gas behaves as foam. The above changes affect only the relative permeability and mobility of the gas phase. Relative permeability and mobility of the liquid phases remain unaffected.

4.4.2   Simulation Model

The physical parameters used in the simulations are summarized in Table 4.10 and discussed below.

Table 4.10 Physical Properties in UTCHEM Simulation

Liquid Properties	Units	DNAPL	Groundwater	4% Surfactant
Viscosity	cp	0.78[a]	1.00	1.5 to 2.5[b]
Density	g/mL	1.38[a]	1.00	1.0045
DNAPL solubility	—	—	none	21.3%[c]
IFT to DNAPL	mN/m	—	9.3[a]	0.002[c]

Gas/foam properties	Units	Value		
Regular gas viscosity	cp	0.5		
Foam mobility reduction factor	--	2 (1 cp effective foam viscosity)		
Foam threshold	v/v	0.2% surfactant		
Gas/liquid IFT	mN/m	30		

Relative permeability	Water	DNAPL	Microemulsion	Gas
Residual saturation	0.20	0.20[d], 0.05[e]	0.20	0.20[f], 0.36[g]
Endpoint relative permeability	0.39	0.43	0.39	0.78
Relative permeability exponent	3.3	2.8	3.3	2

[a] Oolman et al., 1995.
[b] Depending on DNAPL content.
[c] At optimum salinity.
[d] To displacing water.
[e] To displacing gas.
[f] Of ordinary non-foamed gas.
[g] Of foamed gas.

4.4.2.1   Surfactant Phase Behavior

Surfactant-oil-water phase behavior in UTCHEM is modeled based on Hand's rule (Nelson and Pope, 1978; Prouvost et al., 1985; Camilleri et al., 1987). At low salinity, there exists a Winsor Type I phase environment where an excess oil phase that is essentially pure oil coexists with a microemulsion phase that contains water plus electrolytes, surfactant, and some solubilized oil. At high salinity, a Winsor Type II phase environment exists where an excess water phase coexists with a microemulsion phase that contains most of the surfactant, oil, and some solubilized water. Furthermore, a three-phase system, called a Winsor Type III phase environment, exists at intermediate salinity where excess oil and water phases and microemulsion all coexist.

The surfactant/foam field demonstration was performed in the Type III phase environment. Due to current simulator limitation, however, the process was assumed to take place in the Type I regime in the simulation.

Experimental data were matched to Hand's equation to obtain phase behavior parameters in UTCHEM. The oil (or DNAPL) solubility predicted from the simulation is plotted as a function

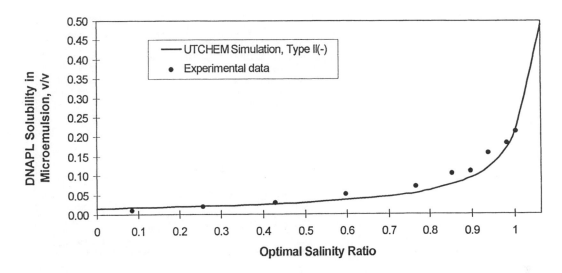

Figure 4.28 Solubility of DNAPL in microemulsion as a function of salinity.

of salinity (Figure 4.28). Circles represent experimental measurements conducted at the University of Texas. In the experiment, excess brine was formed above 0.9 optimal salinity ratio (Type III region). Due to the assumption of Type I behavior in the simulator, the excess brine is combined into the simulated "microemulsion" phase. The maximum oil solubility in this combined microemulsion phase is 21.3%, corresponding to a maximum solubilization ratio (V_o/V_s, volume of solubilized DNAPL per volume of surfactant) of 6.6.

A critical micelle concentration (CMC) of 0.2% in volume fraction of surfactant is used in the simulation. This value is about 20% to 25% of the measured CMC at the operating salinity to account for the reduction in CMC in the presence of oleic-phase contaminants. The 0.2% CMC is consistent with an earlier surfactant-enhanced remediation study using a similar system (Brown et al., 1999; INTERA, 1997b).

4.4.2.2 Liquid Properties

The DNAPL and groundwater properties in the simulation were obtained from measurements at OU2 by Oolman et al. (1995). TCE solubility in groundwater was measured at 1100 mg/L. In the simulation, however, DNAPL is assumed to be insoluble in water, due to current simulator limitation for systems that include gas.

The viscosity of the microemulsion phase is calculated in UTCHEM using the same parameters reported in an earlier study (Brown et al., 1999; INTERA, 1997b). The viscosity ranges from 1.5 cp at low DNAPL content to 2.5 cp at 20% DNAPL concentration.

Interfacial tension between DNAPL and water or microemulsion is modeled using Chun-Huh's relation (Huh, 1979). Figure 4.29 shows the interfacial tension curve used in the simulation, along with experimental data measured at the University of Texas (Brown et al., 1999). The interfacial tension ranges from 9.3 mN/m between DNAPL and groundwater, to 0.002 mN/m with microemulsion at its maximum solubility ratio.

4.4.2.3 Gas/Foam Properties

Gas properties depend on whether the gas exists as foam. A threshold surfactant concentration for foam formation of 0.2% (v/v) equal to the CMC is used in the simulation. At

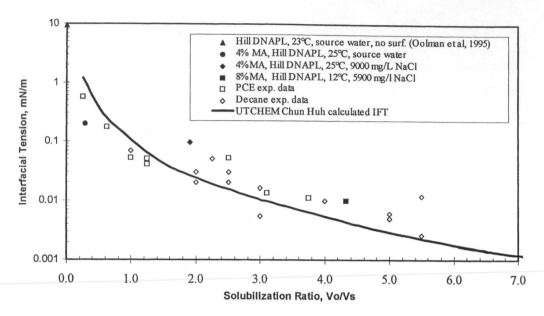

Figure 4.29 Interfacial tension between DNAPL and water or microemulsion.

surfactant concentrations below the threshold, effective viscosity is equal to that of the gas. Above the threshold, its effective viscosity is equal to that of the gas, divided by the mobility reduction factor.

Gas viscosity is assumed to be 0.5 cp instead of the actual viscosity of air (0.02 cp) to avoid numerical problems brought about by high mobility contrast. The high gas viscosity also accounts for weaker foam that exists at surfactant concentrations far below the CMC.

An effective foam viscosity of 1 cp is used in the simulations to best-fit the simulated injection pressures and gas breakthrough times at the observation wells with those observed during the field test. This corresponds to an MRF of 2 with respect to the simulated gas viscosity, and accounts for a 50-fold reduction from the actual air mobility.

The simulations discussed herein use a foamed-gas residual saturation of 36% and ordinary gas residual saturation of 20% at low capillary numbers. For the higher range of capillary numbers attained during the process (~10^{-4}), the foamed-gas residual saturation is around 12% to 18%. These values are low compared with trapped foam saturations reported in the literature. However, they are selected to give the best history fit.

Gas/liquid interfacial tension in the simulation is 30 mN/m, based on experimental measurements of the surface tension of the surfactant solution.

A drainage capillary pressure option for gas was added into the project's working version of UTCHEM. The drainage capillary pressure curve measured by Leverett for unconsolidated sand (Rose and Bruce, 1949) was used.

4.4.2.4 Relative Permeability

The residual saturations and relative permeability parameters for the three liquid phases (water, DNAPL, ME/microemulsion) are identical to those reported in Brown et al. (1999) and INTERA (1997b). A typical gas relative permeability curve is assumed.

Table 4.10 reports the residual saturation and relative permeability endpoint for each phase at low capillary number. At high capillary number, the residual saturation is reduced by capillary desaturation and the relative permeability of each phase approaches unity.

4.4.2.5 Aquifer properties

The 3-D aquifer model is $30.5 \times 10.4 \times 5.5$ m ($100 \times 34 \times 21$ ft) in size, divided into $30 \times 14 \times 18$ grid blocks (total of 7560 grid blocks). The aquifer cells inside the well pattern at the bottom of the aquifer are the smallest, $1 \times 2 \times 0.5$ ft, with a pore volume of 2 gal. The aquifer cells at the periphery of the aquifer model are larger, with the largest one being $12 \times 4 \times 3$ ft with a 291-gal pore volume.

Figure 4.30 compares the clay structure contour originally obtained from kriging with the clay structure in the simulated aquifer model. Control points are shown as symbols on the figure. The upward direction of the vertical axis is 16° west of north, which is the direction along the trench. Kriging results based on the available data may have missed some small-scale features of the aquitard characteristics that are important in the simulation. Therefore, the following modifications were made to the clay structure to obtain Figure 4.30:

- The clay aquitard is deepest along a line in the direction of the vertical axis of the figure that passes through MW-2, east of IN-2/EX-2. The aquitard along this line is generally 15.24 cm (6 in. or 1 grid block) deeper than the depth suggested from kriging. For comparison, kriging results indicated MW-2 as a single deepest point in the aquifer. This modification allows more DNAPL to be present around MW-2, and for a deep trench to exist near IN-2 and HC-2. Together with the next modification, it allows the simulation to match the observed DNAPL production during the test.
- The deep trench just east of IN-2 and HC-2 was assumed to extend 1 grid block (2 ft) farther into the pattern. As discussed later in this section, this trench served as a site for accumulation of mobile DNAPL. A hump at the northern end of the trench prevented the mobile DNAPL from reaching MW-2. Indeed, as discussed below, mobile DNAPL was found in the vicinity of HC-2 and IN-2, but not at MW-2.
- The channel extends farther south than suggested from kriging. Details of the aquitard structure in this area are unknown, due to the lack of control points there. However, more recent measurements in the fall of 1997 indicated that the channel extends at about the same depth much farther south of the pattern than shown in Figure 4.30(A).

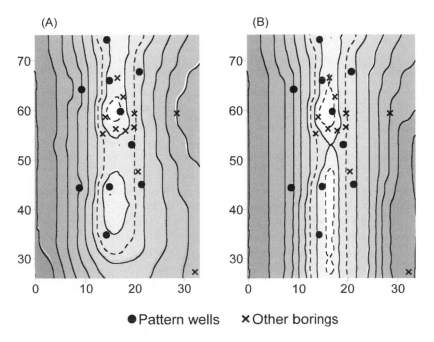

Figure 4.30 A comparison between (a) the clay structure obtained from kriging and (b) the modified clay structure used for the simulation.

Figure 4.31 shows a north-south vertical cross-section along the deepest part of the aquifer along with simulation grid blocks. The vadose zone was included in the simulation model. Gradual change in gas saturation in the vicinity of the water table was employed to avoid numerical difficulties. The initial gas saturations in the top four layers were 90%, 19%, 5%, and 1%, respectively.

No-flow boundaries were specified. The east and west boundaries of the simulation model coincided with the aquitard surface. The north and south boundaries, however, did not all coincide with the aquitard. Fluid was allowed to enter and leave the simulation model through six artificial wells along the north and south boundaries. Four horizontal, constant-pressure wells were placed on the first and second layers of the 3-D model to maintain constant water tables at the boundaries. Two vertical, constant-pressure wells on the north and south boundaries extended from the third layer to the aquitard to allow aquifer fluids to enter or exit the model. The pressure at the south boundary was specified to be 0.76 psi greater than at the north boundary, which corresponds to a natural pressure gradient of 0.0076 psi/ft in the direction of the trench. The porosity is 0.27 in the aquifer and 0.01 in the clay aquitard. The simulated aquitard permeability is 0.001 darcy (10^{-8} m/s). The aquifer permeability distribution is discussed below.

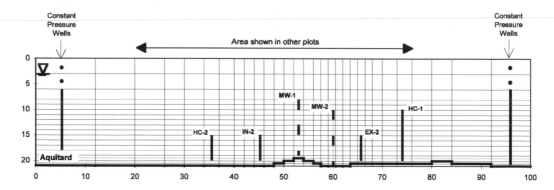

Figure 4.31 North-south vertical cross section along the deepest part of the aquifer (with simulation grid blocks and perforated intervals of wells near the cross section).

4.4.2.6 Permeability and DNAPL Distribution

For simplicity, permeability and DNAPL saturations in the 3-D model were specified on a layer basis. Stochastic permeability distribution, although more realistic, is more difficult to control and causes numerical difficulties in complex systems.

Permeability and DNAPL distributions were chosen to qualitatively match the soil boring measurements. Figure 4.32 shows the horizontal permeability and contaminant saturation at each vertical layer. Several modifications were made to this basic distribution, as discussed below.

The permeability ranged from 10 to 140 darcy. Four high-permeability, thief-zone layers were present. Two thief-zone layers at the fourth and seventh layers connected the IN-1/EX-1 well pair. IN-2/EX-2 and IN-3/EX-3 were connected by two thief zones in layers 11 and 13. Permeability near the bottom of the aquifer ranged from 10 to 25 darcy, which is consistent with the results from the borings.

The horizontal thief zones probably either disappeared or changed direction south of the well pattern. Therefore, the permeability of the thief-zone layers near and south of HC-2 was reduced to 15 to 20 darcy.

Figure 4.33 shows the permeability profile along the deepest vertical cross section of the aquifer that passes through MW-2. Black represents clay, while lighter shades represent areas of higher permeability. To allow earlier breakthrough, as observed at the lowest level of MW-2A, the permeability profile near the monitoring point was modified so that an additional thief zone of 100 darcy extended from the thief zone in the 13th layer to MW-2A.

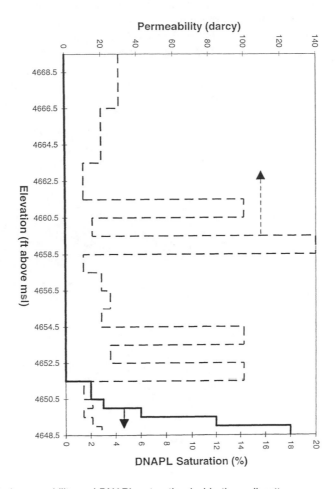

Figure 4.32 Simulated permeability and DNAPL saturation inside the well pattern.

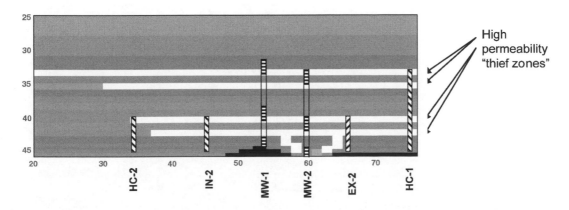

Figure 4.33 Permeability distribution along the deepest cross section of the aquifer.

An anisotropy factor of $k_v/k_h = 0.1$ was used, where k_v and k_h are the vertical and horizontal permeabilities, respectively. Although this anisotropy factor is within reasonable range, it is on the low side. The low value was selected to best match foam breakthrough times in the simulation. Longitudinal dispersivity of 150 cm (5 ft) was used to account for dispersion due to horizontal heterogeneity.

Figure 4.34 compares the simulated DNAPL saturations with soil sample measurements. Figure 4.35 gives an aerial view of the initial DNAPL distribution, shown as the highest DNAPL volume fraction at each location.

Higher DNAPL saturation was specified in the vicinity of HC-2 to match soil sample measurements from that well. DNAPL saturation in the area east and south of IN-2 was specified at 60%

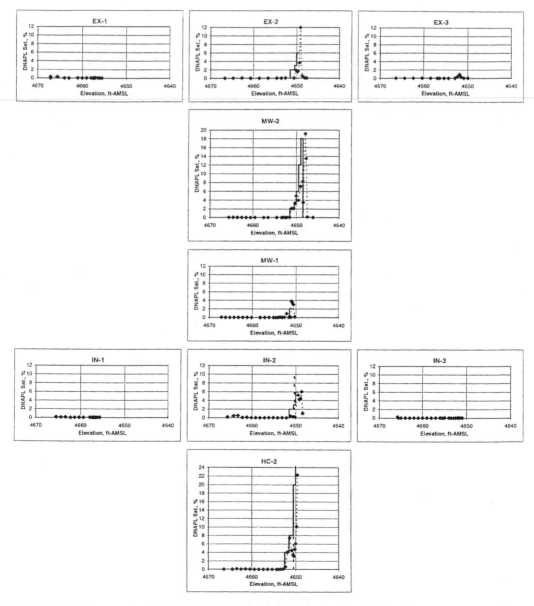

Figure 4.34 Comparison between the simulated initial DNAPL saturation and soil sample measurements.

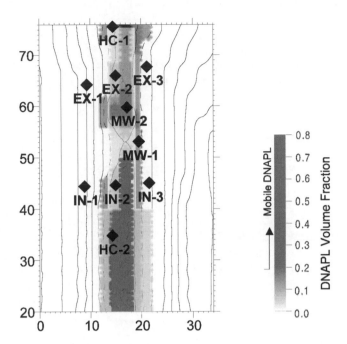

Figure 4.35 Aerial view of the initial DNAPL distribution.

to 80%. The high saturation accounts for the mobile DNAPL detected at HC-2 and IN-2 at the end of the field demonstration. Although no effort to detect mobile DNAPL was made prior to the test, the 22% DNAPL saturation at the deepest soil sample from HC-2 is at the residual saturation. As any soil sample with DNAPL above its residual saturation is likely to have drained before it was measured in the laboratory, it is probable that mobile DNAPL was present in the lower levels of HC-2 prior to the test, where it might form a pool of mobile DNAPL in static equilibrium. A hump in the aquitard just north of IN-2 prevented the mobile DNAPL from spilling into the pattern. When the static equilibrium is disturbed because of injections at HC-2 and IN-2, the pool serves as a source from which DNAPL can enter the pattern.

The basic DNAPL distribution shown in Figure 4.32 gave 18 gal residual DNAPL in the pattern between the lines of injectors and extractors. With the mobile DNAPL pool that extends a little bit into the pattern, however, there is a total of 24 gal DNAPL in static equilibrium before the test. As can be seen in Figure 4.34, soil-boring data were very well respected in this estimation of initial DNAPL distribution.

4.4.3 Computational Method

One-point upstream weighting was used in the finite difference computation. In such surfactant/foam flood simulations, both phase compositions and fluid properties varied considerably from one grid block to another. Therefore, the use of the more stable and robust one-point upstream weighting was necessary. Two-point upstream weighting and three-point spatial discretization resulted in excessive noise and numerical instability.

The time step was selected automatically based on concentration changes. The maximum changes in concentrations for all components were specified. The time-step size was governed by the change in surfactant concentration during surfactant-only injection, and by the change in gas saturation during foam injection.

A summary of initial condition UTCHEM simulations in the foam flood test area is given below:

- A modification was added to UTCHEM to account for foam flow in porous media and this was used to model the surfactant/foam floods.
- Permeability and DNAPL distribution in the 3-D site model were chosen to match both the permeability and contaminant saturation profiles obtained from soil borings as well as the pre-flood partitioning interwell tracer test results.
- Simulations suggest that the surfactant/foam test area contained 24 gal DNAPL before any fluids were injected.

4.5 FINAL FIELD DEMONSTRATION DESIGN

During the first drilling program, it was observed that the AATDF test area had DNAPL only along the bottom 5 ft or less of the 20-ft thickness of the saturated zone. Furthermore, the DNAPL saturation was found to be relatively low where present. It was recognized that the surfactant/foam process would have to be evaluated on the amount of DNAPL remaining after the remediation, rather than on the volume of DNAPL removed. Also, sieve analysis indicated the presence of thief zones that had permeabilities greater than 100 darcy, while DNAPL was present in sands with permeabilities of 30 darcy. This heterogeneity would provide a good demonstration of the effectiveness of foam for mobility control.

4.5.1 Surfactant Solution Composition

The surfactant used was dihexyl sulfosuccinate (product name MA-80I from Cytec). No isopropanol was added because it was shown to be detrimental to foam stability. The surfactant concentration was specified to be 4% active material. This concentration was more than a factor of 4 greater than the CMC. The surfactant/foam's performance was not expected to improve by further increasing the surfactant concentration. The specified salinity was 10,250 ppm NaCl. This is the optimal salinity with Hill OU2 DNAPL at a temperature of 12°C. This optimal salinity is where the microemulsion phase solubilizes an equal amount of DNAPL and brine. At the optimal salinity, the sum of the brine/microemulsion and oil/microemulsion interfacial tensions is a minimum. Thus, the surfactant/foam process was designed for maximum solubilization and mobilization. The surfactant was to be preceded by a 1 pore volume (PV), or 24-h, slug of 10,250 ppm NaCl brine to precondition the aquifer to the optimal salinity. The surfactant slug was to be followed by a 0.5 PV slug of 8000 ppm NaCl brine so that mixing with fresh water would not degrade the effectiveness of the surfactant.

4.5.2 Pattern Volumes and Rates

Laboratory tests in heterogeneous sand packs indicated that the DNAPL could be removed with 1 to 2 pore volumes (PV) of surfactant solution throughput. It was decided to use 3 PV surfactant in the field to be conservative. The test pattern size and rates were specified to be 1 PV per day, so that the test could be completed within the budget and yet be long enough to collect data and make modifications to the process operation. The distance between the rows of injection and extraction wells was 20 ft. The distance was modified slightly when it was discovered during the drilling of the injection wells that the center of the channel (i.e., the deepest point) was not where it was originally expected. The width of the pattern was designed to be 12 to 13 ft, based on the narrowness of the channel and the DNAPL saturation observed in the reconnaissance borings. As the wells were being installed, it was discovered that the width of the channel and DNAPL saturated region were even narrower. Additionally, the aquifer was deeper than the initial simulations, which were based on data from the initial field investigation, predicted. Finally, the DNAPL-saturated region extended to the hydraulic control well (HC-2) upgradient of the injection wells. Well HC-2 had the highest DNAPL saturation, 22%, observed at this site. The DNAPL outside the injection/extraction pattern raised concerns about migration of DNAPL into the pattern.

The pumping rates of the injection and extraction wells were specified such that the total injection and extraction were balanced (see Table 4.11). It was deemed necessary to have balanced rates to maintain a constant water table since the OU2 site was enclosed by a barrier wall. The disadvantage of balanced rates was that the water from hydraulic control well, HC-2, would flow through the pattern, rather than provide a pressure barrier to confine the fluids from the pattern injection wells.

Table 4.11 Design Pumping Rates

Well(s)	Extraction Rate (gpm)	Injection Rate (gpm)
Extraction wells	$3.3 \times 3 = 10$	—
Injection wells	—	$2 \times 3 = 6$
HC-1	—	3
HC-2	—	1
Total	**10**	**10**

4.5.3 Air Injection

The laboratory experiments showed that it was desirable to inject air at a specified pressure corresponding to the degree of mobility reduction desired. Similarly, it was advantageous to inject air intermittently but simultaneously with the surfactant solution so that the surfactant solution would always be flowing through the lower fraction of the screened interval (5 ft) of the injection well. The injection of air was designed to rotate among the three injection wells with a period of 2 h air injection per well. The first 0.5 PV of surfactant solution was to be added without air injection to allow the surfactant to displace DNAPL from high-permeability zones before the foam occluded them.

4.5.4 Monitoring Wells

The monitoring wells were installed to permit observation of the compositions and phases flowing through the formation. This information would not be available from the extraction wells because their fluids would be commingled from many streamlines, including dilution by fresh water from outside the pattern. Each monitoring well was continually sampled at a low rate from screened intervals. Information expected to be gained from the monitoring wells included (1) activity and phase behavior of the surfactant, (2) sweep of the formation from observation of breakthrough times in the different sampling intervals, and (3) extent of foam propagation.

4.5.5 PITTs and Borings

The partitioning interwell tracer tests (PITTs) and soil borings were used as two measures of surfactant/foam process performance. The PITT gives a global or average quantity of DNAPL in the region swept by the process. The borings provide information about the spatial distribution of DNAPL and were the only means to measure the DNAPL that may have been in low-permeability regions not swept by the PITT.

4.5.6 Summary of Design of Surfactant/Foam Demonstration

The final design of the field test is summarized below:

- A 4% (active) solution by weight of MA-80I with 10,250 ppm NaCl was selected as the surfactant solution. This was an optimal surfactant solution designed to exhibit a Winsor Type III behavior at 12°C.
- Surfactant injection was to be preceded by 1 PV injection of 10,250 ppm NaCl and followed by 0.5 PV of 8000 ppm NaCl to improve the effectiveness of the surfactant performance.
- Balanced rates of water injection and extraction were specified to prevent fluctuations in the water table, as OU2 was confined by a slurry wall. However, with balanced flow rates, water from the hydraulic control wells might enter the pattern rather than just provide hydraulic control.
- The first 0.5 PV of surfactant was to be added without air injection to remove DNAPL from the higher permeability zones and allow surfactant to saturate these zones.
- Air injection by a constant-pressure source and rotation of air injection among the three injection wells would occur every 2 h.
- Monitoring wells with multilevel samplers were to be used to sample fluid from Intervals A, B, and C to obtain information about the relative breakthrough times of the various chemicals and contaminants.
- Post-flood PITTs and borings were to be used to determine and confirm the final volume of DNAPL after surfactant/foam flooding.

CHAPTER 5

Field Operations

5.1 FIELD DEMONSTRATION TIMELINE

A graphical timeline of the major field tasks conducted for the surfactant/foam flood is presented in Figure 5.1. The field effort began with mobilization to the OU2 site, grading of the site, and installation of protective tents and other surface facilities, including secondary containment liners, "frac" (waste storage) tanks, tank heaters, electrical and plumbing systems, pumps, sampling devices, and instrumentation. Preparation took approximately 2 weeks.

A summary of the injection/extraction operations during the demonstration is given in Table 5.1. The surfactant/foam flood field test began with a 2-d water flood to establish a forced-gradient flow field between the extraction and injection wells. Potable water was injected into the three injection wells at a rate of 2.0 gpm per well, while extracting at a rate of 3.3 gpm from each of the three extraction wells. To provide positive hydraulic control during the flood, potable water was injected at a rate of 3.0 gpm into hydraulic control well HC-1, and at 1.0 gpm into HC-2. These rates changed during the surfactant/air injection period to focus the remediation effort on the deepest and most contaminated portion of the aquifer in the center of the channel (see Table 5.2).

The initial PITT was performed for 1 week, and consisted of a tracer injection period for 0.5 days, followed by 6 d of water injection to transport the tracers across the zone of interest. During the last 26 h, 10,250 ppm sodium chloride was added to the injected water to exchange cations with the clays before surfactant addition (to raise the salinity of the test zone in preparation for the surfactant/foam flood.)

A 4% by weight Aerosol MA-80I surfactant solution was added to the 10,250 ppm NaCl injectate to begin the surfactant/foam flood. The addition of brine and surfactant continued for 8 h to sweep the high-permeability zones of the aquifer with surfactant solution before generating foam *in situ*. Foam was generated by simultaneously injecting the surfactant solution into the lower section of each injection well screen, while periodically injecting slugs of air into each well consecutively. After approximately 2 h of gas injection at a well, the air line was disconnected and moved to the next injection well so that the gas injection process could continue.

Two days into the surfactant/foam flood, an 18,000 ppm bromide tracer was substituted for the sodium chloride in the surfactant solution. The bromide tracer was used to evaluate the sweep efficiency of the flood in the presence of foam and for comparison with the sweep efficiency of the surfactant alone before foam generation. After 1.5 h of bromide injection, the injectate was returned to 10,250 ppm NaCl.

Surfactant and air injection ended after 3 d (3 PV). For the following 12 h, the NaCl concentration in the injectate was reduced to 8000 ppm to displace the surfactant with a salinity gradient. Afterward, a water flood was conducted for 9.5 d to transport the remaining surfactant solution across the test zone.

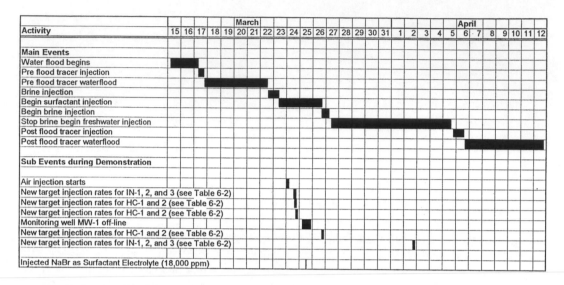

Figure 5.1   Field demonstration timeline.

Table 5.1   Summary of Surfactant/Foam Flood Demonstration Events

Test Segment	Fluids Injected	Pore Volumes (days)
Water injection	Water	1.6
Tracer injection	Water + tracers	0.5
Water injection	Water	4.9
Electrolyte injection	Water + NaCl	1.1
Surfactant/Foam injection	Water + surfactant + NaCl + air	3.25
Electrolyte injection	Water + NaCl	0.5
Water injection	Water	9.8
Tracer injection	Water + tracer	1.0
Water injection	Water	6.3

Table 5.2   Summary of Target Injection and Extraction Rate Changes During the Surfactant/Foam Flood Demonstration at Hill OU2

Date	IN-1	IN-2	IN-3	EX-1	EX-2	EX-3	HC-1	HC-2
3/15/1997	2.0[a]	2.0	2.0	3.3	3.3	3.3	3.0	1.0
3/24/1997	1.5	3.0	1.5	2.7	4.0	3.3	2.0	2.0
3/25–26/1997	1.0	3.0	2.0	2.7	4.0	3.3	3.0	1.0
4/3/1997	2.0	2.0	2.0	2.7	4.0	3.3	3.0	1.0

[a] Rates are in gpm.

The tracer injection of post-flood partitioning tracers lasted for 1 d, and was followed by 6 d of water injection to transport the tracers across the zone of interest. The final PITT was followed by groundwater extraction for 5.25 d to recover any injectate that might have remained in the aquifer.

5.2 SURFACE FACILITIES

The demonstration layout was divided into three process areas: (1) an injection system, which included injectate batch mixing and storage in the tankage area, on the east side; (2) an extraction flow control and sampling system in the well field area on the west side; and (3) an effluent sediment

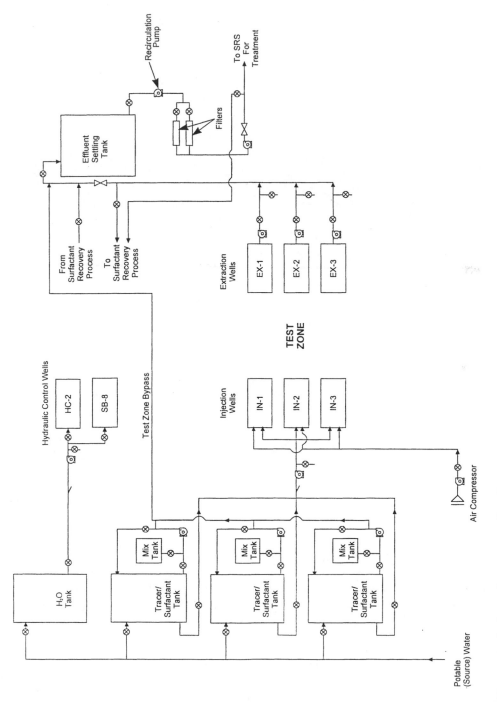

Figure 5.2 Surfactant/foam process flow diagram.

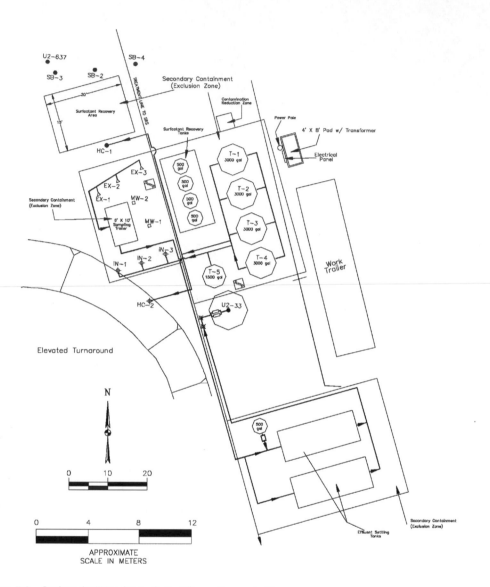

Figure 5.3 Surface facilities for surfactant/foam demonstration.

control system that used filtering and settling equipment to the south of the AATDF demonstration area. These process areas were all secondarily contained using high-density polyethylene liners. The containment areas provided protection from any potential incidental spills or leaks from the mixing/storage tanks and associated plumbing. Figure 5.2 is a process flow diagram and Figure 5.3 shows the surface facilities.

5.3 INJECTION SYSTEM

The fluid injection system consisted of the mixing/storage tanks and the injectate delivery and control system (see Figure 5.4). The fluids were all prepared and staged in a total of five large mixing/storage tanks. Tanks T1, T2, and T3 were dedicated to water/tracer and water/surfactant/electrolyte solutions. Tanks T4 and T5 were used for water injection into both the injection wells and

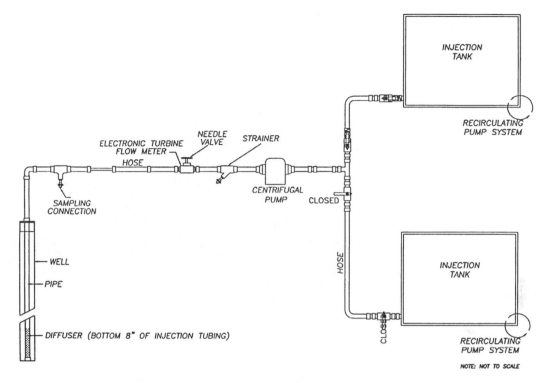

Figure 5.4 Injection system configuration.

the hydraulic control wells. Tank T5 was also used as a backup surfactant solution storage tank during the foam flood. Each tank was equipped with an electric heating element capable of maintaining the tank's contents at >15°C. Centrifugal pumps were used to recirculate and mix injectate batches in the mixing/storage tanks.

A single centrifugal pump drew the injectate from the mixing/storage tanks to each of the three injection wells, where the flow rate was monitored and adjusted separately. Water injection into the hydraulic control wells, HC-1 and HC-2, was controlled by a separate centrifugal pump, with a separate plumbing system similar to that used for the injection wells. A Y-strainer on each line immediately upstream of the centrifugal pumps protected the injection system from potential debris. Injectate samples were collected from a sampling port downstream from the pump on the main injection line before the manifold that diverted the flow to each of the injection wells.

5.3.1 Injectate Preparation

Batches of surfactant solution were prepared by first filling a tank with 1800 gal water and then mixing in 185 lb salt (NaCl). Next, two 475-lb drums of surfactant were added to the tank. Additional water was then added to bring the water level to 2150 gal. Finally, the fluid in the tank was recirculated to mix the surfactant solution. A water heater in the tank maintained the injectate batch temperature at 15°C. The heater was turned off when the fluid level in the tank fell below 400 gal to prevent the element from damaging the tank or itself. All of the fluids injected into the aquifer during the foam flood were preheated, including the potable water injected into the hydraulic control wells.

5.3.2 Injection Well Configuration

The injection well heads were fitted with a manifold that threaded onto the top of the existing casing (see Figure 5.5). The manifold allowed simultaneous injection of both liquid and gas and

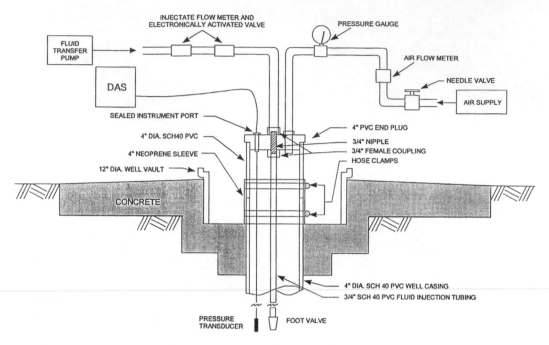

Figure 5.5 Injection well head configuration.

provided a sealed access port for the downhole pressure transducer. Liquid and gas flow control consisted of a series of valves and flowmeters, along with a diverter attached to the bottom of the fluid injection tubing to prevent sediment dispersal in the well. A pressure gauge and a pressure transducer on the air supply line measured the gas pressure. Fluid levels in the well were calculated by subtracting the gas pressure from the total pressure measured by the transducer at the bottom of the well.

5.3.3 Air Injection

A 100-cfm air compressor supplied air for the field demonstration. The air flowed through a filter to remove lubricant before it was injected into the borehole. Air injection rotated among the three injection wells on a 2-h cycle, while surfactant solution continued to be injected at the base of the injection well screen. The air hose had to be disconnected from the wellhead for the rotation. The air flow was shut off several minutes before disconnection to allow the pressure to decline gradually before the sudden pressure decline that occurs when the air hose is disconnected. The pressure setpoint was usually between 8 and 11 psi, except for the last day, when it became apparent that a greater pressure gradient was needed to displace DNAPL from MW-2A. There was no automatic controller, and the rates and pressures were dependent on the operator adjusting the air valve. A difference of 2.2 psi between the bottom-hole pressure and the gas pressure was required for injection of air into the 5-ft screened interval. A short test of the air injection system was made before air injection into IN-1 began. The maximum air injection pressure and flow rate specified to prevent damage to the well or process were 20 psi and 4 scfm, respectively.

5.4 EXTRACTION SYSTEM

Submersible bottom-loading, low-shear pneumatic pumps were installed in each of the three extraction wells, EX-1, EX-2, and EX-3. A portable air compressor supplied the air for the operation of the pumps. The air pressure delivered to all three pumps was controlled by a single-stage regulator. The fluid flow discharged from each of the pumps was adjusted with flow-controlling needle valves. Digital flowmeters installed on each of the discharge lines were used to meter the fluid flow and maintain the desired flow rates.

The fluid discharged from each of the extraction wells was first routed through a metering and sampling system, as shown in Figure 5.6, and then joined in a common effluent line. A manifold at the junction between the pump discharge lines and the main effluent line allowed for independent bucket tests to be performed on each extraction pump. A diversion valve in the effluent line allowed the AATDF Surfactant Recovery Project (Harwell et al., 1999) team to collect batches of the surfactant-laden and VOC-contaminated discharge from the well field during the foam flood for use in their treatment process. The hoses and piping used in the extraction system were rated for burst pressures greater than the expected shut-in pressures that can be generated by the pumps. Multiple valves were installed on all lines to control flow and to minimize fluid release when the system was dismantled.

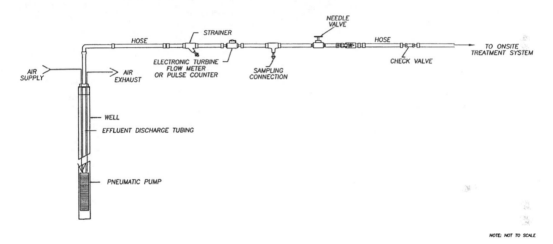

Figure 5.6 Extraction system configuration.

5.5 EFFLUENT SEDIMENT CONTROL SYSTEM

The main effluent line was piped to the effluent sediment control system, which consisted of two large-capacity settling tanks (see Figure 5.2). During the initial and final PITTs, and during the later portion of the post-surfactant/foam water flood, the AATDF effluent was routed directly through two replaceable 5-μm cartridge filters and then onto the SRS treatment system. The filter cartridges were replaced every 12 to 24 h during these segments of the demonstration.

Prior experience led to the expectation that some dispersed clay would be produced from the aquifer as the surfactant solution altered the geochemistry of the swept volume. The SRS treatment system was protected from this sediment by diverting the effluent to one of the two large-capacity settling tanks. To flocculate the clays and promote settling, a chemical feed pump added a brine (20% NaCl by weight) solution at 0.24 gpm as the effluent from a 50-gal storage tank was pumped into the settling tanks. After the flocculated clays settled, the settling tank fluid was pumped through in-line cartridge filters to the SRS treatment line.

The connection to the SRS treatment line was made at the existing connection port at SRS recovery well U2-33R. A check valve was installed at the connection to the fixed discharge line to preclude the possibility of back flow from the SRS treatment line to the extraction wells.

CHAPTER **6**

Field Monitoring, Sampling, and Analytical Methods

6.1 AQUIFER MATERIAL SAMPLE COLLECTION

During soil sampling operations, a string of drill rods was used to seat and retrieve the auger's center bit and the sampling tools through the 4.25-in. ID annulus of the 10.25-in. OD hollow-stem auger. Sediments were classified visually on the basis of their apparent dominant grain size using the Unified Soil Classification System (ASTM D2488-84).

In the upper sections of the alluvium, and in cases where sand flow dictated, a 2.0-ft long California modified split-spoon sampler with an ID of 2.5 in. was seated in the annulus of the lead auger and hammered to depth to collect soil samples ahead of the lead auger. The auger was then advanced to the next planned sampling depth prior to seating the split-spoon sampler for retrieval of the next core. The base of the aquifer, containing the zone of highest DNAPL, was sampled using a large-diameter (3.5-in. ID), 2.0-ft long, custom-built split-spoon sampler. This sampler, driven ahead of the auger bit with the rig hammer, can collect and retain alluvium containing large cobbles, which are commonly found in the basal sections of the OU2 alluvium.

Soil cores were first subsampled and field-extracted for volatile organic analysis (VOA). Most of the samples collected from the soil sampler weighed approximately 50 g. Slightly larger samples were collected for intervals containing large cobbles and gravel. The majority of the samples were collected in 4-oz wide-mouth jars with Teflon gaskets and a septum. For samples consisting of coarse gravel, 8-oz jars were used. The weight of each sample jar was determined before it was filled.

Each sample container was labeled with the sample number, project number, date, time, boring ID, depth, preservative, and sampler's initials. The label and marking medium used were resistant to toluene. The label was then covered with transparent tape. As a precaution, the sample number was also marked on the cap of each jar. One field replicate, one field blank, and one equipment blank (rinse) were taken for every 20 soil VOA samples. A trip blank filled with Type II reagent water was included in each sample shuttle shipped to the laboratory. Samples shipped off-site were packaged securely and sent with a chain-of-custody form in the shipping container. Samples were shipped using DOT-approved packing materials and following the necessary transport requirements. A full discussion of QA/QC for this project, including sample custody and documentation procedures, and definitions for the different types of QA/QC samples used during the initial field work can be found in the Quality Assurance Project Plans (QAPPs) (Rice University, 1995; INTERA, 1997a).

Subsamples from the spoons were also placed in aluminum foil-covered jars and the resulting headspace screened with a photoionization detection (PID) meter as a semiquantitative measurement of volatile contaminants. Finally, the remaining soil in the split spoon was photographed. Each photograph included a scale card identifying the borehole, the sampled interval, and the original orientation of the sample.

The sampler, the sampling tools, and any other equipment used in the sampling procedure were properly decontaminated before reuse to minimize cross-contamination of samples. The decontamination procedure involved washing with detergent, rinsing with water, rinsing with toluene, rinsing with deionized water, and allowing the equipment to air dry. The possibility of cross-contamination during sampling activities was minimized using a combination of newly purchased and unused equipment and sampling materials. Washable materials such as viton gloves were used during the preparation and handling of samples, and systematic cleaning procedures were employed to clean contaminated equipment. All of the soil and fluids resulting from this decontamination procedure were placed in 55-gallon drums to allow the sediments to settle. The supernatant was then pumped into the SRS phase separators for treatment, while the sediments left in each drum were consolidated with the soil waste generated during drilling operations for proper disposal.

6.2 INJECTION SYSTEM

6.2.1 Sampling of Mixing Tanks

Batch samples from the mixing/retention tanks were collected manually from in-line sampling tees on the injection lines to quantify the input concentrations of each injection segment. Batch samples were also collected from different levels in the tanks themselves with a dipper or long-handled scoop.

6.2.2 Flow Rate and Pressure Monitoring

Flow rates for each injection well were controlled by using flow-controlling needle valves. The flow rates were monitored manually every 6 h using an in-line digital flowmeter for each injection well. The injection rates were periodically checked using the autosampler.

The air pressure was regulated by the operator and was measured with transducers at the bottom of the well and in the air flow line. Pressures (including differential pressure, which indicated liquid level in the well) were continually recorded and displayed on a simulated strip chart. The air flow rate was measured with a gas mass flowmeter.

6.3 LIQUID LEVEL MONITORING

Liquid levels were monitored in all the wells to ensure that the water table of the aquifer was not being affected by the injection/extraction operations during the field demonstration. Manual measurements were made using a water-level meter at least every 6 h. The fluid levels were noted to the nearest 0.01 ft. In addition, pressure transducers in the injection and extraction wells measured the pressure at the bottom of the well. The liquid levels were calculated from the bottom-hole pressure, assuming that the air pressure above the liquid level was 1 atm. During the time air was injected, the air pressure in the well increased and the reported liquid level is the total head in feet of water.

6.4 EXTRACTION SYSTEM

6.4.1 Extraction Flow Rate Monitoring

The extraction rate was monitored and recorded every 6 h. Flowmeters on the extraction wells were capable of measuring the cumulative and batch volumes; thus, the differences in the cumulative and batch volumes between successive recording intervals were used for determining the average flow rate. Digital totalizers on the flowmeters were used to track the volume of fluids produced from each well over time, thereby allowing flow rates to be calculated.

The extraction rates were also periodically checked using bucket tests (measurement of the time required to fill a known volume). A digital flowmeter on the effluent line at the manifold was used to log the total amount of fluid extracted from the well field.

6.4.2 Extraction Well Sampling

The fluids produced during the surfactant/foam flood were analyzed for tracer alcohols, TCE, surfactant, calcium, chloride, and bromide (methods given in Chapters 6.8–6.12). Aqueous samples from the extraction and the test-zone monitoring wells were collected automatically using an auto-collector. In addition to the autocollected samples, confirmation or backup samples were collected manually at regular intervals. Exterior monitoring wells were sampled manually using existing in-line ports. The influent and effluent of the SRS were periodically sampled manually to determine the stripper treatment efficiency, and to verify that the treatment effluent discharge limits were being met.

The autocollector was trailer-mounted and contained a series of multiposition actuated valves connected to multiple flow-through sampling loops. Each sample system consisted of a 16-position actuated valve with stainless steel tubing. The stainless steel tubing was maintained at 4°C. The sample line from the well was connected to an actuated valve that directed the flow to one of 16 sample loops. The positions were changed at timed intervals by a personal computer running LabVIEW software, trapping the sample in the stainless steel tubing sample loop. Duplicate samples were collected by setting the activated valve to capture samples at two loop positions, 5 min apart. This allowed the duplicate loop to be adequately flushed before the sample was trapped.

The sampling system also included a valve panel for selection of either manual or automatic operation. The manual sampling port was located upstream of the automated collection system. To manually recover a sample from the sample loop, the computer program was placed on standby, and flow to the entire sample collection system was bypassed. No autocollected samples were missed because their sampling interval was 1 h. The tubing system was flushed first with tap water to clean the line's retrieval port, followed by an air flush to rid the lines of the tap water flush. The actuated valve was then advanced to a sampling position and the collected sample was released into 40-mL sample vials. Finally, the sample loop was flushed with water and air to clean the sample lines. The actuated valve was then advanced to the next position for the second sample retrieval. Manual samples were collected at a frequency of 33% of the autocollector sampling frequency during the tests and surfactant/foam flood. The valves and pipe volumes were purged prior to sampling during collection of manual samples to prevent sample-to-sample cross-contamination, and to minimize the loss of volatile contaminants. The protocol for filling vials from the autocollector sample loops was essentially the same as described above for manual sample collection.

Each sampling system was equipped with a flowmeter, checkvalve, peristaltic pump, and air and water flushing valves. The flowmeter ensured that the system had an adequate flow rate to accommodate automated sampling. The checkvalve and peristaltic pump prevented back-flow contamination, the air flush valve retrieved the sample, and the water flush valve cleared the sample retrieval port.

6.6 MULTILEVEL MONITORING WELL SAMPLING

Every multilevel sampler port module consisted of a screened sampling port, a pneumatic double-valve pump, and a tubing system used to deliver a fluid sample to the surface. The ports were also outfitted with vibrating wire pressure transducers to monitor pressure changes in the aquifer at each screened interval. A packer system isolated each sample port. Samples were collected with the autocollector.

6.7 PITT ANALYSIS

The theoretical foundation for the method of first temporal moment analysis of partitioning tracer tests can be found in Jin et al. (1995) and Jin (1995). This method can be used to estimate the tracer swept volume (the volume of the aquifer through which the tracer solution has flowed), the average DNAPL saturation in the tracer swept volume, and the total DNAPL volume. For a partitioning tracer test with multiple extraction wells, the following equations are applied to each individual extraction/injection well pair.

The average DNAPL saturation in the tracer swept volume (S_n) was calculated using Equation 6.7.1:

$$S_n = \frac{(R_f - 1)}{R_f + K - 1} \tag{6.7.1}$$

where K is the partition coefficient of a partitioning tracer, and the retardation factor, R_f, is defined as

$$R_f = \frac{\bar{t}_p}{\bar{t}_n} \tag{6.7.2}$$

where $\bar{t}_p$ and $\bar{t}_n$ are the first temporal moments of the partitioning tracer and nonpartitioning tracer, respectively, and are calculated using the following equations:

$$\bar{t}_p = \frac{\int_0^{t_f} t C_p(t)dt}{\int_0^{t_f} C_p(t)dt} - \frac{t_s}{2} \tag{6.7.3}$$

and

$$\bar{t}_n = \frac{\int_0^{t_f} t C_n(t)dt}{\int_0^{t_f} C_n(t)dt} - \frac{t_s}{2} \tag{6.7.4}$$

where t_s is the slug size (i.e., the time period in which the tracer mass was injected during the tracer test), t_f is the tracer test cutoff time, and $C_p(t)$ and $C_n(t)$ represent the partitioning and nonpartitioning tracer concentrations as a function of time, respectively.

The average DNAPL saturation was estimated by initially calculating the first moments of the partitioning and nonpartitioning tracers using Equations 6.7.3 and 6.7.4 by numerically integrating the corresponding tracer response curves. Next, Equation 6.7.2 was used to calculated the retardation factor. Equation 6.7.1 was then used to estimate the average DNAPL saturation in the swept volume.

With S_n and $\bar{t}_n$ known, the tracer swept pore volume of a particular extraction well (V_p) was calculated as:

$$V_p = \frac{m}{M} \frac{Q \bar{t}_n}{1 - S_n} \tag{6.7.5}$$

where M is the total mass of tracer injected, m is the total mass of tracer produced from the particular extraction well, and Q is the total injection rate.

To estimate the DNAPL volume accurately, the tracer response curves should be complete, because much of the information is contained in the tails of the response curves. Unfortunately, the tracer response curves are often incomplete, either due to dilution of the tracer concentration below the detectable limit in the latter part of the tracer test, or for some other reason. However, the tracer response curves can be extrapolated with an exponential function, provided the duration of the test is sufficient to establish this decline. The first moments of the tracer response curves can be obtained by dividing the data into two parts. The first represents the data from time zero to the time t_b where the decay becomes exponential, and the second covers the exponential part in which time goes from t_b to infinity.

After time t_b, the tracer response is assumed to follow an exponential decline given by:

$$C = C_b e^{-\left(\frac{t - t_b}{a}\right)}$$

(6.7.6)

where $1/a$ is the slope of the straight line when the tracer response curves are plotted on a semilog scale, and C_b is the tracer concentration at time t_b, when the curve becomes exponential.

The first moment $(\dot{t})$ (Equations 6.7.3 and 6.7.4) of the partitioning and nonpartitioning tracers can then be calculated as:

$$\dot{t} = \frac{\int_0^{t_b} tC\,dt + a(a + t_b)C_b}{\int_0^{t_b} C\,dt + aC_b} - \frac{t_s}{2}$$

(6.7.7)

6.8 MEASUREMENT OF NONPARTITIONING TRACERS

The chloride and bromide concentrations were determined with a titration method. The chloride and bromide ion concentrations were used, among other things, to assess the amount of mixing between injected fluids and fresh water from outside the pattern. The detailed methods are presented in Appendix 1.

6.9 MEASUREMENT OF CATIONS

Multivalent cations, such as calcium and magnesium, are important components of interfacially active surfactant solutions used in remediation. Calcium and magnesium may be added to recovery solutions by design to replace sodium. More often, however, they inadvertently find their way into recovery solutions during field applications. Sources of alkaline-earth metal cations are mixing with source or connate waters at flood fronts and cation exchange. These processes are potentially detrimental to process performance. Reservoir sands and clays in freshwater aquifers are often strongly loaded with calcium and magnesium since fresh waters are usually hard. A sodium-rich flood front can displace calcium and magnesium from mineral surfaces and produce high concentrations of calcium and magnesium in solution. Because these multivalent cations contribute more strongly to electrolyte strength than does sodium, floods can easily be driven to an overoptimum electrolyte condition by cation exchange. The calcium concentration was used to monitor the surfactant/DNAPL phase behavior.

The above considerations provided incentive to develop a method to monitor calcium and magnesium in recovery solutions and in reservoir waters. The basis of the method is potentiometric titration. Here, the appropriate electrode combination was a calcium- and magnesium-specific electrode and an Ag|AgCl reference electrode. The detailed methods are included in Appendix 1.

6.10 MEASUREMENT OF SURFACTANT IN EXTRACTION FLUIDS

To monitor the progress and efficiency of a surfactant-enhanced remediation process, one must analyze for the surfactant of interest. In the present case, the surfactant was an anionic surfactant, sodium dihexyl sulfosuccinate, offered by Cytec Inc. under the trade name Aerosol MA-80I. To determine anionic surfactant by potentiometric titration, one precipitates the surfactant with a cationic surfactant titrant. An electrode combination that responds to anionic surfactant in solution serves as an indicator. A dilute solution of benzethonium chloride was used as the titrant. The detailed method is given in Appendix 1.

6.11 ANALYSIS OF DNAPL

6.11.1   Hill DNAPL Composition

Hill DNAPL at OU2 comprises TCA, TCE, and PCE, with TCE being the major component. Radian (1992) and INTERA (1997b) reported that Hill DNAPL contains approximately 10% 1,1,1-trichloroethane, 70% trichloroethene, and 10% tetrachloroethene. Also, about 10% is minor components, such as other chlorinated hydrocarbons and toluene, along with oil and grease. Rice University analysts determined the TCA, TCE, and PCE contents of two samples of Hill DNAPL. One was from the SRS and one was from monitoring well MW-2, Level A, collected during the pilot. The goal of these analyses was to provide composition data to formulate synthetic DNAPL for current and future experiments. Samples were analyzed by GC; the methods are provided in Appendix 1.

6.11.2   Measurement of DNAPL Constituent Concentration in Surfactant Solutions

During surfactant-enhanced remediation processes, one must analyze production streams for DNAPL. The reasons are threefold:

- In order to establish flood performance, one must determine the material balance. Produced DNAPL — compared with initial DNAPL — is a straightforward measure of flood performance. This, in combination with post-flood cores and tracer surveys, tells how well the flood did.
- Analyses of DNAPL components in phases sampled at observation wells allow one to determine phase compositions and, thus, to determine the characteristics of the flood. For example: Were produced phases those expected? What were their Winsor types? Did cation exchange play a role?
- To satisfy regulatory requirements and also to know when to stop, one must establish DNAPL toward the end of the flood. Here, one expects surfactant to be dilute, but still present in concentrations that might interfere with DNAPL analyses.

The challenge in gas chromatographic (GC) analysis is that the produced DNAPL is in water solutions that also contain high concentrations of nonvolatiles, namely, inorganic salts and anionic surfactant. The preferred procedure depends on available equipment and circumstances.

Direct injection is possible, but not usually recommended. On-column injections of waters containing nonvolatiles foul columns and may not be as sensitive as other methods for low concentrations of DNAPL. Headspace methods (methods that measure analyte in the headspace above solutions) are also used but are subject to constraints. As long as Henry's law applies, headspace methods for volatile analytes are accurate, although they may not be sensitive to low analyte concentrations. DNAPLs solubilized in surfactant solutions above the surfactant CMC do not obey Henry's law. Consequently, concentrations of DNAPL components in solution headspaces are not linear with respect to concentrations in solutions.

Methods that rely on purging solutions with GC carrier gas, trapping analytes on adsorbing columns, and then desorbing concentrated analyte into a GC are preferred for dilute analytes. Purge-and-trap methods are more sensitive than other methods, but one encounters foaming when purging surfactant solutions. Consequently, liquid can be carried over into the adsorbing trap. Weighing all alternatives, a purge-and-trap GC method was chosen to measure DNAPL in surfactant solutions. The solutions were diluted before analysis to minimize foaming. The detailed method is given in Appendix 1.

6.11.3 Analysis of DNAPL in Soil Cores

A novel method was developed for determining the saturation of chlorinated hydrocarbons in soils to preserve the volatile organic compounds present in the soil. In the field, the sample jars were filled with up to 80 mL toluene (Fisher Optima, #T291-4 or equivalent). This procedure reduces the vapor pressures of volatile soil contaminants, thus reducing their evaporation during shipping and storage.

Once the toluene had been placed in the jar, the weight of the solvent and the jar together was recorded. Soil samples were then carefully placed into the toluene in the jar. Finally, the lid was firmly sealed, and the final weight and volume added to each jar was recorded. This sample preservation methodology was designed to minimize the loss of the volatile components from the soil samples. Field blanks were also prepared with toluene. In addition, a 40-mL sample from each lot of toluene was collected to analyze the solvent for background interferences arising from impurities.

The receiving laboratory subsampled the toluene solvent in the sample jars and analyzed for specified VOCs using GC with flame ionization detection (FID). Knowing soil weights, volumes of solvent, and VOC concentrations in the solvent, VOC content in soils was calculated. If one knows, or assumes, soil porosity and grain density, one can also calculate VOC saturations.

Toluene is a convenient solvent because it has a modestly high boiling point (110°C) and dissolves common soil contaminants, such as chlorinated hydrocarbons. However, other solvents might also be used. This method replaces, or augments, more laborious and expensive procedures, which involve freezing cores, refluxing with toluene solvent, and determining NAPL content by weight differences between virgin and extracted samples. The analytical method focused on the principal contaminants in OU2 DNAPL, TCE, PCE, and TCA. For a detailed description of the analytical method, see Appendix 1.

6.12 MEASUREMENT OF TRACER ALCOHOLS

The main objective in the development of GC procedures for partitioning tracer analysis was to obtain adequate chromatographic separation of the VOCs and the individual alcohol partitioning tracers. This was necessary to adequately characterize the tracer tail concentrations required for accurate estimation of the residual DNAPL saturation. Several analytical procedures were developed for simultaneously analyzing the partitioning tracers and VOCs during the initial and final PITTs. The procedures are a modification of EPA method SW8015A. An SRI and a Buck Scientific GC using DYNATECH autosamplers were used by INTERA with the modified SW8015 procedures for continuous analysis of effluent samples. The detailed experimental procedures for tracer and VOC analysis, which involved direct injection of sample onto the column, are given in Appendix 1.

CHAPTER 7

Results

7.1 INJECTED SURFACTANT AND ELECTROLYTE

The operational plan called for injecting 3 pore volumes (PV) of surfactant solution over a period of 3 d. The injectate was intended to be a 4% solution of active MA-80I material at the optimum salinity of 10,250 ppm NaCl at 15°C.

As Figure 7.1 illustrates, the surfactant and chloride concentrations in the injectate samples from each batch of solution were consistently low. When cleaning out the tanks after the surfactant injection was complete, it was noted that some undissolved surfactant was present on the bottom of the tanks. The low delivery concentrations were likely the result of the inefficient mixing process used on site.

During the surfactant/foam flood, 26,600 gal (3.2 PV) surfactant solution were injected. In Table 7.1, the volumes for each injection well are presented. The volumes injected at each well were different, due to the changes in injection rates during the test.

The mass of MA-80I and the mass of NaCl injected during the surfactant/foam flood are also shown in Table 7.1. The total mass of surfactant was 7900 lb and the total NaCl mass was 2800 lb. These numbers were calculated by multiplying the volume injected into each well during each tank's injection period (recorded in field logs) by the measured surfactant and chloride concentrations in the injectate samples. The error margins were calculated based on a 5% error in the laboratory analyses and an error of 10 gal of production for each time step.

Table 7.1   Mass and Volume of Injected Chemicals

Well	Solution Volume (gal)	Surfactant Added (lb)	NaCl Added (lb)	NaBr Added (lb)
IN-1	6800	2000 ± 7%	770 ± 7%	17.7 ± 7%
IN-2	11,700	3500 ± 6%	1140 ± 6%	34.7 ± 6%
IN-3	8100	2400 ± 7%	870 ± 7%	17.7 ± 7%
Total	26,600	7900 ± 7%	2800 ± 7%	70 ± 7%

Approximately 9100 lb (24 drums × 475 lb/drum × 80% MA-80I) surfactant were purchased for the field test. Virtually all of the surfactant was transferred from the drums into the mixing tanks, but the injected mass was 13% less than the mass added to the batch mixing tanks. Approximately 3100 lb NaCl was used to prepare the batches (185 lb × 16 tanks + 120 lb × 1 tank), but the calculated mass of NaCl delivered to the aquifer was only 2800 lb. In this case, the injected value differed from the prepared quantity by 10%.

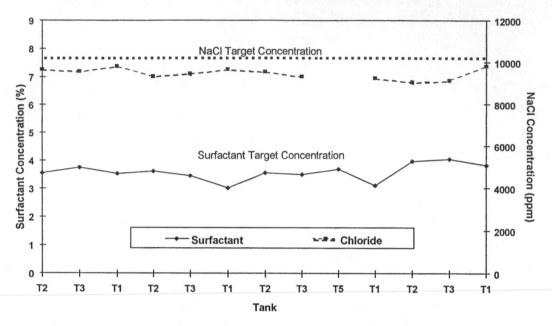

Figure 7.1   Concentrations of chemicals injected during the surfactant/foam flood.

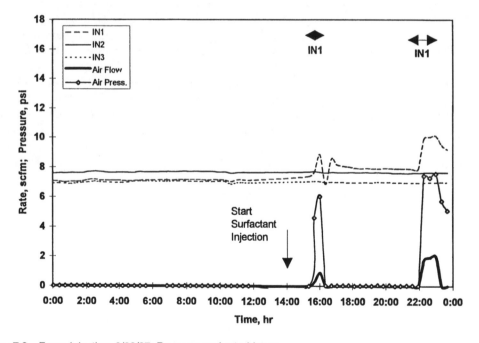

Figure 7.2   Foam injection, 3/23/97: Pressure and rate history.

7.2 PRESSURE AND LIQUID LEVEL RESPONSE TO AIR INJECTION

Pressure and rate histories are shown in Figures 7.2 through 7.5; the liquid level and rate histories are shown in Figures 7.6 through 7.9, along with the liquid levels of the extraction wells and the hydraulic control well, HC-2. Before air injection began, the differences in liquid level or head between the injection and extraction wells were less than the measurement error.

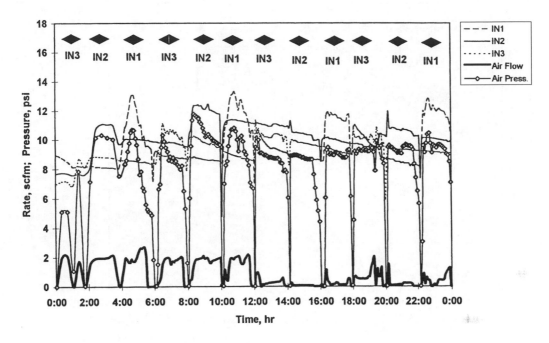

Figure 7.3   Foam injection, 3/24/97: Pressure and rate history.

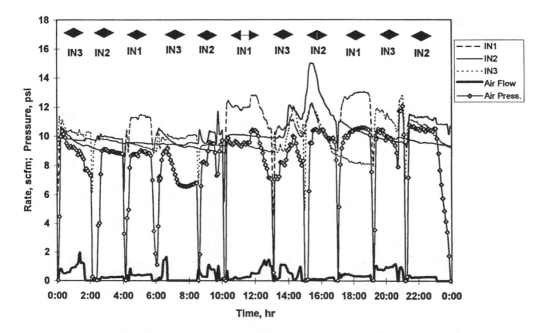

Figure 7.4   Foam injection, 3/25/97: Pressure and rate history.

The computation of liquid head at the bottom of the well from the measured liquid levels was based on the assumption that the air pressure was atmospheric. The actual liquid levels in the wells were lowered when air was injected. After each cycle of air injection, the pressure or liquid

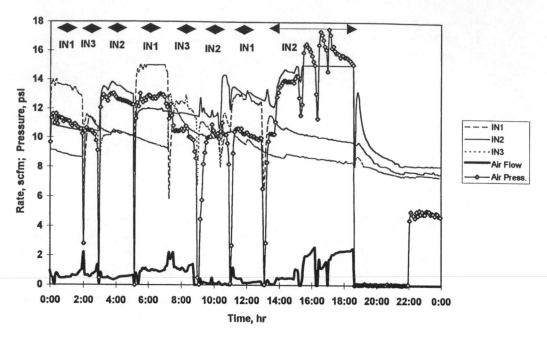

Figure 7.5   Foam injection, 3/26/97: Pressure and rate history.

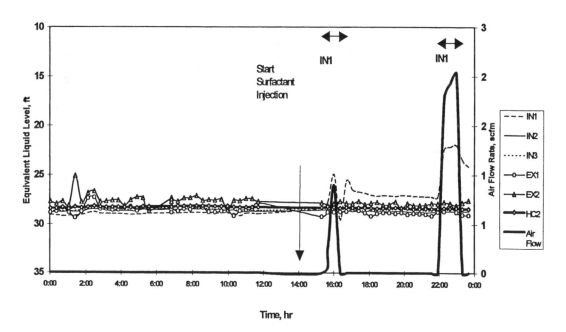

Figure 7.6   Foam injection, 3/23/97: Liquid level and rate history.

level decayed but remained significantly greater than before air injection This result implies that there was enough trapped air or foam to reduce the mobility of the surfactant solution between air injection cycles. The liquid levels quickly returned to the original values after surfactant

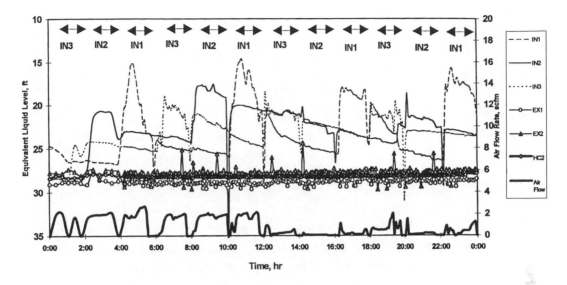

Figure 7.7 Foam injection, 3/24/97: Liquid level and rate history.

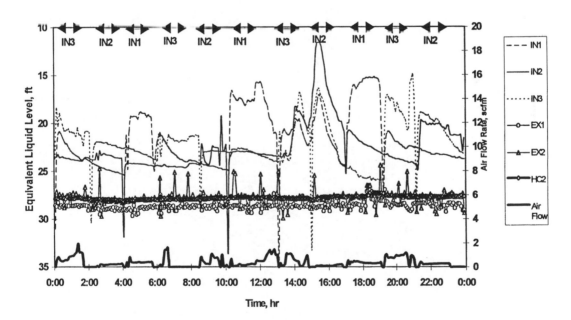

Figure 7.8 Foam injection, 3/25/97: Liquid level and rate history.

solution was switched to 8000 ppm NaCl solution and air injection ceased. Figure 7.10 depicts the long-term liquid levels for the entire project.

The pressures in the monitoring wells are shown in Figure 7.11. The pressures were set to zero at the time surfactant injection began, and the plots represent changes in pressure since that time. It is difficult to attribute these pressure changes to foam because the pressure was changing before air injection began.

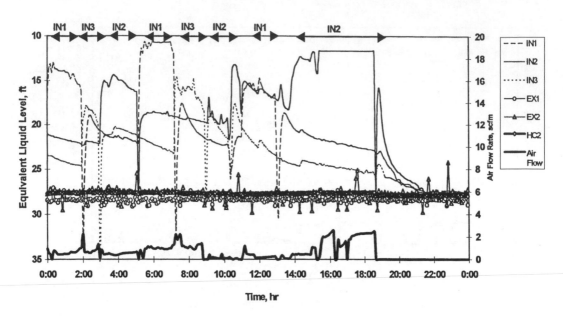

Figure 7.9 Foam injection, 3/26/97: Liquid level and rate history.

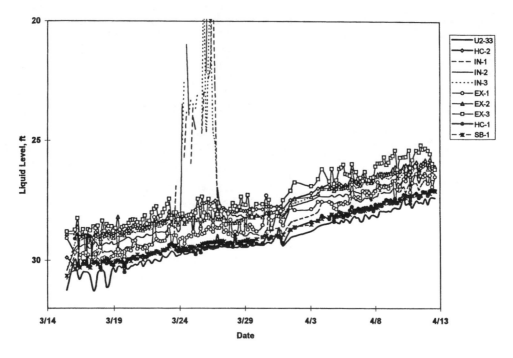

Figure 7.10 Long-term liquid levels for entire project.

7.3 FOAM PROPAGATION AND BREAKTHROUGH

The observation of foam in the monitoring wells indicated that foam was propagating through the formation. The arrival times of foam at different intervals of the monitoring wells are listed in

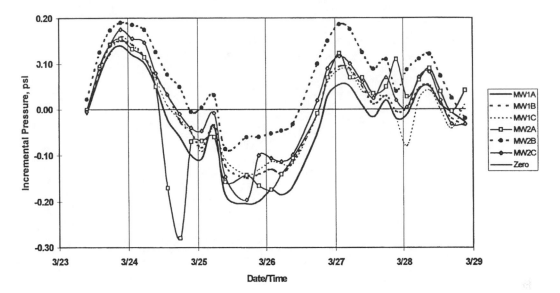

Figure 7.11 Monitoring well pressure response.

Table 7.2   Foam Breakthrough Times

Monitor Well Interval	Time to Breakthrough (h)
MW-1C (top)	0.6
MW-1B (middle)	0.6
MW-1A (bottom)	14
MW-2C (top)	2.5
MW-2B (middle)	8
MW-2A (bottom)	23

Table 7.2. Foam was eventually observed at all levels of the monitoring wells. Once breakthrough occurred, foam persisted in the upper two levels for the duration of air injection. The breakthrough of foam to the top and middle (C and B) intervals of monitoring well MW-1 occurred in less than 0.6 h. The foam was stiff, similar to shaving foam. MW-1 is located approximately 6 ft from IN-1. When air injection was temporarily discontinued in IN-1, the foam at the shallowest interval became wetter, while the foam in the middle remained unchanged.

The breakthrough of foam in MW-2, first in the top interval near the water table and then later in the middle interval where a high-permeability zone was expected, was an indication that a significant amount of foam was both rising to the vadose zone and propagating horizontally, perhaps through the thief zone. The foam in the deepest interval of MW-2 was never a stiff foam, but rather a froth that would quickly break because it contained DNAPL and/or microemulsion with solubilized DNAPL.

7.4 PERFORMANCE OF EFFLUENT TREATMENT SYSTEM

As part of the surfactant/foam field test, Radian operated a source removal system to treat extracted DNAPL and groundwater. During the surfactant/foam flood injection and extraction operations, approximately 23 gal DNAPL was recovered from the well-field effluent and transferred to the SRS product storage tank. An additional 22 gal was transferred after the demonstration had been completed, bringing the total amount of DNAPL recovered by the SRS during the surfac-

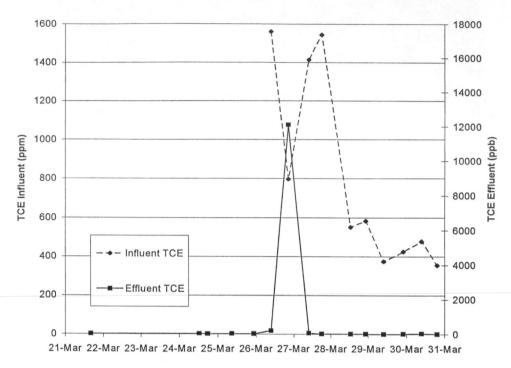

Figure 7.12 TCE influent and effluent concentrations at the SRS.

tant/foam flood to approximately 45 gal. Because the product indicator system was inaccurate, the volume of DNAPL recovered by the SRS actually ranged from 30 to 50 gal. This volume of DNAPL compares well with the volume calculated through the analysis of the extraction well effluent (37 to 48 gal).

In addition to the data collected during the routine operation of the SRS (e.g., liquid flow rates, tank levels, temperatures, pressures), Radian also collected water samples from the stripper influent, stripper effluent, and T-105 (the storage tank for treated groundwater prior to its release to the base industrial waste treatment plant). The samples were analyzed for chlorinated solvents using the on-site gas chromatograph (see Figure 7.12). The chlorinated solvent concentrations in the influent to the SRS seemed to peak on March 26 and 27, and then steadily decline. Except for a slight excursion on March 26, the sum of the chlorinated solvent concentrations in the stripper effluent remained below 100 ppb. The T-105 effluent chlorinated solvent concentrations were generally similar to the stripper effluent, but groundwater from extraction wells upgradient from the OU2 slurry wall may have impacted the results. In general, the SRS performed quite well during the foam flood study, achieving an average TCE removal efficiency of greater than 99%. In addition to the chemical data collected during the foam flood study, Radian noted that low concentrations of antifoam agent were effective in preventing foaming within the stripper column.

Because personnel at the IWTP at Hill AFB were concerned that solvent and/or surfactant loads could cause the IWTP to exceed its chemical oxygen demand (COD) limits, Radian collected and analyzed samples of the T-105 effluent and the IWTP effluent for COD (see Figure 7.13). Using the measured flow rates of the streams, T-105 effluent accounted for one half to three quarters of the total mass of COD discharged from the IWTP during the period. By comparing the mass of surfactant extracted with the resultant COD load in T-105 effluent (see Figure 7.14), 1 lb surfactant exerted approximately 2 lb COD. The COD load created by the AATDF surfactant/foam flood did not exceed the limits permitted at the IWTP. However, COD loads should be addressed in any future scale-up design.

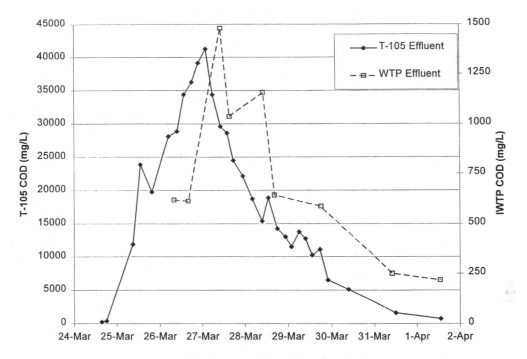

Figure 7.13 COD of samples of the T-105 effluent and the IWTP effluent.

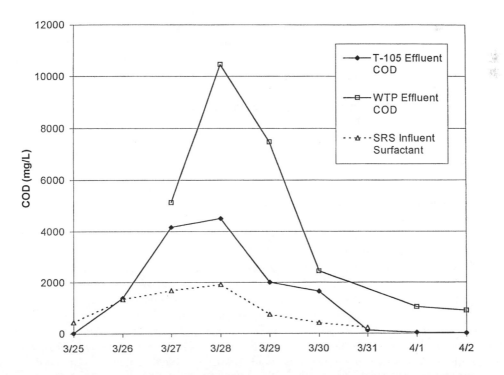

Figure 7.14 Comparison of the mass of surfactant extracted with the resultant COD load in T-105 effluent.

7.5 DNAPL ESTIMATION BY PARTITIONING TRACERS

7.5.1 Approach to Data Analysis

The first step in the data analysis process was quality assurance/quality control (QA/QC) evaluation of the data. Tracer data that did not meet QA/QC criteria were eliminated from the database. The second step was to evaluate the available field data and select a pair of nonpartitioning and partitioning tracers to use for DNAPL volume and saturation estimation. Theoretically, data from each pair of nonpartitioning and partitioning tracers can give an independent estimate of DNAPL volume and saturation. Practically, however, the retardation factor should be greater than 1.2 in order to increase the estimation accuracy (Jin, 1995). The retardation factors of pentanol and hexanol from this tracer test are much smaller than that of heptanol. Therefore, all the calculations for the pre-flood PITT are based on the tracer response data of IPA and heptanol.

The third step was to fit the tracer response data with smooth curves and estimate the DNAPL volume and saturation as a function of tracer cut-off time. The estimated DNAPL volume and saturation should approach a plateau as the tracer test approaches completion. Otherwise, the tracer data need to be extrapolated using the exponential decline function described in Chapter 6.7 to increase the estimation accuracy. For this tracer test, the analysis was done by fitting the tracer response data of IPA and heptanol with two smooth curves and extrapolating using the exponential decline function. DNAPL volume estimation was based on the smooth curves.

The extraction rates in some of the extraction wells were not very stable during this PITT. Because the tracer concentration and the flow rate data were not recorded at the same time, a separate program was used to convert the tracer response data (which were recorded as a function of time) into a function of total volume of water extracted. The program first read in the actual cumulative volume of fluid injected/produced for each well as a function of time, based on the information obtained from the injection/ extraction logs. These data were then used as a lookup table. When the time for which tracer concentration was measured was read in as the input, the program interpolated the corresponding volumes of water injected/produced from the lookup table.

7.5.2 Pre-flood Extraction Well Tracer Data Analysis

The tracer concentration histories of the shallow (IN-1/EX-1), intermediate (IN-3/EX-3), and deep (IN-2/EX-2) well pairs and their corresponding fitted curves are shown in Figures 7.15 through 7.17, respectively. These figures show normalized concentrations of tracers at the extraction wells. Normalized concentrations, which are dimensionless, are calculated by dividing the measured sample concentrations by the average tracer concentration in the injectate measured by the same GC. EX-1, EX-2, and EX-3 samples were analyzed by the same GC, and the monitoring well samples were analyzed using a different GC. The normalizing concentrations of all the alcohol tracers are also indicated in Figures 7.15 through 7.17.

The first time that the concentrations of alcohol tracers were measured in the injectate samples, the concentrations exceeded the calibration range. This exceedance was not noticed until the 7-day holding times expired for these samples. Dilutions of these injectate samples were analyzed approximately 14 d after sample collection and suggested much lower concentrations. Based on integrated tracer mass recoveries at the extraction wells, the initial measurements of injectate concentrations were believed to be more accurate, so they were used to normalize the tracer concentrations. The lower measurements would have implied that tracer mass recoveries had considerably exceeded 100%. Accurate measurement of the concentrations of the injectate samples are important for tracer recovery and swept volume calculations, but they are not important for first-moment calculation of DNAPL saturations.

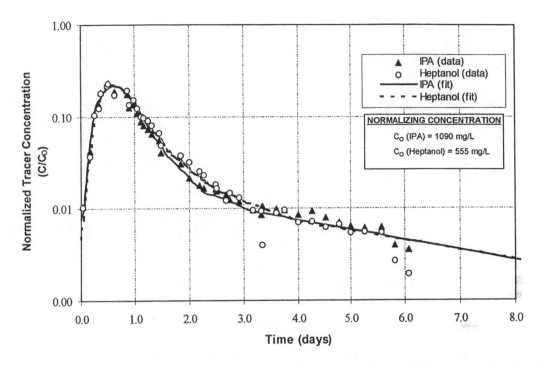

Figure 7.15 Pre-flood PITT shallow well pair (IN-1/EX-1) tracer response data and their corresponding fitted curves.

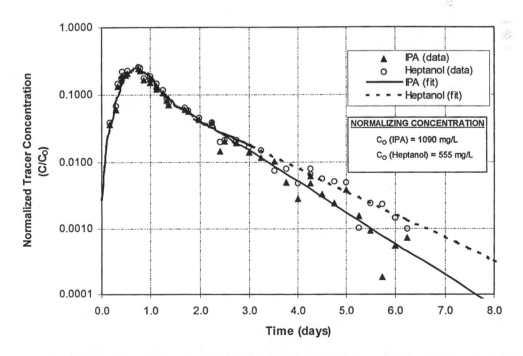

Figure 7.16 Pre-flood PITT intermediate well pair (IN-3/EX-3) tracer response data and their corresponding fitted curves.

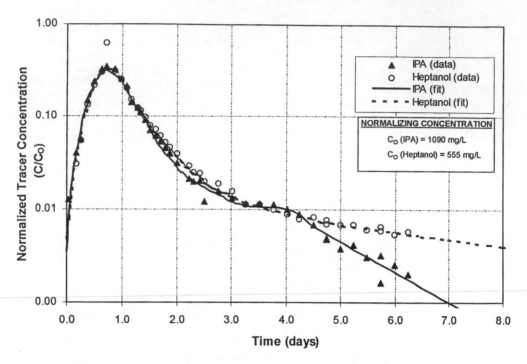

Figure 7.17 Pre-flood PITT deep well pair (IN-2/EX-2) tracer response data and their corresponding fitted curves.

Data for only one conservative tracer (IPA) are presented in Figures 7.15 through 7.17. This is because the IPA and bromide data were almost identical within experimental uncertainty. The separation of the tracer response between IPA and 1-heptanol in the deep well pair (IN-2/EX-2) clearly indicates the presence of DNAPL in the pore space swept by the partitioning tracers. However, because the separation appears only in the tails of the tracer response curves, it can be inferred that the DNAPL is not uniformly distributed in the pore space swept by the partitioning tracers. This also means that most of the pore space is uncontaminated. The lack of tracer separation in the shallow well pair indicates that negligible quantities of DNAPL are present in the pore space swept by partitioning tracers for this well pair.

The tracer curves were analyzed using the method of temporal moments as presented in Chapter 6.7. The resulting estimates of the DNAPL volume for each well pair are summarized in Table 7.3. The pore volume (shown as swept volume) of the aquifer swept by the tracers for each well pair as determined by the moment analysis is also shown in this table. The total aquifer pore volume swept by the tracers was 8180 gal, as determined by adding up the swept volumes calculated for each well pair. The volume of DNAPL in this swept volume was estimated to be about 21 gal,

Table 7.3 Summary of Pre-flood PITT Results

Well Name	Isopropanol Recovery (%)	Heptanol Recovery (%)	Swept Volume (gal; L)	DNAPL Volume (gal; L)	Avg. DNAPL Content of Soil (% saturation; mg/kg)
EX-1	29	29	3,760 gal 14,200 L	0.7 ± 0.7 gal 2.6 ± 2.6 L	0.02 ± 0.02% 50 mg/kg
EX-2	33	35	2,700 gal 10,200 L	19 ± 6 gal 71 ± 23 L	0.7 ± 0.3% 1800 mg/kg
EX-3	21	21	1,720 gal 6,500 L	1.4 ± 0.4 gal 5.3 ± 1.5 L	0.08 ± 0.2% 200 mg/kg
Total	**83**	**85**	**8,180 gal 31,000 L**	**21 ± 7 gal 79 ± 27 L**	**0.26 ± 0.1% 670 mg/kg**

corresponding to an average saturation of 0.26% (670 mg/kg). The tracer recovery for IPA was about 83%. The recoveries of the other tracers used were approximately the same, in the range of 83% (±5%).

7.5.3 Error Analysis

There are two main sources of error associated with the analysis of partitioning tracer data that may contribute to the uncertainty in the estimates of average DNAPL saturation. The first error source is an uncertainty in the estimation of the retardation factor based on the actual tracer data. The second error source is data extrapolation.

The first source can be estimated based on error analyses of numerous laboratory results. It was found that the average error in the retardation factor is expected to be ±0.035 (Dwara-kanath, 1997). Since the average saturation is calculated using the retardation factor (see Equation 6.7.1), the higher the retardation factor, the less uncertainty in the DNAPL estimate. This has also been illustrated on a theoretical basis by Jin (1995). As noted above, a small retardation factor would result in a very large estimation error. In practice, the retardation factor has to be at least 1.2 in order to have a reliable DNAPL volume and saturation estimate. Since the retardation factor is proportional to the tracer partition coefficient, tracers with higher partition coefficients will lead to larger retardation factors and improved accuracy in DNAPL volume estimation. If the tracers with large partition coefficients, such as heptanol and octanol, still yield retardation factors in the range of 1.0 to 1.1, it means that there is little — if any — DNAPL present in the subsurface being tested.

The second source of error is from the extrapolation of experimental data. Extrapolation of experimental data is required when the tails of the tracer concentration histories are not fully characterized. This can cause underprediction of the average DNAPL saturation, thereby causing estimation errors. These errors could be large or small, depending on the quality of the tracer data. The data extrapolation technique is very simple and sound in principle. However, if there is significant scattering in the tracer concentration tail due to the effect of analysis errors and/or low tracer concentrations (on the order of the detection limit), engineering judgment and subjectivity may be required to pick the correct exponential decline of the tracer tail. The average DNAPL saturations are highly sensitive to the changes in the slope of the exponential decline curve and this can cause a relatively large uncertainty in the average DNAPL saturation estimates. On the other hand, if the tracer data are of good quality and a linear decline in tracer concentrations is observed on a semilogarithmic plot, the extrapolation errors will be minimal.

For the pre-flood PITT, the main source of error for all the monitoring wells was the uncertainty in the estimation of the retardation factors. Data extrapolation was not needed, as the estimated DNAPL saturation/volume was observed to reach a constant value using all the measured tracer data. Similar results were observed for extraction wells EX-1 and EX-3. The retardations were 1.02 and 1.11 for wells EX-1 and EX-3, respectively. These retardations are extremely small, suggesting that there was a negligible volume of DNAPL in the area swept by the tracers for extraction wells EX-1 and EX-3. Using an uncertainty of ±0.035 in the retardation estimates of partitioning tracers, the estimated DNAPL saturations in the swept pore volume for wells EX-1 and EX-3 were 0.02 ± 0.02% and 0.08 ± 0.02%, respectively. Although these errors seem high, it should be noted that the actual DNAPL volumes corresponding to these saturations were 0.7 ± 0.7 gal and 1.4 ± 0.4 gal, respectively, indicating that there was little DNAPL in the swept pore volume for wells EX-1 and EX-3.

For well EX-2, the heptanol tail was incompletely characterized, thus necessitating extrapolation. However, as shown in Figure 7.16, the tail of the tracer response curves of extraction well EX-2 was well behaved and clearly established an exponential decline. The data extrapolation technique was used to estimate the DNAPL saturation/volume. To estimate the uncertainty of DNAPL volume estimation due to data extrapolation, the tail portion of the tracer response curve was fit by both linear

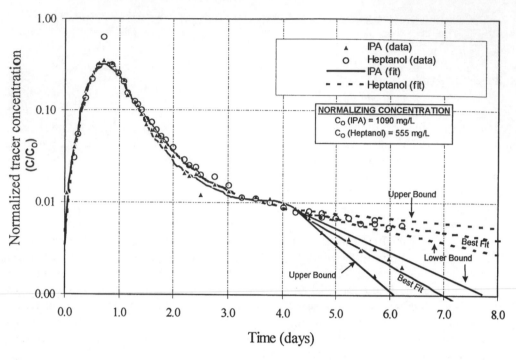

Figure 7.18 EX-2 IPA and heptanol tracer data and their corresponding fitted curves.

regression and visual observation with straight lines with different slopes on a semi-logarithmic scale. The estimated DNAPL volume calculation results ranged from 12 gal to 24 gal. The lower and upper bounds of this DNAPL volume estimation were determined by fitting the tracer response data of IPA and heptanol with four straight lines of different slopes so that the measured data were bounded by these straight lines (see Figure 7.18). The best fit of the tracer response curve, as well as the straight line extrapolation, yielded a DNAPL volume of 19 gal. The R values from the least-squares regression analysis were 0.67 and 0.75 for IPA and heptanol, respectively. The estimated DNAPL volume in the tracer swept volume of extraction well EX-2 was, therefore, approximately 19 ± 6 gal.

The practical quantitation limit (PQL) is essentially the lowest concentration believed to be reliably measurable within 20% of actual values. In this project, the PQL for each tracer was determined as the concentration of the lowest non-zero concentration standard used in the calculation of calibration factors and calibration curves. This definition is in agreement with the nominal definition of the estimated quantification limit (EQL) stated in Chapter One of the EPA SW-846 test methods. Based on this definition, the PQLs (in mg/L) in the pre-treatment PITT for IPA, pentanol, hexanol, and heptanol were 11.6, 1.2, 1.3, and 6.8, respectively, on the SRI GC, and 11.6, 1.2, 1.3, and 0.7 on the Buck GC. For the post-treatment PITT, the PQLs (in mg/L) for propanol, hexanol, heptanol, and octanol were 12.4, 12.3, 9.7, and 2.7, respectively, on the SRI GC, and 2.7, 2.7, 2.1, and 0.6 on the Buck GC. Although rigorous determinations of the method detection limits and PQLs were not performed, previous experience suggested that actual PQLs for these tracers are in the range of 1 to 5 mg/L. Thus, the tails of the tracer breakthrough curves in this report may be unreliable when concentration fell below 5 mg/L. The uncertainties of the average DNAPL saturation and DNAPL volume associated with the pre-flood PITT for the extraction wells are shown in Table 7.3. A similar summary of the uncertainties associated with the average DNAPL saturations for monitoring well MW-2 is presented in Table 7.4. Data from MW-1 were of limited value due to an operational problem that was corrected in time for the post-flood PITT. Because it was necessary to account for layering (heterogeneities), the monitoring well data were analyzed by a different method that is not described herein.

Table 7.4 Summary of the DNAPL Saturation Estimates in Monitoring Wells from the Pre-Flood Tracer Test

Well Name	DNAPL Saturation (%)	DNAPL Concentration (mg/kg)
MW-2A, high-permeability layer	0.5 ± 0.02	1285
MW-2A, low-permeability layer	0.3 ± 0.02	771
MW-2B	0.3 ± 0.02	771
MW-2C	0 ± 0.02	0

7.5.4 Post-Surfactant/Foam Flood PITT Analysis Results

The procedures for data analysis and error analysis used for the post-flood PITT are the same as those used for the pre-flood PITT. Only the final results of the analyses are presented in this section.

The tracer concentration histories of shallow (IN-1/EX-1), intermediate (IN-3/EX-3), and deep (IN-2/EX-2) well pairs are shown in Figures 7.19, 7.20, and 7.21, respectively. Both heptanol and octanol tracer data and their corresponding fitted curves are shown in these figures. In Figure 7.19 for IN-1/EX-1, anomalous behavior of the tracer concentrations was observed 4 d after the start of the tracer test. The cause of this anomalous behavior is unknown. Therefore, the tracer data for this extraction well were extrapolated from the general trend observed during the pre-flood PITT. The extrapolation, however, should not have any significant effect on the DNAPL saturation and volume estimates since there was negligible separation between the heptanol and octanol response curves. It can be inferred that very little DNAPL existed in the pore volume swept by the tracers.

The estimates of the DNAPL volume for each well pair are summarized in Table 7.5. The tracers swept 8310 gal aquifer volume during the post-flood PITT, as compared with the 8180 gal swept in the pre-flood PITT. Although the total swept volume was approximately the same for both PITTs, the individual tracer swept pore volume for each extraction well pair was different. This

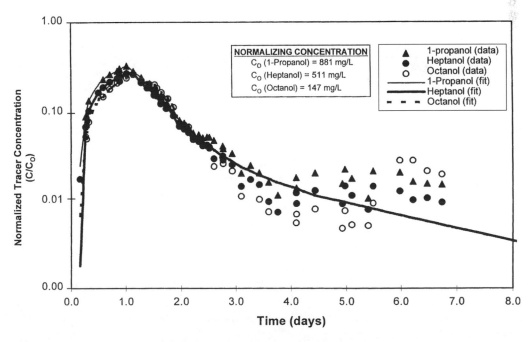

Figure 7.19 Post-flood PITT shallow well pair (IN-1/EX-1) tracer response data.

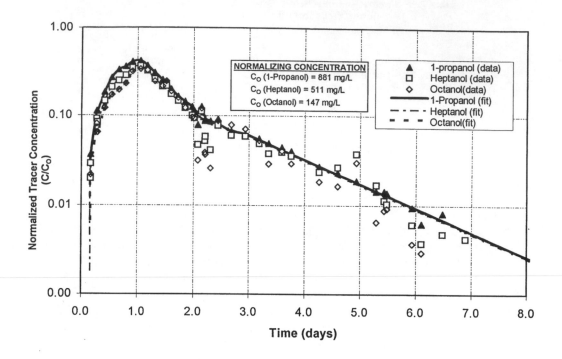

Figure 7.20 Post-flood PITT intermediate well pair (IN-3/EX-3) tracer response data.

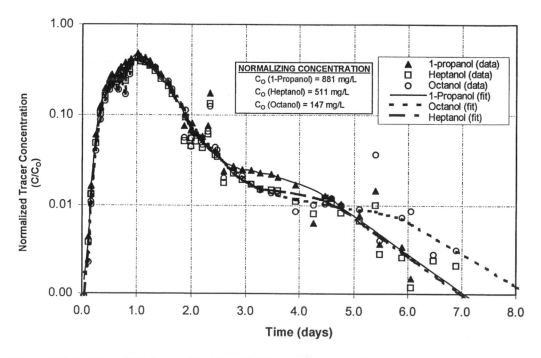

Figure 7.21 Post-flood PITT deep well pair (IN-2/EX-2) tracer response data.

Table 7.5 Summary of Post-Flood PITT Results

Well Name	Isopropanol Recovery (%)	Octanol Recovery (%)	Swept Volume	DNAPL Volume	Avg. DNAPL Content of Soil
EX-1	26	25	2,730 gal 10,300 L	0.9 ± 0.9 gal 3.4 ± 3.4 L	0.03 ± 0.03% 80 mg/Kg
EX-2	27	26	2,140 gal 8,100 L	0.7 ± 0.4 gal 2.6 ± 1.5 L	0.03 ± 0.02% 80 mg/Kg
EX-3	31	28	3,440 gal 13,000 L	1.0 ± 0.7 gal 3.8 ± 2.6 L	0.03 ± 0.02% 80 mg/Kg
Total	**84**	**79**	**8,310 gal 31,500 L**	**2.6 ± 2.0 gal 10 ± 8 L**	**0.03 ± 0.02% 80 mg/Kg**

could be attributed to the highly variable flow rates in the post-flood PITT. However, the change in each individual swept volume had no effect on the total NAPL volume or average DNAPL saturation estimation. The volume of DNAPL in the tracer swept volume was estimated to be 2.6 gal, corresponding to an average saturation of 0.03% or a concentration of 77 mg/kg.

The retardation factors calculated for the post-flood partitioning tracers varied between 1.01 and 1.14 at the three extraction wells. As discussed in the previous section, the uncertainty in the estimated DNAPL saturation and volume is large. The final volume of DNAPL was 2.6 gal (approximately 10 L), indicating that much of the DNAPL was removed. The mass recovery of the tracers ranged between 26% and 31%, for a total of 84% recovery. These recoveries were similar to the pre-flood PITT.

The absence of chromatographic separation of the partitioning tracers in all levels of MW-1 and the two uppermost levels of MW-2 indicated that essentially all of the DNAPL in the shallow and intermediate portion of the aquifer was removed. The separation of the tracers in the deepest level of monitoring well MW-2, MW-2A, indicated the presence of DNAPL at the base of the aquifer (see Figure 7.22). Analysis of the tracer data indicated an average saturation of 0.05%.

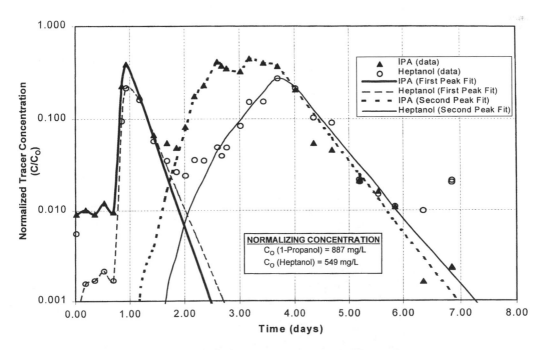

Figure 7.22 Post-flood PITT monitor point MW-2A tracer response and their fitted curves.

7.5.5 Summary of PITT Results

The pre-flood PITT at Hill OU2 estimated a DNAPL volume of 21 ± 7 gal, and the post-flood PITT estimated a DNAPL volume of 2.6 ± 2.0 gal. This corresponds to a DNAPL displacement of 88%. However, the main concern in this demonstration was the final DNAPL saturation, which was 0.03% ± 0.02%. This suggests that very little DNAPL was left in the swept volume of the aquifer after surfactant/foam flooding. The final residual DNAPL saturation of 0.03% ± 0.02% was of the same order of magnitude as the final residual saturation after the AFCEE surfactant flood at Hill OU2, which was 0.03% ± 0.03%.

7.6 RECOVERY OF INJECTED CHEMICALS

Although the main objective of the surfactant/foam flood was the recovery of DNAPL, another objective was to maximize the recovery of the injected chemicals. In order to determine this recovery, the cumulative volume of fluid produced by each of the extraction wells for each time segment was determined from the field production logs. The cumulative volume of fluid produced in each time segment and the average measured concentrations of surfactant, TCE, and sodium chloride were then used to determine the mass of chemicals recovered for that time segment. This information was later used to determine the mass of chemicals recovered during the entire test.

No evidence of foam was observed in the extraction wells, probably because any foam reaching the extraction wells was broken down there. The deterioration of the foam can be attributed to dilution of surfactant caused by the addition of fresh water from the hydraulic control well (Szafranski, 1997). Also, the extraction wells were designed to pump liquid from the bottom of the well. The foam would move upward and away from the pump. Moreover, air released by foam breakage could escape from the top of the wells. For all the above reasons, foam was not present in fluid samples from the extraction wells.

The recovery of surfactant, chloride, and bromide during the surfactant/foam flood is presented in Table 7.6. The errors in estimating surfactant and NaCl recoveries were determined from laboratory analyses and fluid production. Analysis of standards led to an estimate of 5% laboratory error. The 30-gal volume error was based on 10 min (at 3 gpm) between the sample time and the

Table 7.6 Cumulative Production of Surfactant, Chloride, and Bromide

Well	Surfactant Produced (lb)	Low Error Margin (lb)	High Error Margin (lb)	Error ± lb
EX-1	1910	1780	2050	130
EX-2	3990	366	4320	330
EX-3	2000	1810	2200	200
Total	7900	7250	8570	
Well	**NaCl Produced (lb)**	**Low Error Margin (lb)**	**High Error Margin (lb)**	**Error ± lb**
EX-1	720	660	790	61
EX-2	1410	1280	1550	140
EX-3	930	840	1020	93
Total	3060	2780	3360	
Well	**NaBr Produced (lb)**			
EX-1	15.2			
EX-2	42.9			
EX-3	10.9			
Total	69.0			

production log history for each time segment. The injected masses of surfactant and NaCl are shown in Table 7.1. The average error in the injected masses was less than 7%.

The corresponding cumulative production histories of surfactant and NaCl for the three extraction wells are shown in Figures 7.23, 7.24, and 7.25 respectively. From Table 7.6, it can be seen that the average error in the measurements was less than 10%. The total mass of surfactant recovered was 7900 ± 660 lb. This is in close agreement with the estimated mass of surfactant injected, which was 7900 ± 510 lb. Similarly, the mass of NaCl produced was 3100 ± 290 lb, which was in good agreement with the mass of NaCl injected (2800 ± 180 lb).

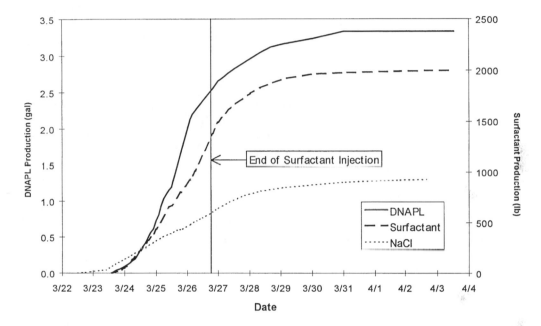

Figure 7.23 Cumulative production for well EX-3.

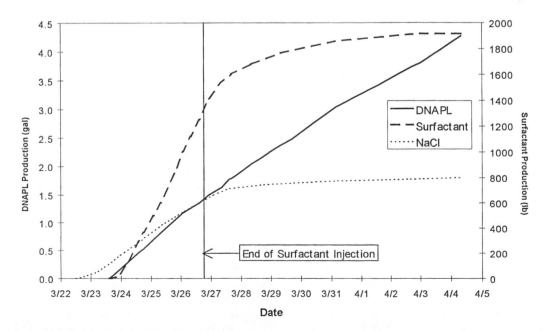

Figure 7.24 Cumulative production for well EX-1.

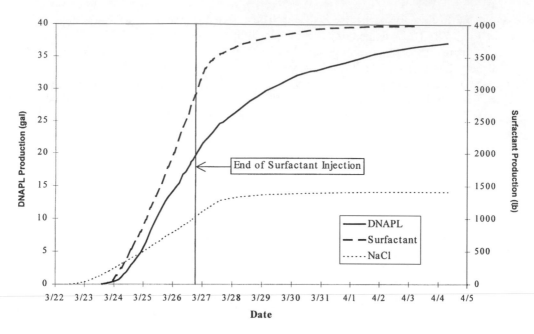

Figure 7.25 Cumulative production for well EX-2.

These data indicate that most of the injected chemicals were recovered. The average surfactant recovery was 100% within experimental error; average NaCl recovery was 110%. Both the ground-water and the source water had small amounts of dissolved NaCl, which may account for the overproduction of NaCl (see also Chapter 7.10.2). The high NaCl and surfactant recoveries indicate good hydraulic control of the injected fluids in the test area. From these results, it can also be inferred that there was minimal retention of surfactant in the aquifer.

Figures 7.26 through 7.28 show that the same trend existed at all three extraction wells for surfactant and chloride concentrations. The surfactant concentration increased to a normalized concentration of between 0.4 and 0.7 for the different extraction wells. The chloride concentration increased earlier because sodium chloride was injected for 26 h prior to surfactant addition. The chloride decline followed the surfactant decline because 8000 mg/L brine was added for 12 h after surfactant injection ended. Both the surfactant and chloride concentrations decreased at the same rate at all the wells.

7.7 DNAPL PRODUCED BY EXTRACTION AND MONITORING WELLS

The extraction well samples from the autocollector were typically single phase, with no evidence of DNAPL discoloration. Some samples had air space above the aqueous phase. Samples collected manually did not have any headspace; headspace in the autocollected samples can be attributed to incomplete sample collection. The cumulative production of DNAPL is listed in Table 7.7.

Since most of the DNAPL was in the middle well pair, IN-2 and EX-2, the contaminant concentration history for this well pair is discussed first. The TCE concentration curve for well EX-2 has three different peaks (see Figure 7.27). The first peak was 1 d after surfactant and air injection began. The second peak is thought to be due to a reduction in the injection rate at hydraulic control well HC-1, which was located at the extraction end of the pattern. The reduced injection rate at HC-1 resulted in less dilution of the fluids entering EX-2. The third peak (on 3/26) is thought to be due to a change in the injection rates some hours earlier. Both the surfactant and the chloride showed behavior similar to that of TCE in that their concentrations peaked at the same times as TCE.

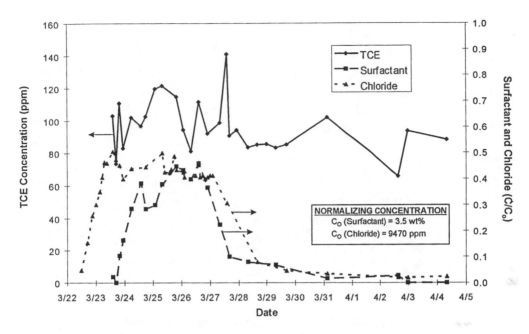

Figure 7.26 Concentration history of TCE, surfactant, and chloride for extraction well EX-1.

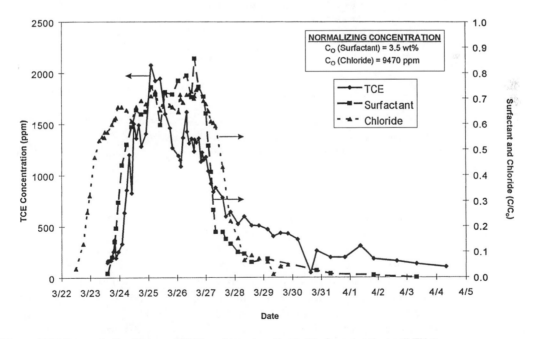

Figure 7.27 Concentration history of TCE, surfactant, and chloride for extraction well EX-2.

An increase in the TCE concentration from 100 mg/L to 2200 mg/L was observed at well EX-2. Similarly, for well EX-3, an increase from 20 mg/L to 700 mg/L was observed (see Figure 7.28). It can be inferred that recoverable DNAPL in the swept pore volume of wells EX-2 and EX-3 resulted in TCE solubilization and, consequently, elevated TCE concentrations in the effluent. For well EX-1 (Figure 7.26), since TCE solubilization was not enhanced, it can be inferred that there

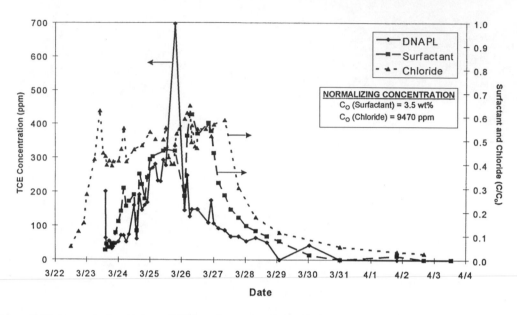

Figure 7.28 Concentration history of TCE, surfactant, and chloride for extraction well EX-3.

Table 7.7   Cumulative Production of DNAPL

Well	DNAPL Produced (gal)	Low Error Margin (gal)	High Error Margin (gal)	Error (± gal)
EX-1	4.3	4.0	4.6	0.3
EX-2	37	34	41	3
EX-3	3.3	3.0	3.7	0.3
Total	45	41	49	4

was no recoverable DNAPL in the swept pore volume. The presence of 100 mg/L TCE might have been caused by infiltration of contaminated groundwater from other zones.

A striking feature of Figure 7.28 for EX-3 is the single data point at 700 mg/L TCE. All other samples had concentrations below 300 mg/L. The single high value at EX-3 occurred shortly after the well was restarted after having been shut down for 3 h, and is almost certainly higher than would have been observed with continuous production.

To mobilize DNAPL upstructure from MW-2A to the bottom of the screened interval in EX-2 (refer to Figure 4.13), the pumping rates were increased in HC-2, IN-2, and EX-2 (see Table 5.2) in an effort to increase the pressure gradient at MW-2A. Figure 7.29 contains several graphs depicting the surfactant/foam flood operating conditions and the responses observed at the monitoring wells and the central extraction well during this time period. As shown in Figure 7.29c, the first monitoring well, MW-1A, which had had a declining TCE concentration, had a rebound in TCE concentration between 2400 h on 3/24/97 and 0200 h on 3/25/97. The second monitoring well, MW-2A, which was producing increasing amounts of foam and excess brine phase, produced samples rich in microemulsion at 2400 h on 3/25/97 (Figure 7.29d). Extraction well EX-2, which had a declining TCE concentration during 3/25/97, rebounded in TCE concentration between 0200 h and 0800 h on 3/26/97 (Figure 7.29e). The increase in TCE in EX-2 was an expected response to the increased pressure gradient in the pattern, but the rebound in MW-1A and MW-2A was not expected. This rebound is interpreted to be a response to the migration of DNAPL from the pool between the hydraulic control well, HC-2, and the injection wells as a result of increasing the injection rates in HC-2 and IN-2.

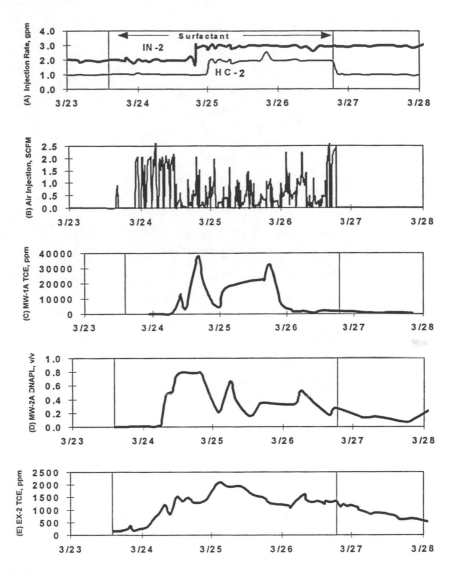

Figure 7.29 Monitoring and extraction well response.

7.8 POST-SURFACTANT/FOAM FLOOD SUBSURFACE CORES

Five borings were drilled after the post-treatment PITT to examine the spatial distribution of remaining DNAPL. The five boreholes — CB-1, CB-2, CB-3, CB-4, and CB-5 — were drilled, logged, and sampled to assess the performance of the surfactant/foam flood in the AATDF test area. The borehole locations were chosen to permit representative samples to be taken from different stratigraphic units within the swept volume of the test area. Because the pretreatment PITT and soil borings showed that the deepest part of the aquifer had the largest volume of DNAPL, and because DNAPL was extracted from MW-2A long after surfactant/air addition, the borings were placed in the vicinity of MW-2. The borings were located on either side of MW-2 along the deepest part and on the sides of the channel (see Figure 7.30). During drilling, injection and extraction with potable water continued to ensure that additional DNAPL hydraulically trapped outside of the test zone to the north and the south did not reinvade the test area and bias the soil sampling results.

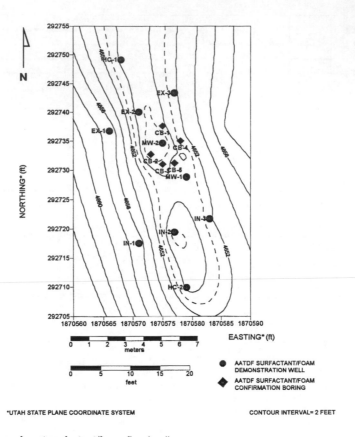

Figure 7.30 Location of post-surfactant/foam flood soil cores.

With the exception of boring CB-5, cores 2 ft in length were obtained at 5-ft intervals to approximately 25 ft bgs for each confirmation borehole drilled. Below 25 ft, the boreholes were sampled continuously to the completion depth of approximately 1 ft into the clay aquiclude. Boring CB-5 was drilled without sampling to 42 ft bgs, and then cored continuously to the completion depth 45 ft bgs. Sediment samples were collected from the cores from each boring using the procedures described previously.

The distribution of DNAPL in the post-treatment borings is compared with the pre-treatment distribution in Figure 7.31. No significant DNAPL was observed in cores CB-1 and CB-4 (with the exception of detectable, >0.02%, but below quantification limits, <0.1%, DNAPL near the water table).

A small amount of DNAPL was found in one to two sampled intervals in cores CB-2, CB-3, and CB-5. The highest DNAPL saturation was found near the level of the clay interface, but DNAPL saturation decreased abruptly below that level. Figure 7.32 gives an expanded scale plot of the DNAPL below the clay surface, expressed as a saturation or volume percent of the pore space. No DNAPL was detected 1 ft into the clay unit. It is possible that DNAPL was present just above the clay because the surfactant/foam bypassed this low-permeability zone. To explore this possibility, the permeability profile in CB-2, the boring with the greatest amount of DNAPL, was determined. The permeability and DNAPL saturation profile for CB-2 are shown in Figure 7.33. The permeabilities of the DNAPL-saturated samples are less than those of the clean samples above, but the difference between them is not enough to conclude that the DNAPL-saturated sample was bypassed because of low permeability.

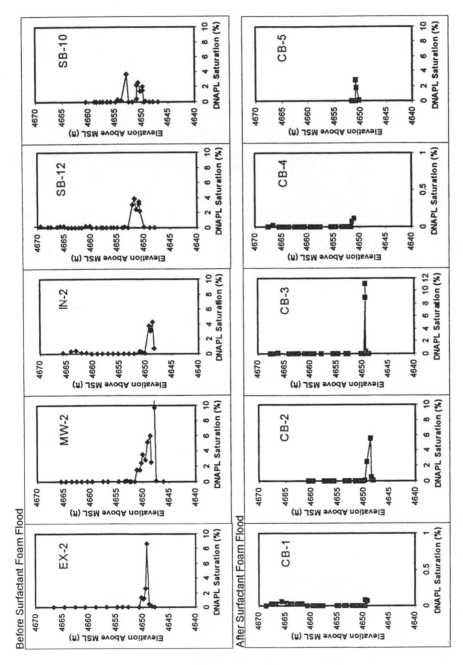

Figure 7.31 DNAPL saturation profiles before and after the surfactant/foam flood.

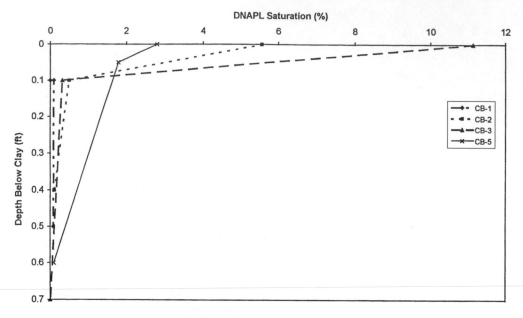

Figure 7.32 DNAPL saturation below clay surface.

CB-2

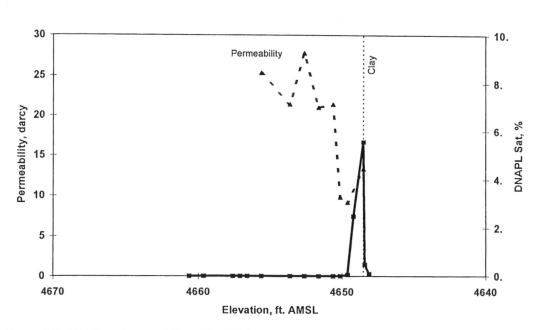

Figure 7.33 DNAPL and permeability profile in CB-2.

7.8.1   DNAPL in Post-flood Subsurface Cores

As indicated previously, DNAPL was detected in CB-2, CB-3, and CB-5, but not in CB-1 and CB-4. At CB-3 and CB-5, a narrow 2- to 3-in. interval was contaminated. At CB-2, DNAPL was present in the bottom foot of the aquifer. Saturation values were assumed for each elevation, based on the post-treatment cores, to determine the volume of DNAPL remaining in the aquifer. CB-2 suggests that DNAPL was present at 6% saturation below the 4649 ft amsl elevation. The borings also indicate that around 4% DNAPL saturation was present between 4649 and 4949.5 ft amsl, and practically none above it. Based on these numbers and the pore volume at each elevation taken from the aquifer model used in the simulations (see Chapter 4.4), the estimated remaining DNAPL volume was 2.8 gal. This number agrees very closely with the volume estimated from the post-treatment PITT, 2.6 ± 2.0 gal.

The post-treatment borings also yielded information about the spatial distribution of the DNAPL. Figure 7.34 is a cross section of the AATDF demonstration area that compares the initial and final DNAPL distributions. This figure is based on results from the borings and the aquifer description used in the simulations. It can be seen that the initial DNAPL level was at 4652 ft amsl. After the surfactant/foam flood, the DNAPL level was on the order of 4649 to 4650 ft amsl. The DNAPL was located in the lowest part of the clay aquiclude surface near IN-2 and MW-2. Some of this remaining DNAPL may be the result of reinvasion from the DNAPL between the injection wells and hydraulic control well, HC-2 (see Chapter 4.4 on simulation).

7.8.2 Cation Exchange Capacity of Post-surfactant/Foam Flood Subsurface Cores

In the post-treatment borings at Hill OU2, it was determined that the alluvium at Hill OU2 had cation exchange capacities ranging from 0.36 to 1.1 meq/100 g (see Table 7.8). These cation exchange capacities are low, but not insignificant. A cation exchange capacity of 1 meq/100 g sand is typical of some clean sands and corresponds to about 0.06 meq/mL of pore space for sand with 33% porosity. It is likely that a substantial portion of the exchangeable cations on the clays were in the calcium form at the start of the test because they were in equilibrium with source water from Hill OU2. This water is quite hard, with total calcium plus magnesium reported to be 0.006 eq/L.

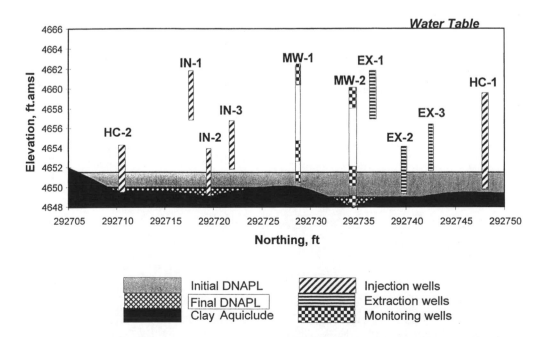

Figure 7.34 Cross section showing the initial and final distribution of DNAPL in the AATDF surfactant/foam demonstration area.

Table 7.8 Cation Exchange Capacities of Sands from Post-Surfactant/Foam Flood Cores

Core	Depth (ft bgs)	CEC[a] (meq/100 g)
CB-2	45.4	0.42
CB-2	46.1	1.1
CB-3	45.2	0.36
CB-3	45.3	0.62

[a] Measurements performed by Shell E&P Technology, 7/7/97.

7.9 OBSERVATIONS ON PRODUCED FLUIDS

The TCE concentration in the effluent from the monitoring wells was measured to determine the performance of the surfactant/foam process. The top levels, B and C, of the closest monitoring well, MW-1, produced a lot of foam; in many samples, there was not enough liquid for analysis. In those samples that could be analyzed, TCE content was low.

The deepest level, A, of MW-1 produced single-phase liquid samples that were usually clear with no visible signs of DNAPL discoloration. Figure 7.35 shows TCE concentration in both microemulsions, where surfactant concentration is above the CMC, and brines, where surfactant concentration is below the CMC, in the effluent from MW-1A. Some microemulsions showed high TCE concentrations, between 20,000 and 40,000 mg/L, confirming the enhanced solubilization capabilities of the surfactant.

Figure 7.35 also shows the amount of gas in samples from MW-1A. After foam breakthrough, significant gas content, which indicated the presence of foam, was seen intermittently until the end of surfactant injection. Gas appeared in most samples during the last day of surfactant injection when gas injection rates were high.

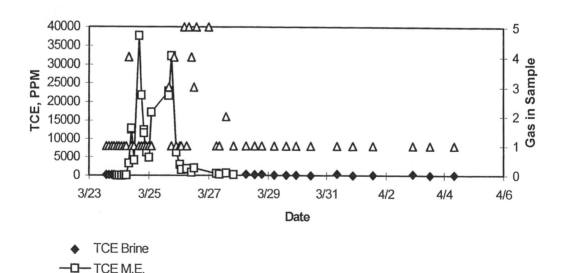

Figure 7.35 Monitoring well MW-1A: TCE in excess brine and microemulsion and gas in sample tube.

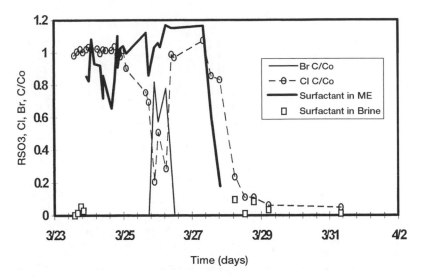

Figure 7.36 Monitoring well MW-1A: produced surfactant, Br⁻, and Cl⁻.

Concentrations of surfactant, chloride ion, and bromide ion are presented in Figure 7.36 for MW-1A. The sum of chloride and bromide concentrations remained near the injected value during the period of surfactant injection. The normalized surfactant concentration increased to a maximum value of 1.16.

Many samples collected from the upper levels, B and C, of MW-2 also contained mostly gas, indicating that foam was present. As happened in MW1, TCE concentrations were low in the samples that were analyzed.

Samples collected from level A of MW-2, the deepest point in the pattern, were mainly liquid. For the most part, they consisted of two phases, one of which was usually a microemulsion rich in DNAPL. However, a few samples had three phases and some had only a single phase. The phases observed in the effluent samples from MW-2A are shown in Figure 7.37. Volumes of the phases, including gas volume, are shown in Figure 7.38. From Figure 7.38, it can be seen that foam was

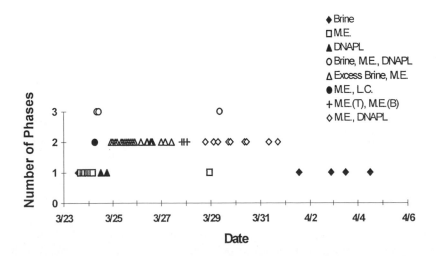

Figure 7.37 Liquid phases produced by monitoring well MW-2A. (Note: ME = microemulsion, LC = liquid crystal.)

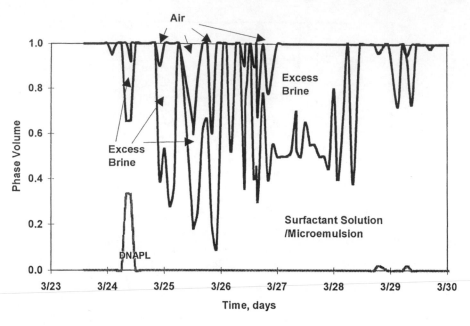

Figure 7.38 Phase volumes (including air) for monitoring well MW-2A.

able to reach MW-2A and reduce the amount of DNAPL-containing microemulsion present at some times during the flood. However, microemulsion volume increased when the gas disappeared.

Initially, DNAPL-rich microemulsions appeared early in the flood as single-phase samples, followed by a single sample containing a microemulsion phase and a liquid crystal phase. These were followed by three-phase samples containing excess brine, microemulsion, and DNAPL. Then, samples containing only DNAPL were seen for about 8 h. Finally, excess brine and microemulsion were produced until after the surfactant injection ended. At the very end of the flood, well MW-2A produced only brine. Because multiple phases were present in nearly all samples from MW-2A, it is useful to have information on the overall volume fraction of TCE+TCA+PCE in the sample as well as the normalized surfactant concentration. This information is presented in Figures 7.39 and 7.40. Further details and interpretation are given in Chapter 8.2.

7.9.1 Composition of Effluent in Drum

During the surfactant/foam flood at Hill OU2, the effluent from monitoring well MW-2 was produced into a drum. After the test, samples collected from the middle and bottom of the drum were analyzed for the presence of DNAPL constituents. Results are presented in Table 7.9. These results suggest that there was a microemulsion containing substantial DNAPL at the bottom of the drum, and surfactant solution with solubilized DNAPL in the middle of the drum. The results are consistent with data from MW-2A, which contained such a microemulsion along with excess brine during much of the test. It is possible that, at various times during the test, some free-phase DNAPL was produced into the drum and subsequently solubilized into the microemulsion.

7.9.2 Post-flood DNAPL Compositions at Hydraulic Control Well HC-2

On completion of the surfactant/foam flood, two samples from hydraulic control well HC-2 were collected. Both samples contained brine and DNAPL. Samples from MW-2A, believed to be mostly DNAPL, were analyzed along with the samples from HC-2. The goal was to compare the composition of DNAPL from HC-2 with that from MW-2A.

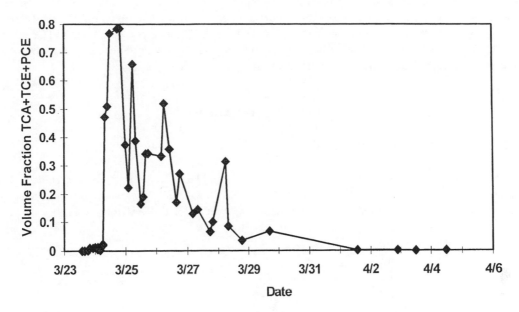

Figure 7.39 Weighted average concentrations of TCA+TCE+PCE in samples from monitoring well MW-2A.

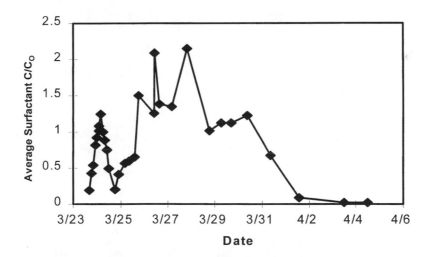

Figure 7.40 Average surfactant C/C_O in monitoring well MW-2A.

Table 7.9 Composition of Contents of Drum Containing Production Fluids from Monitoring Wells

Location	TCA (vol %)	TCE (vol %)	PCE (vol %)
Middle of drum	0.18	0.93	0.08
Bottom of drum	4	19	4

A comparison of these compositions is shown in Figures 7.41 and 7.42. The compositions are plotted in two ways: (1) each major component as a percent of total sample components (greater than 0.2%), and (2) the major components TCA, TCE, and PCE as a percent of TCA+TCE+PCE. Sample HC-2-1 was richer in higher boiling components than the other bailer samples. Otherwise, there were no significant differences among the samples of DNAPL analyzed.

7.10 MASS BALANCES

7.10.1 Surfactant Mass Balance

The quantities of surfactant added to the subsurface and recovered are in close agreement. The amount of surfactant injected was calculated as 7900 lb (±7%). The surfactant produced was determined to be 7900 lb (±8%), so that surfactant recovery was 100% within experimental error.

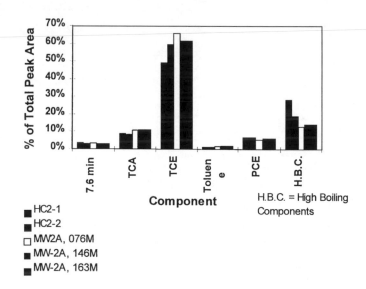

Figure 7.41 Comparison of total peak areas of DNAPL components in MW-2A and HC-2.

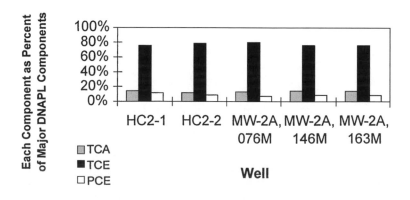

Figure 7.42 Comparison of DNAPL compositions of samples from MW-2A and HC-2.

Thus, very little surfactant escaped, adsorbed, or remained behind in the ground. The subsurface cores were not analyzed for residual surfactant.

7.10.2 Sodium Chloride Mass Balance

The mass balance calculation on the electrolyte had more error than the surfactant mass balance. The mass of sodium chloride (NaCl) injected was 2800 lb ($\pm$7%), while the calculated production was 3060 lb ($\pm$10%). Background chloride may have come from freshwater injection into the hydraulic control wells and from freshwater flooding after the surfactant/foam flood. However, the chloride content of the potable water source (Cl$^-$ ~ 0.0007 M) was too low for it to have contributed significantly.

A more likely explanation is that the aquifer contributed chloride. The background chloride concentration was measured at 0.0045 M both before the brine flood and at the end of the freshwater flood. Based on a production volume of 168,000 gal during the NaCl flood, the surfactant/foam flood, and the water flood, it was calculated that 370 lb NaCl (maximum) could have been contributed by the aquifer water. (This value is an upper limit.) If this 370 lb NaCl is subtracted from the total mass of chloride recovered (3060 lb) it is seen that 2690 lb of the injected NaCl was recovered. Based on the mass of 2800 lb NaCl added, the minimum recovery of NaCl was 96%. The calculated recovery is within the measurement error margin and an indicator of good hydraulic control.

7.10.3 DNAPL Mass Balance

The pre-flood partitioning interwell tracer test indicated that 21 $\pm$ 7 gal DNAPL was initially present in the test area. This value compared well with the estimate of 24 gal, which was based on cores of aquifer material (see Chapter 4.1.2). The post-treatment tracer test suggested that 2.6 gal DNAPL remained in the test area at the end of the field test. The difference between the two tracer measurements indicated that 18 $\pm$ 9 gal DNAPL had been removed from the test site. The post-treatment soil cores estimated that 2.8 gal DNAPL remained; approximately 21 gal DNAPL was removed according to the pre- and post-treatment soil borings.

Part of the DNAPL collected at the extraction wells was transported as a dissolved contaminant in the aquifer water. Since this volume would not be measured by the tracer tests or soil cores, the quantity of dissolved DNAPL was estimated. During the final tracer test, the concentration of TCE (DNAPL) in the effluent streams remained quite steady. These steady-state TCE concentrations for wells EX-1, EX-2, and EX-3 were taken to be 100 mg/L, 90 mg/L, and 20 mg/L, respectively. By scaling the TCE concentration to DNAPL volume fraction and multiplying the result by the total volume extracted from each well, it was determined that EX-1 produced 4.6 gal dissolved DNAPL, EX-2 produced 5.7 gal, and EX-3 produced 0.8 gal. The total dissolved DNAPL produced was 11 gal.

The DNAPL production curves presented in Figures 7.23 through 7.25 indicate that a total of 45 gal DNAPL was produced from the extraction wells over the course of the surfactant/foam flood. Laboratory testing of contaminated effluent from MW-2A showed that it contained 27% DNAPL, resulting in the production of an additional 2.8 gal DNAPL. The total DNAPL produced was thus estimated to be 48 gal. Because the 11 gal of dissolved DNAPL would have been extracted from the wells without the surfactant/foam flood, this quantity was subtracted from the total contaminant produced to give 37 gal. Thus, the amount of DNAPL recovered from the surfactant/foam test area exceeded by more than 50% the amount of DNAPL estimated to be present by both partitioning tracers and soil cores before treatment. The excess DNAPL production suggests that either DNAPL migrated into the area between the injection and extraction wells during the field test, or that the initial estimates of DNAPL contamination were low. This matter is discussed further in section 8.1.

CHAPTER 9

Conclusions

The surfactant/foam process for remediation of DNAPL-contaminated aquifers was successfully demonstrated at Operable Unit 2 (OU2), Hill Air Force Base, Utah. The process was designed to remove DNAPL by solubilization and mobilization. Foam was used for mobility control to divert the surfactant solution from high-permeability thief zones and shallower sands and direct it to the base of the aquifer where the DNAPL was located. Foam generation was possible with a modest increase in injection pressure. The multilevel monitoring wells showed that, as desired, foam first swept the upper intervals and thus diverted surfactant to the lower intervals where DNAPL was present.

The success of this project was due to extensive laboratory testing and experimentation to design the surfactant/foam process. Phase behavior experiments were used in the initial screening experiment to identify surfactants and their active region. Two-dimensional experiments (with a $20 \times 3 \times 0.75$ in. sand pack) showed that a pressure-regulated air injection scheme was required to propagate foam with a desired mobility through the heterogeneous porous medium. Heterogeneous sand packs using soils contaminated with DNAPL indicated that about one pore volume of surfactant was sufficient to remove all the DNAPL using the surfactant/foam process. Finally, experiments in a large 2-D model (8 ft $\times$ 5 ft $\times$ 2 in.) showed that foam could be propagated through a distance of about 2.4 m (8 ft).

Borings both before and after surfactant treatment were useful in characterizing the hydro-stratigraphy, DNAPL distribution, and permeability distribution. The DNAPL was confined to the region near the base of the aquifer and was absent in the high-permeability thief zones. Post-treatment borings indicated that the process could clean the sand to at least the detection level of 0.02% saturation. However, some borings still had a small amount of DNAPL at the base of the sand, indicating that DNAPL continued to migrate into the wellfield from parts of the alluvium farther up-dip. DNAPL did not appear downgradient of the depression containing MW-2, indicating that flushing with surfactant and foam and pumping of the monitoring well displaced or captured most of the DNAPL migrating into the well field.

The initial field investigation gave no indication of the DNAPL pool discovered between IN-2 and HC-2, outside the flood pattern. The flood pattern would have been placed to enclose the DNAPL pool if this information had been available before the well field was drilled. Similarly, an extraction well would have been located at the local depression in the aquitard near MW-2 had the existence of this depression been known. Thus, there is a need for more extensive characterization. Possible use of the cone penetrometer, shallow seismic methods, or ground-penetrating radar should be explored.

Partitioning interwell tracer tests conducted both before and after the surfactant/foam flood were useful in quantifying the volume of DNAPL and its spatial distribution in the treatment region. The initial volume was estimated to be between approximately 0.09 m^3 (24 gal), based on the borings, and 0.079 ± 0.027 m^3 (21 ± 7 gal), based on the PITT. This corresponds to an average saturation in the swept volume of 0.3% (670 mg/kg). The pre-treatment PITT confirmed that the DNAPL was localized between the central injector-extractor well pair in the deepest part of the

161

channel. This observation led to increasing the fraction of liquid injected in the central well pair during the last half of the surfactant injection period. The relatively high uncertainty in the PITT estimates was due mainly to errors in extrapolation of the field tracer data. These errors were also caused by the relatively low retardation of the partitioning tracers, due in part to the presence of only a small amount of DNAPL in the test area. Partitioning tracer errors can be minimized by (1) lowering the detection limit of the partitioning tracers to reduce errors in extrapolation, and (2) identifying and using tracers with higher partition coefficients to reduce errors caused by low retardation of the partitioning tracers.

The estimated post-demonstration volumes of DNAPL from the borings and the PITT are in agreement and show that about 0.01 m³ (2.8 gal) of DNAPL remained in the well field. The latter value corresponds to 0.03% average saturation (80 mg/kg) in the swept volume, a very low value. It is thought that this remaining DNAPL is the result of continued migration of DNAPL into the well field or due to formation of a dense microemulsion phase. Thus, the project successfully met the objective of using PITTs to determine the DNAPL volume in the test area.

The net DNAPL recovery of 0.14 m³ (37 gal) is attributed to the surfactant/foam process. The process recovered more DNAPL than was initially present, thus suggesting that the migration into the well field mentioned above had, in fact, occurred. Migration of even a small amount of DNAPL had a significant effect on the material balance since the initial average DNAPL saturation was only 0.3% of the swept pore volume. This initial DNAPL saturation is only 1% to 2% of what one would expect in a DNAPL pool that had been water-flooded to residual saturation. The material balance indicated that virtually all of the injected surfactant and at least 96% of the injected NaCl had been recovered.

Injection of surfactant solution at a rate corresponding to one pore volume per day was accomplished with little pressure differential between the injection and extraction wells.

Air injection to generate foam *in situ* was accomplished with only a modest increase in injection pressure. The injection pressure and air flow rate gave an instant indication of the resistance being generated by the foam. The multilevel sampling wells indicated the portion of the aquifer that was being swept by foam. The foam served its intended purpose of occupying the thief zone and upper parts of the aquifer so as to divert the surfactant solution to the base of the aquifer where the DNAPL was most likely to remain.

The multilevel monitoring wells provided a valuable window into the foam sweep and the process chemistry. The phase behavior in MW-2A showed that the microemulsion had a higher solubilized DNAPL concentration and thus a higher density than designed. This is likely due to the high calcium concentration in the microemulsion that probably occurred as a result of ion exchange between the clays in the formation and the surfactant micelles. Its detrimental effects can be avoided in future tests by modifying the amount and composition of the NaCl pre-flush preceding surfactant injection and using a lower salinity in the surfactant slug.

This high-density microemulsion tended to accumulate in the deep spot around MW-2A, which is deeper than the bottom of the screened interval in EX-2. It was swept by the foam during the periods of high air injection. However, it was not completely eliminated, possibly because DNAPL continued to enter the pattern.

The chemical flooding simulator, UTCHEM, was modified to simulate foam flow. It models the gas phase to have a specified, increased apparent viscosity if the conditions exist for foam, and has been used to aid in the interpretation of the test. While additional work on modeling and simulating foam flow is needed, the present simulator was able to match observed behavior during the test reasonably well, including overall contaminant production and the migration into the pattern mentioned above.

Steam stripping proved to be a viable treatment process for the effluent from a surfactant flood. It reduced TCE levels in the effluent from approximately 8000 mg/L to 0.1 mg/L.

In summary, it is concluded that the project successfully met the objective of confirming that the surfactant/foam process can remove residual contaminant from a heterogeneous aquifer.

CHAPTER 10

Section I References

Anderson, M.R., Johnson, R.L. and Pankow, J.F. (1992). Dissolution of dense chlorinated solvents into groundwater. 3. Modeling contaminant plumes from fingers and pools of solvent. *Environ. Sci. Technol.* 26(5):901.

Baran, J.R., Jr., Pope, G.A., Wade, W.H., Weerasooriya, V. and Yapa, A. (1994). Microemulsion formation with chlorinated hydrocarbons of different polarity liquids. *Environ. Sci. Technol.* 28:1361–1366.

Bear, J. (1972). *Dynamics of Fluids in Porous Media* Elsevier, New York.

Bernard, G.G., Holm, L.W. and Jacobs, W.L. (1965). Effect of foam on trapped gas saturation and on permeability of porous media to water, *SPEJ*, December 1965, 295–300.

Bhuyan, D, Pope, G.A. and Lake, L.W. (1991). Simulation of high-pH coreflood experiments using a compositional chemical flood simulator. *Proceedings of the SPE International Symposium on Oilfield Chemistry*, Anaheim, CA.

Bourrel, M. and Schechter, R.S. (1988). *Microemulsions and Related Systems*. Marcel Dekker, Inc., New York and Basel.

Bouwer, H. and Rice, R.C. (1976). A slug test for determining hydraulic conductivity of unconfined aquifers with completely or partially penetrating wells, *Water Resour. Res.* 12(3):423–428.

Britton, L.N. (1996). Personal communication

Brown, C.L. (1997). Modeling and Design of Surfactant-Enhanced Aquifer Remediation Field Tests. Ph.D. dissertation, The University of Texas, Austin.

Brown, C.L., Delshad, M., Dwarakanath, V., Jackson, R.E., Londergan, J.T., Meinardus, H.W., McKinney, D.C., Oolman, T., Pope, G.A. and Wade, W.H. (1999). Demonstration of surfactant flooding of an alluvial aquifer contaminated with DNAPL, Innovative Subsurface Remediation: Field Testing of Physical, Chemical, and Characterization Technologies. M.L. Brusseau, D.A. Sabatini, J.S. Gierke, M.D. Annable, eds. *ACS Symposium Series,* 725, American Chemical Society, Washington, D.C.

Brown, C.L., Pope, G.A., Abriola, L.M. and Sepehrnoori, K. (1994). Simulation of surfactant-enhanced aquifer remediation. *Water Resour. Res.* 30(11):2959–2977.

Camilleri, D., Fil, A., Pope, G.A. and Sepehrnoori., K. (1987). Improvements in physical-property models used in micellar/polymer flooding. *SPE Reservoir Eng.* Nov.

Cayais, J.L., Schechter, R.S. and Wade, W.H. (1975). The measurement of low interfacial tension via the spinning drop technique, in adsorption at interfaces. *ACS Symposium Series*, 8, J.R. Gould, Ed., American Chemical Society, Washington D.C., 234.

Core Laboratories (1996). Soil Analysis Test Program. Final report prepared for INTERA, Inc., Austin, Texas, by Core Laboratories, Rock Properties Laboratory, Carrollton TX, dated January 11, 1996.

Corey, A.T. (1994). Mechanics of Immiscible Fluids in Porous Media. Water Resources Publications, Colorado.

Cullum, D.C. (1994). *Introduction to Surfactant Analysis*. Blackie Academic and Professional, New York.

deBlanc, P.C., McKinney, D.C., Speitel, G.E., Sepehrnoori, K. and Delshad, M. (1996). A three-dimensional, multicomponent model of non-aqueous phase liquid flow and biodegradation model in porous media. In *Non-aqueous Phase Liquids (NAPLs), Subsurface Environment: Assessment and Remediation Proceedings of the Specialty Conference Held in Conjunction with the ASCE National Convention*, Washington D.C., Nov. 12–14, 1996), Lakshmi N. Reddi, Ed., ASCE, New York.

Delshad, M., Pope, G.A. and Sepehrnoori, K. (1996). A compositional simulator for modeling surfactant-enhanced aquifer remediation: 1. Formulation. *J. Contam. Hydrol.* 23:303–327.

Dwarakanath, V. (1997) Characterization and Remediation of Aquifers Contaminated by Nonaqueous Phase Liquids Using Partitioning Tracers and Surfactants. Ph.D. dissertation, The University of Texas, Austin.

Engineering-Science (1982). Installation Restoration Program Phase II: Records Search, Hill AFB.

Fountain, J.C., Starr, R.C., Middleton, T., Beikirch, M., Taylor, C. and Hodge, D. (1996). A controlled field test of surfactant-enhanced aquifer remediation. *Ground Water* 34(5):910–916.

Hand, D.B. (1939). Dimeric distribution. I. The distribution of a consolute liquid between two immiscible liquids. *J. of Physics and Chem.* 34:1961–2000.

Harwell, J.H., Sabatini, D.A., Chang, C.L., O'Haver, J.H., and Simpkins, T.J. (1999). Reuse of Sufactants and Cosolvents for NAPL Remediation. Lowe, D.F., Oubre, C.L.,a nd Ward, C.H., Eds., CRC Press, Boca Raton, FL.

Hirasaki, G.J. (1982). Ion exchange with clays in the presence of surfactant, *SPEJ*, April, 181–192.

Hirasaki, G.J. (1982b). Interpretation of the change in optimal salinity with overall surfactant concentration, *SPEJ*, December, 971–982.

Hirasaki, G.J. (1995). *Surfactant/Foam Process for Aquifer Remediation*, Rice University, Houston, TX.

Hirasaki, G.J., van Domselaar, H.R., and Nelson, R.C. (1983). Evaluation of the salinity gradient concept in surfactant flooding, *SPEJ*, June, 486–500.

Hirasaki, G.J. and Lawson, J.B. (1986). An electrostatic approach to the association of sodium and calcium with surfactant micelles, *SPE Reservoir Engineering*, March, 119–130.

Hirasaki, G.J., Miller, C.A., Szafranski, R., Lawson, J.B. and Akiya, N. (1997). Surfactant/foam process for aquifer remediation. Paper presented at the *1997 SPE International Symposium on Oilfield Chemistry*. Houston, TX, February 18–21.

Huh, C. (1979). Interfacial tensions and solubilizing ability of a microemulsion phase that coexists with oil and brine, *J. Colloid Interface Sci.* 71(2):408.

INTERA. (1997a). Phase IIb Work Plan — Field Work: AATDF Surfactant/Foam Process for Aquifer Remediation. Prepared for The Advanced Applied Technology Demonstration Facility, Rice University, Houston, TX, March.

INTERA. (1997b). Demonstration of Surfactant-Enhanced Aquifer Remediation of Chlorinated Solvent DNAPL at Operable Unit 2, Hill AFB, Utah. Draft final, prepared for the Air Force Center for Environmental Excellence, Technology Transfer Division, Brooks Air Force Base, San Antonio, TX, June.

Jackson, R.E. and Mariner, P.E. (1995). Estimating DNAPL composition and VOC dilution from extraction well data, *Ground Water* 33(3):407–414.

Jawitz, J.W., Annable, M.D., Rao, P.S.C. and Rhue, R.D., (1998). Field implementation of a Winsor Type I surfactant/alcohol mixture for *in situ* solubilization of a complex LNAPL as a single-phase microemulsion. *Environ. Sci. Technol.* 32(4):523–530.

Jin, M. (1995). Surfactant Enhanced Remediation and Interwell Partitioning Tracer Test for Characterization of NAPL Contaminated Aquifers. Ph.D. dissertation, University of Texas, Austin.

Jin, M., Delshad, M., Dwarakanath, V., McKinney, D.C., Pope, G.A., Sepehrnoori, K., Tilburg, C.E. and Jackson, R.E. (1995). Partitioning tracer test for detection, estimation and remediation performance assessment of subsurface nonaqueous phase liquids, *Water Resour. Res.* 31(5):1201–1211.

Johnson, R.L. and Pankow, J.E. (1992). Dissolution of dense chlorinated solvents into groundwater. 2. Source functions for pools of solvent, *Environ. Sci. Technol.*, 26(5):896–901.

Khatib, Z.I., Hirasaki, G.J. and Falls, A.H. (1988). Effects of capillary pressure on coalescence and phase mobilities in foams flowing through porous media. *SPE Reservoir Engineering* August 1988, 919–926.

Kim, H. (1995). A Simulation Study of Gel Conformance Treatments. Ph.D. dissertation, The University of Texas, Austin.

King, S. (1996). The Critical Micelle Concentrations of MA-80I. Research project report, Rice University Chemical Engineering Department, Houston, TX.

Kueper, B.H., Redman, D., Starr, R.C., Reitsma, S. and Mah, M. (1993). A field experiment to study the behavior of tetrachloroethylene below the water table: spatial distribution of residual and pooled DNAPL *Ground Water* 31(5):756–766.

Liang, Z., Miller, M.A. and Sepehrnoori, K. (1995). Implementation of a Dual Porosity Model in UTCHEM, Category A Research. 12th Annual Report, Enhanced Oil and Gas Recovery Research Program, Center for Petroleum and Geosystems Engineering, The University of Texas, Austin, (March).

Liu, J., Delshad, M., Pope, G.A. and Sepehrnoori, K. (1994). Application of higher order flux-limited methods in compositional simulations. *J. Transp. in Porous Media* 16:1–29.

Mackay, D.M. and Cherry, J.A. (1989). Ground water contamination: Pump-and-treat remediation. *Environ. Sci. Technol.* 23(6):630–636.

Mariner, P., Jin, M., and Jackson, R.E. (1997). An algorithm for the estimation of NAPL saturation and composition from typical soil chemical analyses. *Ground Water Monitoring and Remediation* 122–129.

Martel, R. and Gelinas, P.J. (1995). *In situ* washing of viscous DNAPL in a contaminated sand aquifer using surfactant solutions: field test at the Thouin sand pit (L'Assumption, Quebec, Canada). Presented at the *XXVI International Congress of the International Association of Hydrogeologists*. June 4–10, 1995, Edmonton, Alberta, Canada.

Mayer, A.S. and Miller, C.T. (1992). The influence of porous media characteristics and measurement scale on pore-scale distributions of residual nonaqueous phase liquids. *J. Contam. Hydrol.* 11:189–213.

Miller, C.T., Poirier-McNeill, M.M. and Mayer, A.S. (1990). Dissolution of trapped nonaqueous phase liquids: mass transfer characteristics, (Paper 9OWRO1964). *Water Resour. Res.* 26(11):2783–2796.

Nelson, R.C. and Pope, G.A. (1978). A chemical flooding compositional simulator. *SPEJ* 18:339–354.

Oolman, T., Godard, S.T., Jin, M. Pope, G.A. and Kirchner, K. (1995). DNAPL flow behavior in a contaminated aquifer: evaluation of field data. *Ground Water Monitoring Remediation* 15(4):125–137.

Pankow, J.F. and Cherry, J.A. (1996). *Dense Chlorinated Solvents and Other DNAPLs in Groundwater.* Waterloo Press, Waterloo, Ontario, Canada.

Pennell, K.D., Pope, G.A. and Abriola, L.M. (1996). Influence of viscous and buoyancy forces on the mobilization of residual tetrachloroethylene during surfactant flushing. *Environ. Sci. Technol.* 30(4):1328–1335.

Pitts, M.J., Wyatt, K., Sale T.C. and Piontek, K.R. (1993). Utilization of chemically enhanced oil recovery technology to remove hazardous oily waste from alluvium. In *Proc. of SPE International Symposium on Oilfield Chemistry.* New Orleans, LA, SPE 25153, 33-44.

Pope, G.A. and Wade, W.H. (1995). Lessons from enhanced oil recovery research for surfactant-enhanced aquifer remediation. In Sabatini, D.A., Knox, R.C., and Harwell, J.H., Eds., *Surfactant-Enhanced Subsurface Remediation: Emerging Technologies.* American Chemical Society Symposium Series 594, ACS Washington D.C., 142-160.

Powers, S.E., Loureiro, C.O., Abriola, L.M. and Weber, Jr., W.J. (1991). Theoretical study of the significance of nonequilibrium dissolution of nonaqueous phase liquids in subsurface systems. *Water Resour. Res.* 27(4):463–477.

Prouvost, L., Pope, G.A. and Rouse, B.A. (1985). Microemulsion phase behavior: a thermodynamic modeling of the phase partitioning of amphiphilic species, *SPE J* Oct., 693–703.

Prouvost, L., Satoh, T., Pope, G.A. and Sepehrnoori, K. (1984). A new micellar phase-behavior model for simulating systems with up to three amphiphilic species, SPE 13031. *Proceedings of the 59th Annual Meeting of the Society of Petroleum Engineers*, Houston, TX, Sept.

Radian. (1996). Operation of Source Recovery System (SRS), Performance Period February 1, 1995 to December 31, 1995. Report prepared for USAF.

Radian. (1995). System Evaluation Report, SRS Commissioning, Startup, and Initial Operation Interim Remedial Action, Operable Unit 2, Hill AFB, Utah. Report prepared in February for OO-ALC/EMR, Hill AFB, Utah.

Radian. (1994). Aquifer Data Evaluation Report, Operable Unit 2. Prepared for OO-ALC/EMR, Hill Air Force Base, Utah.

Radian. (1993). Draft Site Health and Safety Plan for Source Removal System, Operable Unit 2, Hill Air Force Base, Utah. Prepared for OO-ALC/EMR Hill Air Force Base, Utah.

Radian. (1992). Final Remedial Investigation for Operable Unit 2 — Sites WP07 and SS21. Prepared for Hill Air Force Base, Utah.

Reed, R.L. and Healy, R.N. (1977). Some physiochemical aspects of microemulsion flooding: a review. In *Improved Oil Recovery by Surfactant and Polymer Flooding.* D.O. Shah and R.S. Schechter, Eds., Academic Press, New York, 383–437.

Rice University. (1995). Surfactant/Foam Process for Aquifer Remediation. Prepared for Advanced Applied Technology Demonstration Facility (AATDF), Rice University, Houston, TX.

Rose, W. and Bruce, W.A. (1949). Evaluation of capillary character in petroleum reservoir rock. *Pet. Trans. AIME*, 186:127.

Rossen, W.R. and Zhou, Z.H. (1995). Modeling foam mobility at the 'limiting capillary pressure.' *SPE Advanced Technology Series* 3(1):146–153.

Schwille, F. (1988). *Dense Chlorinated Solvents in Porous and Fractured Media*. Lewis Publishers, Boca Raton, FL, 146.

Simpkin, T.J., et al. (1999). *Surfactants and Cosolvents for NAPL Remediation*. Lowe, D.F., Oubre, C.L., Ward, C.H., Eds., CRC Press, Boca Raton, FL.

Smith, J.C., Delshad, M., Pope, G.A., Anderson, W.G. and Marcel, D. (1988). Analysis of unsteady-state displacements using a capacitance-dispersion model. *In Situ* 12(1 & 2).

Szafranski, R. (1997). Laboratory Development of the Surfactant/Foam Process for Aquifer Remediation. Ph.D. dissertation, Rice University, Houston, TX.

Tang, J.S. and Harker, B. (1991a). Interwell tracer test to determine residual oil saturation in a gas-saturated reservoir. I. Theory and design. *J. Canadian Petroleum Technology* 30(3):76.

Tang, J.S. and Harker, B. (1991b). Interwell tracer test to determine residual oil saturation in a gas-saturated reservoir. II. Field applications. *J. Canadian Petroleum Technology* 30(4):34.

Tham, M.J., Nelson, R.C. and Hirasaki, G.J. (1983). Study of the oil wedge phenomena through the use of a chemical flood simulator. *SPEJ* October, 746–758.

U.S. EPA. (1992). Estimating Potential for Occurrence of DNAPL at Superfund Sites. Publication 9355.4-07FS, Office of Solid Waste and Emergency Response (OSWER), January 1992.

PART II

Field Demonstration of Single-phase Microemulsions for Aquifer Remediation

James W. Jawitz
University of Florida

R. Dean Rhue
University of Florida

Michael D. Annable
University of Florida

P.S.C. Rao
University of Florida

EXECUTIVE SUMMARY

This section documents laboratory and field investigations of the use of surfactant/alcohol mixtures to solubilize a complex nonaqueous phase liquid (NAPL) as a single-phase microemulsion (SPME). The first phase of this study was to select a surfactant and cosurfactant that together form the microemulsion precursor. The precursor would produce a low-viscosity, single-phase micro-emulsion on contact with the complex, multicomponent NAPL found at the field site. Eighty-six surfactants and a number of alcohols were screened, with maximum NAPL solubilization and low-viscosity (<2 cp) as the main acceptance criteria. The viscosity of the precursor solution was limited to preclude large hydraulic gradients across the test cell and excessive drawdown around the extraction wells. The precursor solution selected was the surfactant Brij 97® (polyoxyethylene (10) oleyl ether) at 3% by weight and n-pentanol at 2.5% by weight in water. This mixture was evaluated in the laboratory in both column and two-dimensional aquifer models using contaminated and uncontaminated media from the field site at Hill Air Force Base.

The second phase of this study was the field implementation of the SPME flushing technology in a test cell constructed at Hill AFB, Utah. Field testing of the SPME technology was part of a larger study coordinated by researchers at the U.S. EPA National Risk Management Research Laboratory in Ada, Oklahoma. The study was funded through the Strategic Environmental Research and Development Program (SERDP) to conduct side-by-side testing of a total of nine technologies. Each test was conducted in an isolation test cell using evaluation methods that provided consistency among the results.

The SPME flushing field study was conducted in a 2.8 m × 4.6 m test cell made of interlocking sheet piles. The sheet piles isolated a section of the surficial aquifer by penetrating a thick clay aquitard at a depth of 8 m below ground surface (bgs). The flow through the test cell was a line-drive arrangement with four injection wells and three extraction wells located on opposite sides of the test cell. The extent of contamination within the test cell was assessed by collecting soil samples during well installation and analyzing for a selected group of analytes present in the NAPL. The test cell was also characterized by conducting a 10-pore volume (PV) tracer test with a group of partitioning and nonpartitioning tracers. The test cell had an average NAPL saturation of about 0.06 prior to SPME flushing. Water samples were also collected and analyzed for target analytes; however, nearly all the contaminants were below detection limits. Finally, before the SPME flood, the test cell was characterized using interfacial tracers. The method estimates the contact area between that of NAPL and water in the test cell. The interfacial tracer method, developed at the University of Florida, was tested for the first time in the SPME test cell at Hill AFB. The result indicated that the NAPL–water contact area can vary more than 1 order of magnitude within the test cell. This variability in NAPL–water contact area could have significant impact on *in situ* flushing extraction efficiency. Flushing efficiency will be reduced if there are mass transfer limitations that are magnified with low NAPL–water contact area, or if the NAPL is simply not uniformly distributed, as evidenced by a low NAPL-water contact area.

The SPME flood was conducted over an 18-d period by pumping nine pore volumes of the precursor solution with some flow interruption periods. During this time, over 7000 samples were collected for target analyte quantification from the network of multilevel samplers and the three extraction wells. The results were used to quantify the mass removed during the flushing phase of the study and were compared with initial estimates based on both core samples and partitioning tracers. The use of multiple methods of evaluation provides a more comprehensive assessment of the technology and, therefore, a better comparison with other technologies tested in the SERDP project. During the SPME flushing study, samples of effluent (combined from the three extraction wells) were collected for a laboratory investigation of methods to minimize and reduce cost of waste management. The use of salt to separate the waste into oil- and water-rich phases for more economic disposal was investigated on a small pilot-scale test.

Following the SPME flood, a post-flushing partitioning tracer test was conducted. The method and tracers used were the same as for the pre-flushing test. The average NAPL saturation in the test cell following the SPME flood was 0.018, producing an NAPL reduction of about 72%. After the tracer test was completed, post-flushing core samples were collected. The core samples were compared with average values obtained during the pre-flush sampling to estimate mass removal for each target analyte. The mass removal based on core estimates ranged from 64% to 96%, but the three largest constituents present in the NAPL — n-undecane, n-decane, and 1,3,5-trimethyl-benzene — were all removed by more than 90%.

The final method used to evaluate the SPME effectiveness was a mass balance on the target analytes. Using estimates of initial mass from the core data and partitioning tracers, the mass recoveries were calculated. Using the partitioning tracers to estimate initial mass, the mass removal estimates ranged from 62% to 82%. When the core data were used to estimate the initial mass, only two constituents could be compared; the mass removals for these were 93% and 105%, respectively. Once again, an apparent difference in the effectiveness is seen between the results based on tracers and soil cores. A pitch fraction model of the NAPL is proposed as a likely explanation for the differences observed. The methods outlined above demonstrate that the SPME process was capable of effecting significant mass removal of the complex multicomponent NAPL.

Simulations of the SPME flushing study were conducted following the experiment. Both the tracer tests and the SPME flushing results were used in the simulator UTCHEM. The model was capable of simulating the processes observed in the experimental data. It was evident that the model required the use of a mass transfer term to capture the observed behavior. While the model simulation may have limited use in helping to design experiments conducted in a controlled test cell, it can be very important for designing hydraulically controlled systems. Of greater utility are simulations to assess cost and efficiencies of the SPME process at larger scales than the test cell.

CHAPTER 11

Introduction

11.1 PURPOSE

The purpose of this section is to document laboratory and field investigations of the use of surfactant/alcohol mixtures to solubilize a complex nonaqueous phase liquid (NAPL) as a single-phase microemulsion (SPME) and satisfy a contract requirement between the University of Florida and the Advanced Applied Technology Demonstration Facility for Environmental Technology Program (AATDF) at Rice University (Contract R115D-23100095).

11.2 PROJECT OVERVIEW

The primary goal of this project was to conduct a series of controlled laboratory and field experiments to evaluate the feasibility of using single-phase microemulsions for enhanced *in situ* flushing of NAPL-contaminated aquifers. The project comprised the following three phases:

1. Laboratory tests for designing SPME precursors involving mixtures of surfactants and cosurfactants
2. Field-scale demonstration of SPME *in situ* flushing in a hydraulically isolated test cell located in a surficial aquifer contaminated with jet fuel and chlorinated solvents at Hill Air Force Base (AFB), Utah
3. Data analysis and numerical modeling of the laboratory and field tests to assess the effectiveness of the remediation technique

Each of these three phases consisted of several tasks. This section is a summary of Phases I, II, and III. In addition, several papers that describe the laboratory and field results (Annable et al., 1998a; Jawitz et al., 1998) have been published in peer-reviewed journals.

CHAPTER 12

Single-Phase SPME *In Situ* Flushing

12.1 *IN SITU* FLUSHING: BACKGROUND

Recently, many investigators have demonstrated that *in situ* flushing techniques show promise for the remediation of aquifers contaminated with nonaqueous phase liquids (NAPLs), such as fuels and chlorinated solvents, in time frames much smaller than would be expected for traditional pump-and-treat methods (Falta et al., 1996; Fountain et al., 1996; Brown et al., 1997; Rao et al., 1997). Similar to enhanced oil recovery (EOR) (e.g., Gatlin and Slobod, 1960; Gogarty and Tosch, 1968; Shah, 1981; Lake, 1989), these flushing technologies generally involve the addition of chemical agents to an injection fluid during either miscible or immiscible displacements to accelerate the dissolution or displacement of NAPLs from contaminated aquifers. Two classes of chemical agents used for *in situ* flushing that have been the focus of much study are cosolvents (Farley et al., 1993; Peters and Luthy, 1993; Augustijn et al., 1994; Imhoff et al., 1995) and surfactants (Abdul et al., 1992; West and Harwell, 1992; Pennell et al., 1993; Brown et al., 1994; Pennell et al., 1994; Fountain et al., 1996; Shiau et al., 1996). Both cosolvent and surfactant solutions can be used to accelerate the removal of trapped residual NAPLs from porous media through either mobilization or enhanced solubilization. Mobilization of NAPL trapped within the porous media by capillary forces is facilitated by a reduction in the interfacial tension (IFT) between the NAPL and the flushing solution as either a cosolvent or surfactant solution displaces the resident groundwater and contacts NAPL ganglia. Similarly, solubilization of the NAPL is enhanced due to the reduced polarity of the flushing solution, compared with that of the resident groundwater.

Several mobilization studies have shown that a large percentage of the NAPL trapped in contaminated porous media, often >95%, can be removed after injecting a small number of pore volumes of flushing fluid (Pennell et al., 1996; Sabatini et al., 1996; Brown et al., 1999). However, mobilization technologies are limited to sites where the flow of the mobilized NAPL can be controlled and the potential for migration of NAPL through layers that previously acted as capillary barriers can be assured. Remediation technologies that rely on solubilization alone generally do not remove the NAPL as efficiently as technologies that also mobilize, but solubilization-only technologies generally pose less risk, with regard to uncontrolled migration of the mobilized NAPLs, and are less complex to design, especially with respect to the number of parameters requiring optimization. It is for these reasons that the focus of this research has been the development of aquifer remediation strategies that solubilize NAPLs without inducing mobilization (Rao et al., 1991; Augustijn et al., 1994; Ouyang et al., 1996; Rhue et al., 1997). In addition, many other researchers have conducted both laboratory- and field-scale NAPL solubilization studies using both surfactants (Abdul et al., 1992; Pennell et al., 1993; Fountain et al., 1996) and cosolvents (Peters and Luthy, 1993; Roy et al., 1995).

The study reported here was one of eight pilot-scale tests funded by the Strategic Environmental Research and Development Program (SERDP) and AATDF, and conducted side-by-side at Hill AFB for the purpose of evaluating innovative remediation technologies for the removal of NAPLs from subsurface environments. In this study, a surfactant/alcohol mixture was used to generate a Winsor Type I system, wherein the NAPL was solubilized and transported as a single-phase microemulsion. For the SPME process, only 5.5 weight percent (wt%) of the flushing solution was composed of chemical additives (3.0 wt% surfactant and 2.5 wt% alcohol).

12.2 DESCRIPTION OF SPME TECHNOLOGY

A microemulsion is an optically transparent dispersion of liquid droplets (<0.1 μm) suspended within a second, immiscible liquid and stabilized by an interfacial film of surface-active molecules (Rosen, 1989). Generally, the stabilizing interfacial film is composed of both a surfactant and a cosurfactant, such as an intermediate chain-length alcohol. Varying a parameter that changes the character of the system from hydrophilic to lipophilic will produce a phase transition from an oil-in-water microemulsion (oil, or NAPL, droplets in a water-continuous phase; Winsor Type I), to a water-in-oil microemulsion (water droplets in an oil phase; Winsor Type II). Winsor Type III systems are formed midway in the transition from a hydrophilic to lipophilic system. At the transition to a Winsor Type III system, the interfacial tensions between the microemulsion phase, also known as the middle phase, and any excess oil or water phases present reach extreme minimums, much lower than those achieved in Winsor Type I and II systems. It is because of these ultralow interfacial tensions that middle-phase microemulsions have been promoted for both enhanced oil recovery processes (Gogarty and Tosch, 1968; Lake, 1989) and aquifer remediation via NAPL mobilization (Baran et al., 1994; Sabatini et al., 1996; Shiau et al., 1996; Brown et al., 1997). However, middle-phase systems can be difficult to design in field settings because of the relatively large number of parameters requiring optimization (e.g., salinity, surfactant concentration, and aquifer geochemical conditions) (Pithapurwala et al., 1986; Sabatini et al., 1996). Martel and collaborators (Martel et al., 1993; Martel and Gélinas, 1996) proposed the use of Winsor Type I microemulsions to solubilize NAPLs without mobilization. These systems have the advantages of high solubilization of NAPLs (although not as high as middle-phase microemulsions) while requiring low amounts of chemical additives. Chun-Huh (1979) showed that, in microemulsions, solubilization of the oil phase into the microemulsion is related to interfacial tension by an inverse-squared relationship. Therefore, remediation systems that rely on Winsor Type I microemulsification will necessarily be less efficient than those that rely on Winsor Type III microemulsions and mobilization, since solubilization will be lower at the higher interfacial tensions required to prevent mobilization.

The mechanisms of aqueous solubilization, micellar solubilization, and microemulsification are contrasted in the conceptual diagram shown in Figure 12.1. In aqueous solubilization, individual molecules of NAPL constituents dissolve in the water (Figure 12.1a). In the presence of surfactant micelles, solubilization is enhanced as NAPL constituent molecules are scavenged from the aqueous solution (Figure 12.1b). Rather than solubilizing individual constituent molecules, microemulsions, through a reduction of interfacial tension, are able to encapsulate tiny droplets of the NAPL itself (Figure 12.1c). The NAPL solubilization capacities of microemulsions are generally higher than can be achieved through micellar solubilization alone.

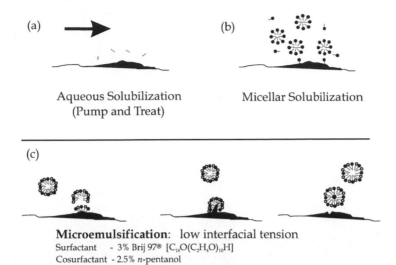

Aqueous Solubilization
(Pump and Treat)

Micellar Solubilization

Microemulsification: low interfacial tension

Surfactant   - 3% Brij 97® [C$_{18}$O(C$_2$H$_4$O)$_{10}$H]
Cosurfactant - 2.5% n-pentanol

Figure 12.1 Conceptual comparison of (a) aqueous solubilization, (b) micellar solubilization, and (c) microemulsification. Solid circles indicate surfactant molecules, open circles indicate cosurfactant molecules, and squiggles indicate NAPL constituent molecules.

CHAPTER 13

Phase I: Laboratory Investigation

13.1 CHARACTERISTICS OF THE OU1 NAPL

The NAPL used in Phase I of this study was free product taken from a monitoring well labeled OU1-202 at Hill AFB in June 1995. It had a viscosity of 10.7 cp and a density of 0.85 g/cm^3 at 24°C. Of the many constituents in the NAPL, nine target analytes were selected for monitoring the presence of NAPL in aqueous microemulsions. These nine compounds are listed in Table 13.1, along with their respective mass fractions in the NAPL, as measured at the University of Florida (UF).

Also listed in Table 13.1 are the mass fractions of the nine target compounds in a free product NAPL sample taken from the monitoring well (labeled MW-2 in Table 13.1) located approximately 2 m north of Well 2852. The spatial heterogeneity of the Hill NAPL composition is indicated by the different compositions of the two samples. While the OU1-202 NAPL was used for Phase I of this project, the MW-2 NAPL was collected very near the field test cell and was thus used for determining NAPL composition for Phase II of the project.

13.2 SURFACTANT SELECTION

Samples of 86 commercial surfactants were obtained directly from manufacturers. They ranged from viscous liquids and solids with essentially 100% active ingredient to aqueous solutions containing <25% surfactant. When calculating how much of a commercial formulation was needed to obtain a given concentration of active ingredient, data were taken from material safety data sheets or lot analyses provided by the manufacturer.

13.2.1 Surfactant Solubility and Viscosity

The solubility, viscosity, and alcohol solubility of each surfactant were evaluated at a surfactant concentration of 3 g/dL. The solubility of alcohol, used as a cosurfactant in these surfactant solutions, was determined by titrating with alcohol until the solubility limit was reached (i.e., phase separation). Of the 86 surfactants evaluated, 28 were insoluble in water at a concentration of 3 g/dL. The remaining 58 surfactants represented six chemical classes of anionic, nine chemical classes of nonionic, and one chemical class of amphoteric surfactants.

The solubility of alcohol in these surfactants, as determined by titration, varied widely. Pentanol was the alcohol used for all but six surfactants that formed stable dispersions, but which did not clear with the addition of pentanol. For these six surfactants, *n*-butanol or isobutanol was used to

Table 13.1 Mass Fractions of Target NAPL Constituents

NAPL Constituent	Mass Fraction (g/100 g NAPL)	
	OU1-202	MW-2
p-Xylene	0.892	0.144
1,2-Dichlorobenzene	1.66	a
1,2,4-trimethylbenzene	1.08	0.438
Naphthalene	0.055	a
Trichlorobenzene	0.62	0.461
n-Decane	2.14	0.477
n-Undecane	1.32	1.573
n-Dodecane	0.680	0.698
n-Tridecane	0.510	0.285

a Not measured.

see if the more water-soluble alcohol would produce water-clear solutions. For three of the six surfactants, water-clear solutions were obtained with n-butanol.

Kinematic viscosities of aqueous surfactant solutions of 3 g/dL were measured in an Ostwald viscometer tube. Viscosities of aqueous surfactant solutions were generally low, most being less than 1.5 cp. The addition of pentanol to two nonionic and two anionic surfactants showed a moderate increase in viscosity, approaching 2 cp at the alcohol solubility limit.

13.2.2 NAPL Solubilization by Precursors

The final 58 surfactants were mixed with alcohol to form microemulsion precursors. The surfactant concentration used was 3 g/dL and alcohol was added whenever possible to give a 1:1 weight ratio of surfactant:alcohol. If this concentration exceeded the alcohol solubility, a lower concentration was used. In all cases, an alcohol concentration was used that resulted in a single-phase microemulsion precursor.

Precursor solutions were equilibrated with OU1 NAPL at 24°C for 1 week. In all but one instance, these oil/precursor mixtures formed Winsor Type I microemulsions. The amount of OU1 NAPL that was solubilized, as measured by gas chromatography (GC), was used to rank the 58 surfactants with respect to solubilization potential.

Chromatograms of OU1 NAPL in hexane and in an aqueous microemulsion obtained with a Brij 97/pentanol solution are compared in Figure 13.1. Peaks associated with the nine target analytes, as well as hexane and pentanol, are identified. Except for the hexane and pentanol peaks, the two chromatograms are similar and indicate little selectivity with respect to the nine target analytes.

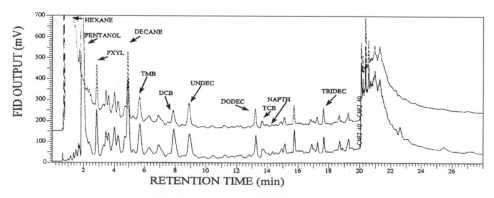

Figure 13.1 Comparison of chromatograms of Hill NAPL dissolved in hexane (upper) and Hill NAPL in aqueous microemulsion (lower).

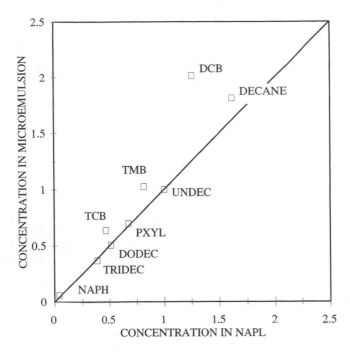

Figure 13.2 Concentration of target analytes relative to that of undecane in OU1 NAPL and in a microemulsion (Brij 97 and pentanol).

Lack of selectivity is also demonstrated in Figure 13.2, where the concentrations of target analytes in the microemulsion, relative to that of undecane, are plotted against relative concentrations in the NAPL. With the possible exception of 1,2-dichlorobenzene (DCB), which tended to be present in the microemulsion phase at a slightly higher concentration (relative to undecane) than it was in the NAPL itself, it appeared that selectivity was negligible and that the concentration of any one of the target analytes, other than DCB, could be taken as representative of the amount of NAPL in a microemulsion. For the laboratory phase of this project, dodecane concentrations were used as an indicator of NAPL microemulsification.

The results of the NAPL solubilization study are presented in Table 13.2. The NAPL concentrations presented in Table 13.2 were determined by dividing dodecane concentrations by the mass fraction given in Table 13.1. For almost all precursors, surfactant concentration was 3 g/dL and the surfactant:alcohol ratio was close to 1. Higher ratios were used where alcohol solubility was limiting. Dodecane concentrations in the aqueous microemulsions ranged from <1 mg/L to over 200 mg/L, corresponding to NAPL concentrations ranging from <0.01 g/dL to almost 3 g/dL.

The ethoxylated alcohols showed the greatest range in NAPL solubilization, from <10 mg/L to over 200 mg/L dodecane. These results are similar to those of Abdul et al. (1991), who chose ethoxylated alcohol ether surfactants for field testing due to their high solubilization potential.

The anionic and sulfosuccinate surfactants were less effective, with maximum dodecane solubilizations of 43 mg/L and 15 mg/L, respectively. Besides the ethoxylated alcohols, the other most promising surfactant was Detaine PB, the only amphoteric surfactant evaluated. The 3 g/dL Detaine PB/0.9 g/dL pentanol solution exhibited a high level of solubilization (108 mg/L dodecane), but the surfactant/alcohol precursor had a viscosity of over 15 cp. However, addition of 5.3 g/dL isopropanol and 1.9 g/dL pentanol to the Detaine PB solution resulted in a solution that was single-phase and had a viscosity of only 2.1 cp. Solubilization with this surfactant/alcohol mixture was 165 mg/L dodecane, which was the highest solubilization of OU1 NAPL observed among surfactant/alcohol solutions having viscosities of about 2 cp or less.

Table 13.2 Hill NAPL Solubilized by Surfactant/Alcohol Solutions at 24°C

		NAPL Solubilization		
Chemical Class	Trade Name	Surfactant/ Alcohol (w/w)	Dodecane (mg/L)	NAPL (g/dL)[a]
ANIONICS				
Alkyl sulfates	Sodium dodecyl sulfate	1.0	35	0.51
	Carsonol SHS	0.9[b]	—[d]	—
	Rhodapon BOS	0.9	—[d]	—
Sulfates of ethoxylated alcohols	Witcolate S-1285C	1.0	20	0.29
	Witcolate SE-5	1.1	29	0.43
	Standapol ES-40	0.9	24	0.35
Sulfates and sulfonates of ethoxylated alkylphenols	Atlas G-7205	—[i]	17	0.25
Alpha-olefin sulfonates	Witconate AOS	1.0	33	0.49
Dialkyl sulfosuccinates	Aerosol AY 65	1.0	7	0.10
	Aerosol MA 80-I	1.1	15	0.22
	Aerosol AOT	—[i]	13[e]	0.19
Sulfonated amines and amides	Emkapon Jel BS	1.0	43	0.63
NONIONICS				
Ethoxylated alcohols	Brij 30	0.9[b]	214	3.15
	Brij 30	0.8[c]	167	2.45
	POE(10) Lauryl	1.0	56	0.82
	Brij 35	1.1	8	0.12
	Brij 56	1.1	100	1.47
	Brij 58	1.1	22	0.32
	Brij 78	1.1	22	0.32
	Brij 97	1.3	104	1.53
	Brij 98	1.1	23	0.34
	Genepol 26-L-45	1.7	77	1.13
	Trycol 5953	1.0	84	1.23
	Hetoxol TD-9	1.1	44	0.65
	Hetoxol OA-5 Special	0.6[b]	<1	<0.01
	Macol LA 790	1.3	74	1.09
	Ritoleth 10	1.4	68	1.00
	Novell II 1216-CO-7	1.5	114	1.68
	Novell II 1216-CO-9	1.0	82	1.21
	Novell II 1216-CO-10.5	1.0	58	0.85
	Syn Fac TDA 92	3.3	54	0.79
	Arapol 0712	1.4	67	0.98
	Ameroxol OE-10	1.3	101	1.48
Ethoxylated fatty acids	Abitec WA-664	1.0	20	0.29
	Myrj 52	1.2	8	0.12
	Nopalcol 2-L	1.7	<1	<0.01
	Pegosperse 400-MO	0.9[b]	29	0.43
	Lipopeg 4 DL	—[i]	—[f]	—
Ethoxylated alkylphenols	Igepal CO-630	2.0	76	1.12
	Igepal CO-730	1.0	37	0.54
Propoxylated/ethoxylated alcohols	Antarox LA-EP 15	6.0	5	0.07
	Antarox LA-EP 25LF	3.3	8	0.12
	Antarox LA-EP-45	1.4	32	0.47
	Antarox LF-224	1.3	—[g]	—
	Anatarox LF-330	3.3[b]	5	0.07
	Rexonic P-5	1.9	27	0.40

Table 13.2 (continued)   Hill NAPL Solubilized by Surfactant/Alcohol Solutions at 24°C

Chemical Class	Trade Name	Surfactant/ Alcohol (w/w)	Dodecane (mg/L)	NAPL (g/dL)[a]
		NAPL Solubilization		
Block polymers	Pluronic L43	2.7	16	0.24
	Pluronic L44	2.7	9	0.13
Polyglycerol esters	Drewpol 10-1-CCK	7.5	20	0.29
	Aldosperse ML-23	1.2	18	0.26
Polysorbates	Tween 20	1.3	31	0.46
	Tween 21	1.0	31	0.46
	Tween 40	1.3	24	0.35
	Tween 60	1.8	16	0.24
	Tween 80	1.3	27	0.40
Sucrose fatty acid esters	Glucamate SSE-20	1.8	15	0.22
	DeSulf GOS-P-70	2.0	32	0.47
Thio and mercapto derivatives	Alcodet SK	—[i]	66	0.97
AMPHOTERICS				
Betaine derivatives	**Detaine PB**	3.3	108	1.59
		1.5[h]	165	2.43

[a] Calculated using a dodecane concentration in the LNAPL of 0.68 g/dL.
[b] Alcohol was *n*-butanol.
[c] Alcohol was isobutanol.
[d] Components in surfactant prevented GC analysis of target analytes in the microemulsion phase.
[e] Surfactant solution consisted of the clear supernatant from a 3 g/dL solution.
[f] Formed a viscous macroemulsion.
[g] Formed a middle-phase microemulsion.
[h] Contained 5.3 g/dL of isopropanol not included in the surfactant:alcohol mass ratio calculation.
[i] No alcohol was added to the surfactant solution.

An empirical parameter that has been related to emulsification efficiency is the hydrophilic–lipophilic balance, HLB (Shinoda and Kunieda, 1983). Figure 13.3 shows NAPL solubilization as a function of HLB for those surfactant/alcohol solutions that had viscosities of about 2 cp or less. There are 37 surfactants represented in Figure 13.3: 34 nonionics, 2 anionics, and the amphoteric surfactant Detaine PB. Thus, the chemical properties of these 37 surfactants vary widely, which probably explains the wide variation in oil solubilization obtained for a given HLB value. However, solubilization of the OU1 NAPL peaked at about 2 g/dL for surfactants with HLB values between 12 and 13, and decreased outside this range. The data in Figure 13.3 suggested that a search for additional surfactants for use at the OU1 field site could be facilitated by focusing on those with HLB values in the 12 to 13 range. In a study of dodecane solubilization by polyoxyethylated cetyl alcohol surfactants, Ballet and Candau (1981) concluded that the HLB parameter played a dominant role, even in the presence of alcohol.

13.3 COLUMN TESTS

Previous sections have documented the results of studies designed to identify single-phase, low-viscosity precursors capable of solubilizing the OU1 NAPL. The most promising candidates were 12- to 18-carbon chain alcohol ethoxylates and a cetyl betaine derivative. Selected precursors were evaluated using 2.5-cm-diameter columns packed with glass beads (0.25 to 0.50 mm in diameter) coated with OU1 NAPL. In addition, one precursor was evaluated using a column packed with contaminated aquifer material from the OU1 site at Hill AFB.

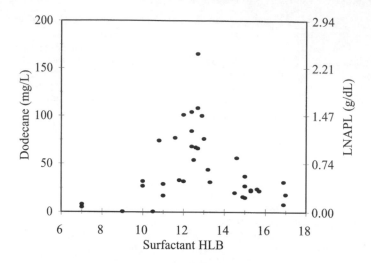

Figure 13.3 Relationship between surfactant HLB and the amount of OU1 NAPL solubilized.

NAPL removal effectiveness was evaluated from the percent change in NAPL saturation as determined from partitioning tracer tests (Jin et al., 1995; Annable et al., 1998b) conducted before and after flushing with precursor. The nonpartitioning and partitioning tracers used were methanol and n-heptanol, respectively. The NAPL/water partition coefficient for m-heptanol, $K_N = C_{NAPL}/C_W$, was measured to be 17.4 using batch equilibration methods.

13.3.1 Glass Bead Column Tests

The three surfactant/alcohol solutions that were tested using NAPL-contaminated glass bead columns are shown in Table 13.3, along with NAPL recoveries estimated from partitioning tracer tests. Dodecane solubilized by these surfactants in batch studies increased in the order GOS P70 (32 mg/L), Brij 97 (104 mg/L), and Detaine PB (165 mg/L). However, Brij 97 was the most effective at removing the NAPL in the column studies. It was observed that each surfactant solution, except Brij 97/pentanol, left a brown, oily residue on the glass beads. The Brij 97/pentanol solution was able to strip all of the NAPL from the glass beads, including the more resistant, difficult-to-solubilize fraction.

13.3.2 Contaminated Soil Column Tests

Based on its effectiveness in glass bead column studies, Brij 97/pentanol was selected for testing on contaminated soil from the OU1 field site. The elution profile for dodecane during flushing of the contaminated soil column is shown in Figure 13.4. Dodecane concentrations in the column

Table 13.3 Recovery of NAPL from Glass Bead Columns as Estimated from Partitioning Tracer Test Results

Flushing Solution Composition				Residual Saturation[a]		
Surfactant (g/dL)		Alcohol (conc in g/dL)		Pre-flush	Post-flush	Recovery (%)
Brij 97	(3.0)	Pentanol	(2.5)	0.047	0.0008	98
Detaine PB	(2.8)	Pentanol	(1.9)	0.075	0.0066	91
		Isopropanol	(5.3)			
GOS P-70	(3.0)	Pentanol	(1.5)	0.068	0.016	76

[a] Residual saturation is defined as the volume of oil divided by the volume of water plus oil.

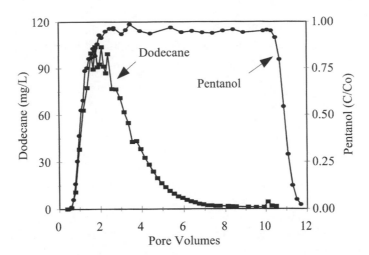

Figure 13.4 Dodecane and pentanol in effluent from a column containing contaminated Hill aquifer material and flushed with Brij 97/pentanol.

effluent were similar to those observed with the glass bead column, peaking around 95 to 100 mg/L and quickly tailing off to low values after several pore volumes. Also shown in Figure 13.4 is the pentanol breakthrough curve (BTC). The pentanol breakthrough coincided with that of the target analytes, as would be expected if the oil were being transported as a microemulsion (vs. immiscibly displaced). Following 10 pore volumes of Brij 97/pentanol flushing, pentanol was flushed from the column using a 3 g/dL Brij 97 solution. The bulk of the pentanol was removed from the column in less than 2 PV.

A post-flushing partitioning tracer test indicated 65% recovery of NAPL from the contaminated Hill column (S_N was reduced from 0.060 to 0.021). However, hexane extraction of this column following the post-flushing partitioning tracer test produced a hexane solution with no brown color, no detectable levels of target analytes, and a chromatogram that was typical of a blank run. These observations suggest that the Brij 97/pentanol solution had solubilized essentially all of the NAPL that had been present in the contaminated Hill aquifer material. However, examination of the mineral grains after flushing showed that they contained a black coating that was not visible on mineral grains from uncontaminated Hill aquifer material. It is believed that this coating is a NAPL weathering product that has accumulated in the contaminated zone at OU1. This substance acts as a sorption site for partitioning tracers, but may have properties more like soil humic matter (i.e., a highly polymerized, high molecular weight solid) than like NAPL.

13.3.3 Two-Dimensional Flow Chamber Test

Parallel-plate aquifer models were constructed for investigating the effectiveness of a single-phase microemulsion (SPME) precursor for removing a nonuniformly distributed NAPL (Figure 13.5). The purpose of these experiments was to evaluate potential problems with displacement of the precursor through a system where the NAPL-contaminated zone may have reduced permeability in comparison with the surrounding aquifer. The potential for enhanced chromatographic separation of the alcohol and surfactant, leading to loss of effectiveness, was also of interest. Conducting experiments in a larger system than previous column experiments provided a better test of this process. All of the aquifer model experiments (two-dimensional flow systems) were aimed at elucidating potential problems not evident at the column scale (one-dimensional flow) prior to conducting the field experiment (three-dimensional flow).

Three complete aquifer model experiments were conducted and the results are summarized here. More details on the results of the laboratory experiments can be found in Ramachandran (1997). Each experiment involved the same steps planned for the field experiment. Following packing of the flow chamber, experiments were conducted in the following order:

1. Chemical analysis of the packed contaminated soil
2. Nonreactive dye tracer experiment
3. Partitioning tracer experiment
4. SPME precursor flushing
5. Post-flushing partitioning tracer test
6. Disassembly and analysis of soil from the contaminated and uncontaminated zones

The first experiment was conducted using a clean quartz sand to which NAPL from Hill AFB was added. The next two experiments were both conducted using contaminated and uncontaminated soil taken from the test cell during well installation. In the third experiment, NAPL was added to the contaminated soil to increase the NAPL saturation. In all three experiments, the contaminated zone was confined to an area of known geometry (see Figure 13.5).

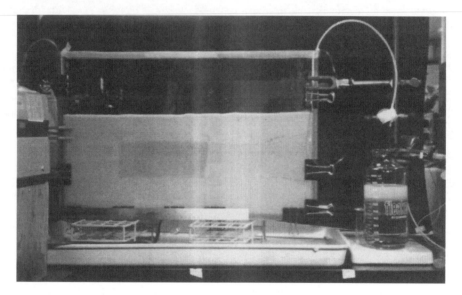

Figure 13.5 Two-dimensional flow chamber with contaminated zone near center of media.

All three experiments indicated that the precursor flushing was very effective in removing target constituents from the flow chamber. Analysis of soils before and after flushing consistently showed >95% removal of the initial mass of target constituents. Significant constituent mass removal was also indicated by the observed decrease in target analyte concentrations in the effluent breakthrough curves (Figure 13.6). These breakthrough curves have been scaled to the maximum concentration measured for each analyte to demonstrate that the removal process was fairly nonselective with respect to components of the NAPL. Separation of the surfactant and alcohol components of the precursor during displacement through the flow chamber was also investigated. Breakthrough curves showed minimal separation (data not shown). These observations support the conclusion that the SPME precursor was acting to microemulsify the NAPL during flushing. All of the observations listed here support the conclusions related to the two major objectives of the aquifer model studies. The SPME flood was capable of removing significant NAPL mass from a heterogeneously contaminated domain, and minimal chromatographic separation of the precursor components was apparently not detrimental to the process.

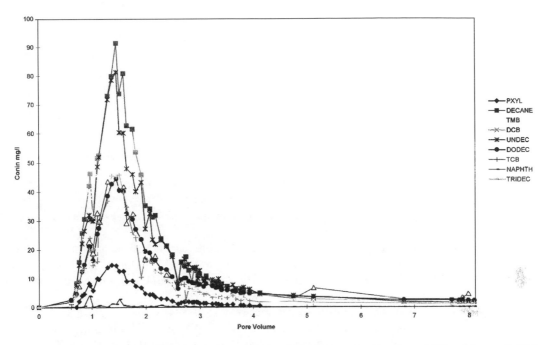

Figure 13.6 Target analyte breakthrough curves from SPME flushing of contaminated Hill aquifer material in the two-dimensional flow aquifer model.

Partitioning tracer test results were less conclusive for determining NAPL removal from the flow chambers. The results from the three experiments are summarized in Table 13.4 (experiments are identified as 1st, 2nd and 3rd, as outlined above). The results of the partitioning tracers tests indicated poor NAPL removal from the chambers. The results from the experiments using aquifer materials from the site in particular indicated very poor removal.

The results led to an investigation of the significance of background retardation on the uncontaminated material used in the models. It was determined that both soil tested prior to packing in the chamber and following the SPME experiment in the chamber exhibited retardation that was equivalent to about 0.5% NAPL saturation. This became very problematic when NAPL was not uniformly present in the flow chamber (generally, only 10% of the volume was contaminated). In the final experiment conducted, the contaminated fraction was increased to 25% to minimize this problem. Still, a large amount of the retardation observed at the extraction well was due to

Table 13.4 Summary of Two-Dimensional Flow Aquifer Model Studies

Experiment	Pre-flushing S_N	Post-flushing S_N	ΔS_N/Initial S_N
1st: based on DMP	0.099	0.004	0.96
2nd: based on DMP	0.0056	0.0099	−0.77
3rd: based on HEPT	0.015	0.012	0.20
4th: based on DMP	0.020	0.012	0.40
	(0.082)[a]	(0.034)	(0.59)
3rd: based on HEPT	0.19	0.014	0.26
	(0.057)	(0.026)	(0.54)

[a] Terms in parentheses are values for S_N in the contaminated zone after correcting for retardation of tracers by uncontaminated aquifer material.

uncontaminated soil within the model. The background retardation associated with the uncontaminated region was then factored out of the calculation used to estimate the NAPL present before and after flushing with SPME precursor. This result is presented as the corrected NAPL saturation. This could only be done because the fraction of the aquifer contaminated was known from the packing process. After this correction, the removal was more reasonable. This effort does demonstrate the potential difficulty of using tracers to quantify mass removal in sparsely contaminated systems. It also stresses the importance of tracer partitioning on uncontaminated media. These issues were further investigated in the SPME field test.

13.4 SUMMARY

The purpose of the first phase of this study was to select a surfactant and cosurfactant, which together form the microemulsion precursor, that would produce a low-viscosity, single-phase microemulsion on contact with the complex, multicomponent NAPL found at the field site. The major findings are presented here.

Eighty-six (86) surfactants and a number of alcohols were screened, with maximum NAPL solubilization and low-viscosity (<2 cp) as the main acceptance criteria (Rhue et al., 1997a). The viscosity of the precursor solution was limited to preclude large hydraulic gradients across the test cell and excessive drawdown around the extraction wells.

Batch solubilization studies identified a number of surfactants suitable for use in the field demonstration phase of the project. The best surfactants had HLB values (hydrophilic-lipophilic balance) between 12 and 14 and solubilized between 1 and 2 g of the Hill NAPL/dL. Flushing tests using NAPL-contaminated glass bead columns showed that the more efficient surfactants could remove >90% of the NAPL in less than 10 pore volumes. The precursor solution selected was the surfactant Brij 97® (polyoxyethylene (10) oleyl ether) at 3 wt% and *n*-pentanol at 2.5 wt% in water.

Greater NAPL solubilization could be achieved with higher concentrations of surfactant and cosurfactant. However, increased solubilization would be accompanied by a concomitant increase in viscosity. Also, the complexity of the NAPL found at the Hill AFB field site should be emphasized. Although only a handful of constituent compounds are discussed here, more than 200 have been identified (Montgomery Watson, 1995). Many contaminated sites have less complex NAPLs that would be more easily solubilized than the NAPL found at OU1 of Hill AFB.

CHAPTER 14

Phase II: Field Investigation

14.1 SITE CHARACTERIZATION

Hill AFB is located approximately 50 km north of Salt Lake City, Utah, at the western foot of the Wasatch Mountain Range. The field site is located in an NAPL source zone within a shallow surficial aquifer on the Weber Delta, approximately 100 m above the Weber River valley. The aquifer was contaminated through the use of two chemical disposal pits in the 1940s and 1950s; migration over the years has caused spreading of the NAPL source zone to an area of approximately 2.8 ha. A variety of liquid wastes, such as spent solvents from degreasing operations, fuels, and waste oils, were disposed in these pits. As part of the SERDP/AATDF project, eight test cells were constructed within the NAPL source zone (Figure 14.1). The SPME field test was conducted in Cell 8, upgradient from the original disposal pit locations. The aquifer material within the test cell area consisted of sand and gravel (Provo Formation), with a thick (>60 m) clay aquitard (Alpine Formation) approximately 8 m below ground surface (bgs). Greater detail can be found in a report written for Hill AFB by Montgomery Watson (1995).

The density of the complex NAPL at this site is less than that of water, and the NAPL is thus generally found at or above the water table. Water table fluctuations since the contamination events created a NAPL smear zone extending up approximately 3 m from the clay layer (i.e., 5 m to 8 m bgs). Although the seasonal water table was approximately 7 m bgs at the time these experiments were conducted (May–August 1996), the water table within the cell was maintained at 4.9 m bgs in order to capture the entire NAPL smear zone within the flow domain (see Figures 14.2 and 14.3). The NAPL collected from wells at the site is a very hydrophobic, multicomponent mixture, and tends to strongly coat solid surfaces. Contaminated aquifer solids collected from within the test cell appeared to be completely coated with the NAPL.

14.1.1 Test Cell Design and Instrumentation

The design of the eight hydraulically isolated test cells at the SERDP site was similar to that of the cell built for the UF/EPA cosolvent flushing project conducted in 1994–1995 within the same NAPL source plume at Hill AFB (Annable et al., 1998b; Rao et al., 1997; Sillan et al., 1998). Each cell was constructed using 9.5-mm-thick, interlocking steel sheet piles with all joints filled with low-permeability grout. The SPME cell enclosed a rectangular area of approximately 3.0 m × 4.8 m and penetrated from the surface to a depth of 10.7 m bgs (2.8 m into the clay confining unit below) (Figure 14.2). The Z-shape of the individual piles resulted in 12 fairly large (0.60 m long × 0.33 m wide) corrugations along the perimeter of the generally rectangular cell. The area bounded by these corrugations represented approximately 17% of the total cell area. The cell instrumentation was similar to that described by Annable et al. (1998b), with four injection wells

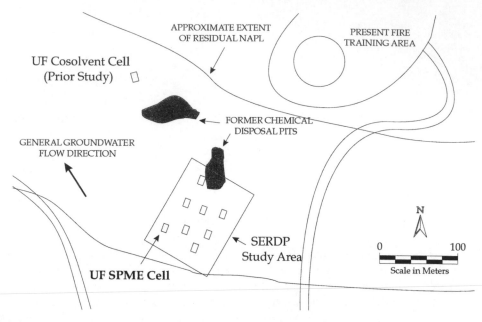

Figure 14.1 SERDP/AATDF test cell locations within the NAPL source zone at Operable Unit 1, Hill AFB, Utah.

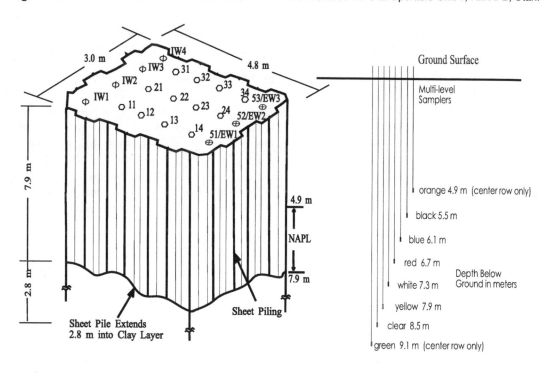

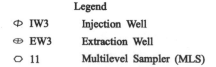

Figure 14.2 Schematic of the SPME test cell.

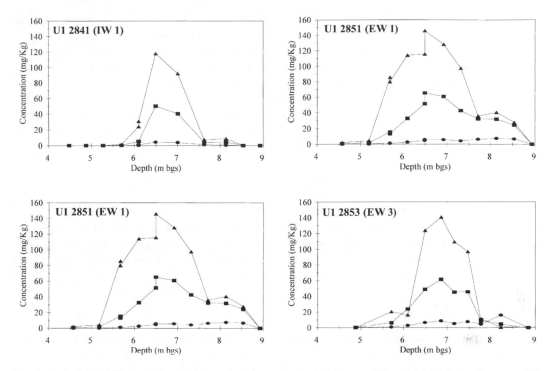

Figure 14.3 Selected pre-flushing soil core profiles for undecane (▲), decane (■), and 1,3,5-trimethylbenzene (●).

(IWs), three extraction wells (EWs), and 15 multilevel samplers (MLSs). Each of the seven wells was screened from 4.9 m to 7.9 m bgs with 0.25-mm slotted stainless steel casing. The MLSs were constructed using 0.32-cm stainless steel tubing that terminated with a 20-μm or 40-μm stainless steel filter. Between five and eight samplers were bundled at each of the 15 MLS locations, for a total of 96 sampling points.

A leak test was conducted in the cell prior to the initiation of flow. Heavy rains at the field site during late spring of 1996 contributed to a rise in the water table within the cell to a level of approximately 1.4 m bgs (greater than 5 m above the water table outside the cell). Water table monitoring over a period of 5 days showed losses of approximately 47 L/day at the extreme head differential, which corresponded to an average leakage of 0.3% of the pore volume per day. Thus, it was deemed that during experiments conducted within the test cell, wherein the head differential across the cell walls was only approximately 2 m, losses from leakage were inconsequential.

14.1.2 Hydrodynamics Characterization

A nonreactive tracer test was conducted to characterize the test cell hydrodynamics prior to partitioning tracer tests and SPME flushing. The nonreactive tracer test followed methods similar to those described by Annable et al. (1998b) using potassium iodide as a tracer. Iodide was selected as a tracer for this preliminary test because bromide was reserved for use as a backup nonpartitioning tracer (in addition to methanol) in the pre-flushing partitioning tracer test.

Steady-state water flow was established in the cell using peristaltic pumps (Masterflex L/S drives, Cole Parmer) to deliver uniform flow distributed equally between the injection and extraction wells. During steady-state water flow, tracer solution (C_0 = 300 mg/L) was delivered to the four injection wells at a flow rate of 4.4 L/min for a period of 2.8 h (i.e., a pulse injection equivalent to 0.12 pore volumes). Subsequent to tracer injection, steady-state water flow was continued for a period of 3.5 days; samples were collected periodically at the three extraction wells and were analyzed for iodide by high-performance liquid chromatography (HPLC). Breakthrough curves were generated, and moment analysis (Valocchi, 1985) was used to determine the effective pore volume of the test cell to be approximately 6000 L. Based on the results of this test, injection and extraction flow rates during all subsequent experiments within the cell were adjusted to 3.6 L/min to deliver an effective flow rate of 0.9 pore volumes per day.

The effective porosity within the test cell was measured to be 0.14 by both hydraulic testing and the nonreactive tracer test. Also, the average saturated hydraulic conductivity across the cell was determined to be 0.01 cm/s by measuring the hydraulic gradient between two piezometers, set 1.6 m apart along the flow direction, at different flow rates.

14.1.3 NAPL Saturation and Morphology

The spatial distribution of residual NAPL mass within the test cell both before and after SPME flushing was determined using two methods: soil cores and partitioning interwell tracer tests (PITTs). Target NAPL constituent BTCs measured at the multilevel samplers during SPME flushing were also used to investigate the spatial distribution of NAPL mass and composition; this analysis was conducted for selected multilevel samplers and is presented in section 14.2.3.

14.1.3.1 Soil Cores

The pre-flushing distribution of target NAPL constituents was determined from 85 soil samples collected from eight borings, and the post-flushing distribution was determined from 40 subsamples collected from six borings. Soil cores were subsampled immediately upon collection, and approximately 10 g soil were placed into 40-mL vials containing 5 mL each dichloromethane and water. The samples were shipped cold overnight to Michigan Technological University (MTU), Houghton, MI, for laboratory analysis by gas chromatography/mass spectrometry (GC/MS). The same protocols were observed for the analysis of 40 post-flushing soil core samples.

Selected pre-flushing soil concentration profiles are presented in Figure 14.3 for the three constituents with the highest soil concentrations: undecane, decane, and 1,3,5-trimethylbenzene (also, see Table 14.6 in section 14.3). The NAPL smear zone is evident where the maximum contaminant concentrations are centered around the approximate position of the regional water table (7 m bgs, or 1 m above the clay confining unit). Note that the center of mass was approximately 0.5 m deeper in the soil cores taken from the west side of the cell (IW 4 and EW 3) than in those from the east side of the cell (IW 1 and EW 1). Pre- and post-flushing soil core data were compared as a measure of the effectiveness of the SPME flood; this analysis is presented in section 14.3.

14.1.3.2 Partitioning Tracers

The amount and spatial distribution of residual NAPL present before and after SPME flushing were estimated using PITTs conducted with methods similar to those described by Annable et al. (1998b). Methanol ($MeOH_2C_0$ = 2700 mg/L) and 2,2-dimethyl-3-pentanol (2,2-DMP; C_0 = 930 mg/L) were used as the nonpartitioning and partitioning tracers, respectively. A tracer pulse of 0.16 PV was delivered to the injection wells during steady water flow of approximately 1 PV per day (3.6 L/min). Samples were collected periodically at the EWs and MLSs and were shipped overnight to laboratories at the University of Florida, Gainesville, and the U.S. EPA National Risk Management Research Laboratory, Ada, OK, for analysis by GC/FID. The analytical methods are described in Appendix 2.

To accurately characterize the tails of the breakthrough curves, where much of the measured retardation is often found, both the pre- and post-flushing tracer tests were continued until tracer concentrations at the extraction wells were below the detection limits (approximately 1 mg/L). The tails of the tracer breakthrough curves were exponentially extrapolated (Jin, 1995). Moment analysis was then performed on the breakthrough curves that were generated to determine the mean arrival times, $\bar{t}_i$ of all tracers at each well. Equations used in calculating $\bar{t}_i$ are presented by Jin (1995). The pre- and post-flushing NAPL saturations within the swept volume of each well were determined using the following relationship (see Jin et al., 1995; Annable et al., 1998b):

$$S_N = \frac{\bar{t}_p - \bar{t}_n}{(K_N - 1)\bar{t}_n + \bar{t}_p} \tag{14.1}$$

where the subscripts n and p refer to the nonpartitioning and partitioning tracers, respectively. The NAPL water partitioning coefficient, K_N, for 2,2-DMP was measured in the UF laboratory to be 10.7.

Pre-flushing partitioning tracer test BTCs for methanol and 2,2-DMP measured at all three extraction wells are presented in Figure 14.4. The delayed arrival of 2,2-DMP in all three BTCs indicates the presence of NAPL within the swept volume of each extraction well.

The mean arrival times for methanol and 2,2-DMP and the calculated value of S_N at each well are presented in Table 14.1. Based on the extraction well data, the average NAPL saturation within the test cell was approximately 0.06.

Note that travel times for the nonreactive tracer through the swept volumes of EW 1 and EW 3 were approximately equal, and both were longer than the EW 2 travel time. Also, the EW 2 BTC showed higher peak concentrations and less spreading than those measured at EWs 1 and 3. These effects are indicative of the geometry of the test cell, because the corrugations of the sheet pile walls along the EW 1 and EW 3 flow paths provided substantially more dead volume than did the center-line flow path of EW 2.

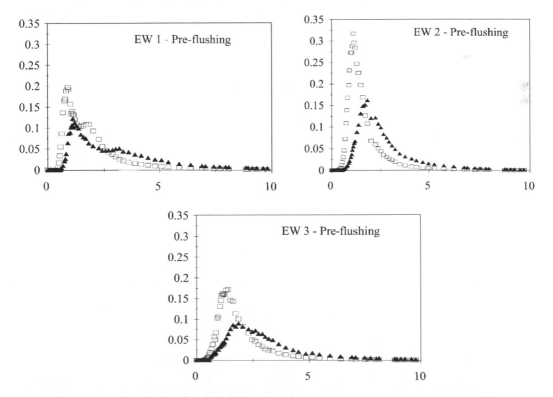

Figure 14.4 Pre-flood partitioning tracer breakthrough curves at each extraction well for methanol (□) and 2,2-dimethyl 3-pentanol (▲).

Table 14.1 Pre-flushing Tracer Arrival Times $\bar{t}_i$, at All Three Extraction Wells with Values of S_N Calculated from Equation 14.1

Tracer	EW-1	EW-2	EW-3
Methanol	30.7[a]	22.5	30.8
2,2-DMP	53.4	39.9	49.3
Initial S_N	0.065	0.068	0.053

[a] Time is in hours.

14.1.3.3 Interfacial Tracers

Interfacial tracers (Saripalli et al., 1997) were used to characterize the spatial distribution of the NAPL morphology within the test cell (Annable et al., 1998a). The ratio of NAPL-water interfacial area to volumetric NAPL content, $a_{Nw}/\phi S_N$, provides a measure of how the NAPL volume is distributed within the porous medium. This ratio was defined as the NAPL morphology index, H_N. High values of H_N indicate that the NAPL is spread out, perhaps as thin coatings on the solid matrix, thus providing large contact area between the NAPL and the aqueous phase. In contrast, a small H_N value indicates that the NAPL is most likely confined to isolated regions and patches of relatively high saturation.

Sodium dodecylbenzene sulfonate (SDBS) (C_0 = 273 mg/L) and bromide (C_0 = 294 mg/L) were used as interfacial and nonreactive tracers, respectively. The tracers were delivered to two of the four injection wells (IW 3 and IW 4) in a pulse of approximately 0.15 PV (4.8 h). Monitoring was focused at the MLS locations that were downgradient of these injection wells. The other wells received tracer-free water to maintain steady-state flow within the test cell.

The retardation of the interfacial tracer with respect to the nonreactive tracer was determined from the tracer mean arrival times, $\bar{t}_i$, calculated using moment analysis. For interfacial tracers, the retardation was defined as (Saripalli et al., 1997):

$$R_{ift} = 1 + \frac{K_0 a_{Nw}}{\theta_w} = 1 + \frac{K_0 H_N S_N}{(1 - S_N)} = \frac{\bar{t}_{ift}}{\bar{t}_n} \qquad (14.2)$$

where the subscript *ift* refers to the interfacial tracer and θ_w is the water content. The interfacial tracer adsorption isotherm is nonlinear, but a linearized adsorption coefficient, K_0, was calculated using the maximum interfacial tracer concentration, C_{max}, observed at each monitoring point. For more details on this approach, see Annable et al. (1998a). Listed in Table 14.2 are the first temporal moments (*M1*) and the calculated interfacial areas (a_{Nw}), NAPL saturations (S_N), and morphology indices (H_N) based on tracer responses for nine locations.

Presented in Figure 14.5 are the breakthrough curves for bromide, SDBS, methanol, and 2,2-DMP measured at MLS 32 Red (located 1.6 m downgradient from IW 3 and IW 4, and 6.7 m bgs). The retardation of SDBS and 2,2-DMP relative to bromide (and methanol) is evident in both BTCs. The interfacial tracer breakthrough curve is quite asymmetric with considerable tailing, which is primarily due to the strong nonlinearity of the SDBS adsorption isotherm.

Figure 14.6 summarizes the interfacial tracer results over the depth profile measured in MLS 32. The NAPL saturation was maximum (0.124) at approximately 6.2 m (20 ft) bgs, which corresponded to the approximate location of the regional water table. The interfacial area measured at this location was a minimum (70 cm²/cm³). These results suggested that the NAPL was located in regions of locally high saturations that did not offer high contact area with the mobile water in this zone. At the next MLS, located at a depth of 6.8 m (22 ft) bgs, the NAPL saturation was lower; however, the interfacial contact area at this location was much higher (870 cm²/cm³). This finding indicates that this zone had a relatively uniformly distributed NAPL, offering good contact with the mobile aqueous phase. The range in calculated H_N over the vertical profile of MLS 32 was approximately one order of magnitude, suggesting that the morphology of the NAPL varied signficantly even within the small test cell. This may be due to the tendency of the NAPL to accumulate above the capillary fringe and smear and trap as the water table fluctuated. Further interpretation of the interfacial tracer test results can be found in Annable et al. (1998a).

The variability in the NAPL morphology index has significant implications for *in situ* remediation technologies. The extraction efficiency of a given *in situ* remediation technique could be greatly reduced where the NAPL distribution is not uniform. If the NAPL is trapped at the pore scale in such a manner that it offers only a small contact area, the performance of remediation technologies could be limited, particularly those with mass transfer constraints. If the heterogeneity measured by tracers is a result of discontinuous "pockets" of NAPL and uncontaminated zones, then any remediation technology that requires the flushing agent (e.g., cosolvent, surfactant, etc.) to contact the NAPL will be inefficient unless the agent can be targeted to the contaminated zones. This information is also critical for comparing performance of remediation technologies at different sites with different NAPL distributions.

14.2 MICROEMULSION FLUSHING

14.2.1 Materials and Methods

The microemulsion flushing field test consisted of pumping 9 PV of precursor solution through the test cell over a period of 18 days, followed by 1 PV of a surfactant-only flood and 6.5 PV of a water flood. The designed duration of the experiment was based on laboratory column studies conducted by Rhue et al. (1997) with contaminated soil from Hill AFB, which showed that after approximately 5 PV of SPME flushing, contaminants were no longer being eluted from the soil. The number of pore volumes was increased approximately twofold to account for greater hydrodynamic dispersion expected as a consequence of field-scale heterogeneities.

The surfactant (Brij 97®) was donated by ICI Americas, Inc. (Wilmington, DE) and the *n*-pentanol was purchased from Union Carbide (Danbury, CT). Approximately 54,000 L (9 PV) microemulsion precursor solution were mixed on site and delivered to the test cell using the fluid mixing and delivery system shown in the schematic diagram of Figure 14.7. Injection and waste tanks were located remotely from the cell, within secondary containment.

The 9 PV of precursor solution were mixed in three batches in two 25,000-L polyethylene tanks (Baker Tanks, Salt Lake City, UT). A 40-L/min sump pump was used for continuous circulation within the tanks during initial mixing and throughout the 18 days of microemulsion flooding. For each precursor batch, the water was added to the tank first, then Brij 97® was slowly (1 L/min) pumped in using a peristaltic pump. After mixing overnight, pentanol was added in the same manner as the surfactant. The two 25,000-L tanks were connected through valved ports at the tank bases. During the experiment, these ports were periodically opened for homogenization of the injection fluids as the fluid level decreased in the tank that was discharging to the test cell.

Fluid samples were collected periodically at the extraction wells and multilevel samplers during the flushing experiment and shipped overnight to UF for target NAPL constituent analysis. A total of 540 extraction well samples and 6500 multilevel samples were collected during the flushing experiment. Analytical methods are given in Appendix 2.

As described previously, the flow rate through the test cell during the partitioning tracer tests was approximately 1 PV per day. It was desired to flush the precursor solution through the same flow domain as that swept by the partitioning tracers. Therefore, because the viscosity of the precursor solution was measured to be approximately twice that of water, the initial flow rate during

Table 14.2 Calculated First Moments (*M1*), Retardation Factors (R_{ift}, R_{pt}), Interfacial Area (a_{Nw}), NAPL Saturation (S_N), and Morphology Index (H_N) for All Monitoring Locations

MLS Location, Depth	M1 Br⁻ (h)	M1 SDBS (h)	R_{ift}	C_{max}/C_0	K_0 cm	a_{Nw} (cm²/cm³)	M1 MeOH (h)	M1 DMP (h)	R_{pt}	S_N (cm³/cm³)	H_N (cm⁻¹)
31 red[a] (6.7 m)	6.47	21.8	3.37	0.50	0.00018	1900	4.54	7.68	1.69	0.061	210,000
32 black (5.5 m)	14.1	18.9	1.34	0.30	0.00027	190	12.1	17.2	1.47	0.054	23,000
32 blue (6.1 m)	26.2	32.2	1.23	0.14	0.00049	70	24.5	61.4	2.51	0.124	3,800
32 red (6.7 m)	14.3	32.3	2.26	0.40	0.00022	870	10.9	19.0	1.75	0.065	91,000
32 white (7.3 m)	17.1	24.3	1.42	0.40	0.00022	290	12.0	24.9	2.08	0.092	21,000
33 red (6.7 m)	24.0	43.4	1.81	0.13	0.00054	230	19.6	41.8	2.13	0.096	16,000
34 red (6.7 m)	40.4	53.6	1.3	0.08	0.00076	59	35.8	75.6	2.11	0.094	4,200
53 red (6.7 m)	22.8	29.2	1.28	0.29	0.00028	150	24.2	32.8	1.36	0.032	32,000
EW3	29.4	65.4	2.23	0.13	0.00052	350	30.8	49.7	1.61	0.054	44,000

[a] See Figure 14.2 for MLS locations; depth is bgs.

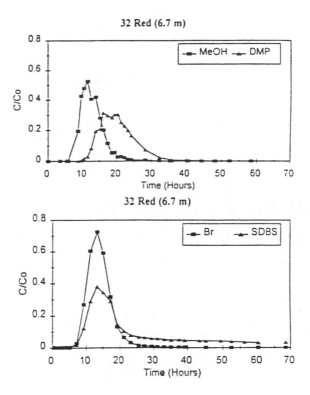

Figure 14.5 Nonreactive and reactive tracer breakthrough curves at MLS 32 Red (6.7 m bgs) for the partitioning tracer test (methanol and 2,2-dimethyl-3-pentanol) and the interfacial tracer test (bromide and sodium dodecylbenzene sulfonate).

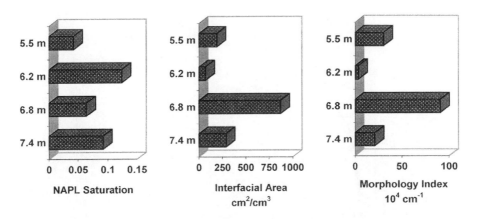

Figure 14.6 Summary of NAPL saturation, interfacial area, and morphology index at MLS 32.

SPME flushing was set at approximately 0.5 PV per day (1.8 L/min). A slower flow rate also served as a precaution against potential mass transfer constraints on NAPL solubilization.

Several flow perturbations were employed to investigate the importance of mass transfer rate limitations during the SPME flood. The flow perturbations included two flow interruptions and three changes in flow rate.

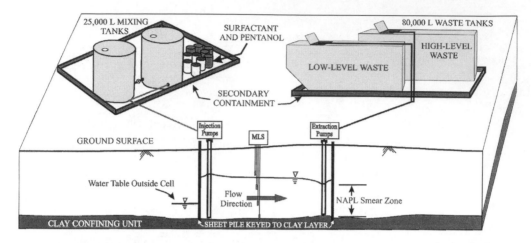

Figure 14.7 Schematic diagram of the SPME field test, including mixing tanks, test cell, and waste tanks.

14.2.2 Extraction Well Results

Fluid samples collected at the three extraction wells during the SPME flood were analyzed for pentanol and the nine target NAPL constituents listed in Table 13.1. Datawill bepresented here for the selected alkanes and aromatics indicated in Table 14.3. The extraction well data for

Table 14.3 Contaminant Mass Removed From the Extraction Wells During SPME Flushing

NAPL Constituent	Total Mass (kg)
p-Xylene	0.301
1,2,3-Trimethylbenzene	1.123
n-Decane	1.121
n-Undecane	3.117
n-Dodecane	1.348
n-Tridecane	0.697

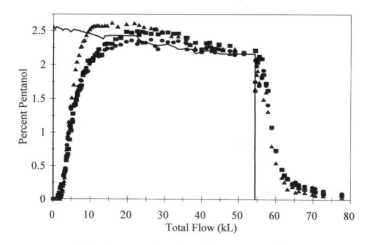

Figure 14.8 Precursor injection profile, averaged over all the injection wells (solid line) and elution profiles for EW 1 (○), EW 2 (▲), EW 3 (□).

trichlorobenzene, 1,2-dichlorobenzene, and naphthalene are not presented here because chromato-graphic peaks led to unreliable results.

14.2.2.1 SPME Hydrodynamic Consideration

Analytical interferences with NAPL constituents precluded the measurement of surfactant concentration in effluent samples. Therefore, the arrival of precursor at the extraction wells was inferred from the arrival of pentanol. This assumption would be invalid if chromatographic separation of the alcohol and surfactant components of the precursor solution occurred during transport through the aquifer; however, laboratory column studies with the precursor solution and soil samples from the field site did not indicate any separation. Also, the arrival time of pentanol at the extraction wells during the SPME flood was equivalent to the arrival time of the nonpartitioning tracer (methanol) during the post-flushing tracer test, indicating no significant retardation of the pentanol. Thus, the precursor injection and elution profiles were characterized by the measured pentanol concentrations (see Figure 14.8).

The breakthrough curves in Figures 14.8 and 14.9 are plotted in terms of the total flow through the entire test cell (Total Flow); all other BTCs presented here are plotted as the flow through an individual well (Total Local Flow). A decreasing trend is evident in the n-pentanol injection profile (averaged over all four IWs) in Figure 14.8, indicating that the n-pentanol concentration varied among the three batches of precursor solution that were mixed during the 18-d experiment. This trend is evidenced in the subsequent n-pentanol BTCs measured at the extraction wells, where concentrations initially reached approximately 2.5%, but then decreased to approximately 2.25%, reflecting the changing injected concentrations. Of perhaps greater significance to the results presented here is the disparity in the shapes of the BTCs between EW 2 and EWs 1 and 3. As noted with the partitioning tracer BTCs (section 14.1.4), the corrugations of the sheet pile walls contributed to longer flow paths in the swept volumes of the two outer wells (EW 1 and EW 3) than in the swept volume of the centerline well (EW 2), leading to greater tailing in the n-pentanol BTCs at the outer wells.

14.2.2.2 Target Analyte Breakthrough Curves

Breakthrough curves for target analytes measured at the extraction wells (Figure 14.9) represent NAPL constituent removal over the entire swept volume of the fully screened wells (Rao et al., 1997). For each target constituent, a significant increase was observed in the effluent fluid concentrations, compared with the pre-SPME flood concentrations. Overlain on the first of the target analyte BTCs (n-decane) is the n-pentanol breakthrough data for the three extraction wells. The increase in n-decane concentration is seen to be coincident with the arrival of the precursor solution; a similar trend was observed for all analytes. The breakthrough curves measured at EW-2 showed higher peak concentrations of all the analytes and less spreading of the curves than those measured at EWs 1 and 3. This behavior suggested that there were differences in hydrodynamic dispersion among these flow zones, as was observed in the n-pentanol and pre-flushing, nonpartitioning tracer breakthrough curves.

The breakthrough curves of the NAPL constituents were integrated to determine the total mass of each contaminant removed from the swept volume of each extraction well (Table 14.3). Each extraction well accounted for approximately one-third of the total mass removed (Jawitz et al., 1998). The breakthrough curves shown in Figure 14.9 indicate that for nearly all NAPL constituents, mass was still being removed when precursor injection ceased.

Continued flushing at this point would have removed additional NAPL mass, but in an inefficient manner. As discussed by Rao et al. (1997), the remaining NAPL was likely located in low-permeability zones and continued flushing would have meant that most of the injected precursor solution would have been displaced through zones where the NAPL had already been solubilized.

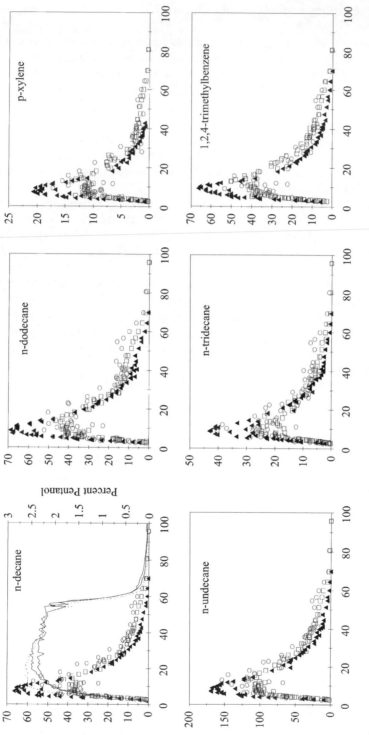

Figure 14.9 Breakthrough curves for target NAPL constituents measured at EW 1 (○), EW 2 (▲), and EW 3 (□). Overlain on the *n*-decane graph are the *n*-pentanol breakthrough data (solid lines for EWs 1 and 3, dashed lined for EW 2).

14.2.2.3 NAPL Microemulsification

During the SPME flood, the physical appearance of the fluids extracted from the test cell followed patterns similar to those evident in the breakthrough curves of Figure 14.9. Both the resident groundwater and the precursor solution were clear and nearly colorless, but as the microemulsified NAPL started to break through at the extraction wells, a light-brown color was evident in the effluent samples. This trend is evident in Figure 14.10, which is a photograph of effluent samples collected throughout the SPME flood. As the concentration of microemulsified NAPL increased, the extracted fluids became darker, reaching a dark-brown color at maximum concentrations and gradually lightening in color as concentrations decreased. Even the darkest extracted fluids remained isotropically clear, with no sign of macroemulsification; in none of the fluids extracted from the test cell was a separate phase, or mobilized, NAPL observed.

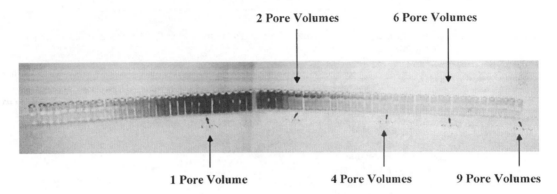

Figure 14.10 Effluent samples collected during the SPME field test.

For microemulsification, where droplets of NAPL are dispersed within the aqueous phase, the relative mole fractions of the NAPL constituents in the microemulsion should be the same as those in the NAPL itself. Petroleum engineers have reported no fractionation (i.e., selective solubilization) in the microemulsification of oil mixtures and crude oils (Salager et al., 1979; Puerto and Reed, 1983), and have successfully modeled the microemulsification of these multiple-component oils as a single "pseudocomponent" (Vinatieri and Fleming, 1979). Recently, Shiau et al. (1996) reported no selective solubilization in the microemulsification of a ternary mixture of chlorinated solvents. Thus, the elution of all constituents was expected to be simultaneous, owing to the nonselective dissolution that is characteristic of microemulsification, while for cosolvent- or surfactant-enhanced solubilization, chromatographic separation according to the relative hydrophobicities of the constituents would be expected (Augustijn et al., 1994). Despite the broad range of hydrophobicities exhibited by the constituents of the Hill NAPL (log K_{ow} values ranged from 3.09 for p-xylene to 7.15 for n-tridecane), they were all eluted from the test cell within a narrow time window. This behavior is demonstrated by the cumulative production curves for each analyte at EW-2, which have been scaled to their respective calculated zeroth moments (Figure 14.11).

While the constituents were all eluted nearly concurrently, a trend in the order of elution can be seen. The first normalized moments of the analyte breakthrough curves shown in Figure 14.9 were calculated in terms of the fluid volume eluted from each well. The EW 2 results are presented in Table 14.4, which lists the constituents in order of arrival time. While only the EW 2 results are presented in Figure 14.11 and Table 14.4, the analytes were eluted in the same order at each extraction well. Generally, the least hydrophobic contaminants were eluted first, followed by contaminants of increasing hydrophobicity.

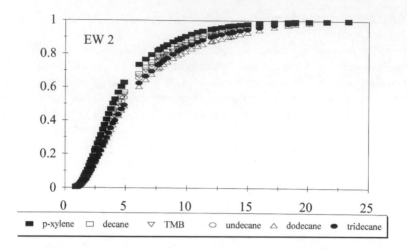

Figure 14.11 Cumulative production curve for the target NAPL constituents, each scaled to its respective zeroth moment.

Table 14.4 First Normalized Moments of the Target NAPL Constituents of Interest at Extraction Well EW 2

NAPL Constituent	EW 2 (kL)
p-Xylene	5.03
n-Decane	5.55
n-Undecane	5.77
1,2,4-Trimethylbenzene	5.83
n-Tridecane	6.17
n-Dodecane	6.44

These data seem to suggest selective dissolution of the less hydrophobic constituents. However, a compositional gradient may exist within the NAPL itself, whereby the more hydrophobic constituents are closely associated with the solid surfaces upon which the NAPL is coated. Thus, as the NAPL is microemulsified, the less hydrophobic constituents that are nearer the NAPL–water interface would be removed preferentially, leading to separation of NAPL constituents in the effluent. However, a detailed analysis of the contaminant breakthrough curves measured at the 96 multilevel samplers is being conducted, and should provide additional information about the relative contribution of NAPL compositional gradients and selective dissolution on the elution behavior of the NAPL constituents. The contaminant breakthrough curves measured at the multilevel samplers are attached as Appendix 3.

14.2.2.4 Elution of Precursor Solution

After 9 PV, the injection fluid was switched from the SPME precursor solution to a 3% solution of surfactant alone. One PV of the surfactant solution was injected to facilitate the removal of *n*-pentanol from the test cell; this was followed by 6.5 PV of water flooding. After water flooding, the *n*-pentanol concentrations at EW 2 had been reduced to 0.5% of C_0, while concentrations were 2% of C_0 at EW 1 and 1% of C_0 at EW 3. Following the post-flushing partitioning tracer test, during which an additional 10 PV of water was flushed through the cell, *n*-pentanol concentrations in all three extraction wells were below 1 mg/L.

14.2.3 Multilevel Sampler Results

Multilevel samplers were used to characterize the performance of the SPME flood on a smaller scale than was possible with the extraction well data. The analysis presented here will focus on the MLSs located at the extraction wells.

Breakthrough curves measured at each extraction well are representative of NAPL dissolution over approximately one third of the cell volume. The swept volume of each well encompassed the entire length of the cell, approximately one-third of the width, and the entire vertical depth, as the wells were fully screened from 4.9 to 7.9 m bgs. Breakthrough curves measured at MLSs attached to each well (five evenly spaced samplers between 5.5 and 7.9 m bgs) are representative of NAPL dissolution over a volume of approximately one-fifth of those measured at the extraction wells.

As described above, the extraction well breakthrough curves were integrated to determine the mass of each constituent removed from within the swept volume of each well (Table 14.5). A similar analysis was performed for the MLSs located at the extraction wells to provide information about the performance of the SPME flood at different depths. However, while the constituent masses were directly determined from the flux-averaged extraction well breakthrough curves, the local fluxes at each multilevel sampler depth were unknown and had to be estimated for determination of the constituent masses from the volume-averaged MLS breakthrough curves.

Fluid flux through each extraction well was equal to one third of the total flux through the cell (3.6 L/min). The initial estimate for the local fluxes at the five samplers at each well was one fifth of the flux through the well. The initial local flux estimate at each MLS was adjusted using the pre-flushing, nonpartitioning tracer (methanol) mean arrival times. At each MLS bundle, the arrival times of methanol at the five depths were arithmetically averaged to provide an estimate of the depth averaged arrival time. The deviation of the local flux from the depth-averaged flux was assumed to be proportional to the deviation of the local mean arrival time for methanol from the depth-averaged methanol arrival time.

The local flux estimates were used to integrate the constituent breakthrough curves measured at each MLS located at the extraction wells. The constituent masses determined from this analysis were summed over all depths and compared with the total mass removed from the extraction well. The results of this comparison at EW 1 are presented in Table 14.5. The results for EW 1 show a good correlation, considering the number of approximations involved in the calculation.

At EW 1, the mass removed from the sampler located at 6.7 m bgs was greater than at any other depth. Therefore, the relative removal of mass with depth was investigated further and the results are presented in Figure 14.12 for each analyte at all three extraction wells. The mass removed at each depth was scaled by the expected value under uniform conditions. The uniform value was determined by dividing the total mass of each constituent removed at an extraction well by the number of multilevel samplers (5 at EWs 1 and 3; 4 at EW 2). Also shown in Figure 14.12, as a horizontal line, is the total removal of each NAPL constituent as a fraction of the total removal as measured at each extraction well vs. depth. The EW 1 results show a strong correlation between the EW and MLS data, while the correlation is not as good for EWs 2 and 3.

Table 14.5 Estimate of Mass of Each Contaminant Removed from EW 1, Based on Multilevel Samplers and Compared with EW 1 Results

NAPL Constituent	Mass Removed (g)	Fraction of EW Values
p-Xylene	90.0	0.87
1,2,4-Trimethylbenzene	390	1.05
n-Decane	353	0.87
n-Undecane	904	0.80
n-Dodecane	546	1.11
n-Tridecane	274	1.06

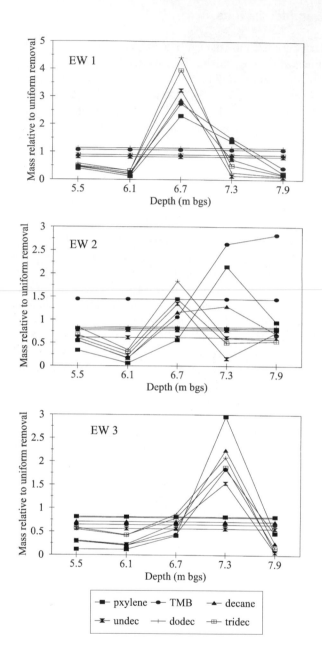

Figure 14.12 Mass removed through multilevel samplers located at the extraction wells. Horizontal lines represent removal based on MLS data relative to EW data

As mentioned above and shown in Figure 14.12, for EW 1, more mass was removed through the sampler located at 6.7 m bgs than through any of the other samplers. This result indicates that the NAPL saturation at this depth was greater than at the other depths. Similarly, at EW 3, based on the MLS data, the NAPL saturation appears to have been greater at 7.3 m bgs than at the other depths; while at EW 2, the highest NAPL saturation may have been between these depths. These results are consistent with the pre-flushing soil core data (Figure 14.3), which showed that the center of mass for the target analytes was slightly deeper on the west side of the cell (IW 4 and EW 3) than on the east side of the cell (IW 1 and EW 1).

14.2.4 Mass Transfer Limitations and Modeling

Under rate-limited conditions, as the residence time of the flushing fluid is increased through either reduced flow rates or flow interruption, solubilization is increased due to greater contact time between the NAPL and the solubilizing agent (Pennell et al., 1993). The SPME precursor residence time was varied by interrupting the flow three times: (1) for a period of 24 h at approximately 4 PV (22.8 kL, or 7.0 d of pumping at 3.6 L/min); (2) for 32 h at 6 PV (35.7 kL, 9.7 d); and (3) for 67 h at 9 PV (54.3 kL, 15.0 d). No significant change in concentration was measured at the extraction wells for the target NAPL constituents following either of the first two flow interrupts, but a rise in the effluent concentrations was measured following the third, and longest, interrupt period (Figure 14.9).

The NAPL constituent breakthrough curves measured at the multilevel samplers (Appendix 3) showed increased solubilization following all three flow interruptions at several monitoring locations (22 Black, 13 Blue, 23 Black, and 14 Yellow; see Figure 14.2 for MLS locations). In addition, many monitoring locations showed increased solubilization following one of the flow interrupts (e.g., 12 Black, 12 Blue, 13 Black, and 34 Blue all showed responses following the final interrupt). These results suggest evidence of mass transfer limitations in the SPME flood.

UTCHEM modeling supported the interpretation that mass transfer limitations were important in the SPME flushing process (results not shown). Nonequilibrium dissolution, with a mass transfer coefficient of 70 day^{-1}, was required to model the laboratory SPME column tests. Additional investigations of the effect of mass transfer limitations on the SPME flood using UTCHEM will continue.

14.3 MEASURES OF PERFORMANCE EFFECTIVENESS

Rao et al. (1997) used pre- and post-flushing data from groundwater samples, soil cores, and partitioning tracer tests and extraction well mass balances to characterize the NAPL removal effectiveness of the UF/EPA *in situ* cosolvent flushing experiment at Hill AFB. The NAPL removal effectiveness of the SPME flood was determined from the percent reduction in NAPL saturation, as measured by partitioning tracer tests, and by the percent reduction in the cell-averaged soil concentrations of several NAPL constituents, as measured by soil core analyses. In addition, breakthrough curves of the NAPL constituents were integrated to determine the total mass of each contaminant removed during the SPME flood. These values were compared with estimates of the initial amount of NAPL determined by both partitioning tracer tests and soil core analyses.

14.3.1 Soil Core Data

Soil concentration profiles, measured from soil core samples collected from within the test cell both before and after flushing, are presented in Figure 14.13 for decane and undecane. A visual comparison of the pre- and post-flushing data indicates a dramatic reduction in the soil concentrations of the target analytes.

The NAPL constituents quantified in the soil samples are presented in Table 14.6, along with the arithmetic average concentration, X_{AVG}, for all samples collected within the flushed zone both before and after SPME flushing. Standard deviations were also calculated, and these data can be found in Jawitz et al. (1998). Data for those constituents with initial values of X_{AVG} less than 0.1 mg/kg are not presented. Note that the initial concentrations of undecane, *n*-decane, and 1,3,5-trimethylbenzene were significantly higher than the other components. The results for these more prevalent constituents are considered to be more reliable than those of the other constituents that were present only in very small amounts.

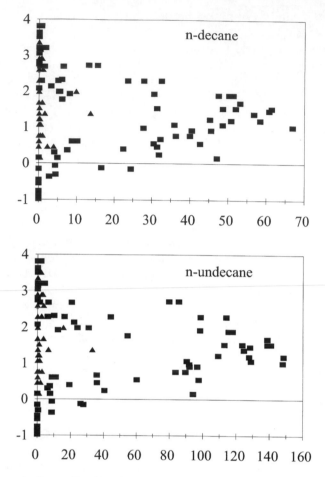

Figure 14.13 Soil concentration profiles for selected NAPL constituents, measured from cores collected before (■) and after (▲) the SPME flood.

The NAPL removal effectiveness, defined here as the percent reduction in X_{AVG}, is presented in Table 14.6 for each analyte. For the NAPL constituents with substantial initial amounts (i.e., initial values of $X_{AVG} > 1$ mg/kg), NAPL removal effectiveness values of greater than 90% were achieved.

Table 14.6 Pre- and Post-Flushing Average Soil Concentrations of Target NAPL Constituents

NAPL Constituent	Pre-flushing Concentration, X_{AVG} (mg/kg)	Post-Flushing Concentration, X_{AVG} (mg/kg)	Percent Reduction
n-Undecane	61.2	2.51	0.96
n-Decane	24.9	1.10	0.96
1,3,5-Trimethylbenzene	2.90	0.25	0.92
Naphthalene	0.91	0.11	0.88
1,2-dichlorobenzene	0.74	0.03	0.96
o-Xylene	0.18	0.06	0.65
m-Xylene	0.12	0.02	0.87

14.3.2 Partitioning Tracers

As shown in the post-flushing partitioning tracer breakthrough curves for methanol and 2,2-DMP at EW 2 (Figure 14.14), there was substantially less retardation of 2,2-DMP than in the pre-flushing test (Figure 14.4). The EW 2 results were representative of all three extraction wells.

The post-flushing mean arrival times for methanol and 2,2-DMP and the calculated values of S_N at each extraction well are presented in Table 14.7. The values reported in Table 14.7 indicate that the cell-averaged reduction in S_N, as determined from partitioning tracer tests, was about 72%. As was described in section 13.4.2, laboratory evidence from Phase I of this study suggested that the complex NAPL found at the Hill AFB site contains a weathering product, or pitch, which is associated with the soil and is not extractable by solvent, but is detected by partitioning tracers. It is suggested from both pre- and post-flushing soil core data that the field SPME flood removed > 90% of the extractable fraction of the NAPL. Therefore, approximately 25% of the measured retardation from the pre-flushing tracer test, and > 90% from the post-flushing tracer test, should be attributed to the non-extractable pitch. This adjustment would bring the NAPL removal effectiveness measures based on pre- and post-flushing partitioning tracer tests into agreement with the estimates based on soil cores.

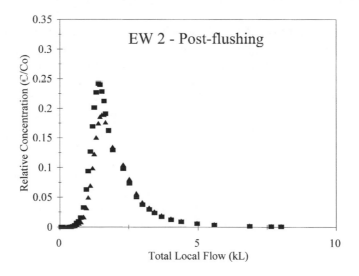

Figure 14.14 Tracer breakthrough curves for methanol (■) and 2,2-dimethyl 3-pentanol (▲) from the post-flood partitioning tracer test.

Table 14.7 Arrival Times for Nonpartitioning and Partitioning Tracers at the Extraction Wells for the Post-flood Tracer Test

	Post-flushing Arrival Time (h)		
Tracer	EW-1	EW-2	EW-3
Methanol	30.2	25.8	28.4
2,2-DMP	37.0	29.7	33.3
Final S_N	0.020	0.014	0.016
Percent Change in S_N	0.69	0.79	0.69

14.3.3 Extraction Well Mass Balance

The volume of NAPL removed from the test cell was estimated by dividing the mass removed for each constituent (Table 14.3) by the NAPL mass fractions (Table 13.1) and the NAPL density (0.85 g/cm^3; measured in the UF laboratory). The initial amount of NAPL present in the cell was estimated from both partitioning tracer and soil core data. Using the pre-flushing partitioning tracer data (Table 14.1), the initial volume of NAPL in the cell was estimated to be 369 L. Soil core data were also used to estimate the initial amount of NAPL present. Of the NAPL constituents that were common to the extraction well and soil core analyses, only n-decane and n-undecane yielded reliable chromatographic peaks in the effluent samples. The total mass of these two constituents initially present was calculated by multiplying the pre-flushing cell-averaged soil concentrations (Table 14.6) by the total mass of soil in the flushed zone (estimated from the cell area, 14.9 m^2), an average saturated depth of 3.0 m, and the soil bulk density. The soil concentrations reported in Table 14.6 are representative of primarily the sand fraction of the aquifer, but the field soil contained a substantial fraction of gravel and cobbles. Therefore, the effective bulk density was estimated to be between 1.7 g/cm^3 — a standard value for sandy soils (Freeze and Cherry, 1979) — and 0.66 g/cm^3 — equivalent to the standard value scaled by the ratio of the measured porosity (0.14) to the expected porosity for sandy soils (0.36).

The extraction well mass balance that used partitioning tracers to estimate the initial NAPL volume yielded NAPL removal effectiveness values of 55% to 75% (Table 14.8). The mass balance obtained using soil core data to estimate the initial amounts of n-decane and n-undecane yielded NAPL removal effectiveness estimates of 60% to 175%. These results demonstrate that determinations of remediation effectiveness made using extraction well breakthrough curves, where the mass removed and the initial mass are determined with different characterization techniques, may exhibit more inconsistency than methods where the initial and final contaminant amounts are measured with the same technique, as with the partitioning tracer and soil core methods.

14.3.4 In-line Analysis

In-line GC analysis was evaluated in the field during the SPME flood and the post-flushing partitioning tracer test because field-scale experiments require the collection, transport, and storage of large numbers of groundwater samples for laboratory analysis. In addition to the expense associated with each of these steps, a considerable amount of time often passes between sample collection and the availability of results for decision-making. In-line analysis offers the potential for immediate availability of results for on-site decision-making while minimizing the expenses associated with sample collection. Also, in-line analysis provides much less opportu-

Table 14.8 NAPL Removal Effectiveness, Determined from Extraction Well BCT Data and Estimates of the Initial Amount of NAPL from the Partitioning Tracer and Soil Core Data

NAPL Constituent	Source of Initial Mass Estimate Partitioning Tracers	Soil Cores
p-Xylene	0.60	
1,2,4-Trimethylbenzene	0.74	
n-Decane	0.68	0.61–1.54
n-Undecane	0.57	0.68–1.72
n-Dodecane	0.56	
n-Tridecane	0.70	

Note: Of the constituents listed, only n-decane and n-undecane were measured in the soil cores. Lower values were obtained with an effective bulk density of 1.7 g/cm^3; higher values were obtained with an effective bulk density of 0.66 g/cm^3.

nity for mass loss through volatilization or degradation than traditional sample collection and analysis schemes.

A portable gas chromatograph (SRI Instruments, Model 8610B) was installed inside the temporary shelter that had been built at the site. A portion of the flow from EW-1 was diverted to the field GC, through 6 m of 3.2-mm Teflon™ tubing, which reduced to 1.6 mm tubing at the GC injection port. The flow rate through this line was approximately 5 L/day (<1/100 of the flow rate at EW-1); but because of the small tubing diameter, the residence time in the line was relatively short (approximately 15 min). Analyses were performed at 8-min intervals for the duration of the post-flushing partitioning tracer test.

The in-line results were compared with laboratory analyses of samples that had been shipped to UF from the field site (Puranik, 1997). The zeroth and first moments of the tracer breakthrough curves were calculated. The zeroth moments were approximately 20% higher for the in-line data, suggesting substantial mass loss during the collection, transport, and storage of the samples analyzed in the laboratory. However, the first moments for both tracers were only approximately 5% higher for the in-line analysis, suggesting that the mechanism of mass loss was linear, occurring over all concentration ranges. For estimation of NAPL saturation by tracer retardation, both in-line and traditional sampling and analysis provide comparable results (note that if both the nonreactive and partitioning tracer first moments are both 5% higher, the retardation calculated will be the same). The in-line method provides rapid realtime information about the test; however, the overall quality of the data might be somewhat lower. The reason for this is that the type of GC typically moved to a field site and installed for a temporary experiment is generally of lower quality (and cost) and is not operated in an environment conducive to stable performance. Despite these limitations, this test has demonstrated that, for the purpose of determining first moments and retardation of tracer breakthrough curves, the low cost system provides good results. It should be noted that a more controlled GC environment can be provided with a mobile laboratory and a technician to provide more controlled QA/QC on-site, but this, of course, adds significant cost. The in-line methods developed as a part of this research project are used routinely in the laboratory in a controlled environment with excellent results.

14.4 WASTE MANAGEMENT AND MINIMIZATION

Disposal and treatment of the waste fluids generated during *in situ* flushing can be a major cost element, especially when scale-up from pilot-scale testing to full-scale site remediation is considered. The SPME field study involved flushing about 54,000 L (about 9 PV) of the precursor solution through the test cell, followed by 1 PV of surfactant alone (Brij 97 at 3 wt%) to displace *n*-pentanol (*n*-pentanol at 2.5 wt%), and then another 6.5 PV of water flood to displace the surfactant and any residual cosurfactant. The total volume of waste fluids produced as a result of this flushing sequence was about 110,000 L. A series of laboratory and on-site experiments were conducted to explore simple and inexpensive, yet effective, techniques to minimize the disposal costs of these waste fluids.

The chemical oxygen demand values for the waste fluids collected in the storage tanks at the site ranged from a low of <1000 mg/L to a high of approximately 120,000 mg/L. The composition of the waste fluids determines, to a great extent, the range of waste disposal or treatment options available, and the costs of waste disposal or treatment are influenced by the method(s) selected.

The project work plan called for separation of the waste fluids into two categories: (1) high-level waste, a composite of fluids generated during the early phases of SPME flood, with COD levels in excess of 100,000 mg/L; and (2) low-level waste, a composite of fluids produced during the water flood, with COD levels in excess of 4000 mg/L. Off-site disposal by incineration of the first category of waste, about 70,000 L, was planned. The waste fluids in the second category, a total of about 60,000 L, were planned to be pumped to the industrial wastewater treatment plant (IWTP) on the Base. The existing IWTP on the Base utilized equalization, clarification, chemical reduction, chemical precipitation, oil-sorbent filtration, and air stripping steps, and a final treatment with granular activated carbon to decrease organic contaminant concentrations to low levels (up to 200 to 300 mg/L) before discharging the wastewater to the county sewer system.

Off-site incineration of all of the waste fluids generated would be perhaps the simplest but also the most expensive option; at about $0.21/lb or $0.42/L, the total cost would be about $46,000. Clearly, the cost of other disposal and treatment options had to be explored to avoid the high cost of off-site incineration. The costs of waste fluid treatment and disposal through the Base IWTP were estimated at a total of about $3300 ($0.03/L for 110,000 L). Associated costs, such as replacement of the activated carbon used at the treatment plant, are not included in this estimate. This, in fact, was the method of disposal of both categories of the waste fluids generated during the pilot-scale SPME test. Waste fluids were stored on-site, and then discharged to the IWTP by metering the fluids at a rate such that the total COD load did not exceed about 700 kg/day (1500 lb/day). This method of disposal may not be an option at sites where access to an IWTP is not available.

The waste disposal and treatment costs can be lowered in at least two ways: (1) decrease the volume of fluids incinerated and (2) decrease the volume and/or the COD load of the fluids sent to the IWTP (thus decreasing activated carbon replacement costs). A simple but effective means of accomplishing both of these objectives was investigated by taking advantage of the phase behavior of the NAPL/water/precursor system. The waste fluid produced during the SPME flushing experiment was a Winsor Type I microemulsion. As described in section 12.2, varying a parameter that changes the character of the system from hydrophilic to hydrophobic will produce a phase transition from an oil-in-water microemulsion (Winsor Type I), to a water-in-oil micro-emulsion (Winsor Type II). The addition of inorganic salts serves to decrease the solubility of hydrophobic organic compounds in water, making the system more hydrophobic. With increasing addition of salt up to a "critical salinity (S1)," organic solubility in the aqueous phase decreases and increasing amounts of NAPL are solubilized in the Type I oil-in-water microemulsion. Salt addition beyond S1 produces a three-phase Winsor Type III middle-phase emulsion, and at the "optimal salinity" equal amounts of oil and water are emulsified (usually the goal for enhanced oil recovery techniques). With further salt addition, a second "critical salinity (S2)" is reached where the system converts to a two-phase system with an oil-rich phase in equilibrium with the aqueous phase (Winsor Type II). Important parameters that define this phase behavior are: the optimal and the critical salinities (S1, S2); the "salinity window" (S2-S1); and the amount of oil solubilized (σ^*, cm^3 oil/cm^3 total) at the optimal salinity.

The treatment strategy for waste minimization was to add enough salt to cause a phase shift from the initial *single-phase* Winsor Type I microemulsion to a *two-phase* Winsor Type II system. Both lab and field experiments showed that the oil:water volume ratio in the two-phase fluid increased with increasing salt addition. For a composite sample of the high-level waste fluids, the water:oil ratio was <1 with 6% salt addition, increased to approximately 10 at 7% salt, and reached a maximum of about 20 for salt additions in excess of about 8%. This type of efficient phase separation could be achieved with salt additions to a composite sample of the high-level waste fluids from the test cell, and for samples of waste fluids collected during various stages of the SPME flood.

These experiments showed that greater than 99% of the NAPL constituent mass was selectively partitioned into the hydrophobic oil-rich phase; and as a result, the associated aqueous-phase concentrations were very low (<1 mg/L). However, the cosurfactant (pentanol) concentrations still remained fairly high (about 3 to 4 mg/mL) and most of the salt (NaCl) was also solubilized in the aqueous phase. The aqueous phase also contained low concentrations (<200 mg/L) of the Brij 97 surfactant. The COD concentration in the aqueous phase waste was reduced to about 3000 to 4000 mg/L after salt addition.

These results suggest that waste disposal and treatment costs can be reduced significantly if the oil-rich phase is disposed by incineration as a fuel-grade waste (<20% water; cost: $0.067/L).

Thus, the first objective of cost minimization via waste volume reduction would be achieved by having to incinerate only the separated oil phase. About 1/25th of the total volume of waste fluids generated need be incinerated at about half the unit cost ($0.21/lb for non-fuel grade vs. $0.115/lb for fuel grade). Furthermore, the decreased COD load in the wastewater would also lead to cost reduction in treating the wastewater discharged to the IWTP. And, since the primary source of COD in the wastewater is pentanol (approximate concentration <3000 mg/L), it can be easily removed either by air sparging or other inexpensive water treatment technologies. This aspect is currently being investigated in the UF laboratory. If pentanol and COD can be decreased below the concentrations acceptable for direct disposal to the county sewer system, then the IWTP costs can also be saved. The cost objective here was to achieve the acceptable level of water treatment at a unit of cost less than that charged by the IWTP.

The foregoing analysis suggests that substantial savings (about a factor of 10 to 20) in waste disposal and treatment costs at Hill AFB could perhaps be accomplished by the simple technique of salt (NaCl) addition to the extraction fluids generated during the SPME *in situ* flushing field test. There are essentially no capital costs associated with this treatment technology because the oil/water phase separation can be readily achieved by simply adding the salt to the waste liquids in the storage tanks, allowing the oil and water phases to separate by gravity alone, and skimming off the oil-rich phase (this concept was demonstrated at the field site on a smaller scale using the extraction fluids from the SPME test cell).

14.5 SUMMARY

A Winsor Type I surfactant/alcohol mixture was used as an *in situ* flushing agent to solubilize a multicomponent NAPL as a single-phase microemulsion in a hydraulically isolated test cell at Hill AFB, Utah. The surfactant (polyoxyethylene (10) oleyl ether) and alcohol (*n*-pentanol) together composed 5.5 wt% of the flushing solution. The NAPL was extremely complex, containing over 200 constituents and a pitch fraction that was strongly associated with the soil and was not extractable with organic solvents. The NAPL-removal effectiveness of the SPME flood was calculated using information from soil cores, partitioning tracer tests, and NAPL constituent breakthrough curves (BTCs) measured at extraction wells. Soil core data for the most prevalent NAPL constituents indicated that approximately 90% to 95% of the target constituents were removed from the cell by the SPME flood. A comparison of pre- and post-flushing partitioning tracer data indicated that about 72% of the measured NAPL volume was removed by the SPME flood. Integration of NAPL constituent BTCs indicated 55% to 75% removal of the target NAPL constituents when partitioning tracer data were used to estimate the initial amount of NAPL present; and 60% to 175% removal of two target constituents when soil core data were used to estimate the amount of NAPL present initially. These results indicate that the SPME flood effectively removed the NAPL constituents of concern, but an insoluble anthropogenic residue was left behind.

CHAPTER 15

Conclusions

This project evaluated single-phase microemulsion solubilization in the laboratory and a pilot-scale field demonstration as a technique for enhanced remediation of a site contaminated with a complex, multicomponent nonaqueous phase liquid waste. The field-scale test was conducted in an isolated test cell installed at the Hill AFB site; hydraulic isolation was deemed necessary to minimize the potential impacts on the remainder of the site. Additional tests of the SPME technology are needed in different hydrogeologic settings with different NAPLs, and in larger test areas, especially without the use of sheet piles for hydraulic confinement. Full-scale testing is also a long-term goal.

The field test involved remediation of an NAPL comprising a mixture of jet fuel, waste chlorinated solvents, and other organic wastes. The compositional complexity of the NAPL confounded several aspects of the field assessment. First, not all of the components in the NAPL could be identified and quantified using the GC/FID method employed, which limited the assessment to the removal of only a few selected constituents. Thus, the total NAPL mass removal was inferred from partitioning tracer tests, but the presence of an insoluble "pitch" component introduced significant uncertainties in interpretation of the data from the post-SPME partitioning tracer tests. It is unlikely that the composition of pitch components can be determined, even using sophisticated GC/MS techniques. Second, significant analytical interference from unidentified components of the NAPL posed problems in measuring the surfactant concentrations in samples from the multilevel samplers and the extraction wells. Thus, an assessment of the surfactant transport behavior (e.g., potential for chromatographic separation of precursor components) and mass balance (i.e., surfactant recovery) was precluded by lack of reliable data. Better GC or LC methods may need to be developed for surfactant concentrations in complex matrices, which would be a significant analytical challenge. However, measured breakthrough curves for n-pentanol confirmed lab column data that suggested minimal partitioning of the precursor components into the NAPL.

Hydraulic testing of the test cell allowed for an estimation of the average hydraulic conductivity of the saturated zone, but an assessment of the spatial variability in hydraulic conductivity was not performed as a part of this site characterization. Even so, data from nonreactive tracer tests did allow a determination of the impact of spatial variability on hydrodynamic processes. Variability in hydraulic properties and NAPL distribution also have a significant impact on NAPL removal effectiveness and efficiency. Analysis of partitioning tracer data from the multilevel samplers did permit characterization of the spatial variability of the NAPL residual saturation within the test cell, both before and after SPME flushing. Parameters such as the variance and correlation length defining the spatial structure of the permeability field and NAPL content are required as inputs for the UTCHEM model. The model simulations were based on best estimates for these parameters. Thus, the spatial variability in the effectiveness of NAPL removal was based on tracer data and analysis of the soil cores collected before and after SPME flushing.

The chosen SPME precursor was quite effective in microemulsifying the complex NAPL at the Hill AFB test site; >90% of several target constituents of the NAPL and >70% of the total NAPL mass was removed after flushing with 9 pore volumes (54 kL). The selection of the surfactant (3% Brij 97) and the alcohol (2.5% n-pentanol) as the SPME precursor components was based on laboratory screening tests of a large number of candidate chemicals. Among the constraints were high NAPL solubilization capacity and low viscosity (<2 cp) of the precursor and the SPME produced during flushing.

Other surfactant/alcohol combinations with much higher NAPL solubilization capacity would have been selected as the SPME precursors if the viscosity constraint had been relaxed somewhat (e.g., <3 cp). Similarly, if waste treatment considerations were also used as screening considerations, other precursors may have been selected. Addition of high concentrations (10% to 12%) of salt was found to be an effective technique for breaking the oil-in-water microemulsions in the extraction fluids. However, other techniques for separation of the NAPL from the extraction fluids (to minimize waste handling costs) and the surfactant and the alcohol (for reuse of the precursor to minimize costs) need to be explored, especially if full-scale implementation is to be cost effective.

The field test of SPME funded by DOD/AATDF was part of a larger study sponsored by EPA-SERDP to evaluate several *in situ* flushing technologies side-by-side in isolation cells installed in Operable Unit 1 at Hill Air Force Base. In eight other test cells of similar design, flushing with other chemical agents (cosolvents, surfactants, steam, cyclodextrin, air) was evaluated. When the data from these tests are available, the relative performance of these flushing technologies should be evaluated.

CHAPTER 16

Section II References

Abdul, A.S., Gibson, T.L., Ang, C.C., Smith, J.C., and Sobczynski, R.E. (1992). *In-situ* surfactant washing of polychlorinated biphenyls and oils from a contaminated site. *Ground Water* 30(2):219–231.

Abdul, A.S., Gibson, T.L., and Rai, D.N. (1991). Selection of surfactants for the removal of petroleum products from shallow sandy aquifers. *Ground Water* 28:920–926.

Annable, M.D., Jawitz, J.W., Rao, P.S.C., Dai, D.P., Kim, H.K., and Wood, A.L. (1998a). Interfacial and partitioning tracers for characterization of effective NAPL-water contact areas. *Ground Water* (36)3:495–502.

Annable, M.D., Rao, P.S.C., Graham, W.D., Hatfield, K., Graham, W.D., Wood, A.L., and Enfield, C.G. (1998b). Partitioning tracers for measuring residual NAPL: Field-scale test results. *J. Environ. Eng.* 124(6):498.503.

Augustijn, D.C.M., Jessup, R.E., Rao, P.S.C., and Wood, A.L. (1994). Remediation of contaminated soils by solvent flushing. *J. Environ. Eng.* 120(1):42–57.

Ballet, F. and Candau, F. (1981). Water-oil solubilization by long chain polyoxyethylene nonionic surfactants. *Coll. Polymer Sci.* 259:548–552.

Baran, J.R., Jr., Pope, G.A., Wade, W.H., and Weerasooriya, V. (1994). Microemulsion formation with chlorinated hydrocarbons of differing polarity. *Environ. Sci. Technol.* 28(7):1361–1366.

Brown, C.L., Delshad, M., Dwarakanath, V., Jackson, R.E., Londergan, J.T., Meinardus, H.W., McKinney, D.C., Oolman, T., Pope, G.A., and Wade, W.H. (1999). Demonstration of surfactant flooding of an alluvial aquifer contaminated with DNAPL. In *Innovative Subsurface Remediation: Field Testing of Physical, Chemical, and Characterization Technologies.* M.L. Brusseau, D.A. Sabatini, J.S. Gierke, M.D. Annable, eds. ACS Symposium Series 725, American Chemical Society, Washington, D.C.

Brown, C.L., Pope, G.A., Abriola, L.M., and Sepehrnoori, K. (1994). Simulation of surfactant-enhanced aquifer remediation. *Water Resour. Res.* 30(11):2959–2977.

Falta, R.W., Brame, S.E., Lee, C.M., Coates, J.T., Wright, C., Price, S., Haskell, P., and Roeder, E. (1996). A field test of NAPL removal by high molecular weight alcohol injection. In *Non-Aqueous Phase Liquids (NAPLs) in Subsurface Environment: Assessment and Remediation.* L.N. Reddi, Ed. ASCE, Washington, D.C., 257–268.

Farley, K.J., Falta, R.W., Brandes, D., Milazzo, J.T., and Brame, S.E. (1993). Remediation of Hydrocarbon-Contaminated Groundwaters by Alcohol Flooding. Hazardous Waste Management Research Fund, Institute of Public Affairs, University of South Carolina, Columbia, SC.

Fountain, J.C., Starr, R.C., Middleton, T., Beikirch, M., Taylor, C. and Hodge, D. (1996). A controlled field test of surfactant-enhanced aquifer remediation. *Ground Water* 34(5):910–916.

Freeze, R.A. and Cherry, J.A. (1979). *Groundwater.* Prentice-Hall, Englewood Cliffs, New Jersey.

Gatlin, C. and Slobod, R.L. (1960). The alcohol slug process for increasing oil recovery. *Trans. AIME* 219:46–53.

Gogarty, W.B. and Tosch, W.C. (1968). Miscible-type waterflooding: oil recovery with micellar solutions. *J. Pet. Tech.* 20:1407–1414.

Huh, C. (1979). Interfacial tensions and solubilizing ability of a microemulsion phase that co-exists with oil and brine. *J. Colloid Interface Sci.* 71:408.

Imhoff, P.T., Gleyzer, S.N., McBride, J.F., Vancho, L.A., Okuda, I., and Miller, C.T. (1995). Cosolvent-enhanced remediation of residual dense nonaqueous phase liquids: experimental investigation. *Environ. Sci. Technol.* 29(8):1966–1976.

Jawitz, J.W., Sillan, R.K., Annable, M.D., and Rao, P.S.C. (1997). Methods for determining NAPL source zone remediation efficiency of *in-situ* flushing technologies. *ASCE In-Situ Remediation of the Geoenvironment Conference.* ASCE, Minneappolis, MN.

Jawitz, J.W., Annable, M.D., Rao, P.S.C., and Rhue, R.D. (1998). Field implementation of a Winsor Type I surfactant/alcohol mixture for *in situ* solubilization of a complex LNAPL as a single-phase microemulsion. *Environ. Sci. Technol.* 32(4):523–530.

Jin, M. (1995). A Study of Nonaqueous Phase Liquid Characterization and Surfactant Remediation. Ph.D. dissertation, University of Texas at Austin, 340 pp.

Jin, M., Delshad, M., Dwarakanath, V., McKinney, D.C., Pope, G.A., Sepehrnoori, K., Tilburg, C.E., and Jackson, R.E. (1995). Partitioning tracer tests for detection, estimation, and remediation performance assessment of subsurface nonaqueous phase liquids. *Water Resour. Res.* 31(5):1201–1211.

Lake, L.W. (1989). *Enhanced Oil Recovery.* Prentice-Hall, Englewood Cliffs, NJ.

Martel, R. and Gélinas, P.J. (1996). Surfactant solutions developed for NAPL recovery in contaminated aquifers. *Ground Water* 34(1):143–154.

Martel, R., Gélinas, P.J., Desnoyers, J.E. and Masson, A. (1993). Phase diagrams to optimize surfactant solutions for oil and DNAPL recovery in aquifers. *Ground Water* 31(5):789–800.

Montgomery Watson (1995). Hill Air Force Base, Utah, Phase I Work Plan for Eight Treatability Studies at Operable Unit 1, Montgomery Watson, Salt Lake City, UT.

Ouyang, Y., Mansell, R.S., and Rhue, R.D. (1996). A microemulsification approach for removing organolead and gasoline from contaminated soil. *J. Hazard. Materials* 46:23–35.

Pennell, K.D., Abriola, L.M., and Weber, W.J., Jr. (1993). Surfactant enhanced solubilization of residual dodecane in soil columns. *Environ. Sci. Technol.* 27:2332–2340.

Pennell, K.D., Jin, M., Abriola, L.M. and Pope, G.A. (1994). Surfactant enhanced remediation of soil columns contaminated by residual tetrachloroethylene. *J. Contam. Hydrol.* 16(1):35–53.

Pennell, K.D., Pope, G.A., and Abriola, L.M. (1996). Influence of viscous and buoyancy forces on the mobilization of residual tetrachloroethylene during surfactant flushing. *Environ. Sci. Technol.* 30(4):1328–1335.

Peters, C.A. and Luthy, R.G. (1993). Coal tar dissolution in water-miscible solvents: experimental evaluation. *Environ. Sci. Technol.* 27(13):2831–2843.

Pithapurwala, Y.K., Sharma, A.K., and Shah, D.O. (1986). Effect of salinity and alcohol partitioning on phase behavior and oil displacement efficiency in surfactant-polymer flooding. *J. Amer. Oil Chem. Soc.* 63(6):804–813.

Puerto, M.C. and Reed, R.L. (1983). A three-parameter representation of surfactant/oil/brine interaction. *Soc. Pet. Eng. J.* 23:669–682.

Puranik, S. 1997. In-line Gas Chromatographic Tracer Analysis: An Alternative to Conventional Sampling & Analysis in Determination of Non-aqueous Phase Liquid (NAPL) Residual Saturation and Remediation Performance Assessment. University of Florida, Gainesville.

Ramachandran, B. (1997). M.S. thesis, University of Florida, Gainesville.

Rao, P.S.C., Annable, M.D., Sillan, R.K., Dai, D., Hatfield, K.H., Graham, W.D., Wood, A.L. and Enfield, C.G. (1997). Field-scale evaluation of *in-situ* cosolvent flushing for enhanced aquifer remediation. *Water Resour. Res.* 33(12):2673–2686.

Rao, P.S.C., Lee, L.S., and Wood, A.L. (1991). Solubility, Sorption, and Transport of Hydrophobic Organic Chemicals in Complex Mixtures. Environmental Research Brief, EPA/600/M-91/009, R. S. Kerr Environmental Research Laboratory, U.S. EPA, Ada, OK.

Rhue, R.D., Annable, M.D., and Rao, P.S.C. (1997). Lab and Field Evaluation of Single-Phase Microemulsions (SPME) for Enhanced *In-Situ* Remediation of Contaminated Aquifers. Phase I: Laboratory Studies for Selection of SPME Precursors. AATDF Report. University of Florida, Gainesville.

Rosen, M.J. (1989). *Surfactants and Interfacial Phenomena.* John Wiley & Sons, New York, 431 pp.

Roy, S.B., Dzombak, D.A., and Ali, M.A. (1995). Assessment of in-situ solvent extraction for remediation of coal tar sites: column studies. *Water Environ. Res.* 67(1):4–15.

Sabatini, D.A., Knox, R.C., and Harwell, J.H. (1996). Surfactant-Enhanced DNAPL Remediation: Surfactant Selection, Hydraulic Efficiency, and Economic Factors. *Environmental Research Brief*, EPA/600/S-96/002, National Risk Management Research Laboratory, U.S. EPA, Ada, OK.

Salager, J.L., Bourrel, M., Schechter, R.S., and Wade, W.H. (1979). Mixing rules for optimum phase-behavior formulations of surfactant/oil/water systems. *Soc. Pet. Eng. J.*, 19(5):271–278.

Saripalli, K.P., Kim, H., Rao, P.S.C., and Annable, M.D. (1997). Measurement of specific fluid–fluid interfacial areas of immiscible fluids in porous media. *Environ. Sci. Technol.* 31(3):932–936.

Shah, D.O., Ed. (1981). *Surface Phenomena in Enhanced Oil Recovery*. Plenum Press, New York, 874 pp.

Shiau, B.-J., Sabatini, D.A., Harwell, J.H. and Vu, D.Q. (1996). Microemulsion of mixed chlorinated solvents using food grade (edible) surfactants. *Environ. Sci. Technol.* 30(1):97–103.

Shinoda, K. and Kunieda, H. (1983). Phase properties of emulsions: PIT and HLB. In P. Becher, Ed. *Encyclopedia of Emulsion Technology*, Volume I. Basic Theory. Marcel Dekker, Inc., New York.

Sillan, R.K., Rao, P.S.C., Annable, M.D., Dai, D., Hatfield, K., Graham, W.D., Wood, A.L. and Enfield, C.G. (1998). Evaluation of *in-situ* cosolvent flushing dynamics using a network of spatially distributed multi-level samplers. *Water Resour. Res.* 34(9):2191–2202.

Valocchi, A.J. (1985). Validity of the local equilibrium assumption for modeling sorbing solute transport through homogeneous soils. *Water Resour. Res.* 21(6):808–820.

Vinatieri, J.E. and Fleming, III P.D. (1979). The use of pseudocomponents in the representation of phase behavior of surfactant systems. *Soc. Pet. Eng. J.* 19(5):289–300.

West, C.C. and Harwell, J.H. (1992). Surfactants and subsurface remediation. *Environ. Sci. Technol.* 26(12):2324–2330.

PART III

Design and Evaluation of Surfactant/Foam and Single-Phase Microemulsion Technologies

Thomas J. Simpkin
CH2M Hill

PART **IIIA**

Design and Evaluation of a Full-scale Implementation of the Surfactant/Foam Process

EXECUTIVE SUMMARY

The Department of Defense (DOD) funded the Advanced Applied Technology Development Facility (AATDF) in 1993 with the mission of enhancing the development of innovative remedial technologies for DOD by bridging the gap between academic research and proven technologies. To accomplish its mission, the AATDF selected 11 projects that involved the quantitative demonstration of innovative remediation technologies. Field demonstrations were completed for each of these projects. To further assist with the potential commercialization of the remediation technologies and to disseminate information on the technologies, the AATDF funded the preparation of a Technology Evaluation for each technology by an independent engineering consulting firm.

This technology design and evaluation is a direct scale-up of the field test at Hill AFB OU2 and is referred to as the Base Case. The field demonstration was designed to be the best technical test of the surfactant/foam concept. Technical issues such as propagation of foam through the subsurface (horizontally), removal of residual DNAPL, comparison of the estimates of DNAPL removal provided by soil cores and the partitioning interwell tracer test, and development of a numerical simulation of the process were paramount. Meeting all these objectives in the first field test was a major accomplishment. Only in the sense that the field test was made to be affordable were economic considerations used in its design. The rationale for directly scaling up the field test in this design and evaluation is that it represents what was known about the technology at the time of writing, but there are problems with this approach. Since this design and evaluation was prepared, a full-scale surfactant-enhanced aquifer remediation has been approved at the Hill AFB, OU2 site, and other surfactant-based remediation demonstrations have been performed. This new information may provide more-accurate estimates of the costs of surfactant and surfactant/foam remediation. The value of the following design and evaluation is that it outlines the technical issues that confront surfactant/foam technology and the elements that most impact the cost of full-scale implementation of the surfactant/foam process.

The surfactant/foam technology involves the injection and extraction of a surfactant solution into the subsurface to solubilize and mobilize nonaqueous phase liquid (NAPL). The processes are enhanced by the injection of air into the injection wells once surfactant is present in the formation to create foam *in situ*. Air injection occurs intermittently during the surfactant injection. The foam fills the pore spaces along the most permeable flow paths, thereby diverting the flow of surfactant to the less permeable flow paths.

Summary of Surfactant/Foam Demonstration

To develop and evaluate the effectiveness of the surfactant/foam process, a series of laboratory studies and a field pilot demonstration were performed. The laboratory tests involved a series of phase behavior tests, critical micelle concentration (CMC) measurements, interfacial tension measurements, foam formation tests, biodegradation studies, one-dimensional column studies, and two-dimensional column studies. These studies were used to select the chemical system and to confirm the potential applicability of the technology. The chemical system selected from these studies and used in the field demonstration was MA-80I (dihexyl sulfosuccinate) at a target concentration of 4% by weight with 10,000 mg/L sodium chloride.

The field demonstration was performed at Hill Air Force Base (AFB) Operable Unit (OU) 2. This site is contaminated with DNAPL, consisting of chlorinated solvents, primarily trichloroethene (TCE). The demonstration involved the following:

- Installation of three injection, three extraction, two hydraulic control, and two monitoring wells over an area 6 m × 3.6 m (20 ft × 12 ft); the deepest wells were 47 ft bgs.
- Construction of the chemical mix system and a piping system to send produced fluids to an existing treatment system
- Injection of chemicals using the following sequence:
  - Water injection (with tracers for part of the time): 6.7 pore volumes (PV)
  - Sodium chloride (NaCl) only: 1.0 PV
  - Surfactant, with periodic air injection: 3.0 PV
  - NaCl only: 0.5 PV
  - Water injection (with tracers for part of the time): 21.8 PV
- Injection of air periodically during the surfactant injection to create *in situ* foam
- Collection of samples during the injection of chemicals
- Collection of soil cores at the completion of the test

The surfactant/foam demonstration removed a large fraction of the DNAPL present. Mass removal estimates ranged from 84% to 95%. This mass removal was achieved in a relatively short time. The majority of the mass removal occurred with less than 3 PV flushed through the aquifer.

Estimates of the volume of DNAPL removed, based on analysis of produced fluids, analysis of soil cores, and partitioning tracer tests, were 140 L (37 gal), 80.2 L (21.2 gal), and 69.7 L (18.4 gal), respectively. The variability of these results in part reflects uncertainties associated with each methodology. Project investigators (INTERA, 1997) attribute production (produced fluids data) in excess of removal (soil cores and tracer tests) to be a result of migration of DNAPL to the target zone during demonstration.

The potential for migration of DNAPL into the test area complicates analysis of the endpoints that were achieved. The majority of the post-surfactant flushing soil cores had TCE concentrations below method detection limits. However, there was a thin layer (7.6 cm [3 in.]) of relatively high concentrations (1600 to 31,600 mg/kg) just above the clay interface. This DNAPL may have been missed by the surfactant solution because the elevations at some of these points may have been lower than the bottom of the extraction wells (i.e., the soil borings were located in depressions of the clay surface). It is also possible that the efforts to flush the DNAPLs from the depressions were successful but that new DNAPL migrated into the depressions during the testing.

The benefits of the air injection (foam production) compared with injection of surfactant alone were evaluated indirectly from the pressure responses at the injection wells, the foam production in the monitoring points, and the swept volume estimated from a bromide tracer conducted during the surfactant/air injection. Taken together, the evidence suggests that the foam did have an impact because it reduced the sweep volume and possibly diverted surfactant to the lower portions of the aquifer. It can be concluded that the injection of air with the production of foam was a better process than surfactant injection alone and is, therefore, a promising addition to surfactant flushing.

Summary of Engineering Design and Cost Estimate

To illustrate the engineering design components that may be required for full-scale application of the surfactant/foam technology, a conceptual design for a hypothetical site using the surfactant/foam technology was prepared. Table E.1 summarizes the conceptual design. The hypothetical site was modeled based on conditions at Hill AFB OU2. However, it is noted that the conceptual design contains a number of assumptions that are not necessarily accurate for OU2. Consequently, the engineering design and cost estimate may not be appropriate for actual remediation of OU2. In addition to preparing the conceptual design that used conservative

Table E.1 Site Characteristics and Base Case System Design Summary for the Surfactant/Foam Site

Site Characteristics

Type of site	Disposal site for waste chlorinated solvents. The site is currently inactive.
Contaminants	Chlorinated solvents TCE, TCA, and PCE present in a multicomponent DNAPL.
Size of target area	0.41 hectare (1 acre).
Hydrogeologic setting	Granular alluvium material consisting of mostly sand with some silt and clay, and occasional stringers of gravel. Relatively homogeneous 12-m thick. Underlying clay that provides barrier to vertical migration. Surrounded by containment wall.
Target depth	DNAPL zone over the bottom 1 to 2 m of granular alluvium.

System design

Remedial objective	95% removal of the recoverable contaminants.
Primary performance measures	Measurement of contaminants (TCE) in the produced fluids. The total mass of recoverable contaminant present will be based on projection of production curves.
Removal mechanism utilized	Solubilization with mobilization using a surfactant foam and sodium chloride. Foam produced *in situ* by injection of air to improve sweep efficiency.
Chemical system	4wt% food grade Aerosol MA-80I (sodium dihexyl sulfosuccinate), 11,000 mg/L NaCl.
Delivery recovery system	Vertical injection and recovery wells arranged in an off-set line-drive pattern with 6-m (20 ft) spacing between wells in a line and 12.2-m (40 ft) spacing between lines. Four injection lines with seven wells each, and five recovery lines with eight wells each. Total length of time to complete is 1 year.
Chemical delivery sequence	• 10 PV of water flooding over 42 d. • 2.5 PV of surfactant and injection over 21 d for each of 34 modules. Air injected at 4 scfm per well. Air injected for 2 h to each well, and then was off for 4 h. • 10 PV of post water flooding over 42 d.
Chemical mix system	• Chemicals delivered in concentrated form, diluted on-site, and mixed in two solution mix and storage tanks. • 187 metric tons (205 tons) MA 80I total for the project. • 201 metric tons (221 tons) NaCl total for the project.
Produced fluids management	Removal of free DNAPL in an oil/water separator, separation of surfactant from contaminants in an air stripper, treatment of off-gas by catalytic oxidation, ultrafiltration to reconcentrate the surfactant, recycle of the majority of the fluid, and discharge of the bleed stream to a WWTP.
Additional studies included in costs	Additional site characterization, bench-scale testing, numerical simulations, field demonstrations in a sheet pile enclosed cell, partitioning tracer tests. (See Appendix 6, Additional Capital Cost sheet for details.)

Table E.2   Unit Cost Comparison

	Base Case		Optimized Cost	
	Quantity	Cost per Unit	Quantity	Cost per Unit
Treatment area				
Hectares (ha)	0.42	$18,050,000/ha	0.42	$11,890,000/ha
Acres	1.03	$7,300,000/acre	1.03	$4,810,000/acre
Treatment volume				
Cubic meters (m³)	16,700	$450/m³	16,700	$300/m³
Cubic yards (yd³)	21,800	$345/yd³	21,800	$227/yd³
Volume of NAPL				
Liters (L)	42,900	$175/L	692,000	$116/L
Gallons (gal)	11,300	$644/gal	183,000	$438/gal

Note: As indicated in Table E.1, the costs are estimates for a one-acre site. Unit costs would be smaller for larger sites.

assumptions from the test (the Base Case), a conceptual design was also prepared for an Optimized Case of the application of the technology.

Cost estimates were prepared for this hypothetical design to illustrate the potential cost of this technology. The cost estimates were order-of-magnitude level and are considered accurate to within + 50% and –30%. For the Base Case, the total cost estimate of $7,523,000 includes $1,893,000 for produced fluids treatment (25% of the total) and $1,813,000 for disposal of residuals. For the Optimized Case, the total cost of $4,957,000 includes $2,013,000 for produced fluids treatment (42%) and $599,000 for disposal of residuals (12%). Treatment costs would be comparable for other processes in which DNAPL is recovered from the source zone. The two most significant cost components for this system were the produced fluids treatment system and the cost for disposal of residuals. Table E.2 summarizes the costs on a unit area and a volume basis. The assumptions that were used to develop these figure are discussed in more detail in Chapter 19.

Summary of Potential Performance and Applicability

The following list summarizes the potential full-scale performance of the surfactant/foam technology. This summary is based on the field demonstration at Hill AFB OU2.

- In the appropriate setting, the surfactant/foam technology has the potential to remove a large fraction of the total DNAPL in relatively short periods of time. Removals in the range of 90% or greater are possible (possibly less than 3 PV for very permeable soils and/or close well spacing).
- The production of foam can reduce the swept volume in the aquifer. It is likely that upper portions of an aquifer will become preferentially filled with foam. This should divert surfactant to the lower portions of the aquifer that are likely to contain more DNAPL, if that is the nature of the contaminant distribution.
- The use of foam should have no adverse impacts on the process. It is also a relatively simple process to install and operate.
- The produced fluids from the surfactant/foam technology are potentially amenable to relatively simple treatment methods (e.g., air stripping) that may allow reuse of the surfactant solution.

Tables 21.1 and 21.2 (see Chapter 21) summarize the site characteristics and design and operating parameters that impact the applicability, performance, and cost of the surfactant/foam technology. The surfactant/foam technology is not yet a fully developed technology. Consequently, additional laboratory and field testing and development may be required on potential sites. Additional laboratory and field-scale tests and demonstrations are needed to address the following issues that may be important in full-scale implementation:

- Definition of the benefits of the mass removal that may be achieved with this technology
- Achievable soil and groundwater endpoint concentrations
- Performance of the technology in other hydrogeologic settings, especially those with heterogeneous soils and an undulating aquitard
- Assessment of whether the surfactant/foam process has the potential to flush all of the depressions in an aquifer
- Potential for vertical mobilization of DNAPL at the particular site
- Cost-effective methods of treating produced fluids
- Improved methods to optimize chemical delivery to targeted contaminated zones of an aquifer
- Methods to optimize the system design, such as well spacing and alternative chemical injection methods
- Stability of foam and the distance it may be propagated in other chemical and hydrogeologic settings

CHAPTER 17

Summary of Technology Demonstration

17.1 INTRODUCTION

The surfactant/foam technology involves the injection and extraction of a surfactant solution into the subsurface to solubilize and mobilize nonaqueous phase liquid (NAPL). The processes are enhanced by the injection of air into the injection wells once surfactant is present in the formation to create foam *in situ*. Air injection occurs intermittently during the surfactant injection. The foam fills the pore spaces along the most permeable flow paths, thereby diverting the flow of surfactant to the less permeable flow paths. Without the foam, the surfactant would contact the less permeable zones at a much slower rate, if at all. This section summarizes the laboratory studies and field pilot-scale demonstration. More details on the studies can be found in Part I.

17.2 SUMMARY OF LABORATORY TESTS AND TREATABILITY STUDIES

The surfactant/foam technology development built on previous work performed by the University of Texas and INTERA for the development of a surfactant-enhanced aquifer remediation process for chlorinated solvents. The team from the University of Texas and INTERA carried out a field demonstration of their process at Hill Air Force Base (AFB), Operable Unit (OU) 2 in 1996 (called the Air Force Center for Environmental Excellence [AFCEE] demonstration [INTERA, 1996; Brown, et al., 1999]). The chemical solution used for the surfactant/foam demonstration, which is the basis for this report, was similar to that used for the AFCEE demonstration, except isopropanol was not included to allow the production of foam. The laboratory tests conducted for the surfactant/foam technology involved a series of phase behavior tests, CMC measurements, interfacial tension measurements, foam formation tests, biodegradation studies, one-dimensional column studies, and two-dimensional column studies. The details of these tests can be found in Chapter 2.

The surfactant selected from the laboratory studies was dihexyl sulfosuccinate, which is commercially available from Cytec Inc. as MA-80I. A target surfactant concentration of 4% by weight was selected. This is lower than that used in the AFCEE demonstration. This surfactant has been approved by the Food and Drug Administration (FDA) as a food-grade surfactant. One modification was made to the solution used for the surfactant/foam technology field trial compared to that used for the AFCEE demonstration: the isopropanol used in the AFCEE project had to be removed from the formulation because the alcohol was found to interfere with foam formation. The physical properties of surfactant solution (e.g., viscosity and interfacial tension) to the DNAPL were not reported in the project report.

One-dimensional (1-D) column tests were used to evaluate the generation of foam. Through this effort, it was learned that the 1-D tests did not scale directly to 2-D conditions; therefore, they were of limited value.

Tests were performed on 2-D laboratory models at two scales. The first test was 20 inches (in.) long, 3 in. high, and 0.75 in. thick. The model was packed with two layers of soil with different permeabilities. The tests demonstrated the benefits of foam injection under controlled conditions. The injection of foam greatly decreased the time to sweep the lower permeability soil with surfactant. The larger 2-D test was conducted in a tank 8 feet (ft) long by 5 ft high by $1^1/_2$ in. thick. This tank was also packed with a layered soil system. Tests in the larger tank demonstrated that the foam could be propagated at least 8 ft across the tank. Different patterns of air injection were also evaluated with this system.

17.3 TEST SITE DESCRIPTION

The surfactant/foam field demonstration was performed at Hill AFB OU2. A brief summary of the field site history and hydrology follows. More detailed information can be found in Part I of this monograph and other references (Radian, 1992).

OU2 is the location of former waste chlorinated solvent disposal pits. A system for removal of the mobile DNAPL produced approximately 87,000 L (23,000 gal) DNAPL since its start-up in 1993. A decision has been made to expand surfactant-enhanced aquifer remediation to the rest of the OU2 site.

The shallow stratigraphic unit at the site is 12 m (39 ft) of the Provo Formation, which consists of sand with silt, clay, and occasional stringers of gravel. The hydraulic conductivity and porosity of the Provo Formation are estimated to be 10^{-2} and 0.27 cm/s, respectively. Underlying the Provo Formation is the 60-m thick (197-ft thick) Alpine Formation. The Alpine is a lacustrine clay that contains minor discontinuous interbedded sand seams. The bulk hydraulic conductivity of the Alpine Formation is estimated to be 10^{-5} to 10^{-8} cm/s. The shallowest regional groundwater aquifer used for domestic purposes is approximately 200 m (657 ft) below the site. The depth to groundwater can be controlled using withdrawal wells within the source zone isolation system. The current depth to groundwater is on the order of 9 m (29 ft) bgs.

The NAPL is a mixture of chlorinated solvents, including trichloroethene (TCE), trichloroethane (TCA), tetrachloroethene (PCE), trichlorotrifluoroethane (Freon 113), and carbon tetrachloride (CTET). The bulk density of the NAPL is 1.4 g/cm^3, making it a DNAPL.

Current site characterization data indicate that the DNAPL is trapped in a north-south trending structural low in the Alpine Formation. These pools are generally in the bottom 2 to 3 m (6.6 to 9.9 ft) of the Provo Formation. There is no information on the presence of DNAPL in secondary features of the clay or on the presence of dissolved DNAPL constituents in the clay. However, TCE had been measured at shallow (0.03 m [0.1 ft]) depths into the clay at concentrations of as much as 1600 mg/kg. The particular location for the field demonstration was located in the southern end of the main channel, which is thought to contain the majority of the DNAPL at the site. Figure 17.1 shows a topographic map of the top of the Alpine clay and the location of the wells used for the demonstration.

Sieve analyses were performed on the soil borings to obtain information on variations in the hydraulic conductivity of the soils. Figures 17.2 and 17.3 show cross sections through the site with a summary of the stratigraphy. Chemical analyses were also performed on the soil borings to measure the contaminant concentrations in the soils. Contamination appeared to be confined to the bottom 1 m (3.3 ft) or less of the channel in the test zone. The "bottom" of the channel was approximately 1.8 m (6 ft) wide in the test area. Figure 17.4 shows the TCE concentrations vs. depth for wells in the test area. Concentrations of TCE ranged from below detection limits to 43,000 mg/kg in the soils. The highest concentrations were found just above the clay.

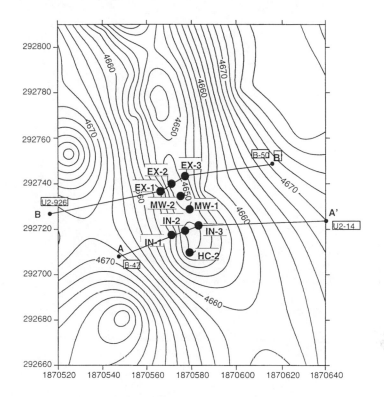

Figure 17.1 Topographic map of Alpine clay surface.

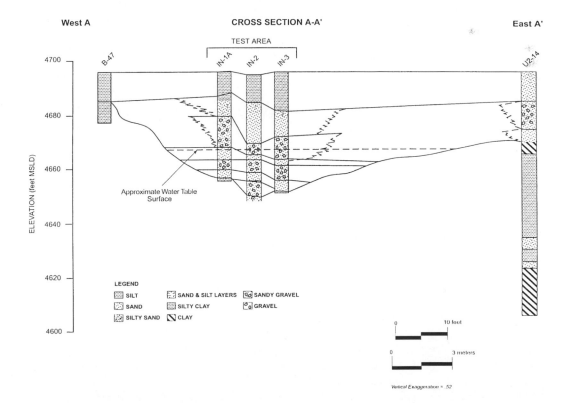

Figure 17.2 Geologic cross section A-A′.

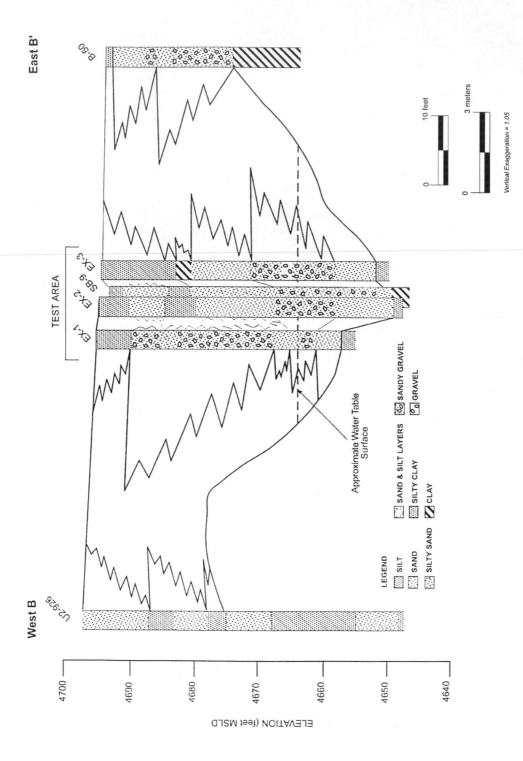

Figure 17.3 Geologic cross section B-B′.

Contamination appeared to be absent in the coarser-grained soils at a greater depth above the clay. It should be noted that TCE concentrations greater than approximately 500 mg/kg may be indicative of the presence of residual NAPL. Concentrations less than 500 mg/kg may be the result of dissolved or sorbed TCE. Concentrations greater than 40,000 or 50,000 mg/kg may be indicative of pooled NAPL.

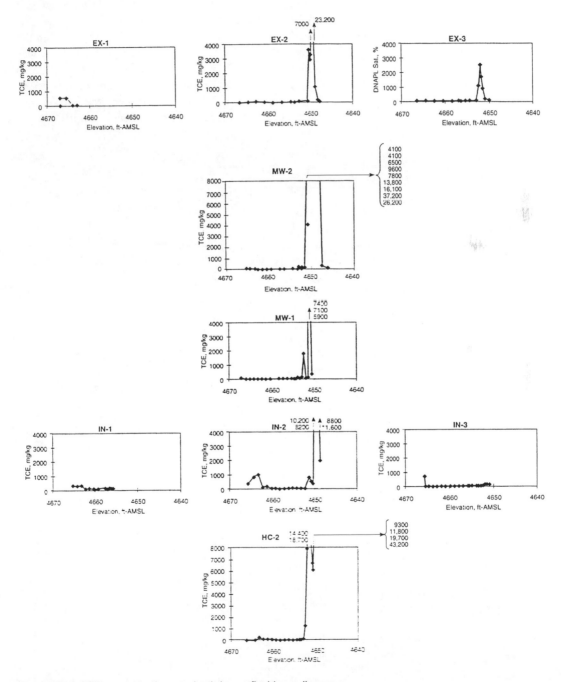

Figure 17.4 TCE concentration vs. depth in pre-flushing soil cores.

17.4 PROCEDURES AND FIELD IMPLEMENTATION

Details on the procedures and field implementation can be found in Part I. A summary of the demonstration is provided below. This summary was prepared before the final analysis of the data by the principal investigators.

Three injection, three extraction, two hydraulic control, and two monitoring wells were installed for the demonstration. Figure 17.5 shows the layout of the wells. The injection and extraction wells were arranged in a line-drive pattern, with the lines spaced 6 m (20 ft) apart. The wells within the line were spaced 2 m (6.6 ft) apart. Therefore, the total area contained between the wells was 6 m × 3.6 m (20 ft × 12 ft). Both the injection and extraction wells were screened over 1.5 m (5 ft) above the Alpine clay. Because the top of the clay was at different elevations, the bottoms of the wells differed in their elevations. Figure 17.6 illustrates the relationship between the elevation of the well screens and the top of the clay down the middle of the test area. Injection well IN-2 and

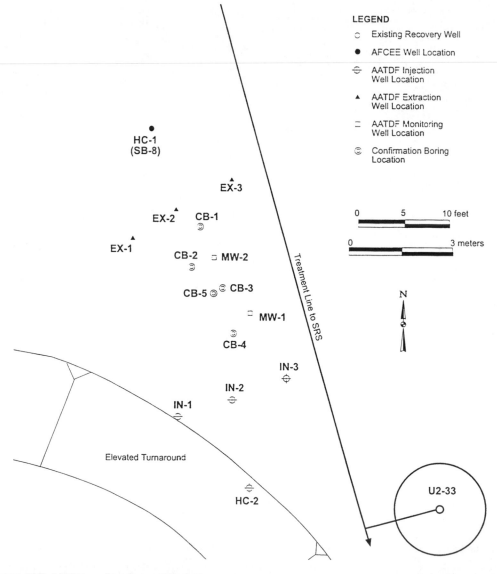

Figure 17.5 AATDF confirmation soil borings.

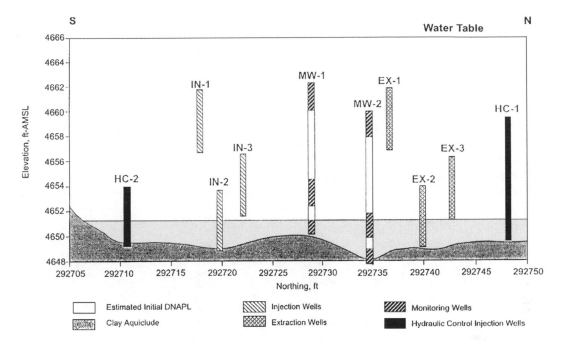

Figure 17.6 Cross section showing the distribution of DNAPL in the AATDF demonstration area.

extraction well EX-2 were the deepest wells because they were in the center of the channel. IN-1, IN-3, EX-1, and EX-3 were all above the estimated DNAPL zone because of the higher elevation of the Alpine clay at these locations. The two monitoring wells were located near the center of the channel at approximately one-third (1.8 m [6 ft] from injection) and two-thirds (3.6 m [12 ft] from injection) the distance between injection and extraction wells. The hydraulic control wells were located outside the pattern, about 10 ft from the central injection and extraction wells.

The fluids injected into the aquifer were prepared and staged in five mixing/storage tanks. Fluids mixed and stored in these tanks included water, water/tracer mixture, or a water/surfactant/salt mixture. Each tank was equipped with an electric heating element capable of maintaining the tank's contents at 15°C. Centrifugal pumps were used to recirculate the contents of the tank to keep them mixed. Chemicals (surfactant, salt, and tracers) were obtained as concentrates and mixed in the tanks. The fluids were delivered to the injection wells by a central centrifugal pump. The flow rate to each well was adjusted using a flow-control needle valve. The flow rate was recorded manually using an in-line digital flowmeter. Air was supplied for the process by a compressor. The air flow rates were measured with a gas mass flow meter.

The monitoring and sampling of the system included measuring fluid injection and extraction rates, sampling injected fluids, sampling and measuring fluids in monitoring wells, sampling extracted fluids, measuring gas injection rates, and measuring pressure in the injection well and at the air source. Fluid injection and extraction rates were recorded twice every 8 h. Transducers were used to monitor the fluid levels in the injection, extraction, and monitoring wells. A transducer was also used to measure the pressure in the air injection line. Grab samples were collected from a variety of points throughout the system. Aqueous samples from the extracted fluids and from the monitoring wells were collected automatically using an autocollector, which collected grab samples. The two monitoring wells inside the test zone were outfitted with multilevel samplers. The levels were set at 0 to 0.3 m (0 to 1 ft) above the Alpine clay (zone containing the most DNAPL), 1.2 to 1.8 m (4 to 6 ft) above the clay (potentially more permeable zone with little contamination), and 3 to 4 m (10 to 13 ft) above the clay (the top of the water table). A pneumatic pump was located at each depth and continuously withdrew fluids at a low rate.

Submersible bottom-loading low-shear pneumatic pumps were used to extract fluids from the extraction wells. The fluid discharged from each of the extraction wells was routed through a metering and sampling system, and then through a common effluent line. The produced fluids from the extraction wells were sent to an effluent sediment control system consisting of two settling tanks and an in-line filter. Most of the time, the produced fluids were routed through this system and then to the existing steam-stripper treatment system. Effluent from the steam stripper was pumped to the Hill AFB industrial wastewater treatment plant (IWTP). Certain produced fluids were diverted to a separate storage tank for use by the surfactant recovery and reuse demonstration (Harwell et al., 1999).

Table 17.1 summarizes the operational phases of operation, while Figure 17.7 summarizes the field activities. The injection rates were approximately 1 pore volume per day (PV/d) for most of the test for all fluids. Consequently, the duration of each solution flushing discussed below is approximately equal to the number of pore volumes flushed.

Table 17.1 Operational Phases of Surfactant/Foam Field Demonstration

Test Segment	Fluids Injected	Time of Injection (days)	Pore Volumes of Fluids (days)
Water injection	Water	1.77	1.7
Tracer injection	Water + tracer	0.52	0.5
Water injection	Water	4.73	4.5
Electrolyte injection	Water + NaCl	1.08	1.0
Surfactant/foam injection	Water + surfactant + NaCl + air	3.19	3.0
Electrolyte injection	Water + NaCl	0.49	0.5
Water injection	Water	9.5	9.5
Tracer injection	Water + tracer	1.0	1.0
Water injection	Water	11.25	11.25

The surfactant/foam field demonstration began with the injection of tap water for 1.8 d to establish a flow gradient in the aquifer. Table 17.2 summarizes the injection and extraction rates during the entire test. The flow rate to IN-2 and from EX-2 were intentionally greater than from the other wells in an attempt to focus more of the remediation effort on the deepest and most contaminated portion of the aquifer in the center of the channel. Water was also injected into the two hydraulic control wells (HC-1 and HC-2) throughout the entire test. An initial partitioning interwell tracer test (PITT) was performed after the initial water flood. A tracer solution was injected for half a day, followed by approximately 6 d of water (with NaCl for part of the time) injection to transport the tracer through the aquifer. The last day of the PITT, prior to the injection of the surfactant solution, a solution of 10,250 mg/L NaCl was flushed through the formation to raise the salinity of the formation. This process was performed to exchange ions off the soil that could interact with the surfactants.

Following the 1 PV NaCl, a solution of 4 wt% (target concentration) surfactant (MA-80I) with 10,250 mg/L NaCl was injected to begin the surfactant/foam flood. The surfactant/NaCl was injected for 3.19 d. The surfactant solution was injected for 8 h prior to starting air injection so that some of the aquifer could be swept with surfactant prior to generating *in situ* foam. Foam was then generated by simultaneously injecting the surfactant solution into each injection well while periodically injecting air. Air was injected into each well individually, for approximately 2 h. The air line was disconnected from the well receiving air and moved to the next injection well. Thus, air was injected into each well for 2 h, and then the air was off for 4 h. Air was injected at rates that varied, but were less than 2 scfm. The pressure of the air injection was typically about 10 pounds per square inch (psi), but was taken up to 17 psi during the end of the test. After 2 d of surfactant solution injection, a bromide tracer at 18,000 mg/L was injected for 1.5 h.

After completion of the surfactant injection period, injection of NaCl was continued at a reduced concentration of 8000 mg/L for 12 h to displace the surfactant with a salinity gradient rather than

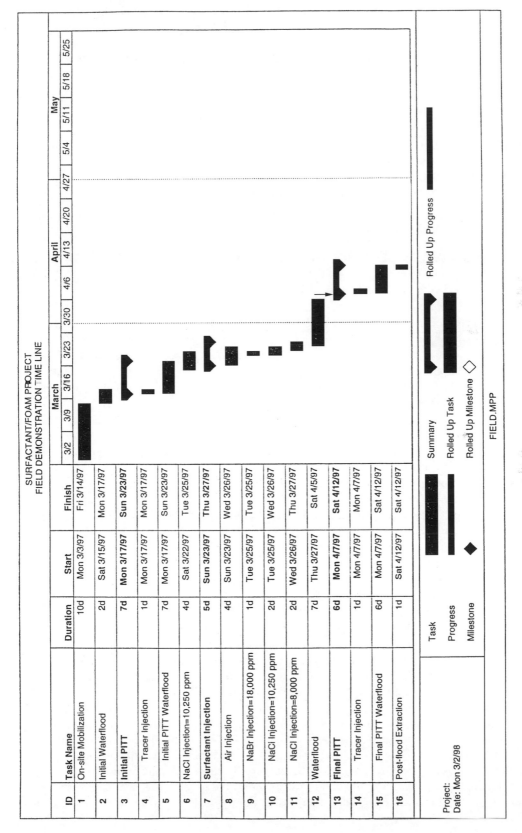

Figure 17.7 Surfactant/foam field demonstration timeline.

Table 17.2 Summary of Target Injection and Extraction Rates During the
Surfactant/Foam Flood Demonstration at Hill AFB OU2

Time (days)	IN-1 (gpm)	IN-2 (gpm)	IN-3 (gpm)	EX-1 (gpm)	EX-2 (gpm)	EX-3 (gpm)	HC-1 (gpm)	HC-2 (gpm)
0	2.0	2	2.0	3.3	3.3	3.3	3	1
9.3	1.5	3	1.5	2.7	4.0	3.3	2	2
11.3	1.5	3	1.5	2.7	4.0	3.3	3	1
19.44	2.0	2	2.0	2.7	4.0	3.3	3	1

fresh water. After the 12 h, water was flushed through the aquifer for another 9.5 d to transport the remaining surfactant solution across the test zone and to sweep the DNAPL out of the aquifer between the test wells.

The final PITT followed the 9.5 d of water flooding. Tracers were injected for 1.0 d and were followed by 11.25 d of water injection to transport the tracers across the aquifer.

17.5 SUMMARY OF RESULTS, MASS BALANCE, AND CONCLUSIONS

17.5.1 Tracer Tests

Conservative tracer tests were used to evaluate the flow paths and travel times of fluids in the aquifer. PITTs were used to estimate the mass of DNAPL present before and after the injection of the surfactant solutions. The volume of DNAPL estimated to be present based on the PITTs is discussed in Section 17.6.

The temporal first moments during the PITTs varied from 22 to 36 h for the three extraction wells. If these first moments can be taken as the average travel time of the fluids, it took approximately 1 d for the fluids to move through the aquifer. This is close to the 1 d estimated from the flow rates and volume of the aquifer. The volume of the aquifer swept by the fluids was calculated to be 31.0 m^3 (8180 gal) during the pre-surfactant PITT to 31.4 m^3 (8300 gal) during the post-surfactant PITT. The volume compares favorably with a volume of 31.8 m^3 (8400 gal) if the swept volume was the total area between the wells (4.3 m [14 ft] by 6 m [20 ft]) times a water depth of 5 m (16 ft) and assuming a porosity of 25%.

During the bromide tracer test, which was conducted after 2 d of surfactant/air injection, the temporal first moments dropped to 11.2 to 14.8 h, and the swept volume dropped to 15.9 m^3 (4200 gal). This result could be explained by the foam's filling approximately one-half of the pore spaces, thereby decreasing the effective swept volume.

17.5.2 Air Injection and Response

Figure 17.8 shows the response of the pressure transducer in the bottom of IN-2 (the center injection well), the air pressure in the air line (the same as the top of the well), and the air flow rates during air injection. This well was selected to illustrate the pressure response. The pressure at the base of the well was approximately 7.5 psi before injection of surfactant and approximately 7.8 psi (18 ft of water) after the start of surfactant injection. The response to the injection of air can be seen by an increase in pressure at the bottom of the well to almost 11 psi during the first period of air injection (at 2 a.m. on March 24, 1997). The air injection pressure was about 10.2 psi at this point. The air flow rate was as much as 2 scfm. The increased pressure in the well forced the liquids originally in the well out of the well until the air could escape through the well screen. Since the well screen is 1.5 m (5 ft) from the bottom of the well, a minimum of 4 m (13 ft) of liquids must have been displaced before air escaped into the formation. Since the pressure at the bottom of the well was always greater than that at the top of the well, some liquids must have

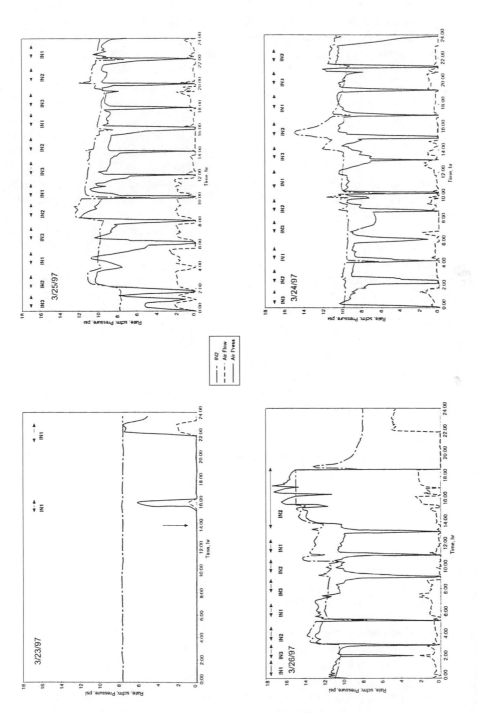

Figure 17.8 Foam injection pressure response.

always remained in the well. The depth of the liquids can be calculated from the difference between the bottom pressure and the pressure at the top of the well.

After the air was turned off, the air line was disconnected to the well so that the top of the well would have been at zero pressure. The pressure at the bottom of the well decreased slightly, but then came back to approximately 10 psi (7 m [23 ft] of water). This suggests that liquids quickly refilled the well bore and came to a level greater than the pre-air injection level by as much as 1.5 m (5 ft). The liquid level gradually declined during the 4 h the air was not being injected into that well.

It should be noted that air flow rates during the second half of March 24 and March 25, 1997, were never as high as the initial two injections. During the first injection period of March 26, 1997, the air was injected at a pressure of as much as 13 psi, but the flow rate only went up to about 0.8 scfm. During the last injection period, the pressure was taken up to as much as 17 psi. The air flow did increase to 2 scfm during this period. The greater pressure required to achieve the same air flow rate may have been caused by the blocking of preferential flow paths by the foam. After the last injection of air stopped at about 5 p.m. on March 26, 1997, the injection fluids were switched from surfactant with NaCl to NaCl alone. The pressure in the bottom of the wells (liquid levels) dropped off after about 3 h to slightly greater than the pre-air injection pressures (between 8 and 9 psi).

The increased liquid levels in the wells after each cycle of air injection compared with pre-air injection may be explained by the foam's filling some of the pore spaces and reducing the mobility of the surfactant solution. The foam is not stable, so the liquid levels dropped slowly while the air was off. They dropped even more rapidly when the injected fluids contained no surfactant, suggesting that the foam was even less stable after surfactant injection stopped. As the foam broke, the air probably escaped upward to the vadose zone as the surfactant solution reentered the soil pores. It is not likely that a significant amount of foam or air remained trapped in the soil pores since the swept volume measured before and after the surfactant flushed remained nearly unchanged.

Instantaneous liquid flow rate information is not available, so it is not clear that the surfactant flow rates remained constant during the increased pressure at the well receiving air. Since all three wells were tied into the same line from one centrifugal pump, it is quite possible that the surfactant flow rate decreased when the pressure in the well receiving air was high. Flow from the pump may have been diverted to the other wells with lower pressure.

To be effective, the foam would need to preferentially fill the pores of the more permeable soils so that the surfactant solution would be diverted to the less permeable soils (at the base of the aquifer in this case). The pressure responses do not provide any direct evidence of this preferential blocking, although the increase in pressure does suggest a decrease in the volume of the target being flushed.

A pressure response in the monitoring wells was not observed in the closest monitoring well, which was 1.8 m (6 ft) from the injection wells. For a typical air sparging system, where air is injected into an aquifer, a pressure response would be observed at this distance. However, it should be noted that the air injection rate in this study was much less than typically used for air sparging.

17.5.3 Monitoring Well Foam Formation

Foam was observed in both monitoring wells at all levels during the test. The times that the foam appeared in the wells since the start of air injection in the first well are presented in Table 17.3. Foam appeared in as little as 0.6 h in the upper portions of the monitoring well 1.8 m (6 ft) from the injection line, and 2.5 h in the well 3.6 m (12 ft) from injection. The time before foam appeared at the deeper depths was much greater. From this information, it appears that the foam or air traveled quite rapidly to the upper portions of the formation. This is partially a result of buoyancy-induced vertical rise of the air or foam.

It should be noted that it is difficult to distinguish between the movement of air (with the production of foam at the point measured) and the movement of foam produced near the injection

Table 17.3 Foam Breakthrough Times

MW Interval	Time to Breakthrough (h)
MW-1C (top)	0.6
MW-1B (middle)	0.6
MW-1A (bottom)	14.0
MW-2C (top)	2.5
MW-2B (middle)	8.0
MW-2A (bottom)	23.0

well. Because the observation of the foam in wells was much more rapid than the movement of liquid as measured in the tracer tests, it could be hypothesized that it was actually the air that was moving and not the foam in the upper formations of the closest monitoring well. The foam that was observed in the deeper monitoring points could have been the result of horizontal movement of foam, since air probably could have moved horizontally. The breakthrough times for the foam in the deep points are also similar to the expected liquid breakthrough times, further suggesting that the foam was possibly moving with the flow of liquids.

Figure 17.8 presents the foam injection pressure response. The foam collected in the upper portions of both monitoring points was very stiff, similar to shaving cream. The foam in the deepest monitoring point was never as stiff, but was a "froth" that would quickly break.

As mentioned above, to be effective, the foam would need to fill the pores of the more permeable zones to divert flow to the less permeable zones. The presence of foam in the monitoring wells does not provide any direct evidence of this occurrence, but it does suggest that the air or foam moved into the upper zones of the aquifer first. Since these are the more permeable zones at this site and the zones with the least contamination, the formation of foam in these zones is desirable. Thus, the observation of foam in the upper monitoring points before the deeper points provides indirect indication that the foam had a beneficial effect.

17.5.4 Produced Fluids Analysis

The composition of the fluids produced in the extraction wells and monitoring wells can be used to evaluate the performance of the system. The samples collected from the extraction wells were typically a single-phase sample with no evidence of free-phase DNAPL, discoloration, or the presence of foam. If any foam was produced at the wells, it may have dissipated because of the dilution with water and DNAPL from outside the test zone, and it may have risen to the surface before it could have been sampled.

Figures 17.9, 17.10, and 17.11 present the trends of TCE, surfactant, and chloride concentrations in the three extraction wells. The surfactant and chloride concentrations have been normalized to the injection concentrations of 3.6% and 9470 mg/L, respectively. The relatively low surfactant and chloride concentrations, compared with the injection concentrations, demonstrate the magnitude of the fluids pulled in from outside the test cell. For example, the normalized chloride concentration for EX-1 varied around 0.5 during the surfactant solution injection, suggesting that 50% of the flow was from outside of the test area. Chloride concentrations from EX-2, on the other hand, were on the order of 0.75, suggesting that less inflow occurred into the middle extraction well. The concentrations of chloride and surfactant dropped to near the background concentrations 2 d after surfactant injection stopped.

Figures 17.12, 17.13, and 17.14 present the cumulative productions of DNAPL, surfactant, and chloride. The DNAPL production was estimated from the TCE concentration, assuming the DNAPL at the site was 72% TCE. The total mass of surfactant produced was 3587 kg (7900 lb), which is in close agreement with the 3609 kg (7950 lb) estimated to be injected. Likewise, the mass of chloride produced (1387 kg [3056 lb]) was close to the amount estimated to be injected (1264 kg

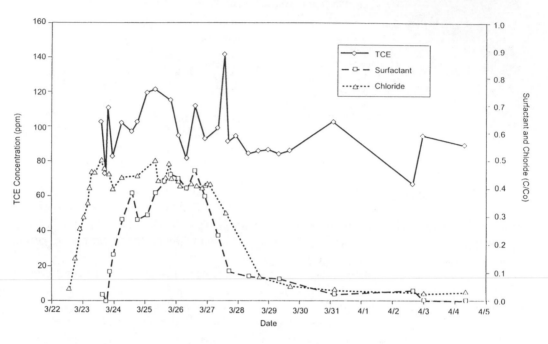

Figure 17.9 Concentration history of TCE, surfactant, and chloride for extraction well EX-1.

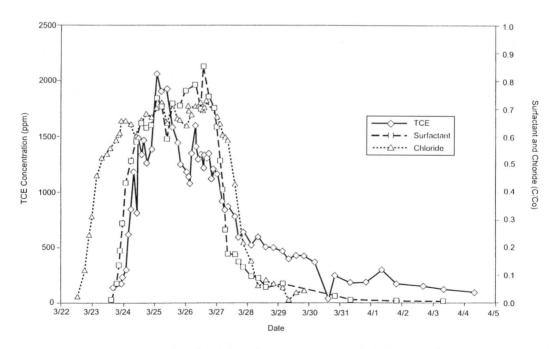

Figure 17.10 Concentration history of TCE, surfactant, and chloride for extraction well EX-2.

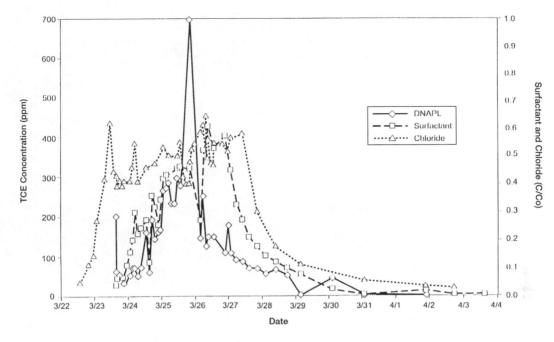

Figure 17.11 Concentration history of TCE, surfactant, and chloride for extraction well EX-3.

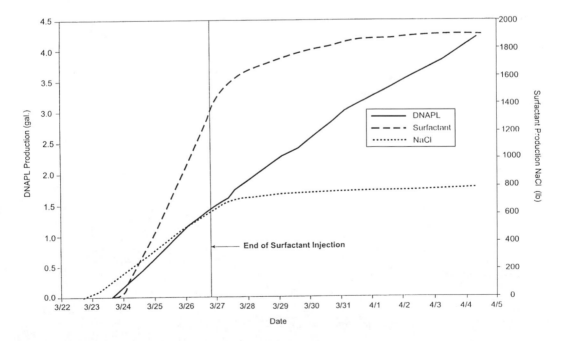

Figure 17.12 Cumulative production for extraction well EX-1.

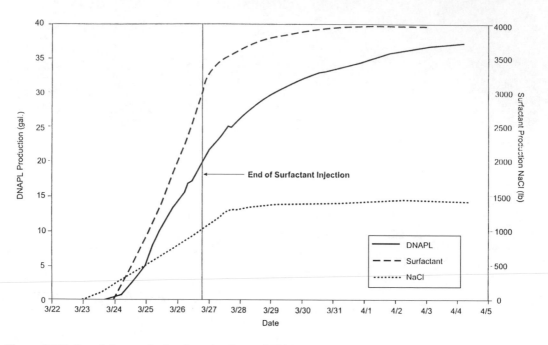

Figure 17.13 Cumulative production for extraction well EX-2.

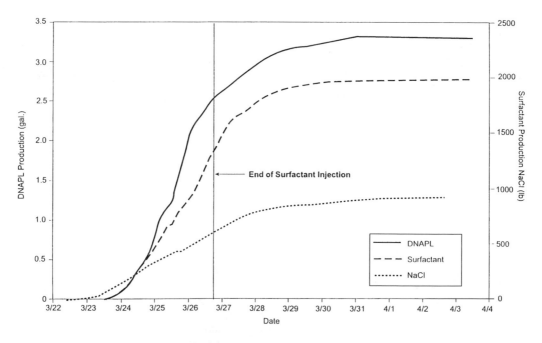

Figure 17.14 Cumulative production for extraction well EX-3.

[2785 lb]). Thus, it appears that virtually all of the injected chemicals were recovered. Little or none of the surfactant appeared to be retained in the aquifer.

The TCE concentration measured in EX-1 remained relatively unchanged throughout the entire test period (approximately 100 mg/L). This suggests that little or no residual or mobile DNAPL was in the flow paths that led to EX-1. However, there was enough DNAPL in the area to produce relatively high dissolved TCE concentrations. TCE concentrations in EX-3 did increase somewhat, suggesting that some DNAPL was present in the flow paths leading to it. TCE concentrations in EX-2 increased dramatically, to a high of 2000 mg/L during the surfactant injection period, suggesting that DNAPL was being contacted. The concentrations had not decreased at the end of the surfactant injection period, suggesting that additional DNAPL could have been removed if the operation had been continued.

From the cumulative production curves, it appears that a total of 169 L (44.7 gal) DNAPL was produced in aqueous form from the extraction wells. Another 10.6 L (2.8 gal) DNAPL was recovered from monitoring well MW-2A during the sampling so that the entire system produced 180 L (47.6 gal). By far, the majority of this was produced from EX-2. As can be seen from the 15 L (4 gal) DNAPL produced from EX-1, a fraction of the DNAPL produced could have been in a dissolved phase and would have been produced anyway.

17.5.5 Monitoring Well Fluid Analysis

As mentioned previously, a significant amount of foam was produced in the upper monitoring points. In many samples, there was not enough liquid in the automatically collected samples to perform an analysis.

The fluids from MW-1A (the deepest point in the well closest to the injection point) were generally single-phase liquid with no visible signs of DNAPL. TCE concentrations ranged from 1500 to 3500 mg/L during the surfactant injection period. This compares with a maximum of 2000 mg/L in the extraction wells. The lower concentrations of TCE in this well compared with MW-2A could have been a consequence of the lower soil DNAPL concentration compared with MW-2.

The samples collected from MW-2A (the deepest point in the well farthest from injection) typically had two distinct phases. However, a few samples had three phases and some a single phase. The phases can be described as a mobile DNAPL phase (which occurred only early during the surfactant flood), a microemulsion (high surfactant, high DNAPL content, and salinity equal to injected concentration), and an excess brine (low surfactant, salinity equal to injected concentration). The microemulsion contained as much as 80% DNAPL, and surfactant concentrations as high as 3 times the injected concentration. The chloride concentration in the microemulsion was the same as that injected. The surfactant concentration in the brine was about 25% of the injected concentration, while the TCE concentration was also much lower than the microemulsion, ranging from 1000 to 2000 mg/L. The microemulsion was typically the largest fraction of the three.

The three phases observed can be explained by classic phase behavior of surfactant solutions. Part I of this book provides a good description of the phase behavior. The sequence of phase behaviors typically observed is called Winsor Types I, III, and II:

- Winsor Type I: Characterized by a predominant water phase with surfactant in solution in the water phase. A pure DNAPL phase may also be present. DNAPLs may be solubilized in the micelles of the surfactant in the water phase.
- Winsor Type III: Three phases may be present: a pure DNAPL, a water phase, and a middle phase containing DNAPL, water, and surfactant micelles. This is typically considered an optimum solution for the purposes of mobilization in conjunction with solubilization.
- Winsor Type II: Characterized by a predominant DNAPL phase with surfactant in the DNAPL phase. This type of solution is considered overoptimum.

The very high DNAPL concentration in the microemulsion was higher than predicted, based on laboratory work for an optimum surfactant solution. Thus, it could be concluded that the microemulsion was overoptimum and characterized by a Winsor Type II system. One implication of the overoptimum system is that the microemulsions were denser than water or the optimum solution. This could be trapped in the topographic lows (depressions) in the formation. The overoptimum system is also likely to have higher interfacial tensions than the rest of the surfactant solution. The higher interfacial tension could result in trapping of the microemulsions in the soil pores and would require a higher pressure to force the microemulsion blobs through the soil pore throats and this was the approach taken in the latter part of the field demonstration. The fact that the nature of the solutions sampled from MW-2A did change over time suggests that the overoptimum microemulsions did move from the area around the monitoring well and that they were not trapped. The overoptimum fluids from around MW-2A, and possibly other depressions, may have made their way to the production wells where they would have been diluted with surrounding waters, changing the overoptimum conditions. No other evidence is available to indicate whether they were trapped.

The investigators provide evidence that the reason the system had a status of overoptimum around MW-2A was the presence of calcium ions in the solution (see Part I). Even low concentrations of calcium can cause an otherwise optimum system to become an overoptimum system with this surfactant. The calcium may have been produced *in situ* by cation exchange from the soil with sodium in the solution.

17.5.6 Soil Core Data

Soil cores were collected from all nine wells installed before the test and from five locations at the end of the demonstration. Figure 17.4 shows the results of the soil borings collected initially, while Figure 17.15 shows the results after surfactant flushing. The initial soil cores demonstrate the low concentrations present on the sides of the test area (IN-1, IN-2, EX-1, and EX-3). In the center of the test area, DNAPL was present, although the concentrations were not extremely high and were located in a very narrow band above the Alpine clay.

The post-surfactant flushing borings were all located in the deepest part of the channel in the Alpine clay, in the vicinity of MW-2. Figure 17.5 shows the location of the soil borings along with the wells. In general, the extent and concentration of TCE decreased substantially after flushing. As Figure 17.15 shows, TCE concentrations were below quantification limits in two of the five borings. Significantly higher concentrations were found at the clay interface (approximately 7.6 cm [3 in] above the clay), in the other three borings. Concentrations as high as 31,600 mg/kg were measured in one boring, while concentrations as high as 12,800 mg/kg were measured in two others. These concentrations are as high as those observed in the initial samples, and correspond to the presence of residual DNAPL.

The estimated DNAPL mass present initially and at the end of the study is very sensitive to the volume of soil assumed for each range of DNAPL concentrations. For example, if it were assumed that the DNAPL was confined to a 1.8-m wide (6-ft wide) trough at the bottom of the aquifer, the distance between the extraction and injection wells was 20 ft, the DNAPL was located in the bottom 0.5 m (1.6 ft), and the average concentration of TCE was 7800 mg/kg, the total volume of DNAPL would be approximately 77 L (20.5 gal). If, however, the average TCE concentration was 12,000 mg/kg, the total volume of DNAPL would be 119 L (31.5 gal). Given the soil core data available, this 77 to 119 L (20.5 to 31.5 gal) is a likely range of DNAPL initially present.

Pockets or pools of differing depths and DNAPL concentrations could drastically impact this initial volume estimate. This is especially true if the DNAPL is at high enough concentrations to be mobile (e.g., a pool 1.2 × 1.2 m [4 × 4 ft] by 0.3 m [1 ft] deep at 412,000 mg/kg TCE would add 10 L [2.6 gal] DNAPL). The presence of such a pool cannot be ruled out.

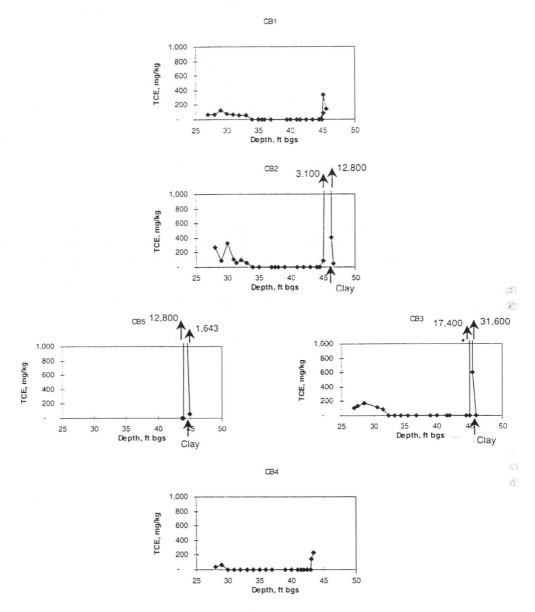

Figure 17.15 TCE concentration in post-flushing soil cores.

The post-surfactant flushing soil cores had DNAPL present over a much narrower band and at lower concentrations. If it is assumed that the average concentration was in the range of 7800 mg/kg TCE over a depth of approximately 0.08 m (0.25 ft), the volume of DNAPL would be 12 L (3.2 gal). If the average concentration was 3900 mg/kg TCE over 0.08 m (0.25 ft) of depth, the volume would be 6 L (1.6 gal).

The soil core data suggested that the initial DNAPL content was reduced from 77 to 119 L (20.5 to 31.5 gal) to 6 to 12 L (1.6 to 3.2 gal), corresponding to a mass removal of 84–95%. The investigators (see Part I) calculated their DNAPL reduction, based on soil core data, as 90.8 L (24 gal) to 10.6 L (2.8 gal), which yields a mass removal of 88%.

17.5.7   Partitioning Interwell Tracer Test Data

The PITT data can also be used to estimate the volume of DNAPL present initially and after the surfactant flushing. Details of the PITTs performed can be found in Section I.

The results of the initial PITT suggest that 79 L (20.9 gal) DNAPL were initially present in the test area. An error analysis was performed and suggests that this number is ±26.5 L (7 gal). This range of values is in general agreement with the soil core data. The post-flushing PITT data suggested that 9.8 L (2.6 gal) ± 7.56 L (2.0 gal) were present. This is also in the range estimated by the soil cores. Thus, approximately 69 L (18.3 gal) DNAPL appeared to be removed (88% of the initial amount). This is less than the 140 L (37 gal) produced from the extraction wells.

One potential problem with PITTs is that they measure only the volume of DNAPL in the volume of soil swept by the tracers. Both the initial and final PITTs estimated the swept volume at around 31,374 L (8,300 gal). This is very close to the pore volume (31,752 L [8,400 gal]) in a soil volume of 4.3 m × 6.1 m × 4.6 m deep (14 ft × 20 ft × 15 ft). Thus, the PITTs may have swept the majority of the aquifer. There is still the potential that the PITTs underestimated the mass of DNAPL present because of rate-limited mass transfer into regions of high DNAPL saturation.

17.6 CONCLUSIONS FROM THE RESULTS OF THE SURFACTANT/FOAM FIELD TEST

The points of greatest interest in assessing the performance of the surfactant/foam project are the contaminant mass removal achieved and the benefits of the foam formation. The ability of the technology to remove DNAPLs from depressions is a critical aspect of the latter point.

17.6.1   Contaminant Mass Removal

The surfactant/foam demonstration removed a large fraction of the DNAPL present. Mass removal was approximately 88%, based on PITT data and soil core data. It should also be noted that this mass removal was achieved in a relatively short time. The majority of the mass removal occurred with less than 3 PV flushed through the aquifer. For this demonstration, this volume was delivered in approximately 3 days. A longer period would be required for larger spacing or less permeable soils.

Estimates of the volume of DNAPL recovered based on analysis of produced fluids, analysis of soil cores, and partitioning tracer tests were 169 L (44.7 gal), 80.2 L (21.2 gal), and 69.2 L (18.3 gal), respectively. The variability of these results in part reflects uncertainties associated with each methodology. Project investigators (INTERA, 1997 and Part I) attribute production (produced fluids data) in excess of removal (soil cores and tracer tests) to be a result of migration of DNAPL to the target zone during demonstration.

The potential for migration of DNAPL into the test area also adds uncertainty as to the endpoints that were achieved during the demonstration. The majority of the post-surfactant flushing soil cores had TCE concentrations below method detection limits. However, there was a thin layer of relatively high concentration just above the clay interface. This DNAPL may have been missed by the surfactant solution since the elevations at some of these points may have been lower than the bottom of the extraction wells (i.e., the borings were from depressions). For example, the aquitard was 1.3 ft deeper at MW-2A than the bottom of the lowest extraction well, EX-2. Attempts were made to get surfactant to flow through these deeper points by increasing the air injection pressures at the end of the test. There was some evidence that MW-2A (the deep monitoring point) was impacted by the surfactant injection, production of foam, and microemulsions.

It is also possible that efforts to flush the DNAPL from the depressions were successful but that new DNAPL migrated into the depressions during the testing. As mentioned above, mobile DNAPL was observed in HC-2 at the end of the test. This DNAPL may have migrated, by gravity or under the increased pressure caused by the injection of water at this point. At IN-2, this DNAPL may have transferred into microemulsions. Some of these microemulsions may have moved to the extraction wells, and others may have accumulated in depressions in the aquitard. As depicted in Figure 17.6, IN-2 and MW-2 appear to be separated by a ridge (approximately 1 ft high) in the aquitard. The microemulsions would have had to fill up the depression around IN-2 before being able to flow by gravity to MW-2. Pressure from the injection well may have also helped to move this microemulsion. DNAPL from HC-2 would have had to move approximately 25 ft to enter MW-2. It does not seem likely that free-phase mobile DNAPL would have migrated this far since the quantity of it was limited, but microemulsions may have. Because the site was not physically or hydraulically isolated, the dissolved phase concentrations of TCE that can be achieved with this technology could not be assessed. The data regarding the achievable endpoint concentration are inconclusive because there was the potential for migration of DNAPL into the surfactant/foam test area.

17.6.2   Benefits of Foam

The benefits of air injection (foam production) over injection of surfactant alone can be evaluated based on pressure responses, foam production in the monitoring points, and the bromide tracer test conducted during the surfactant injection. As discussed above, the pressure response data showed an increase in pressure (head) in the injection wells after the injection of air. This suggests that the foam did fill some pore spaces and impeded the flow of the liquids. However, the pressure response data provide no direct evidence that the foam preferentially filled the more permeable air spaces, which is necessary for the foam to force surfactant into the less permeable soils.

Foam was produced rapidly in the upper monitoring wells. It is quite possible that this foam was formed at or near the monitoring wells as a result of the rapid propagation of air to that point. Foam was noted at much later times in the deeper monitoring points. The preferential vertical movement of air is expected in an air injection system like this. For this site, the vertical movement of air and the likely formation of foam in the upper sections of the aquifer are favorable because these are the more permeable and less contaminated sections of the aquifer. Thus, the observation of foam in the upper monitoring points provides evidence that the air was preferentially moving to the upper portions of the aquifer and producing foam.

The bromide tracer data provide evidence that the foam filled a significant fraction of the pore space, thereby decreasing the sweep volume of the aquifer. The bromide tracer showed that the sweep volume decreased by approximately 50% during the surfactant/foam injection, as compared with the sweep volume both before and after the surfactant/foam injection. These data, combined with the occurrence of foam in the upper monitoring points first, suggest that the upper portions of the aquifer were those that were filled with foam. This should have diverted the surfactant solution to the lower portions of the aquifer where the majority of the DNAPL was located.

Thus, taken together, this evidence suggests that the foam did have an impact because it reduced the sweep volume and possibly diverted surfactant to the lower portions of the aquifer. It can be concluded that the injection of air with the production of foam was a better process than surfactant injection alone and is, therefore, a promising addition to surfactant flushing.

Another positive aspect of the surfactant/foam technology is that the chemical solution is potentially amenable to treatment and recycling of the surfactant. The use of air stripping and a hollow fiber stripper are discussed in more detail in the companion monograph on the surfactant recovery/reuse project (Harwell et al., 1999).

CHAPTER 18

Measurement Procedures

A number of measurement procedures should be used to characterize the site before implementation of the technology. Specific procedures should also be used to monitor the system during operation, as well as after operation. Many of these procedures are discussed in Simpkin et al., 1999. Tables 18.1, 18.2, and 18.3 provide references for performing these analytical tests. The Federal Remediation Technologies Roundtable (FRTR, 1995) also provides a list of measurement procedures.

Table 18.1 Methods for NAPL Characterization

Parameter	Method Description	Method
Density	Hydrometer	ASTM D 1298
	Pycnometer	ASTM D 1480
Viscosity	Capillary (kinematic)	ASTM D445
	Rotational (Brookfield)	
Interfacial tension	Ring (DuNouy)	ASTM D971
	Pendant drop (drop weight)	ASTM D2285
	Spinning drop	Cayias et al., 1975
NAPL composition	Volatile organics	SW846-8270[a]
	Semivolatile organics	SW846-8080[a]
	Pesticides/PCBs	EPA418.1,[b] SW846-9071[a]
	Total petroleum hydrocarbons	SW846-8015M[a]
	Gasoline range organics	SW846-8015M[a]
	Diesel range organics	SW846-8260[a]

[a] U.S. EPA. Test Methods for Evaluating Solid Waste, SW-846, 3rd edition. November 1986, as amended.
[b] U.S. EPA. Methods for Chemical Analyses of Water and Wastes, EPA600/4-79/020. March 1983, as amended.

Table 18.2 Methods for Water Characterization

Parameter	Method Description	Method
Fluid properties		
Viscosity	Capillary (kinematic)	ASTM D445
Density	Pycnometer	ASTM D1480
Inorganic Chemistry		
Calcium	Atomic adsorption	APHA 301A II (APHA-AWWA-WPCF)
Magnesium	Atomic adsorption	APHA 301A II
Iron	Atomic adsorption	APHA 301A II
Potassium	Atomic adsorption	APHA 317A
Sodium	Atomic adsorption	APHA 320A
Barium	Atomic adsorption	APHA 301A II
Strontium	Atomic adsorption	APHA 321A
Chloride	Mercuric nitrate	APHA 408 B
Sulfate	Turbimetric precipitation	APHA 427C
Carbonate and bicarbonate	Alkaline titration	APHA 403
pH	Electrode	APHA 424
Conductivity	Electrode	APHA 205
Total dissolved solids	Gravimetric	APHA 208
Organic chemistry		
Volatile organics	Mass spectrometry	SW846-8260[a]
Semivolatile organics	Mass spectrometry	SW846-8270[a]
Pesticides/PCBs	Gas chromatography	SW846-8080[a]
Total petroleum hydrocarbons	Infrared spectrometry	EPA418.1[b]
Gasoline range organics	Gas chromatography	SW846-8015M[a]
Diesel range organics	Gas chromatography	SW846-8015M[a]

[a] U.S. EPA Test Methods for Evaluating Solid Waste, SW-846, 3rd edition. November 1986, as amended.
[b] U.S. EPA Methods for Chemical Analyses of Water and Wastes, EPA600/4-79/020. March 1983, as amended.

Table 18.3 Methods for Soil Characterization

Parameter	Method Description	Method
Physical properties		
Dry bulk density	Physical method	ASTM D2937
Grain size distribution	Physical method	ASTM D422
Permeability and porosity	Physical method	ASTM D5084
Wettability (capillary moisture retention)	Physical method	ASTM D3152
Total organic carbon	Infrared carbon dioxide	SW846-9060[a]
Displacement pressure	Physical method	Bear (1972), Corey (1986)
Cation exchange capacity	Chemical method	SW846-9081[a]
Organic contaminants		
Volatile organics	Mass spectrometry	SW846-8260[a]
Semivolatile organics	Mass spectrometry	SW846-8270[a]
Pesticides/PCBs	Gas chromatography	SW846-8080[a]
Total petroleum hydrocarbons	Infrared spectrometry	EPA418.1,[b] SW846-9071[a]
Gasoline range organics	Gas chromatography	SW846-8015M[a]
Diesel range organics	Gas chromatography	SW846-8015M[a]

[a] U.S. EPA Test Methods for Evaluating Solid Waste, SW-846, 3rd edition. November 1986, as amended.
[b] U.S. Methods for Chemical Analyses of Water and Wastes, EPA600/4-79/020. March 1983, as amended.

CHAPTER 19

Engineering Design

To illustrate the engineering design components that may be required for full-scale application of the surfactant/foam technology, a conceptual design for a hypothetical site using the surfactant/foam technology was prepared. Other designs are possible and may prove to be more cost effective. This conceptual design also includes the air stripping technology that was tested for the Surfactant Recovery/Reuse Project (Harwell et al., 1999). Consequently, it also serves as a conceptual design for that technology. Cost estimates that were prepared using these conceptual designs are discussed in Chapter 20.

The conceptual design was prepared for a "hypothetical" site that was modeled after an actual site — Hill AFB OU2. This model site was selected because it was the location of the field demonstration conducted for this project (as previously discussed). As a result, information is available on the performance of the technology at the site. In the preparation of the conceptual design, OU2 was modeled as closely as possible. However, complete information is currently not available on the application of this technology at OU2, so some assumptions had to be made. A key assumption concerned the extent of the site that has mobile or residual NAPL that would require remediation using the surfactant/foam technology. The actual site area that might require remediation may deviate from that assumed here, so this conceptual design and the cost estimate should not be considered an estimate of what the remediation would consist of or cost for OU2. Further site investigations may be necessary, as well as a site-specific design and cost estimates.

Two conceptual designs and cost estimates were prepared for this site. The first design, the Base Case, is a direct scale-up of the field test. For example, the volume and concentration of fluids injected were kept the same as those used in the field demonstration. Successful application of the technology may be possible with lower volumes and concentrations of chemicals, thereby resulting in potentially lower costs. A conceptual design for an Optimized Case was also prepared to evaluate the potential costs under more optimum conditions. A number of assumptions were made on methods to optimize the system that may reduce the cost. The costs for the Base Case and the Optimized Case may be on the extremes of the cost scale, with the actual costs for many sites lying somewhere between.

19.1 BASE CASE

Table 19.1 summarizes the conceptual design of the surfactant/foam system coupled with surfactant recovery and reuse, and Figure 19.1 is a process flow diagram of the system. Other design material is included in the appendices, as follows:

- Appendix 4: Drawings
  - Site plan
  - Well detail
  - Process instrumentation and control diagrams

- Appendix 5: Mass balance and design spreadsheets
  - Base Case design worksheet
  - Base Case mass balance
  - Optimized Case design worksheet
  - Optimized Case mass balance
- Appendix 6: Cost estimates
  - Base Case additional capital costs
  - Base Case O&M cost
  - Base Case cost summary
  - Optimized Case additional capital costs
  - Optimized Case O&M cost
  - Optimized Case cost summary
  - Capital cost detail

Table 19.1 Site Characteristics and Base Case System Design Summary

Site Characteristics

Type of site	Disposal site for waste chlorinated solvents. The site is currently inactive.
Contaminants	TCE, TCA, and PCE present in a multicomponent DNAPL.
Size of target area	0.41 hectare (1 acre).
Hydrogeologic setting	Granular alluvium material consisting of mostly sand with some silt and clay, and occasional stringers of gravel. Relatively homogeneous 12-m thick. Underlying clay that provides barrier to vertical migration. Surrounded by containment wall.
Target depth	DNAPL zone over the bottom 1 to 2 m of granular alluvium.

System Design

Remedial objective	95% removal of the recoverable contaminants.
Primary performance measures	Measurement of contaminants (TCE) in the produced fluids. The total mass of recoverable contaminant present will be based on projection of production curves.
Removal mechanism utilized	Solubilization with mobilization using surfactant/foam and NaCl. Foam produced *in situ* by injection of air to improve sweep efficiency.
Chemical system	4 wt% Aerosol MA-80I (sodium dihexyl sulfosuccinate); 11,000 mg/L NaCl.
Delivery recovery system	Vertical injection and recovery wells arranged in an offset line-drive pattern with 6 m (20 ft) spacing between wells in a line and 12.2 m (40 ft) spacing between lines. Four injection lines with seven wells each, and five recovery lines with eight wells each. Total length of time to complete is 1 year.
Chemical delivery sequence	• 10 PV of water flooding over 42 days. • 2.5 PV of surfactant. Injection over 21 days for each of 34 modules. Air injected at 4 scfm per well. Air injected for 2 h to each well and off for 4 h. • 10 PV of post water flooding over 42 days.
Chemical mix system	Chemicals delivered in concentrated form, diluted on site, and mixed in two solution mix and storage tanks. • 187 metric tons (205 tons) MA-80I total. • 201 metric tons (221 tons) NaCl total.
Produced fluids management	Removal of free DNAPL in an oil/water separator, separation of surfactant from contaminants in an air stripper, treatment of off-gas by catalytic oxidation, ultrafiltration to reconcentrate the surfactant, recycling most of the fluid, and discharging the bleed stream to a WWTP.
Additional studies included in costs	Additional site characterization, bench-scale testing, numerical simulations, field demonstrations in a sheet-pile-enclosed cell, partitioning tracer tests. (See Appendix 6, Additional Capital Cost sheet for details.)

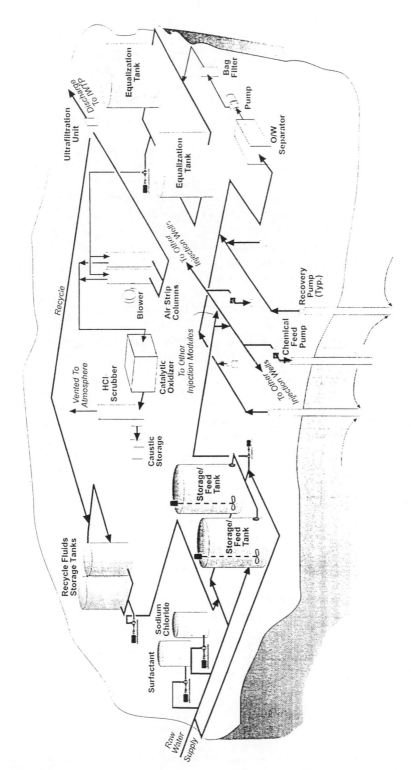

Figure 19.1 Process flow diagram for Base Case design of surfactant/foam remediation.

19.1.1 Site Setting

The site is the location of former chlorinated solvent waste disposal pits. A system for removal of the mobile DNAPL had been in operation and removed 113,400 L (30,000 gal) DNAPL. The primary ongoing site activity is operation of the source zone isolation system, which consists of a 300-m long (986.8-ft) vertical cutoff wall and a groundwater extraction system that sustains an inward and upward hydraulic gradient. As fas as is known, no other activities at the site will conflict with site remediation.

The site contains 12 m (39.5 ft) of granular alluvium, consisting of mostly sand with some silt and clay and occasional stringers of gravel. The hydraulic conductivity and porosity of the granular alluvium are estimated to be 10^{-2} cm/s and 0.27, respectively. The depth to groundwater can be controlled by pumping and is on the order of 9 m (29.6 ft) bgs. Underlying the granular alluvium is a 60-m thick (197.4-ft) lacustrian clay formation, which contains minor, discontinuous, inter-bedded sand seams. The bulk hydraulic conductivity of the lacustrian clay is estimated to be 10^{-5} to 10^{-8} cm/s. The shallowest regional groundwater aquifer used for domestic purposes is approximately 200 m (657.9 ft) below ground surface.

The NAPL is a mixture of chlorinated solvents, including TCE, TCA, PCE, Freon 113, and CTET. The density of the NAPL is 1.4 g/cm³, making it a DNAPL. The viscosity, DNAPL-water interfacial tension, and exact DNAPL composition are not known.

Current site characterization data indicate that the DNAPL is trapped in a north-south trending structural low in the clay. The major pools of mobile DNAPL present at the site prior to DNAPL extraction are shown in Figure A5.2. It can be inferred from high dissolved TCE concentrations that residual DNAPL may exist in an area larger than the pooled DNAPL. The area, which is the target DNAPL zone assumed for this example, is within the well array shown in Figure A5.2. This area covers 0.41 hectare (1 acre).

The residual DNAPL is generally in the bottom 2 to 3 m (6.6 to 9.9 ft) of the granular alluvium. There is no information on the presence of DNAPL in secondary features of the clay or on the presence of dissolved DNAPL constituents in the clay. Although little information is available at this time on actual upward gradients caused by operation of the source zone isolation system, it will be assumed for this example that upward gradients exist and, therefore, that downward mobilization of DNAPL is not a major concern.

19.1.2 Remedial Objective

For the purpose of the Base Case design it was assumed that a Record of Decision for this site will be written with the following elements. The surfactant flushing system will be operated to a point that 95% of the recoverable contaminants are removed. The mass of recoverable contaminants present will be based on extrapolation of production curves during the operation of the system. TCE will be used to judge the extent of recovery because it makes up the bulk of the DNAPL.

It should be noted that 95% of the recoverable contaminants criteria is very different from removing 95% of the mass of contaminant present. The 95% of the recoverable contaminants criteria takes into account the fact that some of the mass present will not be recoverable.

19.1.3 Up-front Activities

It was assumed for the Base Case that a number of up-front activities had not been completed for this site and needed to be included in the costs. These activities were:

- Additional site characterization: it was assumed that although this site had been relatively well characterized, 20 additional soil borings will be necessary to verify the locations of the wells, evaluate the extent of NAPL, and collect soils and NAPL samples for laboratory testing. Including site characterization costs with the costs of the remediation may lead to a higher unit cost for the surfactant/foam technology as compared to other remediation technologies where these costs may not have been included in their estimate.

- Laboratory testing: laboratory testing consists of component analysis of the NAPL, interfacial tension screening tests, solubilization tests, phase behavior tests, testing of the interactions of the solution with the site soils, a series of columns tests to evaluate the potential effectiveness of the system, and a series of produced fluids treatability tests (ultrafiltration and stripping).
- Numerical simulations: numerical simulations will be performed to aid the design and to predict the performance of the system.
- Field Demonstrations: The field demonstration for this site will include the installation of a sheet pile wall ($5 \times 5 \times 12$ m [$16.4 \times 16.4 \times 36$ ft]) for containment, four injection wells, three extraction wells, and four nested monitoring points. A sheet pile enclosure for a pilot-scale test may not be necessary. it was not required for the surfactant/foam field demonstration, which saved the project considerable expense. Two soil samples will be collected from each monitoring and operations well and analyzed for contaminants of concern. Fluids will be monitored at a frequency of approximately 3 times that suggested for the full-scale system. Pilot tests of the produced fluids treatment system will be conducted. The bulk of the produced fluids will be concentrated by ultrafiltration and sent off-site for incineration. The permeate will be sent to the industrial wastewater treatment plant (IWTP).

A detailed design of the facility will not be required. Because the size of this site is moderate, it can be assumed that a design/construct/operate turnkey type of approach would be used to deliver the project.

19.1.4 Chemical System and Injection Sequence

The chemical systems and sequence used in this conceptual design are identical to that used for the surfactant/foam field demonstration. The surfactant chemical system is made of 4 wt% surfactant (Aerosol MA-80I, sodium dihexyl sulfosuccinate) and 11,000 mg/L NaCl. This system reduces DNAPL/water interfacial tension and enhances the effective solubility of the DNAPL constituents, resulting in the potential to recover DNAPL through both displacement and solubilization processes. Furthermore, periodic *in situ* injection of air produces a high-viscosity, *in situ* foam that improves solution sweep efficiency by plugging higher permeability layers.

The following sequence of injection will be used:

- Water flooding (10 PV water) for mobile DNAPL recovery. Anticipated benefits of water flooding include reduced overall produced fluids treatment costs and reduced overall chemical requirements.
- Electrolyte flood consisting of 11,000 mg/L NaCl for 1 PV.
- 2.5 PV of the surfactant/NaCl solution. Air will be intermittently injected at the delivery wells during this time to generate and drive surfactant/foam to the recovery systems.
- 10 PV water flood to displace the surfactant/NaCl solution and any remaining recoverable DNAPL and dissolved DNAPL constituents.

19.1.5 Injection/Recovery Approach

Fluids will be injected into and recovered from the DNAPL zone using vertical, 4-in., low-carbon-steel wells. Injection and delivery wells will be constructed in a line-drive pattern as shown in Figure A5.2. One injection line and two recovery lines will make up one treatment module. Adjacent modules will share a recovery line. There will be a total of four modules. Spacing between lines of delivery and recovery wells will be 12.2 m (40.1 ft). This spacing is based on success in driving foam a distance of 6 m (19.7 ft) during the 1997 field demonstration and estimates of what should be achievable on a larger scale. Spacing between wells in each line will be 6 m (19.7 ft) to minimize stagnation zones between wells and to maximize total system flow rates.

The initial water flooding will be conducted in all four modules at the same time. Water produced during the water flood will be passed through an oil/water separator and then reinjected. Injection of the surfactant/foam will be performed in each module individually. Sequential delivery to the modules will allow for reuse of the chemicals. During the injection of the surfactant, air will be periodically injected into the wells to generate an *in situ* surfactant/foam. Air will be injected at a frequency of 2 h on, 4 h off, with an air flow rate of 2 scfm per well and a pressure of 20 psi. The final water flood will be performed on each module individually.

Flow calculations for this system are presented in the attached design spreadsheet (see Appendix 5). The flow calculations assumed uniform, steady-state flows in a homogeneous, isotropic aquifer. This assumption may not be accurate, but it is suitable for designing the system. The formation at this site is capable of transporting water at a very high rate so that the chemical flushing could theoretically be completed in a very short time. However, the rate of injection will be intentionally limited to reduce the size of the chemical feed and produced fluids handling systems and to prevent the operations from occurring too rapidly. Sufficient time is desired during the operations to allow collection of information and modification of operations. The flow rates were set so that the surfactant flushing for each module will be completed in 21 days. Completion of all four modules and the initial pre- and post-water floods will require approximately 1 year.

19.1.6 Chemical Mix System

The surfactant will be purchased and delivered as a concentrated liquid and stored on-site. The NaCl will be delivered in a dry form in bags. A concentrated NaCl solution will be prepared and stored in a storage tank. All of the concentrated chemical tanks will be provided with mixers. The concentrated chemicals will be diluted, mixed, and stored in two feed/storage tanks. It is assumed that the surfactant system has been designed for site tap water and that the tap water has reasonably low hardness so that it can be used without softening. The surfactant will be recycled after air stripping, but it will be assumed that 22% of the surfactant is either lost in the formation, or comes out late in the post-water flush so that it cannot be economically recycled.

The surfactant solution will be pumped from the storage tanks through a bag filter to the main injection header by two variable-speed pumps. The bag filter will remove any particulates that may be in the fluids. Chemical metering pumps will be used to pump and meter the fluids from the main header to the injection wells. The pumps will be adjusted manually to maintain a constant head (water level) in the injection wells. Using the metering pumps will allow the flow to be varied between wells as necessary to optimize operations. During water flooding, the metering pumps will be bypassed.

The pumping systems of the chemical makeup system will be constructed within a temporary building. The storage tanks will be located outside, and will be heated and insulated to prevent freezing in the winter. The chemical makeup system will be enclosed by a secondary containment system consisting of a lining and earthen berms. The surfactant piping outside the secondary containment berms will be aboveground and will be heat traced and insulated. Because the piping will be aboveground, any leaks will be visible so that double-walled pipes will not be necessary.

19.1.7 Produced Fluids Handling

As with all extractive remediation technologies, the produced fluids from a surfactant/foam process will require treatment to remove the contaminated fraction. Recovered fluids will be treated using packed-column air strippers to remove the chlorinated solvents from the surfactant solution. Ultrafiltration will be used to reconcentrate the surfactant. The bulk of the treated fluids will be reused in the process. A bleed stream will be sent to an IWTP located at the facility.

The fluids will be recovered from the system using individual pumps at each recovery well. The recovery rate will be 1.1 times the injection rate to contain the fluids. This relatively low recovery rate compared to the injection rate can be assumed because the site is isolated within a containment wall. From the recovery wells, the produced fluids will be piped in aboveground, heat-traced insulated pipes to a central handling area. This area will be within secondary containment, and all major equipment will be within a temporary building.

The produced fluids will be stored in two insulated equalization tanks. The fluids will then be treated in a packed-column air stripper. The off-gas from the air stripper will be treated in a catalytic oxidation system with an HCl scrubber. The effluent from the air stripper will be passed through an ultrafiltration unit to reconcentrate the surfactant before recycling. This process will reduce the amount of surfactant used, as well as the amount of activated carbon consumed at the IWTP. The water from the ultrafilter will be pumped to the IWTP. The treated effluent to be reused will be stored in two tanks, capable of storing approximately one-third of a pore volume. This storage capacity should allow some flexibility in the operation from one module to the next. The bleed stream will be piped to the IWTP.

The final pore volume of surfactant that cannot be reused will be concentrated in the ultrafiltration unit by an assumed 50%. It was assumed that this material must be sent off-site for incineration. The water passing through the ultrafilter will be sent to the IWTP.

19.1.8 Operations, Maintenance, and Monitoring

Table 19.2 summarizes the labor required. Some of the staffing level is driven by the Occupational Safety and Health Administration (OSHA) requirement to have two people on-site at all times. The operators will keep all of the systems running and will monitor functions. The maintenance technician will be required to keep the equipment running. It was assumed that a significant amount of engineering time will be required to monitor the system, evaluate the operating data, suggest changes to the operations, and write reports because the surfactant/foam process is a relatively new technology, the operating period is short, and decisions must be made quickly. It was also assumed that a laboratory technician will be hired to perform much of the operational laboratory analysis.

Table 19.2 Labor Requirements

Staff Description	Day Shift	Swing Shift	Night Shift	Total Staff	Total Hours per Month (h)
Operator[a]	1	2	1	4	224
Supervisor[a]	1			1	56
Lab Technician[a]			1	1	56
Maintenance Technician	0.5			0.5	20
Engineering	1.5			1.5	60
Total	4	2	2	8	416

[a] 7 days per week (56 hours per week).

The utility costs for this operation include the costs to run the pumps, blowers, and mixers. The bleed water from the air stripper will be sent to the IWTP. It is assumed that the IWTP charges $2.91/m^3 ($11 per 1000 gal). The IWTP does not have a surcharge for chemical oxygen demand (COD), but it was assumed that the project will pay for the amount of activated carbon that is consumed as a consequence of the discharge. Consequently, costs were included for replacement of activated carbon, assuming a carbon weight pickup of 10% and a cost for disposal and replacement of carbon of $1.50/lb.

The laboratory analyses required for both operational and performance monitoring were assumed to be relatively substantial. Since the operating period of each module is so short, monitoring must be performed frequently so that corrective actions can be taken before the operating period is over. Many of the measurements can be performed on-site by the laboratory technician, but it was assumed that analysis requiring gas chromatography (GC) or gas chromatography-mass spectrometry (GC/MS) instruments would be sent to a fixed-base laboratory. The analyses are summarized in Table 19.3.

Table 19.3 Analytical and Modeling Requirements

Parameter	Fluid to be Analyzed:				
	Injected Fluids	Produced Fluids	Air Stripper Effluent	Treated Produced Fluids	Monitoring Well Fluids
Viscosity	Used for quality control (3 times/week)	Used for quality control (3 times/week)			Used for quality control (3 times/week)
Visual observation of NAPL and phase behavior		Indicator of performance (3 times/day)			Indicator of performance (3 times/day)
Interfacial tension (NAPL to water)	Useful for understanding performance (7 times/week)	Useful for understanding performance (e.g., dramatic increase suggests chemical changes) (7 times/week)			Useful for understanding performance (7 times/week)
Surfactant concentration	Quality control (5 times/week)	Indicator of surfactant breakthrough (3 times/week)			Indicator of surfactant breakthrough (3 times/week)
Chloride concentration	Quality control (5 times/week)	Indicator of surfactant breakthrough (3 times/week)			Indicator of surfactant breakthrough (3 times/week)
COD		Indicator of surfactant breakthrough (7 times/week)	Indicator of treatment effectiveness (5 times/week)	Indicator of treatment effectiveness (5 times/week)	Indicator of the breakthrough of injected chemicals (1 time/week)
Volatile organics		Indicator of performance (7 times/week)	Indicator of treatment effectiveness (7 times/week)	Indicator of treatment effectiveness (7 times/week)	Indicator of performance (7 times/week)

Other monitoring activities include monitoring water levels in the formation and in all tanks. Monitoring wells will be installed at the locations shown on the site plan. Each well will be set up to collect samples from three discrete depths. These locations were selected to provide information sufficient for operations over most of the site, plus additional information on the performance of the system in critical areas of highest contamination. The monitoring wells will be standard 2-in. PVC wells.

As discussed previously, the production of the contaminants of concern in the produced fluids will be the primary method of judging the success of the system. The cumulative production of the contaminants will be plotted with time or pore volume produced. A projection of the maximum recoverable quantity of the contaminants will be made by extrapolating the production curves to infinite time. The system will be considered complete when 95% of the maximum recoverable is reached. Only a limited number of soil cores (approximately 10) will be collected at the completion of the remediation to verify the performance of the system. The number will be limited to reduce cost, since soil cores will not be the primary factor used to judge success. Partitioning tracer tests will also be performed to verify the performance of the system.

19.1.9 Miscellaneous Cost Components

Site preparation includes construction of a secondary containment system, construction of the buildings, installation of utilities to the site (i.e., water, electrical, and sewer), and installation of site facilities, including office trailers and decontamination facilities.

19.2 OPTIMIZED CASE

The conceptual design discussed above was prepared for the surfactant/foam and surfactant recovery/reuse technologies as a scale-up of the completed field demonstrations. A number of optimizations could be made to the technologies to reduce their cost and to improve their effectiveness. Many of these are currently under investigation by the developer of the technology. In addition, less monitoring will be needed as more experience is gained with the technology. The following is a list of optimizations that were assumed for the Optimized Case:

- The number of pore volumes (PV) of flushing required is as follows:
  - 5 PV tap water
  - 1.5 PV surfactant and alcohol
  - 5 PV tap water
- The up-front activities discussed above (Chapter 19.1.3) have been conducted by others, and thus do not need to be accounted for here.
- The air:water ratio of the air stripper can be reduced to 20:1.
- The oil/water separator removes 40 wt% of the DNAPL recovered.

In addition, it was assumed for the Optimized Case that the goal will be operation of the system until 70% of the recoverable contaminants are removed.

CHAPTER 20

Cost and Economic Analysis

This chapter presents cost estimates developed for the conceptual designs presented in Chapter 19. An economic analysis is also presented; it looks at the factors that influence the cost and ways the cost for full-scale applications of the technology could be reduced. To facilitate this analysis, a cost estimate for the Base Case application of the technology was prepared, along with a cost estimate for the Optimized Case.

The capital cost estimates were developed using an estimating program called Hard Dollar Estimating Office System (EOS). The capital cost estimate was developed by entering subcontractor and equipment vendor quotes, combined with detailed estimated costs for labor to install the system. Operations costs were based on assumptions or calculations of operating requirements (see Chapter 19) and common unit costs for labor and materials. Costs for chemicals and disposal were obtained from chemical suppliers and a hazardous waste disposal facility.

It should be noted that the cost estimates developed here are at what would typically be considered an order-of-magnitude level. Estimates of this type are generally accurate within +50% or –30%. The accuracy of the estimate is subject to substantial variation because the details of the specific designs will not be known until the projects are actually implemented. Factors that may impact the cost include site conditions, final project scope and schedule, design details, the bidding climate and other competitive market conditions, changes during construction and operations, productivity, interest rates, labor and equipment rates, tax effects, and other variables. As a result, actual costs will likely vary from these estimates. Other designs for a full-scale implementation of a surfactant/foam process will also yield different estimates. Additional cost information for surfactant-enhanced remediation can be gained from a full-scale implementation scheduled at Hill AFB OU2 and other locations in the U.S.

Table 20.1 summarizes the cost estimate for the Base Case. The costs in this table are the present-worth costs for the project. Appendix 6 presents more details of the cost estimate. The estimate summary is organized according to the Work Breakdown Structure (WBS) suggested by the Federal Remediation Technologies Roundtable (1995). This WBS is designed to facilitate comparison between remediation technologies. Not all of this WBS matched the way the conceptual design and cost estimate were prepared; thus, certain costs had to be subdivided to fit into the WBS categories. For example, the conceptual design for the system operations was developed with labor requirements and utilities for the entire project. These labor and utilities costs were divided equally between the "groundwater collection and treatment" (which includes the operation of the chemical injection and pumping system) and "physical/chemical treatment" (which includes the operation of produced fluids treatment system) WBS categories for the summary tables.

20.1 MOST SIGNIFICANT COST COMPONENTS

The total cost for the Base Case application of the surfactant/foam technology for this conceptual design is $7,523,000. Of this total, just under 50% is for operation of the system. Costs for disposal

Table 20.1 Present Worth Costs of Base Case Design

WBS No.	Standard Description	Cost	Subtotal	% of Total	
321	Pre-Construction and Project Management Activities		$1,112,000		14.8
32190	Additional Site Characterization and Laboratory Testing	$284,000		3.8	
32191	Numerical Simulations	$24,000		0.3	
32192	Field Demonstration	$580,000		7.7	
32193	Design	$224,000		3.0	
331	Remedial Action (Construction)		$2,244,000		29.8
33101	Mobilization and Preparatory Work	$97,000		1.3	
33103	Site Work	$25,000		0.3	
33106	Groundwater Collection and Control	$669,000		9.2	
33113	Physical Treatment	$1,893,000		25.2	
33121	Demobilization	$15,000		0.2	
33190	Salvage Value of Equipment	(455,000)		6.1	
332	Engineering During Construction	$100,000	$100,000	2.4	2.4
333	Construction Management	$399,000	$399,000	5.3	5.3
342	Operations and Maintenance (Post-Construction)		$3,588,000		47.7
34202	Monitoring, Sampling, Testing, and Analysis	$326,000		4.3	
34203	Site Work	$121,000		1.6	
34206	Groundwater Collection and Control	$456,000		6.1	
34213	Physical Treatment	$456,000		6.1	
34218	Disposal (other than commercial)	$1,813,000		24.1	
34290	Chemicals	$363,000		4.8	
33120	Site Restoration	$54,000		0.7	
	Total Project Cost		$7,523,000	100.0	100.0

Note: These estimates are considered "order-of-magnitude" level and are accurate within +50% or −30%. Assumed 5% interest rate.

of residuals are the largest part of the operation costs (24%). The majority of this cost is to pay for activated carbon at the IWTP. For the short duration of this project, activated carbon was assumed to be the most cost-effective form of treatment. No other alternative were investigated. A large amount of activated carbon will be consumed because of the high COD of the water that will be discharged to the IWTP (18,700 mg/L). Much of this discharge is surfactant, and some is contaminant. Other methods of removal of the COD could be evaluated (e.g., biological treatment) and may prove to be more cost-effective.

The cost for chemicals is a relatively small percentage of the total cost, which is a consequence of the recovery and reuse of the surfactant. The labor costs are a significant fraction of the total costs. Labor costs were split for Table 20.2 into the physical treatment and groundwater collection and control categories. Labor costs make up the majority of the costs in these two categories, and are about 12% of the total costs.

Of the capital costs, the costs of physical treatment (the chemical mix system and the produced fluids treatment system) make up the largest cost categories (25% of the total). Major pieces of equipment that fall into this category include the air stripper, catalytic oxidation unit, pumps, and tanks. It should be noted that there will be a salvage value for some of this equipment, so the capital costs are somewhat higher than the true costs of this equipment. Salvage value is discussed below.

The total cost of the upfront activities is fairly significant (12% of the total), as is likely to be the case with a relatively small site. In other words, the upfront activities are likely to be the same, regardless of the size of the site; thus, they will be a larger percentage of the total costs with a small site.

20.2 OPTIMUM COSTS

Table 20.2 presents a summary of the cost estimate for the Optimized Case. The total cost of $4,957,000 is significantly lower than that of the Base Case. The capital cost for the optimized system is the same as the Base Case since no major capital systems were optimized. The capital cost for the Physical Treatment category is the most significant category in the Optimized Case. This suggests that the costs could be reduced further if more efficient treatment, chemical injection, and recovery methods were used. For example, performing in smaller modules over a greater total time period — thereby reducing the size of the equipment — could potentially further reduce the costs for the Optimized Case.

The majority of the cost reduction can be attributed to the reduced number of pore volumes of surfactant that were assumed to be required (1.5 PV vs. 2.5 for the Base Case). The reduced duration of surfactant injection reduced almost all operating costs, especially the costs for disposal of the effluent (cost for activated carbon at the IWTP).

Based on the data from the field demonstration at Hill AFB OU2, a significant amount of contaminant removal could have been achieved with a total of 1.5 PV, so that 1.5 PV might be applicable at some other sites. However, the 1.5-PV number is likely to be the bare minimum that could ever be used to achieve significant contaminant reductions.

20.3 OTHER AREAS OF SENSITIVITY

Other factors that could be considered in an economic analysis are salvage value, interest rate assumed for present-worth calculations, and contractor margins and profits. For this analysis, a salvage value of 25% for all the equipment was assumed. This resulted in a decrease in cost of $455,000. The salvage value will depend on the value of the equipment to others after use (market value), the durability of the equipment with use, and the rate of return required by the contractor. Most contractors will try to recoup the cost of capital equipment in less than 2 years. Therefore,

Table 20.2 Present-Worth Costs of Optimized Case Design

WBS No.	Standard Description	Cost	Subtotal	% of Total
321	Pre-Construction and Project Management Activities			0.0
32190	Additional Site Characterization and Laboratory Testing			0.0
32191	Numerical Simulations			0.0
32192	Field Demonstration			0.0
32193	Design			0.0
331	HTRW Remedial Action (Construction)		$2,376,000	47.9
33101	Mobilization and Preparatory Work	$97,000		2.0
33103	Site Work	$25,000		0.5
33106	Groundwater Collection and Control	$669,000		13.5
33113	Physical Treatment	$2,093,000		42.2
33121	Demobilization	$15,000		0.3
33190	Salvage Value of Equipment	$(523,000)		−10.6
332	Engineering During Construction	$190,000	$190,000	3.8
333	Construction Management	$399,000	$399,000	8.0
342	HTRW Operations and Maintenance (Post-Construction)		$1,992,000	40.2
34202	Monitoring, Sampling, Testing, and Analysis	$224,000		4.5
34203	Site Work	$92,000		1.9
34206	Groundwater Collection and Control	$241,000		4.9
34213	Physical Treatment	$241,000		4.9
34218	Disposal (other than commercial)	$599,000		12.1
34290	Chemicals	$536,000		10.8
33120	Site Restoration	$59,000		1.2
	Total Project Cost		$4,957,000	100.0

Note: These estimates are considered "order-of-magnitude" level and are accurate within +50% or −30%. Assumed 5% interest rate.

it is likely that for a project that is less than 1 year old, the salvage value could be significant. If the same contractor is performing a large number of these types of projects, and can reuse the equipment many times, the capital costs may be significantly reduced. However, at this point, a large number of these projects are not being performed, so reuse cannot be assumed.

Because this project is to be completed in less than 1 year, interest rates do not enter into this analysis. If the project were to be performed in smaller increments over a longer time to reduce the capital costs, then interest rates could be significant.

The contractor margins and profits will have some impact on the total cost of the project. Because this technology is somewhat specialized, some contractors may try to obtain relatively high profit ranges. As the technology matures, the profit range will drop. The difference is still likely to be less than 10% of the total project cost.

A number of other factors will impact the project costs for the surfactant/foam technology. These factors could include site characteristics (e.g., soil heterogeneity) and design and operating parameters (e.g., injection flow rate). A complete quantitative analysis of these types of factors was beyond the scope of this report. A qualitative assessment of these factors is provided in the tables in Chapter 21.

20.4 COMPARISON WITH THE COSTS FOR OTHER TECHNOLOGIES

It is difficult to compare the cost estimates developed here for the surfactant/foam technology for application to a hypothetical site with costs for other technologies. First, the cost must be brought to a common unit basis (e.g., dollars per acre or cubic yard). This can be difficult because of the variations in site depth if a unit area is used, and uncertainty in the volume of contaminated soil to use in the calculation. The size of the site used for the development of the cost estimate will affect the design, total cost, and the unit cost (especially by treatment area), with smaller sites (and the the technology used at them) being penalized by higher cost overall. Many fixed costs, such as engineering design efforts and equipment costs do not increase in proportion to the site area. Second, cost estimators often prepare cost estimates differently, making direct comparison difficult. For example, some estimators may not include laboratory testing of the technology in their total costs. Many estimators would include not site characterization in their remediation. The best way to compare costs of different technologies is to have the same estimator prepare estimates for different technologies for the same site. This was beyond the scope of this document, so only rough comparisons can be made here.

Table 20.3 summarizes the cost estimates developed here, expressed as unit area, unit volume, and unit quantity of NAPL removed for the Base Case and the Optimized Case. These estimates can be compared with the cost estimates for surfactant/cosolvent flushing reported by Simpkin et al., 1999. Figure 20.1 is a summary of the costs presented by Simpkin et al., 1999. The Base Case

Table 20.3 Unit Cost Comparison

	Base Case		Optimized Case	
	Quantity	Cost per Unit	Quantity	Cost per Unit
Treatment Area				
Hectares	0.42	$18,050,000	0.42	$11,890,000
Acres	1.03	$7,300,000	1.03	$4,810,000
Treatment Volume				
Cubic meters (m³)	16,700	$450	16,700	$300
Cubic yards (yd³)	21,800	$345	21,800	$227
Volume of NAPL				
Liters (L)	42,900	$175	692,000	$116
Gallons (gal)	11,300	$644	183,000	$438

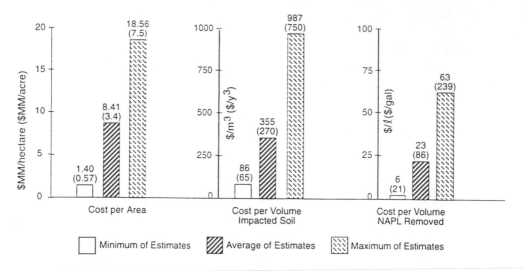

Figure 20.1 Average and range of cost estimates for surfactant/cosolvent technologies. (From Simpkin et al., 1999.)

cost for surfactant/foam is on the high end of the costs reported for other surfactant/cosolvent flushing technologies. The optimum costs are more in the middle of costs developed by Simpkin et al., 1999 and others.

As will be seen in Part IIIB, the SPME design and evaluation was prepared for a much larger site than the one used for the surfactant/foam technology. The larger site prompted costs saving features that were not utilized for the sufactant/foam design. For these and other reasons, the costs for the costs for the surfactant/foam technology presented here are higher on a per-unit basis (e.g., dollars per acre) than the cost for the SPME technology developed (see Chapter 26). The cost per unit acre for the SPME technology was approximately $5,000,000, compared with approximately $7,300,000. Some of this difference is likely a result of the economy of scale for the SPME technology, which was assumed to be performed on an 11-acre site, compared with the 1-acre site for the surfactant/foam technology study.

These unit costs for surfactant/foam technology might seem to be on the high side for common *ex situ* remediation technologies (after excavation of the soil). For example, the Department of Defense (DOD), Remediation Technologies Screening Matrix and Reference Guide (DOD, 1996), states that costs for low-temperature thermal desorption (LTTD) range from $40 to $100 per ton ($60 to $148 per cubic yard for typical soil). For deep sites, an additional $5 to $20 per yard may be required for excavation (including removal of clean overburden). An additional $10 per cubic yard may be required to replace the contaminated soil, for a total cost ranging from $75 to $178 per cubic yard. These costs for LTTD are somewhat old, and market conditions appear to have reduced them somewhat. Excavation and LTTD are possible alternatives for remediation of the hypothetical site developed here. Other, possibly lower cost, *ex situ* technologies, such as pile air stripping, might also be used.

In summary, the cost for the surfactant/foam technology could be in the same range as other competing technologies if the cost of a few key aspects can be kept low. Reducing the costs of the produced fluids treatment is key to reducing the overall cost of this technology. The Optimized Case described in this report is a good starting point in working to produce a potentially cost-effective approach. With further developments, it might be possible to achieve these lower costs.

CHAPTER 21

Performance and Potential Application

21.1 SUMMARY OF POTENTIAL PERFORMANCE

The field demonstration at Hill AFB OU2 provided valuable information on the potential effectiveness of the surfactant/foam technology. However, since the conditions at OU2 were unique in a few specific ways, it may be difficult to extrapolate these results to sites that are significantly different. OU2 was unique in terms of the relatively homogeneous and permeable soils, and the presence of a clay barrier beneath the zone of contamination. Below is a summary of the potential performance of the technology.

- In the appropriate setting, the surfactant/foam technology is likely to remove a significant mass of DNAPL. Removals in the range of 90% or greater are possible. These removals may be achieved in a relatively short time (possibly less than 3 PV) for very permeable soils and/or close well spacing.
- The production of foam can reduce the swept volume in the aquifer, and it is likely that upper portions of an aquifer will become preferentially filled with foam. This should divert surfactant to the lower portions of the aquifer that are likely to contain more DNAPL, if that is the nature of the contaminant distribution.
- The use of foam should have no adverse impacts on the process. It is also a relatively simple process to install and operate.
- Data were not available from the demonstration, suggesting that the surfactant/foam process can effectively flush all of the depressions in an aquifer.
- Data from the field trial on achievable soil endpoint concentrations may have been impacted by inflow of contaminants into the treatment area.
- Care must be taken when applying this technology to sites not underlain by a strong capillary barrier and with the potential for pooled DNAPL.
- As discussed in *Reuse of Surfactants and Cosolvents for NAPL Remediation* (Harwell et al., 1999), the produced fluids from the surfactant/foam technology are potentially amenable to relatively simple treatment methods (e.g., air stripping) that may allow reuse of the surfactant solution.

The surfactant/foam process is not yet a fully developed technology. Consequently, additional laboratory and field testing and development are required on potential sites. Additional laboratory and field-scale tests and demonstrations are needed to address the following issues that may be important in full-scale implementation:

- Definition of the benefits of the mass removal that may be achieved with this technology
- Achievable soil and groundwater endpoint concentrations
- Performance of the technology in other hydrogeologic settings, especially those with heterogeneous soils and an undulating aquitard

- Potential for vertical mobilization of DNAPL at a particular site
- Cost-effective methods of dealing with produced fluids
- Improved methods to optimize chemical delivery to targeted contaminated zones of an aquifer
- Methods to optimize the system design, such as well spacing and alternative chemical injection methods

21.2 SITE CHARACTERISTICS AFFECTING APPLICABILITY, PERFORMANCE, AND COST

As with any *in situ* remediation technology, the surfactant/foam technology has a certain niche of site characteristics where it is likely to be applicable. Chapter 2 of Simpkin et al. (1999) provides a description of the niche of surfactant/cosolvent flushing technologies in general, but it is also applicable to the surfactant/foam technology. Table 21.1 provides a summary of the site characteristics affecting applicability, performance, and cost of the technology. Because the technology has had limited field testing, some of the following information is speculation based on experience at other sites with similar technologies.

21.3 DESIGN AND OPERATING PARAMETERS AFFECTING PERFORMANCE, APPLICABILITY, AND COST

Table 21.2 provides a summary of the most important design and operating parameters, and how they might affect performance, applicability, and cost. Because the surfactant/foam technology has only been field-tested at one site under a limited number of design and operating conditions, some of the discussion in Table 21.2 is speculation based on the results of the demonstration and on other similar technologies.

21.4 MATERIAL HANDLING REQUIREMENTS

The material handling requirements for the surfactant/foam technology should be relatively straightforward. However, because many of the chemicals that are being used may be at relatively high concentrations and may cause environmental contamination if released, special care is needed to ensure that there are no releases during the operation.

Piping and other materials of construction in contact with either the injected or recovered fluids must be selected to be compatible with the chemicals in the solution. Excessive corrosion may lead to leaks or failures of the materials. This may not be a critical issue for short-term projects.

It is possible that double containment with leak detection may be required for both injected and recovered fluids. Double containment will avoid contamination of the environment from leaks or breaks. Leak detection with automatic system shutdown can also be used. For exposed piping where leaks can be seen, it may be possible to not use double containment as long as the system is monitored by on-site personnel 24 hours per day so that it can be manually shut down in case of leaks.

21.5 REGULATORY REQUIREMENTS FOR PERFORMANCE AND COMPLIANCE CRITERIA

The regulatory requirements for performance and compliance criteria are very complicated and are based on the remedial objective. Simpkin et al. (1999) provide a discussion of possible remedial

Table 21.1 Site Characteristics Affecting Applicability, Performance, and Cost

Parameter	Comments	Effects on Applicability	Effects on Performance	Effects on Cost
Remedial objective, endpoint required	Critical factor for selection of the technology	Potentially applicable at sites where mass removal is the objective; may not be applicable where restoration of DNAPL zone to drinking water is required		Less stringent cleanup objectives are likely to require less time and to cost less
Presence of NAPL	If a significant amount of mobile DNAPL is present, it should be removed first; this will reduce costs and reduce the risk of vertical DNAPL mobilization	Applicable to residual and possibly pooled DNAPL; not likely to be applicable to dissolved or sorbed contaminants; great care must be taken when applying to sites with pooled DNAPL since there is the potential to increase the vertical mobilization of the DNAPL		Higher contaminant concentrations will increase the chemicals required and their cost, and the cost for produced fluids treatment
Chemical composition of NAPL	Specific chemical constituents and LNAPL or DNAPL are of importance	Demonstrated on chlorinated solvent DNAPL, but potentially applicable to other DNAPLs or LNAPLs; application to an LNAPL may be complicated by the vertical rise of the foam, which may plug the zone being targeted	Chemical composition will greatly impact performance; different surfactant formulations may be needed; the stability of the foam will depend on the NAPL present and the chemicals used	Surfactant formulation and other design/operating parameters will be impacted, which will also impact costs
Permeability of soils/soil classification	Requires flushing of multiple pore volumes, thus impacted by permeability; demonstration at Hill AFB OU2 in relatively high permeability soils	Most applicable to sites with high permeability (sands and gravel)	Performance not likely to be as good on low permeability soils, or in fractured clay and rock settings	Durations will be longer with low permeability soils, which will increase costs
Heterogeneity of soils	Highly layered or heterogeneous soils will impact ability to get surfactant chemicals to contaminated soils	Use of foam is likely to have its greatest benefit in heterogeneous soils compared with surfactant systems without foam	Removals may still be poorer in very heterogeneous soils; the time required is likely to be greater in heterogeneous soils since more pore volumes may be required	More pore volumes may be required for very heterogeneous soils, which will increase the costs
Depth of contamination	May not be a theoretical limitation	Most applicable to sites with contamination less than 22 to 32 m (70 to 100 ft)	At very deep sites, it may be difficult to control injection of fluids and to accurately locate the NAPL	Costs are likely to be greater for deeper depths; injection of air for foam production will be more expensive at greater depths
Presence of a competent confining layer	May be important for DNAPL sites with significant amounts of pooled DNAPL	May not be applicable to sites without a competent confining layer and with pooled DNAPL		

continued

Table 21.1 (continued) Site Characteristics Affecting Applicability, Performance, and Cost

Parameter	Comments	Effects on Applicability	Effects on Performance	Effects on Cost
Depressions and variability in confining layer(s) elevations		May not be applicable to sites with a large number of depressions and/or variability in confining layer elevation	Depressions in confining layer may be difficult to access, thus reducing the performance	Expensive site characterization may be needed to fully define the depressions
Area of site	Large sites may require a phased approach			Large sites will have higher costs, although there will be some economies of scale, so the cost per unit area may decrease with increased area
Current land use	Active sites will be more complicated to work on	May not be applicable at active sites where vertical or horizontal wells cannot be installed	Horizontal drilling methods may have to be used at active sites with many obstructions; these may reduce the performance, depending on the specifics of the drilling method	Costs will be higher for active sites
Monitoring and reporting required by regulators	May vary with the regulatory agencies involved			Extensive monitoring may increase the cost slightly
Permits required	May vary with the regulatory agencies involved	Permit limitations on chemicals that may be injected may limit the applicability in certain locations		Extensive permitting requirements could increase the cost
Availability of discharge facilities for liquid wastes	Final disposal of liquid waste streams may be difficult in remote settings since they may not have access to either a POTW or industrial WWTP	Most applicable at sites with existing wastewater treatment facilities that can treat the wastes		Cost for treatment of produced fluids will vary with receiving stream discharge requirements
Availability of residual management facilities	Some residual waste may require off-site incineration or other management	Most applicable at sites near low-cost residual management facilities		Cost for disposal of residuals will have a major impact on the project cost
Need for post-treatment or long-term care	Post-treatment or long-term care may be needed if the remediation goal is to achieve concentration-based standards	Most applicable at sites where costly post-treatment and/or long-term care are not required		Post-treatment and long-term care may have a significant impact on the total costs
Site geochemistry	Water inorganic chemistry, ion exchange of soil, and surfactant/cosolvent sorption characteristics		As was found at OU2, calcium may be detrimental to certain surfactants. The design of the chemical system may need to be adjusted to account for this. A high degree of sorption of the surfactant and/or cosolvent may reduce performance	Water used to make up surfactants may require softening, which will increase the cost. A high degree of sorption will increase the mass of chemicals required and thus the cost

Table 21.2 Design and Operating Parameters Affecting Applicability, Performance, and Cost

Parameter	Comments	Effects on Applicability	Effects on Performance	Effects on Cost
Injection/extraction system	Pattern of wells or horizontal drain lines		Choice of wells or horizontal drain lines will impact performance; the pattern of wells will influence dead zones that may not be remediated	Cost must be optimized on a case-by-case basis; trade offs between performance and optimized injection/extraction system
Spacing of injection/extraction system	Smaller spacings generally required in lower permeability settings		Well or drain line spacing may have an impact on performance and time to complete	Spacing will impact time to complete and cost
Screened interval or depth of injection point	Chemicals must be injected at appropriate depths		Inappropriate depth of injection could result in chemicals missing contaminated zones	
Surfactant type and concentration			The appropriate surfactant will have an impact on performance of NAPL removal and foam formation; the concentration will also impact removals; the surfactant type and concentration will influence the performance of the produced fluids treatment system	The cost of different surfactants varies; higher concentrations will increase the cost to purchase the surfactant as well as the cost for treatment of the produced fluids and residual disposal
Cosolvent type and concentration			Cosolvents may not be desirable for the surfactant/foam technology since they may impact foam stability	
Electrolyte type and concentration	An electrolyte may be needed to optimize the phase behavior of the NAPL solution created		The phase behavior of the NAPL/surfactant solution may impact the removal of the solution from the subsurface	The cost of the electrolytes is relatively low
Number of pore volumes	Measure of the volume of chemicals flushed		Critical to the NAPL removal achieved; although the majority of NAPL may be removed in the first one or two pore volumes, many more may be needed to achieve high degrees of removal	The number of pore volumes will dictate the quantity of chemicals required and thus the cost for the chemicals and the treatment and disposal of the produced fluids
Sequence of chemical floods	A pre-water flood, electrolyte flood, surfactant flood with air injection, electrolyte flood, and post-water flood were used for the surfactant/foam demonstration		The sequence of flooding used may have some influence on the NAPL removal achieved	

continued

Table 21.2 (continued) Design and Operating Parameters Affecting Applicability, Performance, and Cost

Parameter	Comments	Effects on Applicability	Effects on Performance	Effects on Cost
Gradient applied	The head used will impact the flow rate of fluids, depending on the soil characteristics		Head, and thus flow rate, will impact duration	The balance between flow rate and duration will have an impact on cost; high flow rates will require larger process equipment
Flow rates			Flow rate will impact duration	Certain materials (e.g., stainless steel) that are more resistant to corrosive chemicals may increase the cost slightly
Materials of construction	Surfactants and other additives, along with the contaminants, will impact the materials of construction required		The physical system may fail if very corrosive chemicals are present and the proper materials of construction are not used	
Complexity and cost of the produced fluids unit processes required	The type and complexity of the produced fluids treatment system unit processes will depend on the chemical solution used and the NAPL; the system at OU 2 required relatively simple unit processes		Failure of the produced fluids system could result in failure of the entire system	The specific unit processes required for the produced fluids treatment system could increase the cost for this system, which may be a significant portion of the overall cost
Degree of automation	The degree of automation will impact the labor required		Automation may improve the potential to understand the system that could improve the performance	Automation may increase the capital cost but could reduce the labor costs of operations
Monitoring, sampling, and analytical methods used			Additional monitoring and sampling may improve the potential to understand the system, which could improve the performance	Additional monitoring and sampling will increase the costs somewhat; the type of analytical method will impact the cost slightly

objectives. The requirements for demonstrating performance or compliance with a remedial objective will be specific to the remedial objective. The following are examples:

- If restoration of the aquifer to drinking water standards is the remedial objective, groundwater data will be the key to demonstrating compliance. Data on the ability of the technology to achieve drinking water standards are not available from this surfactant/foam project. However, it may be very difficult to achieve drinking water standards with any technology where NAPL and/or low-permeability media are present.

- If mass reduction is the remedial objective, a number of methods can and should be used to measure compliance. Soil cores before and after flushing are probably the best true measure, but they are prone to very high variability as a result of variability in soil conditions. Partitioning tracer tests can also be used to measure mass reduction. The analysis of the produced fluids can also be used to estimate the amount of mass removed from the system, but this analysis alone will not provide any indication of the amount of mass still remaining in the formation.

- The remedial objectives might also be based on achieving as much mass removal as is practical with a given technology. This is similar to a best-available or -practical technology approach, as is used for some air and wastewater treatment criteria. With this type of criterion, the produced fluids can be analyzed, and a certain percentage of the "maximum removable" contaminants would have to be removed. The "maximum removable" level is likely to be different from the total mass present; it can be estimated based on extrapolation of produced fluids production curves to infinite time.

CHAPTER 22

Section IIIA References

American Public Health Association, (1996) *Standard Methods for the Examination of Water and Wastewater*, 19th edition. APHA, Washington, D.C.

American Society for Testing and Materials. (1995). *Standard Guide for Developing Conceptual Site Models for Contaminated Sites*. Designation: E 1689-95, Philadelphia, PA.

Bear, J. (1972). *Flow Through Porous Media*. Elsevier, New York.

Brown, C.L., Delshad, M., Dwarakanath, V., Jackson, R.E., Londergan, J.T., Meinardus, H.W., McKinney, D.C., Oolman, T., Pope, G.A., and Wade, W.H. (1999). Demonstration of surfactant flooding of an alluvial aquifer contaminated with DNAPL. In *Innovative Subsurface Remediation: Field Testing of Physical, Chemical, and Characterization Technologies*. M.L. Brusseau, D.A. Sabatini, J.S. Gierke, M.D. Annable, Eds. ACS Symposium Series 725, American Chemical Society, Washington, D.C.

Cayais, J.L., Schechter, R.S., and Wade, W.H. (1975). The measurement of interfacial tension via the spinning drop technique. *Adsorption at Interfaces*. American Chemical Society Symposium Series, Vol. I, Washington, D.C. pp. 234-248.

Corey, A.T. (1986). *Mechanics of Immiscible Fluids in Porous Media*. Water Resources Publications.

Federal Remediation Technologies Roundtable. (1995). Guide to Documenting Cost and Performance for Remediation Projects. EPA-542-B-95-002.

Hirasaki, G.J., Miller, C.A., Szafranski, R., Lawson, J.B., and Akiya, N. (1996). Surfactant/foam process for aquifer remediation, SPE paper No. 37257. Presented at the *SPE International Symposium on Oil Field Chemistry*, Houston, TX, February 18–21, 1997.

INTERA, Inc. (1997). *Surfactant /Foam Process for Aquifer Remediation*. Advanced Applied Technology Demonstration Facility, Rice University. September.

Simpkin, T.J. et al. (1999). *Surfactants and Cosolvents for NAPL Remediation*, Lowe, D.F., Oubre, C.L., Ward, C.H., Eds. CRC Press, LLC. Boca Raton, FL

Harwell, J.H., Sabatini, D.A., Chang, C.C., O'Haven, J.H., Simpkin, T.J. (1999). *Reuse of Surfactants and Cosolvents for NAPL Remediation*, Lowe, D.F., Oubre, C.L., Ward, C.H., Eds. CRC Press, LLC. Boca Raton, FL

Radian. (1994). Aquifer Data Evaluation Report, Operable Unit 2. Hill Air Force Base, Utah, RCN 279-100-26-41.

U.S. Department of Defense. (1996). *Remediation Technologies Screening Matrix and Reference Guide*. In http://clu-in.com, NTIS PB95-104782

U.S. EPA. (1986). *Test Methods for Evaluating Solid Waste*, SW-846, 3rd edition, as amended.

U.S. EPA. (1983). *Methods for Chemical Analyses of Water and Wastes*. EPA 600/4-79/020, as amended.

Design and Evaluation of a Full-scale Implementation of the Single-Phase Microemulsion Process

EXECUTIVE SUMMARY

Introduction

The Department of Defense (DOD) funded the Advanced Applied Technology Development Facility (AATDF) in 1993 with the mission of enhancing the development of innovative remedial technologies for DOD by bridging the gap between academic research and proven technologies. To accomplish its mission, the AATDF selected 11 projects that involved the quantitative demonstration of innovative remediation technologies. Field demonstrations were completed for each of these projects. To further assist with the potential commercialization of remediation technologies and to disseminate information on the technologies, the AATDF sponsored the preparation of a Technology Evaluation Report by an independent engineering consulting firm.

The technology design described herein is a direct scale-up of the field demonstration at Hill AFB and is referred to as the Base Case. The rationale for directly scaling the field test is that the field test represents what is known about the technology. The hazards in doing so warrant mention here. The field test of the single-phase microemulsion (SPME) process was designed to be the best technical demonstration of the concept. Therefore, the surfactant selected for the field test was the one that best fulfilled technical performance criteria — rather than a less expensive, second-best performer. The impact of surfactant cost is shown in Chaper 26. The SPME field test was part of a large, side-by-side demonstration of eight remediation technologies funded by the Strategic Environmental Research and Development Program (SERDP). The number of pore volumes of flushing fluid used in the SPME test was set by the larger demonstration program. It will also be seen in Chapter 26 how the use of this number affects cost, although one can easily understand that the more surfactant used, the higher the cost. The value of this technology design and cost evaluation lies less in the total cost and unit cost estimates that were developed than in outlining the technical issues that confront the SPME process and the elements that most impact the cost of full-scale implementation of the SPME technology.

SPME technology involves the injection and extraction of a surfactant/alcohol solution into the subsurface to solubilize nonaqueous phase liquids (NAPLs). In theory, microemulsion refers to the dispersion of very small (less than 0.1 micrometer [μm]) liquid droplets that result when the surfactant/alcohol solution contacts an NAPL. "Single phase" refers to the fact that the microemulsion remains dispersed in the surfactant solution and does not form a separate phase. NAPL constituents will concentrate in the microemulsion (solubilization), which can result in very high water phase concentrations of the NAPL constituents, thereby accelerating their removal from the subsurface as the solution is flushed through the contaminated zone.

Summary of SPME Field Demonstration

To develop the technology and evaluate the effectiveness of the SPME technology, a series of laboratory studies and a field pilot demonstration were performed. The laboratory tests screened surfactants and cosolvents for their NAPL solubilization, viscosity, and phase behavior of the solutions when mixed with NAPL, and included one-dimensional (1-D) column studies and two-dimensional (2-D) flow chamber tests. These studies were used to select the chemical system and to confirm the potential applicability of the technology. The chemical system selected from these studies and used in the field demonstration was Brij 97 (polyoxyethene (10) oleyl ether) at 3 percent by weight (wt%) with 2.5 wt% n-pentanol.

The field demonstration was performed at Hill Air Force Base (AFB) Operable Unit (OU) 1. This site is contaminated with an NAPL that is a complex mixture of weathered JP-4, waste oils, and chlorinated solvents. The demonstration involved the following:

- Installation of a sheet pile wall to hydraulically isolate the test cell. The test cell covered an area 2.8 m × 4.6 m (9.2 ft × 15.1 ft) and was approximately 10 m (32 ft) deep.
- Installation of four injection wells, three extraction wells, and 12 multilevel samplers
- Construction of the chemical mix system, and tank and piping system to store produced fluids and pump them to the base IWTP and the sampling system
- Injection of chemicals using the following sequence:
  - Water injection (with tracers for part of the time): 20 pore volumes (PV)
  - Surfactant/alcohol solution: 9 PV
  - Surfactant only: 1 PV
  - Water injection (with tracers for part of the time): 15.8 PV
- Collection of water samples during the injection of chemicals
- Collection of soil cores at the completion of the test

The SPME technology demonstrated at Hill AFB OU1 was able to remove a large fraction of the NAPL present. Mass removal ranged from 73% to 96%, depending on the method used to assess the removal. The majority of the contaminant removal occurred in a relatively short time. Most of the removal occurred in less than 4 PV, which took from 4 to 8 d in this test cell. Longer periods may be required for larger spacings.

Evaluation of the soil endpoint concentrations achieved during the demonstration was complicated by the complex mixture of compounds that make up the NAPL at the site and the fact that most of the NAPL constituents used to evaluate the process are not regulated or do not pose significant human health risk. For example, the post-flushing average soil concentrations of n-undecane were 2.5 mg/kg (down from 60 mg/kg initially). No criteria are available with which to judge the adequacy of this value. If the average n-undecane concentration is used to estimate a total NAPL soil concentration (similar to a total petroleum hydrocarbon [TPH]), using the mass fraction of n-undecane in the site NAPL, the total NAPL would be approximately 160 mg/kg (down from 3800 mg/kg). An estimate of the average NAPL soil concentration from the post-flushing partitioning tracer data is 1100 mg/kg. Groundwater data were not available from the investigation.

Summary of Engineering Design and Cost Estimate

To illustrate the engineering design components that may be required for full-scale application of the SPME technology, a conceptual design for a hypothetical site using the SPME technology was prepared. Hill AFB OU1 was used as a model for the hypothetical site, but the conceptual design and cost estimate are not specifically for Hill AFB OU1. Table E.3 summarizes the conceptual

Table E.3 Site Characteristics and Base Case System Design Summary for the SPME Site

Site Characteristics

Type of site	Waste disposal site for waste oil, fuels (JP-4), and solvents. The site is currently inactive.
Contaminants	Petroleum hydrocarbon from waste oils and fuels, with small amount of chlorinated solvents, an LNAPL.
Size of target area	4.8 hectares (11.9 acres).
Hydrogeologic setting	Granular alluvium material, consisting of poorly sorted gravel intermixed with sand and cobbles, 8 to 10 m thick. Historic water table from 5 to 8 m. Underlying unfractured lacustrian clay that provides a barrier to vertical migration.
Target depth	LNAPL smear zone over the bottom 1.5 m of alluvial material.

System Design

Remedial objective	95% removal of the recoverable contaminants
Primary performance measures	Measurement of contaminants (1,2-dichloroethene; 1,2-dichlorobenzene; tetrachloroethene; toluene; chlorobenzene; total xylene; and naphthalene) in the produced fluids. The total mass of recoverable contaminant present will be based on projection of product curves.
Removal mechanism utilized	Solubilization using a single-phase microemulsion.
Chemical system	3 wt% Brij 97® (polyoxyethylene(10)oleyl ether), 2.5 wt% *n*-pentanol.
Delivery recovery system	Horizontal injection and recovery drain lines spaced 31 m on centers. Constructed in units consisting of injection and recovery drain lines of various lengths. Total length of time to complete is 3 years.
Chemical delivery sequence (flows for each drain line)	• 2 PV water flooding over 22 days • 9 PV surfactant and alcohol injection over 200 days • 1 PV surfactant over 22 days • 6.5 PV post-water flooding over 72 days
Chemical mix system	Chemicals delivered in concentrated form, diluted on-site, and mixed in two solution mix and storage tanks. • 4762 metric tons (5200 tons) Brij total for the project. • 3900 metric tons (4300 tons) *n*-pentanol total for the project.
Produced fluids management	Concentration of surfactant and NAPL via ultrafiltration, steam-stripping to remove alcohols, biological treatment with SBRs to further reduce the COD, off-site incineration of the wastes, and WWTP discharge of the effluent.
Surfactant reuse	Not used
Additional studies included in costs	Additional site characterization, bench-scale testing, numerical simulations, field demonstrations in a sheet pile enclosed cell, partitioning tracer tests (see Appendix 9, Additional Capital Cost sheet, for details)

Note: COD = chemical oxygen demand; SBR = sequencing Batch Reactor; WWTP = wastewater treatment plant.

design. In addition to preparing the conceptual design based on the process used in the SPME field test (the Base Case), a conceptual design was also prepared for an Optimized Case of the application of the technology.

Cost estimates were prepared for this hypothetical design to illustrate the potential cost of this technology. The cost estimates were order-of-magnitude level and are accurate to within +50% and −30%. The total cost of the Base Case system was $58,600,000, which includes $29,280,000 for the cost of the Brij 97 and pentanol (50% of the total) and $17,300,000 (30%) for the cost of disposal. The cost dropped to $24,000,000 for the Optimized Case, which includes $14,100,000 for the chemical costs (59%) and $2,170,000 (9%) for disposal costs. The two most significant cost components of this system were the cost for the chemicals and the cost for disposal of residuals. Table E.4 summarizes the costs on a unit area and a unit volume basis. These costs are on the high side for competing technologies but within the range of other surfactant floods.

Table E.4 Unit Cost Comparison

	Base Case		Optimized Cost	
	Quantity	Cost per Unit	Quantity	Cost per Unit
Treatment area				
Hectares	4.8	$12,200,000	4.8	$5,000,000
Acres	11.9	$4,930,000	11.9	$2,000,000
Treatment volume				
Cubic meters (m³)	144,000	$400	144,000	$170
Cubic yards (yd³)	188,000	$311	188,000	$130
Volume of NAPL				
Liters (L)	692,000	$85	692,000	$35
Gallons (gal)	183,000	$320	183,000	$130

Summary of Potential Performance and Applicability

Based on the pilot-scale demonstration of the SPME technology at Hill AFB OU1, the following conclusions regarding the potential performance and applicability of the technology can be made:

- In the appropriate setting, the SPME technology is likely to remove a significant mass of NAPL: removals from 75% to greater than 90% are possible.
- Soil endpoint concentrations that are potentially achievable will depend, in part, on the nature of the NAPL being removed. For the Hill AFB OU1 NAPL, average concentrations of individual compounds of less than 2.5 mg/kg were achieved. However, most of the compounds analyzed are not priority pollutants, so it is difficult to evaluate the merits of these endpoints. Estimated endpoint TPH concentrations ranged from 160 to 1100 mg/kg.
- In the appropriate site setting, operational problems should not be major. Heterogeneities can, however, lead to longer periods of operation.

Tables 26.2 and 26.3 (in Chapter 26) summarize the site characteristics, as well as the design and operating parameters that impact the applicability, performance, and cost of the surfactant/foam technology. The SPME technology is not yet a fully developed technology. Consequently, additional laboratory and field testing and development may be required for each potential site. The additional laboratory and field-scale tests and demonstrations are needed to address the following issues that may be important in full-scale implementation:

- Definition of the benefits of the mass removal that may be achieved with this technology
- Achievable soil and groundwater endpoint concentrations
- Performance of the technology with other contaminants, especially dense nonaqueous phase liquids (DNAPLs)
- Performance of the technology in other hydrogeologic settings, especially those with heterogeneous soils
- Performance of the technology with fewer pore volumes of chemicals
- Cost-effective methods of dealing with produced fluids, including possible methods of recovery and reuse of the surfactant and cosolvent
- Improved methods to optimize chemical delivery to targeted contaminated zones of an aquifer
- Methods to optimize the system design, such as well spacing and alternative chemical injection methods

CHAPTER 23

Summary of Technology Demonstration

23.1 INTRODUCTION

The SPME technology involves the injection and extraction of a surfactant/alcohol solution into the subsurface to solubilize nonaqueous phase liquids (NAPLs). In theory, microemulsion refers to the dispersion of very small (less than 0.1 μm) liquid droplets that result when the surfactant/alcohol solution contacts an NAPL. Single phase refers to the fact that the microemulsion remains dispersed in the surfactant solution and does not form a separate phase. NAPL constituents will concentrate in the microemulsion (solubilization), which can result in very high water phase concentrations of the NAPL constituents, thereby accelerating their removal from the subsurface as the solution is flushed through the contaminated zone.

To develop the technology and evaluate the effectiveness of the SPME technology, a series of laboratory studies and a field pilot demonstration were performed. This section summarizes these studies. More details on the studies can be found in Part II of this volume.

23.2 SUMMARY OF LABORATORY TESTS AND TREATABILITY STUDIES

The laboratory tests for this project began with the screening of surfactants and cosolvents to be used in the chemical solution. High NAPL solubilization and a low solution viscosity were the primary selection criteria. Eighty-six surfactant samples were evaluated. They were first screened for their aqueous solubility and the solubility of pentanol in the resulting surfactant solution. The viscosities for the solution at 3% surfactant were also measured. Fifty-eight surfactants with pentanol had acceptable surfactant solubilities (at least 3%) and viscosities (less than 2 cp). These surfactant solutions were further evaluated for their solubilization of Hill AFB OU 1 NAPL and the phase behavior of the resulting mixtures. Solubilization was determined by measuring the concentration of compounds most prevalent in the NAPL with gas chromatography after mixing the NAPL with the surfactant solution.

The solubilization of the NAPL constituents varied significantly among the various surfactants. For example, dodecane concentrations in the aqueous microemulsions ranged from less than 1 mg/L to over 200 mg/L. The surfactant/alcohol solution eventually selected was able to solubilize 100 mg/L of dodecane, which is equivalent to approximately 14,700 mg/L of NAPL. Detailed results from these studies can be found in Part II.

Three of the more promising surfactant solutions were further evaluated in column studies with glass beads and Hill AFB OU1 NAPL. It was noted that two of the three surfactant solutions left a brown, oily residue on the glass beads, although the total removal was good. The surfactant/alcohol solution that was selected and that did not leave the oily residue by these studies was Brij 97 (polyoxyethene (10) oleyl ether) at 3 wt% with 2.5 wt% *n*-pentanol. Removal of NAPL from the glass beads with this surfactant was 98%, as measured by a partitioning tracer test.

This surfactant/alcohol solution was further tested with site-contaminated soil in columns. The removal of NAPL, as measured by partitioning tracer tests, was 65% with the site soils. A hexane extraction of the soil after the flushing showed no measurable NAPL constituents. Visual observation of the mineral grains of the soils revealed that they were coated with a black material that was not obvious on clean site soil. It is believed that this coating is an NAPL weathering product that has accumulated at OU1. This coating can act as a sorption site for partitioning tracers, but may have properties more similar to soil humic matter than NAPL.

Three 2-D flow chamber tests were also performed on the selected surfactant/ alcohol solution. The first test was conducted on clean quartz sand to which NAPL from Hill AFB OU1 was added. The second two tests were conducted with both contaminated and uncontaminated site soils. Contaminated soil was packed into a central zone within the chamber. The contaminated portion of the test cell was from 10% to 25% of the total volume of the chamber. All three experiments indicated that the chemical solutions were effective in removal of the contaminants, as measured by soil concentrations before and after the test. Greater than 95% of the initial mass of target contaminants was removed. However, as with the column tests, the results of the partitioning tracer test were not as promising; measured removals ranged from 26% to 96%. Some of the poor performance, as measured by the partitioning tracer test, could be attributed to the coating on the soil that was noted in the column test. In addition, it was noted that some partitioning occurred on the clean site soil.

23.3 TEST SITE DESCRIPTION

The field demonstration was performed at Hill AFB, OU1, the location of two chemical disposal pits, two fire training areas, and two landfills. Most of the NAPL contamination at this site originated in the chemical disposal pits. Disposal of JP-4, waste oils, and chlorinated solvent wastes occurred at this site.

The hydrostratigraphy of the site includes 8 to 10 m of poorly sorted alluvial sand intermixed with gravel and cobbles (Provo Formation). The hydraulic conductivity and porosity of the Provo Formation are estimated to be 10^{-1} to 10^{-2} cm/s and 0.14 to 0.16, respectively. Underlying the Provo Formation is a 60-m thick lacustrian clay (Alpine Formation) that contains minor discontinuous interbedded sand layers. The bulk hydraulic conductivity of the Alpine clay is estimated to be 10^{-5} to 10^{-8} cm/s. Based on borehole information, the Alpine clay appears to be unfractured. The most shallow regional groundwater aquifer used for domestic purposes is approximately 200 m (640 feet [ft]) below the site.

Groundwater occurs at the base of the Provo Formation, perched by the Alpine clay. Based on an observed 2- to 3-m (6.4 to 9.6 ft) thick NAPL smear zone in the Provo, directly over the Alpine clay, historic water levels have ranged from 5 to 8 m (16 to 25.6 ft) below ground surface (bgs). Groundwater flow follows local topography and is generally to the west through a series of erosional channels cut into the Alpine clay. There may be a downward component to the groundwater flow, although the difference in the hydraulic conductivity between the Provo and Alpine clay is so large that it is thought that the vast majority of the flow in the alluvium is horizontal.

The NAPL is a complex mixture of weathered JP-4, waste oils, and chlorinated solvents. Table 23.1 presents a summary of the mass percent of the most prevalent compounds measured in two samples of the site NAPL. It should be noted that the sum of the mass fractions listed is a relatively small fraction of the total NAPL. The rest of the NAPL is a complex mixture of hydrocarbon compounds that are difficult to determine using standard techniques. The bulk density of the NAPL is 0.85 g/cm³, making it an LNAPL. The viscosity of the LNAPL is 11 cp at 24°C. Information is not available on the interfacial tension of the NAPL with water.

Table 23.1 Mass Fractions of Target NAPL Constituents

NAPL Constituent	Mass Percent (g/100 g NAPL)	
	OU1-202	MW-2
p-Xylene	0.892	0.144
1,2-Dichlorobenzene	1.66	a
1,2,4-Trimethylbenzene	1.08	0.438
Naphthalene	0.055	a
Trichlorobenzene	0.62	0.461
n-Decane	2.14	0.477
n-Undecane	1.32	1.573
n-Dodecane	0.680	0.698
n-Tridecane	0.510	0.285

a Not measured.

23.4 PROCEDURES AND FIELD IMPLEMENTATION

The field demonstration of the SPME technology was one of eight field demonstrations performed at Hill AFB OU1 as part of an overall program of technology evaluations funded by SERDP and AATDF. The tests were coordinated by the EPA National Risk Management Research Laboratory in Ada, Oklahoma. The procedures used for this demonstration were similar to the other eight demonstrations to allow comparison among the technologies.

The demonstrations were performed in hydraulically isolated test cells. Each cell was constructed using 9.5-mm (0.37-in. [in.]) thick interlocking steel sheet piles with all joints filled with low-permeability grout. The cell enclosed a rectangular area of approximately 2.8 m (9.2 ft) × 4.6 m (15.1 ft). The sheet piles were driven to a depth of 10.7 m (35.2 ft) bgs, which was approximately 2.8 m (9.2 ft) into the Alpine clay. As a result of the Z-shape of the individual piles, the sides of the cell were corrugated, with 17% of the total cell area inside the corrugations.

Four injection and three extraction wells were installed in the cell, as shown in Figure 23.1. The injection wells were spaced 0.71 m (2.3 ft) apart and 4.13 m (13.6 ft) from the extraction wells. The extraction wells were spaced 0.76 m (2.5 ft) apart. Each of the seven wells was screened from 4.9 to 7.9 m (16.1 to 25.9 ft) bgs, for a screen length of 3 m (9.9 ft). The wells were 2-in. polyvinyl chloride (PVC) with 0.25-mm slotted stainless steel screens.

Twelve multilevel samplers (MLSs) were installed in the test cell — in three rows, four per row. An additional three MLSs were located at each extraction well. Each MLS consisted of either five or eight samplers. The MLSs were constructed from 0.32-cm stainless steel tubing, terminating with a 20-μm or 40-μm stainless steel filter. The MLSs started at 4.9 m (16.1 ft) bgs in the center row, 5.5 m (18 ft) bgs on the outer rows, and were spaced approximately every 0.6 m (2 ft) down to, and into the clay, in the case of the middle row.

Peristaltic pumps were used to both inject and recover fluids from the test cells. When the system was operated at 1 PV per day, the injection wells were operated at 0.90 liters per minute (L/m) (0.24 gpm) each, and the extraction wells at 1.2 L/m (0.317 gpm) each, for a total flow of 3.6 L/m (0.95 gpm).

Samples were collected from the MLSs using a vacuum pump system. The vacuum pump was connected to the manifold of the sampling tubes, and each sampling point was sampled independently. Water knockouts were installed between the pump and the manifold to protect the pump. Samples were also collected from the discharge of the extraction pumps. A total of 540 extraction well samples and 6500 MLS samples were taken during the demonstration.

Figure 23.2 is a schematic diagram of the SPME field test. The surfactant/alcohol solution was mixed in two 25,000-L (6614-gallon [gal]) polyethylene tanks. A sump pump was used to recirculate

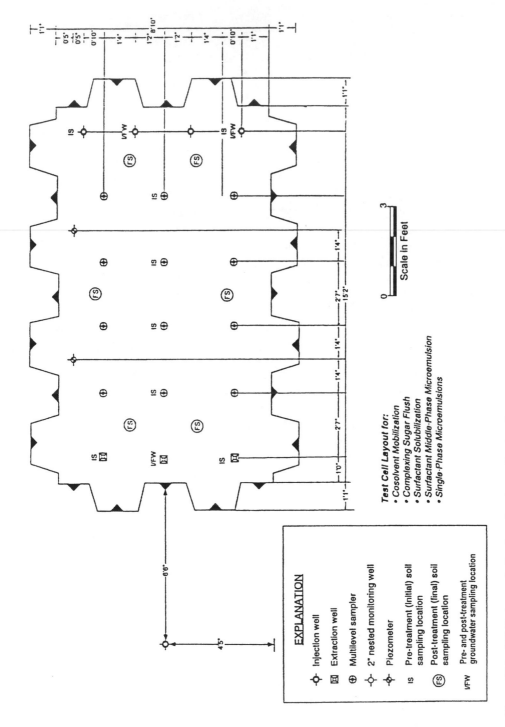

Figure 23.1   SPME demonstration layout and sampling locations.

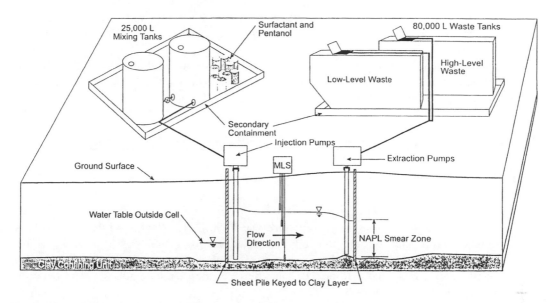

Figure 23.2 Schematic diagram of SPME demonstration.

and mix the fluids; three batches of chemical were prepared. Water was added to the tank, followed by slow addition of the surfactant and alcohol. The tanks were within secondary containment.

Fluids generated during the flushing were stored on-site in 80,000-L (21,164-gal) storage tanks. The fluids had chemical oxygen demands that ranged from 1000 mg/L to over 120,000 mg/L. The fluids were metered at a slow rate to the Hill AFB industrial wastewater treatment plan (IWTP), where they were treated with granular activated carbon.

Table 23.2 summarizes the operational phases of the field work. The injection rates were approximately 1 PV per d during the tracer tests and 0.5 PV per d for the surfactant/alcohol injection, although the injection rate was varied, as discussed later.

A nonreactive tracer test was conducted to characterize the hydrodynamic behavior of the test cell prior to other tests. After flushing the nonreactive tracer with water, the partitioning tracer test was performed to estimate the mass of NAPL present. Methanol and 2,2-dimethyl-3-pentanol (DMP) were used as the nonpartitioning and partitioning tracers, respectively. An interfacial tracer test was also performed to estimate the surface area of the NAPL. Sodium

Table 23.2 Operational Phases of Field Work

Test Segment	Fluids Injected	Time of Injection (days)	Pore Volumes of Fluids
Nonreactive tracer	Water + iodide	0.12	0.12
Water injection	Water	3.5	3.5
Partitioning tracer injection	Water + tracer	0.16	0.16
Water injection	Water	9	9
Interfacial tracer injection	Water + tracer	0.2	0.15
Water injection	Water	7	7
Surfactant/alcohol solution	Water + surfactant + pentanol	18	9
Surfactant only	Surfactant	2	1
Water injection	Water	6.5	6.5
Partitioning tracer injection	Water + tracer	0.16	0.16
Water injection	Water	4	4
Interfacial tracer injection	Water + tracer	0.2	0.15
Water injection	Water	5	5

dodecylbenzene sulfonate (SDBS) and bromide were used as interfacial and nonreactive tracers, respectively. A partitioning tracer test was also performed after the completion of the surfactant/alcohol flood.

The monitoring and sampling of the system included measuring fluid injection and extraction rates, sampling injected fluids, sampling the MLSs, and sampling extracted fluids. Fluid levels were measured in two piezometers located in the test cell.

23.5 SUMMARY OF RESULTS, MASS BALANCE, AND CONCLUSIONS

23.5.1 Hydraulic Performance

The swept volume of the aquifer and the effective porosity were estimated from the nonreactive tracer test. The test suggested that the pore volume was 6000 L (1587 gal), and the porosity was 0.14 (assuming uniform flow through the cell). The relatively low porosity is possibly the result of the large number of cobbles, which have zero porosity present, present. The hydraulic conductivity across the cell was estimated to be 0.01 cm/s by measuring the hydraulic gradient between two piezometers at various flow rates.

The breakthrough curves (BTCs) for pentanol can also be used to provide some indication of the hydraulic performance of the system. Analytical interference with NAPL constituents precluded analysis of effluent surfactant concentrations. Figure 23.3 presents the average injected pentanol concentration, as well as the pentanol concentration in the three extraction wells. The injected pentanol concentration decreased during the study from approximately 2.5% to 2.25%. This decrease is likely due to variations in the batches of solution prepared. It is not likely that this decrease in pentanol concentration was enough to impact the performance of the system. Also evident from the data are the differences in BTCs among the three extraction wells. The curve from EW-2 (the inside well), is steeper and has somewhat less tailing than the outside wells. This is

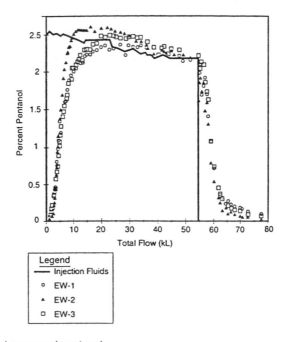

Figure 23.3 Injected and recovered pentanol.

likely a result of the presence of the large dead volumes along corrugations of the sheet pile wall that were in the flow paths of the outer wells.

23.5.2 Produced Fluids Analysis

The samples collected from the extraction wells (produced fluids) were always single-phase samples with no evidence of NAPL, either as a separate phase or as a macroemulsion. The produced fluids developed a light-brown color as microemulsified NAPL was produced. This light-brown color changed to a dark-brown color and then gradually lightened in color again. The color of the fluids correlated very well with the concentration of contaminants.

Breakthrough curves (BTCs) for four of the target NAPL constituents are presented in Figure 23.4; also shown is the breakthrough of pentanol in the three extraction wells. A significant increase in NAPL constituent concentration is obvious in these curves (e.g., a maximum n-undecane concentration of 170 mg/L was measured in EW 2, up from a background of less than 1 mg/L). The increase in n-undecane concentration, and other analytes, coincided with the arrival of pentanol. The concentrations dropped off quickly after reaching their peak. The peak concentrations in EW 2, the center extraction well, were considerably higher than those of the outer wells, and the outer well data also showed more spreading. This is likely a result of the dead volume along the walls of the sheet piles.

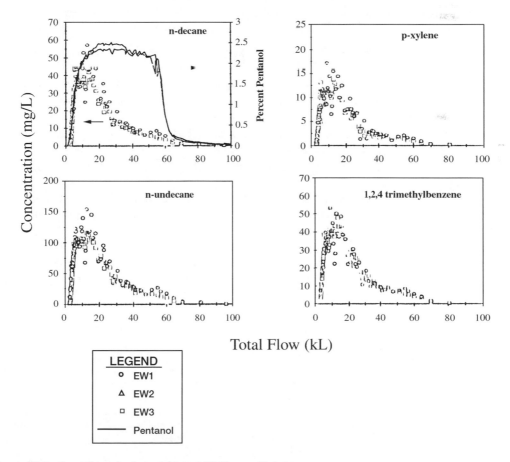

Figure 23.4 Breakthrough curves of target NAPL constituents.

Cumulative production curves for a number of the NAPL constituents are provided in Figure 23.5. This figure and the BTCs suggest that mass was still being removed when chemical injection was stopped. Continued flushing of the chemicals would have likely removed some additional NAPL mass, but not very efficiently. The investigators of the SPME technology (see Part II) speculate that the remaining NAPL was likely located in low-permeability zones, and continued flushing would have resulted in most of the injected chemicals being flushed through the more permeable zones where the NAPL had already been removed.

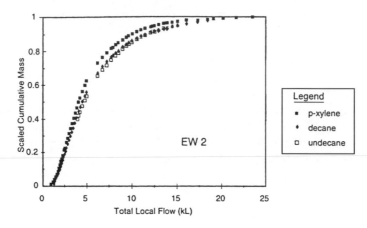

Figure 23.5 Cumulative production of target NAPL constituents.

The total mass of the major NAPL constituents removed is summarized in Table 23.3 (integrated from the BTC). This table also includes estimates of the volume and mass of NAPL removed, extrapolated from the mass fraction of each compound measured in a site NAPL sample. For example, assuming n-undecane is present in the NAPL at 1.57 g/100 g, the total removal of NAPL was 233 L (61.6 gal) based on the n-undecane in the produced fluids. The range in estimated NAPL removal was from 227 to 302 L (60 to 80 gal), depending on the compound that was assumed to represent the NAPL.

Information on the total production of surfactant is not available because of analytical problems with the surfactant method. Consequently, the losses of the injected surfactant in the subsurface cannot be estimated.

To evaluate the potential for rate-limited dissolution processes during the SPME flushing, a number of flow changes were performed during the demonstration. If rate-limited dissolution is a factor for SPME flushing, concentrations of the contaminants should change as the contact time changes with flow rate. The following flow changes were made:

Table 23.3 NAPL Constituent Removal

NAPL Constituent	Mass Fraction of NAPL Sample (g/100 g NAPL)	Mass of Constituent Removed Based on Produced Fluids (kg)	NAPL Removal Based on Produced Fluids (L)	Mass of Constituent Removed Based on Pre and Post Soil Cores (kg)	NAPL Removal Based on Soil Cores (L)
1,2,4-Trimethylbenzene	0.438	1.12	302		
n-Decane	0.477	1.12	276	1.77	438
n-Undecane	1.573	3.12	233	4.42	331
n-Dodecane	0.698	1.35	227		

- Doubled to 1 PV/day after about 2 PV (12.6 kiloliter [kL]) had been flushed
- Flow interrupted for 24 h after 4 PV (22.8 kL) had been flushed
- Flow interrupted for 32 h after 6 PV (35.7 kL) had been flushed
- Reduced back to 0.5 PV/d after 6.8 PV (40.8 kL) had been flushed
- Increased to 1 PV/d after 8.2 PV (49.2 kL) had been flushed

Only very minor responses were obvious in the measurement of NAPL constituents in the extraction wells. There was a slight change in the slope of the curve at 6.8 PVs (40.8 kL). This lack of change in concentration in response to flow rate suggests that nonequilibrium dissolution processes were not significant during the SPME flood.

23.5.3 Multilevel Sampler Data

Data from the MLSs provide some indication of preferential flow paths and the location of the more contaminated zones. Figure 23.6 presents the variation in mass removed with depth in the MLSs located at the extraction wells. The mass removals shown in Figure 23.6 were scaled to the average removal from each MLS. These data suggest that there were specific depth ranges in the aquifer that produced the majority of the NAPL. These depths correlated fairly well with the depths that had the highest soil concentrations (shown in Figure 23.7).

23.5.4 Soil Boring Data

Eighty-five soil core samples were collected during the installation of the wells and MLSs and were analyzed for NAPL constituents. Forty soil core samples were also collected after the surfactant/cosolvent flushing.

Figure 23.7 presents the data for 1,3,5-trimethylbenzene (TMB), n-decane, and n-undecane (the three most abundant compounds in the soil samples). The relatively high concentrations from 0 m to 2 m above the Alpine clay are indicative of the smear zone of the LNAPL. It should be noted that the position of the Alpine clay varied (± 0.47 m) so that the points shown below the clay in Figure 23.7 may not have really been in the clay.

These figures indicate a dramatic reduction in the soil concentrations after the surfactant/cosolvent flushing. Two post-flushing samples at 1.5 m above the clay had higher concentrations of all three compounds, although the concentrations were still much lower than the initial concentrations. There was no other obvious pattern for the remaining locations. For example, although there were a few samples with elevated concentrations at the clay interface, this does not appear to be a clear trend.

Table 23.4 presents the average concentrations for all samples collected within the flushed zone before and after the SPME flushing. The table only includes those compounds with initial average concentrations greater than 0.1 mg/kg. In addition, some compounds were not measured in the soils (e.g., dodecane). The post-flushing soil concentrations of these NAPL dramatically decreased. Percent reductions were greater than 90% for the three most abundant compounds. Percent reductions for the less abundant compounds were somewhat lower, but they were also more sensitive to sampling variability.

Estimates of the pre-flushing NAPL concentration from the n-undecane and n-decane concentrations and their mass fractions in the NAPL were 3800 mg/kg and 5000 mg/kg, respectively. The estimated post-flushing NAPL concentration ranged from 160 mg/kg to 230 mg/kg using these same two compounds. These values represent a percent removal of 95.5 and 95.8, respectively.

Estimates of the mass of the constituents removed based on the soil core data were presented in Table 23.3. This table also include the removals estimated from the analysis of produced fluids. The estimates of removals by these two methods were fairly close for n-decane and n-undecane (e.g., 3.1 kg based on produced fluids, and 4.4 kg based on soil cores for n-undecane).

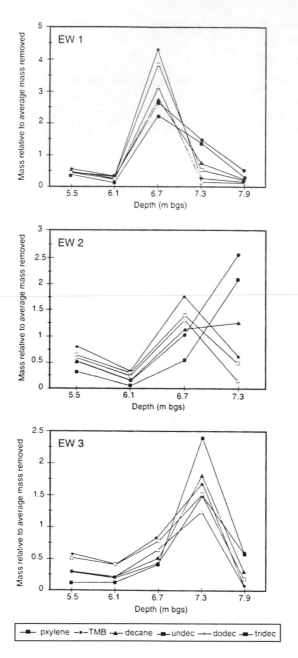

Figure 23.6 Mass removed through multilevel samplers located at the extraction wells.

The volume of total NAPL removed was also estimated from these data (see Table 23.3). The volume of NAPL removed based on the produced fluids analysis varied from 230 to 302 L (60.8 to 79.4 gal). This is a relatively close range. The values based on soil cores were 340 and 438 L (90 to 115.9 gal) for *n*-undecane and *n*-decane, respectively. The soil core estimates were slightly greater than the produced fluids but probably within the errors of the estimates. It should also be noted that the values based on soil cores are very sensitive to the volume of soil that is assumed to be swept. Variability in soil concentrations also has a significant impact on these results.

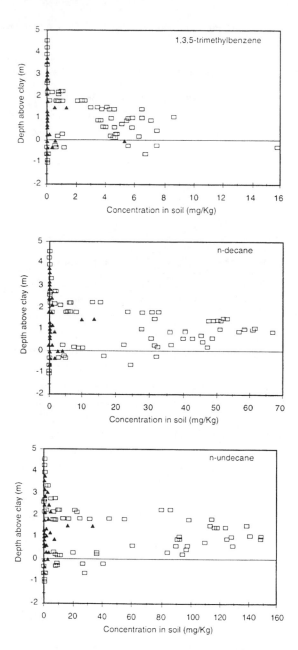

Figure 23.7 Soil concentration profiles for selected NAPL constituents before (□) and after (▲) SPME flushing.

23.5.5 Partitioning Tracer and Interfacial Tracer Data

Partitioning tracer tests were performed before and after the surfactant/alcohol flushing to estimate the volume of NAPL present in the test cell. Table 23.5 summarizes the percent saturation estimates from the pre-and post-flushing data. The saturation is the percentage of pore space occupied by the NAPL. These data also suggest a significant reduction in the mass of NAPL present. The reduction in NAPL based on the partitioning tracer data is 73%. This is less than the 90% or greater estimated from the soil cores for the most abundant NAPL constituents. The lower removals estimated from the partitioning tracer tests may be a result of the heavy "pitch" fraction of the NAPL observed in the

Table 23.4 Pre- and Post-Flushing Average Soil Concentrations of Target NAPL Constituents

NAPL Constituent	Pre-flushing Concentration (mg/kg)	Post-flushing Concentration (mg/kg)	Percent Reduction (%)
n-Undecane	59.55	2.50	96
n-Decane	23.98	1.09	95
1,3,5-TMB	2.73	0.24	91
Naphthalene	0.86	0.11	87
1,2-DCB	0.74	0.03	96
o-Xylene	0.17	0.06	64
m-Xylene	0.12	0.02	87

Table 23.5 Partitioning Tracer Test Results

	% Saturation	
Well	Pre-flushing	Post-flushing
EW 1	0.065	0.02
EW 2	0.068	0.014
EW 3	0.053	0.016
Average	0.062	0.017

laboratory studies that could not be removed, even with laboratory hexane extraction. The soil samples from the field demonstration also had a visible black staining, possibly this pitch fraction of NAPL.

The volume of NAPL present was estimated from the partitioning tracer data to be 380 L (100.5 gal) and 100 L (26.5 gal) from the pre-flushing and post-flushing tests, respectively. The NAPL removal from these data (280 L [74 gal]) is well within the range of the removal estimated from produced fluids production (230 to 300 L [60.8 to 79.4 gal]), but less than that from the soil core data (340 to 438 L). Thus, given the inherent error in all of these types of measurements, these data are in relatively good agreement. Given the confined cell nature of this test cell and the relatively close spacing of the injection and recovery wells, the partition tracer test should provide a good estimate.

The interfacial tracer test provided information on how the NAPL is distributed within the porous medium. Figure 23.8 summarizes the results of the interfacial tracer data from one of the MLSs. The NAPL saturation was at a maximum at about 6.1 m (20 ft) bgs. The interfacial area measured at this same depth was the lowest level measured. This suggests that the NAPL was located in regions where the NAPL filled the pore space in relatively large globs, with little surface area. At the next depth (6.7 m [22 ft]), the NAPL saturation was lower, but the interfacial area was much higher. This suggests that this zone had NAPL distributed as a thinner layer, possibly coating the soil particles or in smaller globs having a higher surface-area-to-volume ratio. The extent of the surface area in a particular zone and the variability in the surface area relative to the NAPL saturation may provide insight into the effectiveness of remediation technologies that depend on mass transfer. A higher surface area should improve mass transfer.

23.5.6   Conclusion for the Results of the SPME Field Test

The points of greatest interest in assessing the performance of the SPME project are the contaminant mass removal achieved and the endpoints achieved. In addition, the occurrence of any operational problems that would limit its application is of interest.

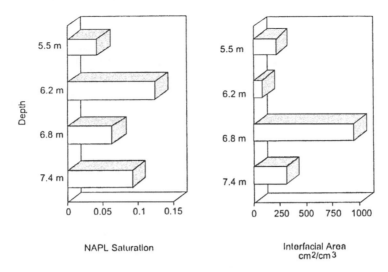

Figure 23.8 Summary of NAPL saturation and interfacial area at MLS 32.

The SPME technology demonstrated at Hill AFB OU1 was able to remove a large fraction of the NAPL present. Mass removal ranged from 73% to 96%, depending on the method used to assess the removal. The lower removals are from methods that measured "total" NAPL removal; the higher numbers were for specific compounds. But, in general, the range of values suggests fairly good agreement among the different measurement producers. It should be noted that the production curves for the NAPL constituents suggested that additional removal was occuring when the chemical injection stopped (after 10 PV). Based on the shape of the curves, the additional removals may have only been on the order of a few percent of the initial mass. It should also be noted that the contaminant removal occurred in a relatively short time. Most of the removal occurred in less than 4 PV, which took from 4 to 8 d in this test cell. Longer periods will be required for larger spacings.

Evaluation of the endpoints achieved during the demonstration is complicated by the complex NAPL mixture at the site and the fact that most of the NAPL constituents used to evaluate the process are not regulated or do not pose significant risk. Thus, it is difficult to judge the adequacy of the soil endpoint concentrations. For example, post-flushing average soil concentrations of n-undecane were 2.5 mg/kg (down from 60 mg/kg initially). No standard risk-based criteria are available with which to judge the adequacy of this value, although concentrations as high as 10,000 mg/kg may be acceptable under certain risk calculations. If the average n-undecane concentration is used to estimate a total NAPL soil concentration (similar to a total petroleum hydrocarbon [TPH]), using the mass fraction of n-undecane in site NAPL, the total NAPL would be approximately 160 mg/kg (down from 3800 mg/kg). This is a relatively low TPH value, although some states in the U.S. may regulate TPH down to less than 100 mg/kg. An estimate of the average NAPL soil concentration from the post-flushing partitioning tracer data is 1100 mg/kg. This value may be higher than desirable by some regulatory authorities. However, some of this total NAPL may not be extractable with the common TPH methods (e.g., freon extraction for Method 418.1), so the measured TPH may actually be lower. Average soil concentrations of less than 0.06 mg/kg were achieved (down from less than 0.74 mg/kg) for compounds that have more risk implications (e.g., 1,2-DCB and xylenes). However, even the initial concentrations for these compounds were less than possible risk-based criteria. For example, EPA Region IX has developed Preliminary Remediation Goals for 1,2-DCB of 17 mg/kg for protection of ground-water. Groundwater data are currently not available, so the impact of this technology on groundwater quality cannot be assessed.

No major operational problems were observed in this field demonstration of the SPME technology. There was some indication of the impact of heterogeneous soil conditions on the rate of contaminant removal (i.e., removal was still occurring when the chemical flushing ended). Except for this one potential problem, this site was relatively ideal (e.g., high permeability), so that information is not available from this demonstration on the performance of the technology at sites with less favorable hydrogeologic conditions.

CHAPTER 24

Measurement Procedures

A wide variety of measurement procedures are associated with implementation of an SPME remediation. These procedures are associated with site characterization, performance monitoring, and achieved endpoints in media of concern. Many of these procedures are discussed by Simpkin et al., 1999. The Federal Remediation Technologies Roundtable (FRTR, 1995) also provides a list of measurement procedures.

Table 25.1 provides a list of suggested monitoring for the hypothetical conceptual design discussed below. This table is a good summary of testing that should be performed during the operation of an SPME system. Tables 24.1, 24.2, and 24.3 provide references for performing these analytical tests.

Table 24.1 Methods for NAPL Characterization

Parameter	Method Description	Method
Density	Hydrometer	ASTM D1298
	Pycnometer	ASTM D1480
Viscosity	Capillary (kinematic)	ASTM D445
	Rotational (Brookfield)	
Interfacial tension	Ring (DuNouy)	ASTM D971
	Pendant drop (drop weight)	ASTM D2285
	Spinning drop	Cayias et al. 1975
NAPL composition	Volatile organics	SW846-8260[a]
	Semivolatile organics	SW846-8270[a]
	Pesticides/PCBs	SW846-8080[a]
	Total petroleum hydrocarbons	EPA418.1,[b] SW846-9071[a]
	Gasoline range organics	SW846-8015M[a]
	Diesel range organics	SW846-8015M[a]

[a] US EPA Test Methods for Evaluating Solid Waste, SW-846, 3rd edition. November 1986, as amended.
[b] US EPA Methods for Chemical Analyses of Water and Wastes, EPA600/4-79/020. March 1983, as amended.

Table 24.2 Methods for Water Characterization

Parameter	Method Description	Method
Fluid properties		
Viscosity	Capillary (kinematic)	ASTM D445[a]
Density	Pycnometer	ASTM D1480[a]
Inorganic chemistry		
Calcium	Atomic adsorption	APHA 301A II (APHA-AWWA-WPCF)
Magnesium	Atomic adsorption	APHA 301A II
Iron	Atomic adsorption	APHA 301A II
Potassium	Atomic adsorption	APHA 317A
Sodium	Atomic adsorption	APHA 320A
Barium	Atomic adsorption	APHA 301A II
Strontium	Atomic adsorption	APHA 321A
Chloride	Mercuric nitrate	APHA 408 B
Sulfate	Turbimetric precipitation	APHA 427C
Carbonate and bicarbonate	Alkaline titration	APHA 403
pH	Electrode	APHA 424
Conductivity	Electrode	APHA 205
Total dissolved solids	Gravimetric	APHA 208
Organic chemistry		
Volatile organics	Mass spectrometry	SW846-8260[a]
Semivolatile organics	Mass spectrometry	SW846-8270[a]
Pesticides/PCBs	Gas chromatography	SW846-8080[a]
Total petroleum hydrocarbons	Infrared spectrometry	EPA418.1[b]
Gasoline range organics	Gas chromatography	SW846-8015M[a]
Diesel range organics	Gas chromatography	SW846-8015M[a]

[a] US EPA Test Methods for Evaluating Solid Waste, SW-846, 3rd edition. November 1986, as amended.
[b] US EPA Methods for Chemical Analyses of Water and Wastes, EPA600/4-79/020. March 1983, as amended.

Table 24.3 Methods for Soil Characterization

Parameter	Method Description	Method
Physical properties		
Dry bulk density	Physical method	ASTM D2937
Grain size distribution	Physical method	ASTM D422
Permeability and porosity	Physical method	ASTM D5084
Wettability	Physical method	ASTM D3152
Total organic carbon	Infrared carbon dioxide	SW846-9060[a]
Displacement pressure	Physical method	Bear (1972), Corey (1986)
Cation exchange capacity	Chemical method	SW846-9081[a]
Organic contaminants		
Volatile organics	Mass spectrometry	SW846-8260[a]
Semivolatile organics	Mass spectrometry	SW846-8270[a]
Pesticides/PCBs	Gas chromatography	SW846-8080[a]
Total petroleum hydrocarbons	Infrared spectrometry	EPA418.1,[b] SW846-9071[a]
Gasoline range organics	Gas chromatography	SW846-8015M[a]
Diesel range organics	Gas chromatography	SW846-8015M[a]

[a] US EPA Test Methods for Evaluating Solid Waste, SW-846, 3rd edition. November 1986, as amended.
[b] US EPA Methods for Chemical Analyses of Water and Wastes, EPA600/4-79/020. March 1983, as amended.

CHAPTER 25

Engineering Design

To illustrate the engineering design components that may be required for full-scale application, a conceptual design for a hypothetical site using the SPME technology was prepared. Other designs are possible and may prove to be more cost effective. Cost estimates that were prepared using this conceptual design are discussed in Chapter 26.

The conceptual design was prepared for a hypothetical site that was modeled after an actual site, Hill AFB OU1. This model site was selected because it was the location of the field demonstration (as previously discussed), so information is available on the performance of the technology at the sites. In the preparation of the conceptual design, OU1 was modeled as closely as possible. However, complete information is currently not available on the application of this technology at OU1, thus, some assumptions had to be made. One was on the extent of the site that has mobile or residual NAPL that would require remediation using the SPME technology. The actual site area that might require remediation may deviate from that assumed here, so this conceptual design and the cost estimate should not be considered an estimate of what the remediation would consist of or cost for Hill AFB OU1. Further site investigations may be necessary at OU1, as well as a site-specific design and cost estimates.

Two conceptual designs and cost estimates were prepared for this site. The first is a Base Case that is a direct scale-up of the field test. For example, the volume and concentration of fluids injected were kept the same as those used in the field demonstration, despite the fact that the volume used in the field test was set to satisfy the requirements of the larger demonstration program. In the field test, most of the NAPL removal occurred by the halfway point. Successful application of the technology should be possible with lower volumes and concentrations of chemicals, resulting in potentially lower costs. A conceptual design for an Optimized Case was also prepared to evaluate the potential costs under more optimum conditions. A number of assumptions were made on methods to optimize the system that may reduce the cost. The costs for the Base Case and the Optimized Case are likely to be on the extremes of the potential cost, with the actual costs for many sites somewhere between the extremes.

25.1 BASE CASE

Table 25.1 summarizes the conceptual design of the Site 1 SPME system, and Figure 25.1 is a process flow diagram of the system. Other design material is included in the appendix, as follows:
•Appendix 7: Drawings
  - Site plans
  - Typical drain line profiles
  - Drain line details
  - Process instrumentation and control diagrams
- Appendix 8: Mass balance and design spreadsheets

- Base Case design worksheet
- Base Case mass balance
- Optimized Case design worksheet
- Optimized Case mass balance
- Appendix 9: Cost estimates
  - Base Case additional capital cost
  - Base Case O&M cost
  - Base Case capital cost detail
  - Optimized Case additional capital cost
  - Optimized Case O&M cost
- Optimized Case capital cost detail

Table 25.1 Site Characteristics and Base Case System Design Summary for the SPME Site

Site Characteristics

Type of site	Waste disposal site for waste oil, fuels (JP-4), and solvents. The site is currently inactive.
Contaminants	Petroleum hydrocarbon from waste oils and fuels, with small amount of chlorinated solvents, an LNAPL.
Size of target area	4.8 hectares (11.9 acres).
Hydrogeologic setting	Granular alluvium material consisting of poorly sorted gravel intermixed with sand and cobbles, 8 to 10 m thick. Historic water table from 5 to 8 m. Underlying unfractured lacustrine clay that provides a barrier to vertical migration.
Target depth	LNAPL smear zone over the bottom 1.5 m of alluvial material.

System Design

Remedial objective	95% removal of the recoverable contaminants.
Primary performance measures	Measurement of contaminants (1,2-dichloroethene; 1,2-dichlorobenzene; tetrachloroethene; toluene; chlorobenzene; total xylene; and naphthalene) in the produced fluids. The total mass of recoverable contaminant present will be based on projection of production curves.
Removal mechanism utilized	Solubilization using a single-phase microemulsion.
Chemical system	3 wt% Brij 97® (polyoxyethylene(10)oleyl ether), 2.5 wt% n-pentanol.
Delivery recovery system	Horizontal injection and recovery drain lines spaced 31 m on centers. Constructed in units consisting of injection and recovery drain lines of various lengths. Total length of time to complete is 3 years.
Chemical delivery sequence (flows for each drain line)	• 2 PV of water flooding over 22 d. • 9 PV of surfactant and alcohol injection over 200 d. • 1 PV of surfactant over 22 d. • 6.5 PV of post-water flooding over 72 d.
Chemical mix system	• Chemicals delivered in concentrated form, diluted on-site, and mixed in two solution mix and storage tanks. • 4762 metric tons (5200 tons) Brij total for the project. • 3900 metric tons (4300 tons) n-pentanol total for the project.
Produced fluids management	Concentration of surfactant and NAPL via ultrafiltration, steam stripping to remove alcohols, biological treatment with SBRs to further reduce the COD, off-site incineration of the wastes, and WWTP discharge of the effluent.
Surfactant reuse	Not used.
Additional studies included in costs	Additional site characterization, bench-scale testing, numerical simulations, field demonstrations in a sheet pile enclosed cell, partitioning tracer tests (see Appendix 9, Additional Capital Cost sheet for details).

Note: COD = chemical oxygen demand; SBR = sequencing batch reactor; and WWTP = wastewater treatment plant.

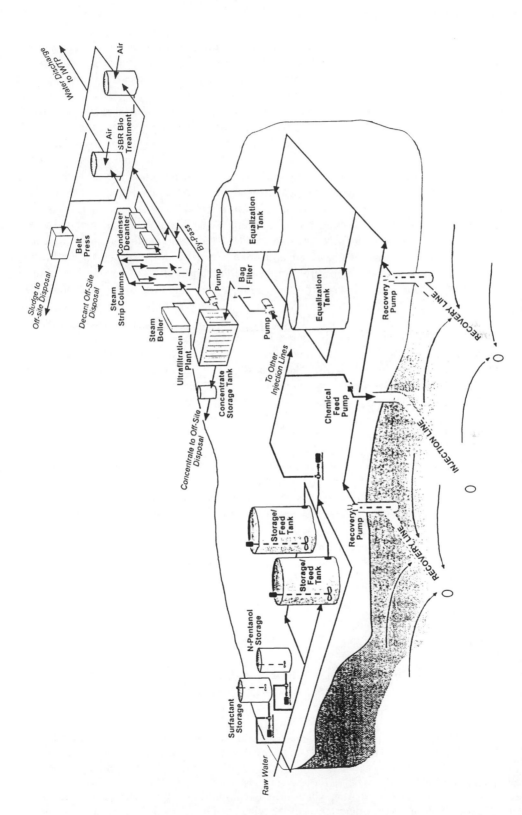

Figure 25.1 SPME flushing remediation process flow design.

25.1.1   Site Setting

The site is the location of two chemical disposal pits, two fire training areas, and two landfills. The majority of the NAPL contamination at this site originated in the chemical disposal pits. Disposal of JP-4, waste oils, and chlorinated solvent wastes occurred at this site. Currently, the site is a level field with no active uses that conflict with site remediation.

The hydrostratigraphy of the site includes 8 to 10 m of poorly sorted alluvial sand intermixed with gravel and cobbles (granular alluvium). The hydraulic conductivity and porosity of the granular alluvium are estimated to be 10^{-1} to 10^{-2} cm/s and 0.16, respectively. Underlying the granular alluvium is a 60-m thick lacustrian clay that contains minor discontinuous interbedded sand layers. The bulk hydraulic conductivity of the lacustrian clay is estimated to be 10^{-5} to 10^{-8} cm/s. Based on borehole information, the lacustrian clay appears to be unfractured. The shallowest regional groundwater aquifer used for domestic purposes is approximately 200 m (658 ft) below the site.

Groundwater occurs at the base of the granular alluvium, perched by the lacustrian clay. Based on an observed 2- to 3-m (6.6- to 9.9-ft) NAPL smear zone in the granular alluvium, directly over the lacustrian clay, historic water levels have ranged from 5 to 8 m (16 to 26 ft) bgs. Groundwater flow follows local topography and is generally to the west through a series of erosional channels cut into the clay. There may be a downward component to the groundwater flow, although the difference in the hydraulic conductivity between the alluvium and the clay is so large that it is thought the vast majority of the flow in the alluvium is horizontal.

The NAPL is a complex mixture of weathered JP-4, waste oils, and chlorinated solvents. All of the NAPL in the soil is thought to be at immobile, residual saturations. The bulk density of the NAPL is 0.85 g/cm^3, making it an LNAPL. The viscosity of the LNAPL is 11 cp. Because the NAPL is less dense than water, vertical mobilization of the NAPL is not a significant concern. The targeted NAPL covers 4.8 hectares (12 acres).

25.1.2   Remedial Objective

It was assumed that a Record of Decision for this site would be written with the following elements. The surfactant flushing system will be operated to a point that 95% of the recoverable contaminants are removed. The mass of recoverable contaminant present will be based on projection of production curves during the operation of the system. Contaminants that will be used to judge the extent of recovery will be those that are most important to the risk at the site, including 1,2-dichloroethene, 1,2-dichlorobenzene, tetrachloroethene, toluene, chlorobenzene, total xylene, and naphthalene.

It should be noted that the 95% removal of the recoverable contaminants criterion is very different from removing 95% removal of the mass of contaminant present. The 95% removal of the recoverable contaminants criterion takes into account the fact that some of the mass present will not be recoverable.

25.1.3   Up-front Activities

It was assumed for the Base Case that a number of up-front activities had not been completed and were included in the costs. Other estimates might not include up-front activities such as site characterization in their preparation of remedial cost estimates. These activities include:

- Additional site characterization: It will be assumed that the site has been relatively well character-ized, but that three additional soil borings will be installed along the alignment of each drain line (for a total of 105 samples) to evaluate the extent of NAPL, confirm the depth of the drain lines, and collect soils and LNAPL samples for laboratory testing.
- Laboratory testing: Laboratory testing will involve solubilization testing to select the most appro-priate surfactant, testing of the interactions of the solution with the site soils, a series of column

tests to evaluate the potential effectiveness of the system, and a series of produced fluids treatability tests (ultrafiltration, steam stripping, and biological treatment).

- Numerical simulations: A standard two-dimensional steady-state groundwater flow model will be used to model the full-scale system to aid in the design of the system.
- Field demonstrations: The field demonstration for this site will include the installation of a sheet pile wall ($5 \times 5 \times 10$ m [$16 \times 16 \times 32$ ft]) for containment, four injection wells, three extraction wells, and four nested monitoring points. A sheet pile enclosed demonstration may not be necessary at most sites, which could reduce costs. Two soil samples will be collected from each monitoring and operations well and analyzed for contaminants of concern. Fluids will be monitored at a frequency approximately 3 times that suggested above for the full-scale system. Pilot tests of the produced fluids treatment system will be conducted. The bulk of the produced fluids will be concentrated by ultrafiltration and sent off-site for incineration. The permeate will be sent to the IWTP.

A detailed design of the facility will be required. Because this is a relatively large site, it can be assumed that design and construction will follow a traditional detailed design/bidding approach.

25.1.4 Chemical System and Injection Sequence

The chemical system and injection sequence used for this conceptual design were identical to that used for the SPME field demonstration. The targeted mass removal mechanism using the SPME process is enhanced solubilization. The SPME chemical system is made up of 3 wt% surfactant (Brij 97, polyoxyethylene (10) oleyl ether) and 2.5 wt% alcohol (n-pentanol). This system results in the formation of a clear microemulsion (Winsor Type I behavior). Recall that this chemical system was optimized for performance only; chemical cost was not a factor as the materials were donated to the field demonstration. An actual full-scale implementation would balance performance and cost of the chemicals.

Because there is no significant mobile NAPL pool at the site, water flood NAPL recovery will not be conducted as a precursor to the chemical flood. Instead, preliminary water flooding will be conducted only to facilitate testing process equipment.

The chemical injection system will involve the following quantities in the following order:

- 2 PV tap water
- 9 PV surfactant and alcohol
- 1 PV surfactant only
- 6.5 PV tap water

Nine pore volumes are used in the full-scale design because 9 PV were used in the field test; 9 PV were used so that all the SERDF tests could be compared from the target zone. The final 6.5 PV of tap water is used to displace any remaining injected chemical.

25.1.5 Injection/Recovery Approach

Because the depth to contamination is relatively shallow at this site, and the site is large, horizontal drain lines were selected as the approach for injecting and recovering fluids because they are likely to be the most cost-effective method. Horizontal drain lines can also achieve a more uniform sweep of the target zone compared to wells. Use of drain lines reduces varying travel times along radial flow paths and stagnation zones that are associated with flow between wells. Use of drain lines also reduces the total number of pumping systems needed. The drain lines will be corrugated high-density polyethylene (HDPE) installed with continuous trenching equipment.

A total drain line length of 785 m (2582 ft) will be used for this example. With the types of soils found at the site, spacing of 31 m (102 ft) between drain lines should provide effective treatment. This spacing could be optimized using process models. A treatment module will consist of an injection drain line and two recovery drain lines. The recovery drain lines will be shared between two back-to-back modules. Assuming this spacing, a total of eight units will be required to cover the site. The length of the drain lines will be limited to 100 m (329 ft). Breaking up the potentially long drain lines will allow variations in chemical injection rates to optimize the system to a certain extent.

Flow calculations for this system are presented in the attached design spreadsheet (see Appendix 8). The flow calculations assumed uniform, steady-state flow in a homogeneous, isotropic aquifer. Although this may not accurately depict the site, it does provide a basis for estimating flow rates. The head loss was constrained to 1 m (3.3 ft) to minimize chemical flooding above the target. Differences between flow rates reflect water and surfactant/NAPL/alcohol solution viscosities of 0.01 and 0.02 poise, respectively. Based on these amounts, times to deliver 1 PV of water and the surfactant/alcohol solution are 11 and 22 days, respectively. Multiplying these rates by the total number of pore volumes, the total time to complete the delivery to one module will be on the order of 1 year.

The modules will be operated in three phases, with each phase having approximately the same length of drain line, and thus the same flow. Operating the system in phases will reduce the total flow being injected or produced at any time, thereby reducing the size of the chemical mix and treatment systems. The first phase will include two modules and will cover the eastern end of the main plume. The experience gained in operating the first phase will be used to optimize the design and operation of the later phases. The first phase will be constructed first, and the second and third the following year. The total amount of time to complete all three phases will be about 3 years, or 3.5 years assuming some downtime.

25.1.6 Chemical Mix System

The surfactants will be purchased and delivered as concentrated liquids and stored in tanks on-site. All of the concentrated chemical storage tanks will be provided with mixers. The concentrated chemicals will be diluted, mixed, and stored in two feed/storage tanks. It is assumed that the surfactant system used is not sensitive to calcium and that the tap water has reasonably low hardness so that it can be used without softening.

The surfactant solution will be pumped from the storage tanks through a bag filter to the main injection header by two variable-speed pumps. The bag filter will remove any particulates that may be in the fluids. Chemical metering pumps will be used to pump and meter the fluids from the main header to the injection drain lines. The pumps will be adjusted manually to maintain a constant head (water level) in the injection drain line. Using the metering pumps will allow the flow to be varied between drain lines as necessary to optimize operations.

The pumping systems of the chemical makeup system will be constructed within a small building. The storage tanks will be located outside and will be insulated and heated to prevent freezing in the winter. The chemical makeup system will be enclosed by a secondary containment system consisting of a lining and earthen berms. The surfactant piping outside the secondary containment berms will be double-walled and will have leak detection to capture any leaks. The piping will be buried to prevent freezing.

25.1.7 Produced Fluids Handling

Recovered fluids will be handled and pretreated as described below before being discharged to the IWTP located at the facility. All extractive remediation technologies require the removal and treatment of contaminants in recovered fluids. It has been assumed for this example that no field-tested technology currently exists that can separate surfactant and alcohol from the predominantly semivolatile contaminants at the site. Solvent extraction and pervaporation are two technologies that have promise in this regard, but, at the time of writing, they have not been fully developed or tested. Consequently, it was assumed that ultrafiltration for concentration of the surfactant/alcohol/NAPL mixture will be used, and that the concentrate will require off-site incineration. The water passing through the ultrafilter will contain relatively high concentrations of the alcohol, which will be stripped out in a steam stripper. Since the effluent from the steam stripper will have as much as 4500 mg/L of COD, which would consume a large amount of granular activated carbon (GAC) at the IWTP, it was decided to use a biological treatment system consisting of two sequencing batch reactors (SBRs) to further reduce the COD before discharge. The costs of other treatment options were not explored for the Base Case design.

The fluids will be recovered from the system using individual pumps at each recovery drain line sump. The recovery rate will be 1.1 times the injection rate to contain the fluids. This relatively low recovery rate compared with the injection rate should be adequate because the site is almost naturally dewatered from the small amount of groundwater inflow. From the recovery drain lines, the produced-fluids will be piped in buried, double-walled pipes with leak detection to a central produced fluids handling area. This area will be within secondary containment, and all major equipment will be within a temporary building. The produced fluids will be stored in two insulated equalization tanks with a retention time of 1.5 d. The fluids will then be treated in a package ultrafiltration unit. The concentrate (or retentate) will be stored in an insulated storage tank before being shipped off-site. The effluent water will be pumped through a steam stripper and then piped to the IWTP. The off-gas from the steam stripper will be condensed and sent off-site for incineration.

The SBRs will be operated on a cycle consisting of fill, react, settle, and decant modes. Flow will be going to one or the other SBR at all times so that when one SBR is in the fill mode, the other is in one of the other modes. The SBRs will be aerated during the fill and react modes. The majority of the biomass (or sludge) that develops in the SBRs will be retained during the decant period. A portion will be sent to an aerated storage tank. Sludge in the storage tank will be periodically dewatered using a filter press. The dewatered sludge will be stored in roll-off boxes until it is shipped off-site.

25.1.8 Operations, Maintenance, and Monitoring

Table 25.2 summarizes the labor required. Some of the staffing level is driven by the Occupational Safety and Health Administration (OSHA) requirement to have two people on-site at all times. The operators will keep all of the systems running and will perform the monitoring functions. The maintenance technician will be required to keep the equipment running. Because surfactant/cosolvent flushing is a relatively new technology and OU1 is relatively large, it was assumed that a significant amount of engineering time will be required to monitor the system, evaluate the operating data, suggest changes to the operations, and write reports. A laboratory technician will be hired to perform much of the operational laboratory analysis.

Table 25.2 Labor Requirements

Staff Description	Day Shift	Swing Shift	Night Shift	Total Staff	Total Hours per Week (h)
Operator[a]	1	2	1	4	224
Supervisor[a]	1			1	56
Lab Technician[a]			1	1	56
Maintenance Technician	0.5			0.5	20
Engineering	1			1	40
Total	3.5	2	2	7.5	396

[a] 7 d per week (56 h per week)

Table 25.3 Analytical and Monitoring Requirements

Parameter	Injected Fluids	Produced Fluids	Ultrafilter and Steam Stripper Effluent	Treated Produced Fluids	Monitoring Well Fluids
			Fluid to be Analyzed:		
Visual observation of produced fluids and NAPL					Indicator of performance
Alcohol	Quality control (once/d)	Quality control (once/d)	Indicator of treatment effectiveness (once/month)	Indicator of treatment effectiveness (once/month)	Indicator of the breakthrough of injected chemicals (once/week)
Surfactant	Quality control (once/week)				
COD			Indicator of treatment effectiveness (5 times/week)	Indicator of treatment effectiveness (5 times/week)	Indicator of the breakthrough of injected chemicals (once/week)
TPH		Indicator of treatment effectiveness (twice/week)	Indicator of treatment effectiveness (once/month)	Indicator of treatment effectiveness (once/month)	Surrogate of contaminant removal (once/week)
Volatile and semivolatile organics		Indicator of treatment effectiveness (twice/week)	Indicator of treatment effectiveness (once/month)	Indicator of treatment effectiveness (once/month)	Indicator of contaminant removal (once/month)

The utility costs for this operation will include the costs to run the pumps and mixers, and to generate steam. The fluids concentrated from the ultrafiltration units will be sent off-site for incineration. The water from the ultrafiltration unit will be sent to the facility IWTP. It is assumed that the IWTP charges $2.91/m^3 ($11/1000 gal). The effluent of the biological treatment system should have a COD of less than 500 mg/L, so there should not be a surcharge on biochemical oxygen demand (BOD) or COD.

The laboratory analyses required for both operational and performance monitoring were also assumed to be substantial. Many of the measurements can be performed on-site by a laboratory technician, but analysis requiring GC or GC/MS instruments were assumed to be performed off-site. The analyses and their frequencies are summarized in Table 25.3.

Another monitoring activity is the measurement of water levels in the formation and in all tanks. Monitoring wells will be installed at the locations shown on the site plans (Figures A7.1 and A7.2). Each well will be designed to allow sample collection from three discrete depths. These locations were selected to provide information sufficient for operations over most of the site, plus additional information on the performance of the system in areas of highest contamination. The monitoring wells will be standard 2-in. PVC wells. Piezometers will also be installed in the gravel pack of the drain lines so that water levels at the drain lines can be measured.

As discussed previously, the production of the contaminants of concern in the produced fluids will be the primary indicator of the success of the system. The cumulative production of the contaminants will be plotted with time or pore volume produced. A projection of the maximum recoverable quantity of the contaminants will be made by extrapolating the production curves to infinite time. The system will be considered complete when 95% of the maximum recoverable quantity is reached. Only a limited number of soil cores (approximately 10) will be collected at the completion of the first phase to verify the performance of the system. Partitioning tracer tests will not be performed because of the time and expense required to perform such tests on an area as large as this site. In addition, partitioning tracers are not specific to a particular contaminant, so they are not relevant to the remediation objective. It should also be noted that the remedial objective is based on the removal of 95% of the maximum recoverable contaminants and not a percentage of the total contaminants initially present. Consequently, it is not critical if the initial mass or final mass is not known.

25.1.9 Miscellaneous Cost Components

Site preparation includes construction of gravel access roads, construction of roads to the units within the site, construction of the secondary containment system, construction of the buildings, installation of utilities to the site (i.e., water, electrical, and sewer), and installation of site facilities, including office trailers and decontamination facilities.

25.2 OPTIMIZED CASE

The conceptual design discussed above was prepared for the SPME technology as a scale-up of the demonstration. A number of optimizations could be made to the technology to reduce its cost and to improve its effectiveness. Many of these are currently under investigation by the developer of the technology. The following is a list of optimizations that were assumed for this Optimized Case. A process flow diagram and mass balance for this Optimized Case are also included in the Appendices.

- The number of pore volumes of flushing required are as follows:
  - 2 PV of tap water
  - 3 PV of surfactant and alcohol
  - 1 PV of surfactant only
  - 3 PV of tap water

- Salt addition, followed by an oil/water separator, will be used in place of the ultrafiltration unit. The salt addition will produce an effluent with less than 200 mg/L of surfactant and 7000 mg/L pentanol. The concentrated NAPL and surfactant will be 6% of the influent volume. A steam stripper will still be used to remove the pentanol, but the biological system will not be needed. The high salt concentration in the final effluent will be assumed not to be a problem with discharge of the effluent. This may only be true if the WWTP receiving the high salt water has a high total flow rate, or the receiving stream is not sensitive to high salt content.
- The upfront activities discussed in Chapter 25.1.3 have been conducted by others, so they do not need to be accounted for here.
- The condensate from the steam stripper can be incinerated as high British thermal unit (BTU) fluids at a reduced cost.

In addition, it will be assumed for the Optimized Case that the removal of the total recoverable NAPL mass is 70% vs. 95%.

CHAPTER 26

Cost and Economic Analysis

This chapter presents cost estimates developed for the conceptual designs discussed in Chapter 25. An economic analysis is also presented that looks at the factors that influence the cost, and ways to reduce the cost for full-scale applications of the technology.

The capital cost estimates were developed using an estimating program called Hard Dollar Estimating Office System (EOS). The capital cost estimate was developed by entering subcontractor and equipment vendor quotes, combined with detailed estimated costs for labor to install the system. Operations costs were based on assumptions or calculations of operating requirements (see Chapter 25) and common unit costs for labor and materials. Costs for chemicals and disposal were obtained from chemical suppliers and a hazardous waste disposal facility.

It should be noted that the cost estimates developed here are at what would typically be considered an order-of-magnitude level. Estimates of this type are generally accurate within +50% or –30%. Other designs for a full-scale implementation of SPME will yield different estimates. The accuracy of the estimate for the design in Chapter 25 is subject to substantial variation because the details of the specific designs will not be known until the projects are actually implemented. Factors that may impact the cost include site conditions, final project scope and schedule, design details, the bidding climate and other competitive market conditions, changes during construction and operations, productivity, interest rates, labor and equipment rates, tax effects, and other variables. As a result, actual costs will likely vary from these estimates.

Table 26.1 summarizes the cost estimate for the Base Case. The costs in this table are the present-worth costs for the project. Appendix 8 presents more details of the cost estimate. The estimate summary is organized according to the Work Breakdown Structure (WBS) suggested by the Federal Remediation Technologies Roundtable (1995). This WBS is designed to facilitate comparison among remediation technologies. Not all of this WBS matched the way the conceptual design and cost estimate were prepared; thus, certain costs had to be subdivided to fit into the WBS categories. For example, the conceptual design for the system operations was developed with labor requirements and utilities for the entire project. These labor and utilities costs were divided equally between the "Groundwater Collection and Treatment" (which includes the operation of the chemical injection and pumping system) and "Physical/Chemical Treatment" (which includes the operation of produced fluids treatment system) WBS categories for the summary tables.

26.1 MOST SIGNIFICANT COST COMPONENTS

The total cost for the Base Case design of the SPME technology at Site 1 is $58,570,000. Of this amount, more than 85% is for system operations. Costs for chemicals and disposal of residuals make up the majority of the operations costs — 50% and 30% of the total project cost, respectively. The high cost for chemicals is a consequence of the inability to recycle the surfactant, the large

Table 26.1 Present-worth Cost Summary of the Base Case Design for SPME

WBS No.	Standard Description	Cost	Subtotal	% of Total
321	Pre-Construction and Project Management Activities		$1,540,000	2.6
32190	Additional Site Characterization and Laboratory Testing	$430,000		0.7
32191	Numerical Simulations	$20,000		0.0
32192	Field Demonstration	$500,000		0.9
32193	Design	$580,000		1.0
331	HTRW Remedial Action (Construction)		$5,830,000	10.0
33101	Mobilization and Preparatory Work	$90,000		0.2
33103	Site Work	$30,000		0.0
33106	Groundwater Collection and Control	$2,200,000		3.7
33113	Physical Treatment	$3,840,000		6.5
33121	Demobilization	$20,000		0.0
33190	Salvage Value of Equipment	(330,000)		-0.6
332	Engineering During Construction	$470,000	$470,000	0.8
333	Construction Management	$400,000	$400,000	0.7
342	HTRW Operations and Maintenance (Post-Construction)		$50,300,000	85.9
34202	Monitoring, Sampling, Testing, and Analysis	$400,000		0.7
34203	Site Work	$440,000		0.7
34206	Groundwater Collection and Control	$1,380,000		2.4
34213	Physical Treatment	$1,380,000		2.4
34218	Disposal (other than commercial)	$17,300,000		29.6
34290	Chemicals	$29,280,000		50.0
33120	Site Restoration	$130,000		0.2
	Total Project Cost		$58,570,000	100.0

Note: These estimates are considered "order-of-magnitude" level and are accurate within +50% or −30%; 5% interest rate assumed.

number of pore volumes used in the direct scale-up (10), and the high cost of the selected surfactant (Brij 97 at $2.07/lb). Other surfactants used for surfactant/alcohol flushing cost approximately $1/lb. A full-scale application would likely balance surfactant performance and cost, unlike the field test presented in Part II.

The high residual disposal costs are driven by the incineration cost for the biological sludge produced in the SBR and the concentrate from the ultrafiltration unit. Both residuals may be classified as listed hazardous wastes because they are derived from a hazardous waste. Therefore, hazardous waste incineration costs were used as a conservative estimate. The SBR was included in the design to reduce the amount of activated carbon that would be consumed (and paid for by the project) at the IWTP. Very roughly, the cost for carbon if the SBR were not used would be $7 million. With the capital cost for the SBR, plus the cost for sludge disposal (incineration as hazardous waste), the cost for the biological treatment is close to the cost for activated carbon. The high cost for the SBR is driven by the high organic content COD of the waters, even after the ultrafilter and steam stripper. The costs for the SBR and residual disposal would decrease substantially if the surfactant or cosolvent could be separated and reused, since much of the COD being treated is made up of the surfactant and cosolvent. Novel separation and recycle technologies might be able to drastically reduce these disposal costs.

Of the capital costs, the costs for the "physical treatment" (the chemical mix system and the produced fluids treatment system) and the "groundwater collection and control" (the horizontal drain lines and associated pumps and pipes) make up the largest cost categories. Much of the physical treatment cost is for the SBR system, most notably the SBR tanks. The size of these tanks is driven by the high COD of the waste. It should be noted that the capital costs are all relatively small compared with the operations costs, so that even major changes in the capital costs will not make dramatic changes in the total cost of this design.

The total cost of the up-front activities is fairly significant ($1,540,000) compared with the capital cost ($5,830,000). The up-front activities could be minimized to save costs; for example, a field test without sheet piles could be conducted as part of the first module. However, as with the capital costs, the cost for the upfront activities is relatively small compared with the total project cost as designed here.

26.2 OPTIMUM COSTS

Table 26.2 presents a summary of the cost estimate for the Optimized Case discussed in Chapter 25. The total cost of $24,000,000 is significantly lower than for the Base Case. The categories with the greatest cost reductions include chemical costs and disposal costs. The chemical costs decreased by about 50% by reducing the total number of pore volumes required for the system. The applicability of reducing the number of pore volumes will be a function of the regulatory requirements for contaminant removal (i.e., the remedial objectives) and the site characteristics. Based on the data from the field demonstration at Hill AFB OU1, a significant amount of contaminant removal could have been achieved with a total of 4 PV; therefore, 4 PV might be applicable at some other sites.

The cost of disposal dropped by over 80% as a consequence of the reduced number of pore volumes and the assumption that a more efficient treatment process — salt addition — could be used. Preliminary studies by the principal investigators during the field demonstration suggested that salt addition would be an effective means of breaking the microemulsion so that a concentrated mixture of NAPL and surfactant could be disposed of by incineration. Steam stripping could be used to remove the alcohol, leaving relatively small concentrations of COD in the effluent requiring final disposal. Consequently, the biological SBR system would not be needed. Further testing of this optimization is needed before this system can be considered implementable.

Table 26.2 Present-worth Costs of the Optimized Case Design for SPME

WBS No.	Standard Description	Cost	Subtotal	% of Total	% of Total
321	Pre-Construction and Project Management Activities		$470,000		1.9
32190	Additional Site Characterization and Laboratory Testing				
32191	Numerical Simulations				
32192	Field Demonstration				
32193	Design	$470,000		1.9	
331	HTRW Remedial Action (Construction)		$4,670,000		19.4
33101	Mobilization and Preparatory Work	$90,000		0.4	
33103	Site Work	$30,000		0.1	
33106	Groundwater Collection and Control	$2,200,000		9.1	
33113	Physical Treatment	$2,580,000		10.7	
33121	Demobilization	$20,000		0.1	
33190	Salvage Value of Equipment	($240,000)		-1.0	
332	Engineering During Construction	$370,000	$370,000	1.6	1.6
333	Construction Management	$400,000	$400,000	1.7	1.7
342	HTRW Operations and Maintenance (Post-Construction)		$18,130,000	0.0	75.4
34202	Monitoring, Sampling, Testing, and Analysis	$200,000		0.8	
34203	Site Work	$450,000		1.9	
34206	Groundwater Collection and Control	$560,000		2.3	
34213	Physical Treatment	$560,000		2.3	
34218	Disposal (other than commercial)	$2,170,000		9.0	
34290	Chemicals	$14,100,000		58.7	
33120	Site Restoration	$110,000		0.5	
	Total Project Cost	$24,000,000		100.0	100.0

Note: These estimates are considered "order-of-magnitude" level and are accurate within +50% or −30%; 5% interest rate assumed.

26.3 OTHER AREAS OF SENSITIVITY

Other factors that could be considered in an economic analysis are salvage value, interest rate assumed for present-worth calculations, and contractor margins and profits. For this analysis, a salvage value of 10% for all the equipment was assumed. This resulted in a decrease in cost of $333,000. The salvage value will depend on the value of the equipment to others after use (market value), the durability of the equipment with use, and the rate of return required by the contractor. Most contractors will try to recover the cost of capital equipment in less than 2 years, so that it is likely that for a 3-year project, 10% is reasonable. The difference between assuming 20% and 10% salvage value changes the total project cost by only 0.7% so that the total project cost is not sensitive to salvage value.

Because the operations cost for the system is such a large portion of the total costs, the interest rate assumed for the calculations could have a significant impact on the overall cost. A 5% "real interest" rate (meaning the difference between the nominal interest rate and inflation is 5%) was used. It might be possible for some organizations to use a 10% real interest rate. In such a case, the costs would decrease by $4,200,000. Although this is a significant amount of money, it is only 7.2% of the total project cost.

The assumed interest rate would have a greater impact on the cost if the project were of longer duration. It might be possible to reduce the total project cost by extending the period of operations for this project from 3 years to 10 years. This is true for this project because so much of the cost is for chemicals and disposal, which could be spread out evenly over a greater time. Some costs, such as labor, cannot be spread out evenly over time so that the impact will not be as great with projects that have relatively high labor costs.

The contractor margins and profits are only likely to have a significant impact on the total project cost if the operations contractor is able to mark up the chemical and disposal costs. Because these two costs make up such a large portion of the total project, a few percentage changes in mark-up on these costs is likely to change the total project cost by almost the same percentage. For this estimate, no mark-up on chemicals or disposal has been assumed. The mark-up of the construction contractor is not likely to have a significant impact on the total project cost because the capital costs are a relatively small fraction of the total cost.

A number of other factors will impact the project costs for the SPME technology. These factors could include site characteristics (e.g., soil heterogeneity) and design and operating parameters (e.g., injection flow rate). A complete quantitative analysis of these types of factors was beyond the scope of this report. A qualitative assessment of them can be found in the tables in Chapter 27. As has been mentioned a number of times in this chapter, any factor that will impact the quantity of surfactant or cosolvent could have a dramatic impact on total costs, since chemical and disposal costs have such a large impact on total costs.

26.4 COMPARISON WITH COSTS FOR OTHER TECHNOLOGIES

A comparison of the cost estimates developed here for the SPME technology for application at the hypothetical site with costs for other technologies is difficult. First, the cost must be brought to a common unit basis (e.g., dollars per acre or cubic yard). This can be difficult because of the variations in site depth if a unit area is used, and uncertainty in the contaminated soil volume to use to calculate a cost per volume of soil if a unit volume is used. Second, cost estimators often prepare cost estimates differently, making direct comparison difficult. For example, some estimators may not include laboratory testing of the technology in their total costs. Other estimates probably don't include site characterization in their total costs. The best way to compare costs of different technologies is to have the same estimator prepare estimates for different technologies for the same site. This was beyond the scope of this document, so only rough comparisons can be made here.

Table 26.3 Unit Cost Comparison

	Base Case		Optimized Cost	
	Quantity	Cost per Unit	Quantity	Cost per Unit
Treatment area				
Hectares (ha)	4.8	$12,200,000	4.8	$5,000,000
Acres	11.9	$4,930,000	11.9	$2,000,000
Treatment volume				
Cubic meters (m³)	144,000	$400	144,000	$170
Cubic yards (yd³)	188,000	$311	188,000	$130
Volume NAPL				
Liters (L)	692,000	$85	692,000	$35
Gallons (gal)	183,000	$320	183,000	$130

Table 26.3 summarizes the cost estimates developed here, expressed as unit area, unit volume, and unit quantity of NAPL removed for the Base Case and the Optimized Case. These estimates can be compared with the cost estimates for surfactant/cosolvent flushing reported by Simpkin et al., 1999. Figure 26.1 is a summary of the costs presented by Simpkin et al., 1999. The Base Case cost for the SPME is on the high end of the costs reported for other surfactant/cosolvent flushing technologies, but well within the range of reported values. The optimized costs are on the low end of costs developed by Simpkin et al., 1999.

These costs for SPME are on the high side for common *ex situ* remediation technologies (after excavation of the soil). For example, the Department of Defense Remediation Technologies Screening Matrix and Reference Guide (DOD, 1996), states that costs for low-temperature thermal desorption (LTTD) range from $40 to $100 per ton ($60 to $148/yd³) for typical soil. For deep sites, an additional $5 to $20/yd³ may be required for excavation (including removal of clean overburden). An additional $10/yd³ may be required to replace the contaminated soil, for a total cost ranging from $75 to $178/yd³. These costs for LTTD are somewhat old, and market conditions appear to have reduced these costs somewhat. Excavation and LTTD are possible alternatives for remediation of the hypothetical site developed here. Other, possibly lower cost *ex situ* technologies, such as biopile bioremediation, could also be used.

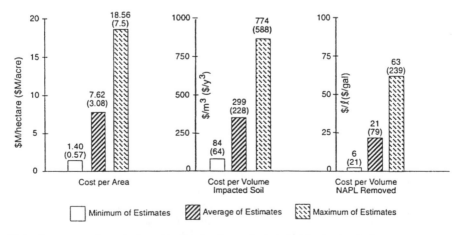

Figure 26.1 Average and range of cost estimates for surfactant flushing technologies.

In summary, the cost for the SPME technology could be in the same range as other competing technologies if the cost of a few key aspects can be kept low. Reducing the costs of the chemicals, residuals disposal, and produced fluids treatment is the key to reducing the overall cost of this technology. The Optimized Case described in this report is a good target to work toward to produce a potentially cost-effective approach. With further developments, it might be possible to achieve these lower costs.

CHAPTER 27

Performance and Potential Application

27.1 SUMMARY OF POTENTIAL PERFORMANCE

The field demonstration at Hill AFB OU1 provided valuable information concerning the potential effectiveness of the SPME technology. However, since the conditions at OU1 were unique in a few specific ways, it may be difficult to extrapolate these results to sites that are significantly different. OU1 was unique in terms of the NAPL present and the relatively homogeneous and permeable soils. Below is a summary of the potential performance of the technology:

- In the appropriate setting, the SPME technology is likely to remove a significant mass of NAPL. Removals from 75% to greater than 90% are possible.
- Soil endpoint concentrations that are potentially achievable will depend on the nature of the NAPL being removed and the initial concentrations. For the Hill AFB OU1 NAPL, average concentrations of individual compounds of less than 2.5 mg/kg were achieved. However, most of the compounds analyzed are not priority pollutants, so it is difficult to evaluate the merits of these endpoints. Estimated endpoint TPH concentrations ranged from 160 to 1100 mg/kg.
- Under the appropriate site setting, operational problems should not be major. Heterogeneities may, however, lead to longer periods of operation.

The SPME technology is not yet a fully developed technology. Consequently, additional laboratory and field testing and development are required on potential sites. Additional laboratory and field-scale tests and demonstrations are needed to address the following issues that may be important in full-scale implementation:

- Definition of the benefits of the mass removal that may be achieved with this technology
- Achievable soil and groundwater endpoint concentrations
- Performance of the technology with other contaminants, especially DNAPLs
- Performance of the technology in other hydrogeologic settings, especially those with heterogeneous soils
- Performance of the technology with fewer pore volumes of chemicals
- Cost-effective methods of dealing with produced fluids, including possible methods of recovery and reuse of the surfactant and cosolvent
- Improved methods to optimize chemical delivery to targeted contaminated zones of an aquifer
- Methods to optimize the system design, such as well spacing and alternative chemical injection methods

27.2 SITE CHARACTERISTICS AFFECTING PERFORMANCE, APPLICABILITY, AND COST

As with any *in situ* remediation technology, the SPME technology has a certain niche of site characteristics where it is likely to be applicable. Chapter 2 of Simpkin et al., 1999 provides a general description of the niche of surfactant/cosolvent flushing technologies. Many of these are also applicable to the SPME technology. Table 27.1 provides a summary of the site characteristics affecting applicability, performance, and cost of the technology. Because the technology has had limited field demonstrations, some of the following is speculation based on experience at other sites with similar technologies.

27.3 DESIGN AND OPERATING PARAMETERS AFFECTING PERFORMANCE, APPLICABILITY, AND COST

Table 27.2 provides a summary of the most important design and operating parameters, and how they might affect applicability, performance, and cost. Because the SPME technology has been field-tested at only one site under a limited number of design and operating conditions, some of the discussion in Table 27.2 is speculation based on the results of the demonstration and other similar technologies.

27.4 MATERIALS HANDLING REQUIREMENTS

The materials handling requirements for the SPME technology should be relatively straightforward. However, since many of the chemicals that are being used may be at relatively high concentrations and may cause environmental contamination if released, special care is needed to ensure there are no releases during the operation.

Piping and other materials of construction in contact with either the injected fluids or the recovered fluids must be selected to be compatible with the chemicals in the solution. Excessive corrosion may lead to leaks or failures of the materials. This may not be a critical issue for short-term projects.

It is possible that double containment with leak detection may be required for both injected and recovered fluids. Double containment will avoid contamination of the environment from leaks or breaks. Leak detection with automatic system shutdown can also be used. For exposed piping where leaks can be seen, it may be possible to not use double-containment as long as the system is monitored by on-site staff 24 hours per day so that it can be manually shut down in case of leaks.

27.5 REGULATORY REQUIREMENTS FOR PERFORMANCE AND COMPLIANCE CRITERIA

The regulatory requirements for performance and compliance criteria are very complicated and are based on the remedial objective. The requirements for proving performance or compliance with a remedial objective will be specific to the remedial objective. The following are examples:

- If restoration of the aquifer to drinking water standards is the remedial objective, then groundwater data will be the key to demonstrating compliance. Data on the ability of the technology to achieve drinking water standards are not available from the SPME project. However, it may be very difficult to achieve drinking water standards with any technology where NAPL and/or low-permeability media are present.

Table 27.1 Site Characteristics Affecting Applicability, Performance, and Cost

Parameter	Comments	Effects on Applicability	Effects on Performance	Effects on Cost
Remedial objective, endpoint required	Critical factor for selection of the technology	Potentially applicable at sites where mass removal is ultimate objective; may not be applicable where restoration of NAPL zone to drinking water is required		Less stringent cleanup objective are likely to require less time and to cost less
Presence of NAPL	If a significant amount of mobile NAPL is present, it should be removed first	Applicable to residual and possibly mobile NAPL; not likely to be applicable to dissolved or sorbed contaminants		Higher contaminant concentrations will increase the chemicals required and their cost, and the cost for produced fluids treatment
Chemical composition of NAPL	Specific chemical constituents and LNAPL or DNAPL are of importance	Demonstrated on LNAPL of mixed composition, but potentially applicable to other LNAPLs or DNAPLs	Chemical composition will greatly impact performance; different surfactant formulations may be needed	Surfactant formulation and other design/operating parameters will be impacted, which will also impact costs
Permeability of soils/ soil classification	Requires flushing of multiple pore volumes, thus impacted by permeability; demonstration at Hill AFB OU1 in relatively high-permeability soils	Most applicable to sites with high permeability (sands and gravel)	Performance not likely to be as good on low-permeability soils, or in fractured clay or rock settings	Durations will be longer with low-permeability soils, which will increase costs
Heterogeneity of soils	Highly layered or heterogeneous soils will impact ability to get surfactant chemicals to contaminant soils; mobility control agents may help reduce impacts on heterogeneities	Most applicable to relatively homogeneous soils; may not be applicable in very heterogeneous soils f contaminants are located in the low-permeability zones	Removals likely to be poorer in very heterogeneous soils; the time required is likely to be greater in heterogeneous soils since more pore volumes may be required	More pore volumes must be flushed for very heterogeneous soils, which will increase the costs
Depth of contamination	May not be a theoretical limitation	Most applicable to sites with contamination less than 22 to 32 m (70 to 100 ft)	At very deep sites, it may be difficult to control injection of fluids and to accurately locate the NAPL	Costs are likely to be greater for deeper depths
Presence of a competent confining layer	May be important for DNAPL sites with significant amounts of pooled DNAPL	Not applicable at sites without a competent confining layer and with pooled DNAPL		

continued

Table 27.1 (continued)   Site Characteristics Affecting Applicability, Performance, and Cost

Parameter	Comments	Effects on Applicability	Effects on Performance	Effects on Cost
Area of site	Large sites may require a phased approach			Large sites will have higher costs, although there will be some economies of scale, so the cost per unit area may decrease with increased area
Current land use	Active sites will be more complicated to work on	May not be applicable at active sites where vertical or horizontal wells cannot be installed	Horizontal drilling methods may have to be used at active sites with many obstructions; these may reduce the performance depending on the specifics of the drilling method	Costs will be higher for active sites
Monitoring and reporting required by regulators	May vary with the regulatory agencies involved			Extensive monitoring may increase the cost slightly
Permits required	May vary with the regulatory agencies involved	Permit limitations on chemicals that may be injected may limit the applicability in certain locations		Extensive permitting requirements could increase the cost
Availability of discharge facilities for liquid wastes	Final disposal of liquid waste streams may be difficult in remote settings since they may not have access to either a POTW or industrial WWTP	Most applicable at sites with existing wastewater treatment facilities that can treat the wastes		Cost for treatment of produced fluids will vary with receiving stream discharge requirements
Availability of residual management facilities	Some residual waste may require off-site incineration or other management	Most applicable at sites near low-cost residual management facilities		Cost for disposal of residuals will have a major impact on the project cost
Need for post-treatment or long-term care	Post-treatment or long-term care may be needed if the remediation goal is to achieve concentration based standards	Most applicable at sites where costly post-treatment and/or long-term care is not required		Post-treatment and long-term care may have a significant impact on the total costs
Availability and cost of local labor				For long-term operations, labor may have a significant impact on the total project costs
Site geochemistry	Water inorganic chemistry, ion exchange of soil, and surfactant/cosolvent sorption characteristics		Calcium may be detrimental to certain surfactants; the design of the chemical system may need to be adjusted to account for this. A high degree of sorption of the surfactant and/or cosolvent may reduce performance.	Water used to make up surfactants may require softening, which will increase the cost. A high degree of sorption will increase the mass of chemicals required and thus the cost.

Table 27.2 Design and Operating Parameters Affecting Applicability, Performance, and Cost

Parameter	Comments	Effects on Performance	Effects on Cost
Injection/extraction system	Pattern of wells or horizontal drain lines	Choice of wells or horizontal well lines will impact performance; pattern of wells will influence dead zones that may not be remediated	Cost must be optimized on a case-by-case basis; trade-offs between performance and optimized injection/ extraction system
Spacing of injection/ extraction system	Smaller spacings generally required in lower permeability settings	Well or drain line spacing may have an impact on performance and time to complete	Spacing will impact time to complete and cost
Screened interval or depth of injection point	Chemicals must be injected at appropriate depths	Inappropriate depth of injection could result in chemicals missing contaminated zones	
Surfactant type and concentration		The appropriate surfactant will have an impact on performance of NAPL removal; the concentration will also impact removals; the surfactant type and concentration will influence the performance of the produced fluids treatment system	The cost of different surfactants varies; higher concentrations will increase the cost to purchase the surfactant as well as the cost for treatment of the produced fluids and residual disposal
Cosolvent type and concentration	Cosolvents may be necessary to achieve the single-phase microemulsion	The type and concentration of cosolvent may be critical to achieving NAPL removal with the SPME technology; the cosolvent may influence the performance of the produced fluids treatment system	Cosolvent is relatively inexpensive, however, it will add to the complexity and cost of produced fluids treatment and residual disposal
Mobility control agents	Agents such as polymer have not been used to date with the SPME technology, although they could be	At sites with significant heterogeneities, the addition of mobility control agents may improve the removal of NAPL from less permeable zones	The cost of polymer is typically not substantial since they are used at relatively low concentrations
Number of pore volumes	Measure of the volume of chemicals flushed	Critical to the NAPL removals achieved; although the majority of NAPL may be removed in the first one or two pore volumes, many more may be needed to achieve high degrees of removal	The number of pore volumes will dictate the quantity of chemicals required and thus the cost for the chemicals and the treatment and disposal of the produced fluids
Sequence of chemical floods	A pre-water flood, surfactant/alcohol flood, surfactant-only flood, and post-water flood were used for the SPME demonstration	The sequence of flooding used may have some influence on the NAPL removal achieved	
Gradient applied	The head used will impact the flow rate of fluids, depending on the soil characteristics	Head and thus flow rate will impact duration	

continued

Table 27.2 (continued)   Design and Operating Parameters Affecting Applicability, Performance, and Cost

Parameter	Comments	Effects on Performance	Effects on Cost
Flow rates	Flow rate will impact duration	Flow rate will impact duration	The balance between flow rate and duration will have an impact on cost; high flow rates will require larger process equipment
Materials of construction	Surfactants and other additives, along with the contaminants, will impact the materials of construction required	The physical system may fail if very corrosive chemicals are present and the proper materials of construction are not used	Certain materials (e.g., stainless steel) that are more resistant to corrosive chemicals may increase costs
Produced fluids treatment system complexity	The type and complexity of the produced-fluids treatment system will depend on the chemical solution used and the NAPL	Failure of the produced fluids system could result in failure of the entire system	The complexity of the produced-fluids treatment system could increase the cost for this system, which may be a significant portion of the overall cost
Degree of automation	The degree of automation will impact the labor required	Automation may improve the potential to understand the system, which could improve the performance	Automation may increase the capital cost but could reduce the labor costs of operations
Monitoring, sampling, and analysis used		Additional monitoring and sampling may improve the potential to understand the system, which could improve the performance	Additional monitoring and sampling will increase the costs somewhat

- If mass reduction is the remedial objective, a number of methods can and should be used to measure compliance. Soil cores before and after flushing are probably the best true measure, but they are prone to very high variability, depending on variability in soil conditions. An important question to be answered with the use of soil cores is which compounds should be measured at sites that have NAPLs that are complex mixtures (i.e., should total NAPL be measured or only specific compounds that drive the risk?). Partitioning tracer tests can also be used to measure mass reduction. However, there are limitations with these tests, as was observed in the SPME demonstration. The analysis of the produced fluids can also be used to estimate the amount of mass removed from the system, but this analysis alone will not provide any indication of the amount of mass still remaining in the formation.
- The remedial objectives can also be based on achieving as much mass removal as is practical with a given technology. This is similar to a best-available or -practical technology approach, as is used for some air and wastewater treatment criteria. With this type of criteria, the produced fluids can be analyzed and a certain percentage of the "maximum removable" contaminants would have to be removed. The "maximum removable" level would be based on extrapolation of production curves to infinite time.

CHAPTER 28

Part IIIB References

American Public Health Association. (1990) *Standard Methods*, 17th edition, APHA, Washington, D.C.

American Society for Testing and Materials. (1995). *Standard Guide for Developing Conceptual Site Models for Contaminated Sites*. Designation: E 1689-95, Philadelphia, PA.

Bear, J. (1972). *Flow Through Porous Media*. Elsevier, New York.

Cayais, J.L., Schechter, R.S., and Wade, W.H. (1975). *The Measurement of Interfacial Tension via the Spinning Drop Technique, Adsorption at Interfaces*. American Chemical Society Symposium Series, Vol. I, Washington, D.C. 234–248.

Corey, A.T. (1986). *Mechanics of Immiscible Fluids in Porous Media*. Water Resources Publications.

CH2M Hill. (1997). Draft Final Feasibility Study for Operable Unit 1, Hill Air Force Base, Utah, May.

Federal Remediation Technologies Roundtable. (1995). Guide to Documenting Cost and Performance for Remediation Projects. EPA-542-B-95-002.

Jawitz, J.W., Annable, M.D., Rao, P.S.C., and Rhue, R.D. (1998). Field Implementaion of a Winsor Type I surfactant/alcohol mixture for *in situ* solubilization of a complex LNAPL as a single-phase microemulsion. *Environ. Sci. Technol*. 32:523–530.

Simpkin, T.J. et al. (1999). *Surfactants and Cosolvents for NAPL Remediation*, CRC Press, Boca Raton, FL

U.S. Department of Defense. (1996). *Remediation Technologies Screening Matrix and Reference Guide*. In http://clu-in.com, NTIS PB95-104782

US EPA. (1986). *Test Methods for Evaluating Solid Waste*, SW-846, 3rd edition, as amended.

US EPA. (1983). *Methods for Chemical Analyses of Water and Wastes*. EPA 600/4-79/020, as amended.

Analytical Procedures Used in the Demonstration of the Surfactant/Foam Process for Aquifer Remediation

A1.1 THE POTENTIOMETRIC TITRATION OF HALIDE IONS

Chloride affects interfacial activity and phase behavior of surfactant solutions. In fact the interfacial tension of an NAPL/surfactant solution is controlled by adjusting the chloride concentration. Chloride and other halides can also serve as nonadsorbed tracers to monitor the transport of surfactant and partitioning tracers. Therefore, analysis of halides, and especially chloride, during a surfactant-enhanced recovery process is essential.

A potentiometric titration method for determining halide ions in surfactant-enhanced remediation solutions is presented. The method was tested using solutions of sodium chloride; an example of analysis of a mixed chloride/bromide solution from the field is presented.

The potentiometric titration for halide ion utilizes an Ag,Ag|AgCl electrode pair. The relationship between electrode potential and halogen ion concentration is described by the Nernst equation. Although it is possible to calculate concentrations directly from measured potential, there are many complicating factors. In practice, one titrates an unknown solution with a standard solution of $AgNO_3$ and records electrode potential as a function of titrant volume. The equivalence point of the reaction $Ag^+ + Cl^- = AgCl$ is the inflection point in the sigmoidal titration curve.

Reagents:

1. Silver nitrate, available as standardized solution or crystalline solid
2. Sodium acetate
3. Acetic acid
4. Sodium chloride
5. Triton X-100. (ethoxylated octyl phenol)

All the above are available from:
Aldrich Chemical, 1001 West Saint Paul Ave., Milwaukee, WI, 53233;
Phone: 1-800-558-9160

Apparatus:

1. A combination electrode, Ag,Ag|AgCl; or a silver bullet electrode and a single or double junction reference electrode. An example of the former is Metrohm Cat. No. 6.0404.100.
2. An automatic titrator, or a pH meter capable of measuring potential and a burette.

3. A balance capable of accurate weight to at least two decimal places. A combination of a top-loading electronic balance, accurate to two decimal places, and an analytical balance accurate to four decimal places is best.
4. A 10-mL pipet.
5. Appropriate beakers and measuring glassware.

Procedure:

1. Prepare a standard solution of sodium chloride, approximately 0.1 M or 0.1 moles/1000 g solution: weigh accurately about 5.9 g NaCl and dissolve in enough deionized water to make 1 L or 1000 g solution. The use of weight or volume as the basis of analysis is the choice of the analyst.
2. Prepare a solution of silver nitrate, approximately 0.1 M, (17 g/L). This concentration is somewhat arbitrary. A titrant should be dilute enough to yield a measurable titration volume with a reasonable sample of unknown. That is the only constraint.
3. Prepare a buffer solution 0.2 M in sodium acetate and 0.2 M in acetic acid.
4. Accurately weigh approximately 5 to 10 g standard sodium chloride solution into a 100-mL beaker and dilute to approximately 60 mL with deionized water.
5. Pipet in 10 mL acetate buffer (item 3 above) and add a couple of drops of a 1% Triton X-100 solution.
6. Immerse the combination electode (or the electrode pair) in the analyte solution. Commence stirring using a magnetic stirrer. If titrating manually with a pH meter and burette, record the voltage.
7. Begin adding titrant in small increments. If titrating manually, observe and record voltage changes and volume changes.
8. Continue titrating past the maximum rate of voltage change. With an automatic titrator that updates the titration curve on a plot or visual display, this presents no problem. A manual titration requires that the analyst plot the data while collecting it, or else be experienced enough to pick an approximate inflection point.
9. Locate the titration endpoint by drawing a tangent to the inflection portion of the curve and calculating a midpoint. Otherwise, accept the endpoint calculated by the automatic titrator.
10. Calculate the normality of the silver nitrate solution according to the following: N_t = meq/g (sodium chloride solution) × wt. (sodium chloride solution)/mL titrant. Repeat at least three times.
11. For unknown determinations, weigh enough sample into a 100-mL beaker to provide approximately 0.5 to 1 meq chloride for titration. This is based on 0.1 N titrant. Other concentrations of titrant will require different quantities of sample. The goal is to provide enough chloride to titrate to an accurate endpoint, but not so much as to make the analysis unnecessarily lengthy. Dilute to approximately 60 mL with deionized water.
12. Repeat steps 5 through 9. Calculate halide ion concentration according to: Cs = N_t × Volume titrant/weight sample, where N_t is the normality of the titrant. An automatic titrator provides a calculation of chloride ion concentration if it is programmed to do so.
13. When titrating samples containing both bromide and chloride ions, volume to the first inflection point is the titer for bromide ion. The total titer minus the first titer is the titer for chloride ion. An example of a titratrion curve for a sample containing both chloride and bromide appears in Figure A1.1.

Notes on the Method:

1. The procedure outlined here is based on using a 20-mL burette to contain the titrant (other sizes of burette may be chosen). The object is to select sample sizes so that repetitive burette filling is not necessary. Sample weights rather than volumes were used. Note, however, that the titration is based on volume; thus, the titrant will be standardized in M (molarity).
2. Silver nitrate titrant should be protected from light and should be restandardized frequently. Light promotes the decomposition of AgCl to Ag and Cl_2. The acid buffer is added to suppress the reaction:

$$3Cl_2 + 3H_2O + 5Ag^+ = ClO_3 + 6H^+ + 5AgCl$$

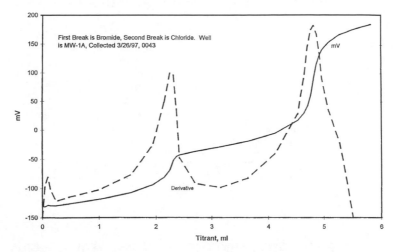

Figure A1.1 Halogen ion titration curve.

which further consumes Ag⁺ and gives high results. However, this is usually not a problem in rapid titrations.

3. As long as one uses deionized water and determines that it in fact contains no chloride, frequent blanks are not strictly necessary. However, blanks are sound practice. When changing water sources, or after preparing fresh reagents, the analyst should run a blank. When using tap water, or source waters, blanks are essential. Table A1.1 contains data illustrating the precision of the method. Data in Table A1.2 determine a detection limit for the chloride analysis. Constraints were a 1-g sample and approximately 0.025 N AgNO₃ titrant.

Table A1.1 Precision of Potentiometric Titration of Halogens Method

Sample Number	Concentration (meq/g)
1	0.058
2	0.058
3	0.0591
4	0.0591
5	0.0590
6	0.0589
7	0.0592
8	0.0588
%RSD	1%
9	0.0311
10	0.0323
11	0.0296
12	0.0294
13	0.0295
%RSD	4%
14	0.0148
15	0.0148
16	0.0149
17	0.0149
18	0.0147
%RSD	1%

Table A1.2 Determination of the Detection Limit of the Chloride Titration

Sample Number	Concentration (meq/g)
1	0.0013
2	0.0018
3	0.0025
4	0.0015
5	0.0029
Mean	0.0020
Standard deviation	0.000678
Mean - 3 Standard Deviation	$3.47E - 05$

Note: Under the constraint of a 1-g sample and 0.025 N $AgNO_3$, the method detection limit is approximately 0.002 meq/g.

A1.2 THE POTENTIOMETRIC TITRATION OF TOTAL CALCIUM AND MAGNESIUM, AND CALCIUM ONLY IN SOLUTION

A method for determining calcium and magnesium is presented first. The method for calcium differs only in type of titrant. All chemicals arc reagent grade.

Reagents:

1. CDTA: *trans*-[1,2-Cyclohexylenedinitrilo]tetraacetic acid monohydrate
2. Ammonium chloride
3. Ammonium hydroxide
4. Calcium chloride for standardization
5. Potassium hydroxide for pH adjustment during solution preparation

Vendor: Aldrich Chemical, 1001 West Saint Paul Ave., Milwaukee, WI, 53233; Phone: 1-800-558-9160

Apparatus:

1. A double- or single-junction reference electrode. Ag/AgCl is preferred.
2. A calcium ion-specific-indicating electrode.
3. An automatic titrator; or a pH meter capable of measuring potential and a burette.
4 A balance capable of accurate weight to at least two decimal places.
5. A combination of a top-loading electronic balance (accurate to two decimal places) and an analytical. balance accurate to four decimal places is best.
6. A 5-mL pipet.
7. Appropriate beakers and measuring glassware.

Procedure:

1. Prepare a standard solution of calcium chloride. For dilute solutions of divalent cations (e.g., fresh waters) 0.025 M is suitable. Weights are 2.77 g anhydrous calcium chloride to be dissolved in deionized water to make 1 L of solution.
2. Prepare a standard solution of CDTA., approximately 0.025 M. The formula weight of CDTA is 364.36 g/mol. Therefore, a 0.025 M solution contains 9.198 g/L. This material may not dissolve readily unless the analyst adjusts the pH to 10 or above. If necessary, do not hesitate to do so. Use KOH to adjust pH. Some sources recommend dissolving CDTA in a 0.1 M KOH solution.

3. Prepare a pH 10 buffer solution as follows: add 300 mL deionized water to 570 mL concentrated ammonium hydroxide. In this mixture, dissolve 60 g ammonium chloride. Bring to 1 L with deionized water.

4. The procedure outlined here is based on using a 20-mL burette to contain the titrant (other sizes of burette can be chosen). The object is to select sample sizes so that repetitive burette filling is not necessary. Sample weights were used rather than volumes. Note, however, that the titration is based on volume; thus, the titrant will be standardized in M (molarity).

5. Accurately weigh approximately 5 to 10 g standard calcium chloride solution into a 250-mL beaker and dilute to approximately 150 mL with deionized water.

6. Pipet in 5 mL of pH 10 buffer (item 3) and add a couple of drops of 1% Triton X-100 solution. This keeps the electrode frits and membranes clean.

7. Immerse the electrode pair in the solution of analyte. Commence stirring, using a magnetic stirrer. If titrating manually with a pH meter and burette, record the voltage. Begin adding titrant in small increments. If titrating manually, observe and record voltage changes and volume changes. Continue titrating past the maximum rate of voltage change.

8. Locate the titration endpoint by drawing a tangent to the inflection portion of the curve and calculating a midpoint. Otherwise, accept the endpoint calculated by the automatic titrator; these are usually based on a maximum in the derivative of EMF vs. volume.

9. Calculate the molarity of the CDTA titrant according to the following:

$$M_t = (\text{Conc. Std} \times \text{Wt. Std})/V_t \tag{A1.2.1}$$

where M_t is molarity of titrant,
  Conc. Std = concentration of the sodium chloride standard
  Wt.Std = the weight of the sodium chloride standard
  V_t = the volume of sodium chloride standard

Repeat at least three times.

10. For unknown determinations, weigh enough sample into a 250-mL beaker to provide approximately 0.1 to 0.2 mm total calcium plus magnesium for titration. Add buffer and Triton X-100 solution, as above, and dilute to approximately 150 mL with deionized water.

11. Repeat steps 3 through 6. Calculate total calcium plus magnesium concentration according to:

$$C_s = M_t \times V_t/W_s \tag{A1.2.2}$$

where W_s = Weight of the unknown sample.
An automatic titrator provides a calculation of total calcium plus magnesium concentration if it is programmed to do so.

The Potentiometric Titration of Calcium Only is identical to the above except that CDTA titrant is replaced by EGTA ([ethylene-bis(oxyethylenenitrilo)tetraacetic acid). The formula weight of EGTA is 380.35 g/mol; therefore, a 0.025 M standard solution contains 9.51 g/L.

Notes on the Method:

1. The procedure here is based on using a 20-mL burette to contain the titrant (other sizes of burette may be chosen). The object is to select sample sizes so that repetitive burette filling is not necessary. Sample weights rather than volumes were used. Note, however, that the titration is based on volume; thus, the titrant will be standardized in M (molarity).

2. As long as deionized water determined to contain no calcium or magnesium is used, frequent blanks are not strictly necessary. However, blanks are sound practice. When changing water sources, the analyst should run a blank. When using tap water, or source waters, blanks are essential.

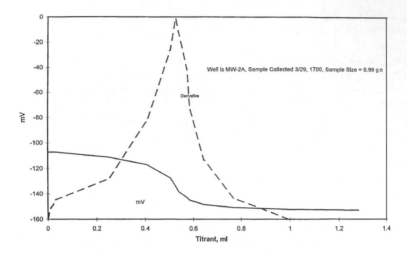

Figure A1.2 Titration curve of a field sample containing calcium.

3. In testing this method, the calcium analysis was stressed. A typical titration curve of a field sample containing calcium appears in Figure A1.2. The instrument used was a Metrohm Titrino Model 716 autotitrator. Table A1.3 presents data to estimate the accuracy of the combined calcium plus magnesium method. Average error was 2%. Data in Table A1.4 illustrate the accuracy of the calcium method in solutions containing both calcium and magnesium. The average error was 3%. Data in Table A1.5 show that surfactant (sodium dihexyl sulfosuccinate) does not interfere with the calcium analysis. Data in Table A1.6 demonstrate that the detection limit of the calcium method using 1-g samples and 0.0310 M titrant is less than 0.0006 M.

Table A1.3 Measurement Accuracy of Total Divalent Cations by Potentiometric Titration.

Calcium Stock[a] (g)	Magnesium Stock[a] (g)	Measured Ca+ Mg (M)	% Error
0.05	0	0.0265	6
0.49	0	0.0264	6
0.50	0	0.0256	2
0.54	0	0.0238	5
1.01	0	0.0255	2
1.02	0	0.0253	1
2.02	0	0.0257	3
2.03	0	0.0255	2
3.01	0	0.0254	2
4.99	0	0.0256	2
5.00	0	0.0253	1
0.53	0.49	0.0244	2
1.05	1.0	0.0247	1
2.01	2.02	0.0248	1
2.99	1.99	0.0248	1
2.01	3.0	0.0246	2
1.01	4.02	0.0245	2
4.03	1.02	0.0247	1
0	5.02	0.0249	0
		Average	2

[a] All stocks were 0.0250 M.

Table A1.4 Measurement Accuracy of Calcium by Potentiometric Titration.

Calcium Stock[a] (g)	Magnesium Stock[a] (g)	Measured Conc. (M)	% Error
4.01	1.0	0.0205	2
3.00	2.01	0.0154	3
2.51	2.48	0.0129	3
2.01	3.0	0.0104	4
1.56	3.52	0.0080	4
1.00	4.12	0.0051	4
		Average	3

[a] Stocks were 0.0250 M in Ca^{2+}.

Table A1.5 Accuracy of the Calcium Method in the Presence of Surfactant MA-80I and Salt.

Sample	Calculated Ca (M)	Measured Ca (M)	% Error	Wt% Surfactant	NaCl (ppm)
1	0.0220	0.0246	12	2	4035
2	0.0220	0.0214	3	1	3134
3	0.0220	0.0222	1	2	3651
4	0.0220	0.0216	2	2	4197
5	0.0220	0.0227	3	2	3938
Standard 1	0.0220	0.0214	3	0	0
Standard 2	0.0220	0.0219	0	0	0
% RSD		5%			
Avg. Error			3%		

Table A1.6 Detection Limit for Ca^{2+} Analysis

Sample	Measured Concentration (M)	Calculated Concentration (M)
Standard	0.0216	0.0220
1	0.0008	
2	0.0007	
3	0.0006	
4	0.0005	
5	0.0006	
Average	0.0006	
Standard Deviation	0.0001	
Average - 3 × Std. Deviation	0.0003	
Detection limit	<0.0006	

A1.3 THE POTENTIOMETRIC TITRATION OF ANIONIC DETERGENTS

The indicating electrode is a combination electrode pair consisting of an Ag|AgCl reference cell and a PVC membrane cell. The composition of the PVC membrane is optimum for surfactant determination. An electric potential, which is a function of analyte concentration, forms at the solution/membrane phase boundary. As with other electrode reactions, electrode response is described by the Nernst equation:

$$E = E_o + S \times \log C \tag{A1.3.1}$$

where E is electrode potential in millivolts, E_o is a constant, S is the electrode slope, and C is concentration. In the ideal case, S is 59 mV per concentration decade of monovalent anions and cations. In practice, lower slopes occur, especially for specific ion indicator electrodes discussed here. Therefore, titrations should always be evaluated using the inflection point of the sigmoidal titration curve. The inflection point corresponds to the reaction equivalence point.

Reagents:

1. Benzethonium chloride (Hyamine 1622), the titrant reagent
   Vendors: Gallord-Schlesinger, Inc., 584 Mineola Ave., Carle Place, NY 11514;
   Phone: 516-33-5600; 0.004 M solution
   Sigma Chemical Co., P.O. Box 355, Milwaukee, WI; Phone: 800-325-3010; crystalline solid, Cat. #B8879;
   Aldrich Chemical, 1001 West Saint Paul Ave., Milwaukee, WI, 53233; Phone: 800-558-9160; crystalline solid, Cat. #B470-8.
   For any large number of analyses, the solid materials are much more economical.
2. Sodium dodecyl sulfate for standardizing the titrant; other surfactants can also be used. (A good alternative is sodium dodecyl benzene sulfonate.)
   Vendor: Aldrich Chemical, 1001 West Saint Paul Ave., Milwaukee, WI, 53233;
   Phone: 800-558-9160
3. Triton X-100 (ethoxylated octyl phenol)
   Vendor: Aldrich Chemical, 1001 West Saint Paul Ave., Milwaukee, WI, 53233;
   Phone: 800-558-9160

Apparatus:

1. Beakers for titration
2. Magnetic stirrer
3. Combination anionic surfactant specific electrode. Available from Brinkman Instruments, Inc. Cat No. 019-01-950-1. Also available from Pheonix Electrode Co., 6103 Glenmont, Houston, TX 77081; Phone: 800-522-7920. Cat. No. SUR1502 or SUR1503.
4. An automatic titrator, or a pH meter capable of measuring potential and a burette. Work reported here was done on a Metrhom Titrino Model 716 available from Brinkman Instruments.
5. A balance capable of accurate weight to at least two decimal places. A combination of a top-loading electronic balance accurate to two decimal places, and an analytical balance accurate to four decimal places is best.

Procedure:

1. Prepare a standard solution of sodium dodecyl sulfate, approximately 0.004 M.
2. Prepare a solution of benzethonium chloride, approximately 0.004 M. This concentration is somewhat arbitrary. A titrant should be dilute enough to yield a measurable titration volume with a reasonable sample of unknown, yet concentrated enough so that one does not use large volumes of titrant. Those are the only constraints. The procedure outlined here is based on using a 10-mL burette to contain the titrant. Other sizes of burette can be used. The object is to select sample sizes so that repetitive burette filling is not necessary. The work was done in sample weights rather than volumes. Note, however, that the titration is based on volume; thus, the titrant will be standardized in M (molarity).
3. Accurately weigh approximately 5 g standard sodium dodecyl sulfate into a 100-mL beaker and dilute to approximately 60 mL with deionized water.
4. Immerse the surfactant-specific electrode in the solution of analyte. Add one or two drops of 1% Triton X-100 solution. Commence stirring using a magnetic stirrer. If titrating manually with a pH meter and burette, record the voltage.
5. Begin adding titrant in small increments. Observe and record voltage changes.
6. Continue titrating past the maximum rate of voltage change. With an automatic titrator that updates the titration curve on a plot or visual display, this presents no problem. Manual titration requires

that the analyst plot the data while collecting it, or else be experienced enough to pick an approximate inflection point.

7. Locate the titration endpoint by drawing a tangent to the inflection portion of the curve and calculating a midpoint. Otherwise, accept the endpoint calculated by the automatic titrator.
8. Calculate the normality of the benzethonium chloride as follows:

$$N_t = C_s \times W_s/V_t \qquad (A1.3.2)$$

where N_t = Normality of titrant
      C_s = Concentration of standard in meq/g
      W_s = Weight of standard in grams
      V_t = Volume of titrant in mL.

Repeat at least three times

9. For unknown determinations, weigh enough sample into a 100-mL beaker to provide approximately 0.02 meq surfactant for titration. Dilute to approximately 60 mL with deionized water.
10. Repeat steps 4 through 7 and calculate surfactant concentration according to:

$$C_{samp} = N_t \times V_t/W_{samp} \qquad (A1.3.3)$$

where C_{samp} = Sample concentration in meq/g
      W_{samp} = Weight of sample in g.

Of course, an automatic titrator provides a calculation of surfactant concentration if it is programmed to do so.

Notes on the Method:

1. Notice that no buffer is recommended. As far as known, none is required for solutions at near-neutral pH. However, strong solutions of acids or bases may interfere or even damage electrodes. There is anecdotal evidence that solvents will destroy surfactant-specific ion electrodes. Samples containing several thousand ppm trichloroethylene and also Hill Air Force Base contaminant were run with no obvious electrode damage. However, samples of field DNAPL larger than about 0.25 g tended to temporarily foul electrodes. The result is slow electrode response and vague endpoints. Of course, these smaller samples reduce accuracy.
2. As with all titrations, sample size and normality of titrant affect method precision and accuracy. When using metered pipets to deliver sample to the titration vessel, it is convenient to use 1-g samples. These are large enough to ensure accuracy, provided one uses a two-place balance to determine sample weights. Automatic titrators perform efficiently when one adjusts titrant normality so that titration volumes are between 1 and 10 mL; 1-mL titers are large enough to ensure accuracy. Titers larger than 10 mL result in frequent burette filling and excessive titrant consumption. A typical surfactant titration curve, determined using an automatic titrator, appears in Figure A1.3. The subject sample is an injectate sample collected 3/24/97 at 6:45 a.m. Sample size was 1.01 g, and titrant concentration was 0.014 N.
3. The overall accuracy and precision of titration for anionic surfactant was found to be about ±5% of the amount of surfactant present (see Tables A1.7–A1.9). The detection limit of the anionic surfactant method was investigated and results appear in Table A1.10. Successively lower concentrations of surfactant were titrated using a fixed sample size (1 g) and a fixed titrant normality (0.014 N). For multiple determinations, the detection limit was taken to have been reached whenever the average concentration minus three times the determined standard deviation was approximately zero. Subject to the above constraints of sample size and titrant normality, 0.001 M surfactant exceeded the method detection limit. Larger samples and more dilute titrant will produce even lower method detection limits. For purposes of reference, the CMC of MA-80I in 11,000 ppm NaCl is approximately 0.025 M.

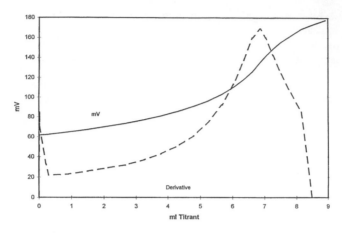

Figure A1.3 Typical surfactant titration curve.

Table A1.7 Precision of the Surfactant Method

Sample No.	Nominal Concentration (wt%)	Measured Concentration (meq/g)
1	5	0.096
2	5	0.093
3	5	0.097
4	5	0.095
5	5	0.096
	Standard Deviation	0.002
	%RSD	2%
1	1	0.0210
2	1	0.0213
3	1	0.0210
4	1	0.0207
5	1	0.0208
	Standard Deviation	0.0002
	%RSD	1%
1	0.17	0.0030
2	0.17	0.0037
3	0.17	0.0034
4	0.17	0.0032
5	0.17	0.0035
	Standard Deviation	0.0003
	%RSD	8%
1	0.08	0.0013
2	0.08	0.0016
3	0.08	0.0015
4	0.08	0.0015
5	0.08	0.0014
6	0.08	0.0016
7	0.08	0.0016
8	0.08	0.0017
9	0.08	0.0016
10	0.08	0.0017
	Standard Deviation	0.0001
	%RSD	8%

Note: Analyte was sodium dihexyl sulfosuccinate; mol. wt. 388. MA-80I
product is 80% active.

Table A1.8 Analyses of Surfactant Solutions Containing Both Sodium and Calcium Chloride

Wt%	NaCl (ppm)	CaCl$_2$ (ppm)	Measured Conc. (meq/g)	Calculated Conc. (meq/g)	Error	% Error
1	1000	0	0.0201	0.0206	0.0005	3
1	5000	1000	0.0208	0.0206	0.0002	1
1	0	2000	0.0217	0.0206	0.0011	5
3	10000	0	0.062	0.062	0.000	0
3	5000	1000	0.065	0.062	0.003	5
3	5000	1000	0.064	0.062	0.002	3
3	5000	1000	0.069	0.062	0.007	11
3	0	2000	0.064	0.062	0.003	4
5	10000	0	0.111	0.103	0.008	8
5	5000	1000	0.112	0.103	0.009	8
5	0	2000	0.113	0.103	0.010	10
				Average		5

Note: Normality based on 80% active ingredient in MA-80I.

Table A1.9 Spiked Field Samples to Investigate Interferences

Sample No.	Field Sample RSO$_3$ (meq/g)	Weight of Field Sample (g)	Weight of Standard (g)	Conc. of Spiked Sample (meq/g)	RSO$_3$ from Field Sample (meq/g)	RSO$_3$ from Spike (meq/g)	Recovery of Spike (%)
229M	0.0646	0.48	0.41	0.0822	0.0348	0.0468	101
229M	0.0646	0.58	0.58	0.0814	0.0323	0.0508	97
222M	0.0633	0.50	0.58	0.0846	0.0293	0.0546	101
222M	0.0633	0.50	0.53	0.0874	0.0307	0.0523	108
204A	0.0476	0.50	0.57	0.0754	0.0222	0.0541	98
204A	0.0476	0.56	0.50	0.076	0.0251	0.0479	106

Table A1.10 Detection Limit of the Anionic Surfactant Titration

Sample 1 g	Measured Concentration (meq/g)
1	0.0012
2	0.0011
3	0.0013
4	0.0010
5	0.0007
6	0.0013
7	0.0013
8	0.0007
9	0.0009
10	0.0010
Mean	0.0011
Standard Deviation	0.0002
Detection Limit; assumed to be reached when mean = 3 × standard deviation	<0.001
Practical Quantitation Limit = 5 × detection limit	0.005

Note: 0.014 *N* titrant, one gram sample size.

4. Cullem (1994) advises rinsing and wiping electrodes with tissue and placing them in a solution of nonionic surfactant for a few seconds before each analysis. This was not found to be necessary. The electrode manufacturer, Phoenix Electrode Co., recommends that pH be controlled between 2 and 12 or the electrode will be damaged. Also, electrodes respond best to sulfates, or sulfonates, in slightly acid solutions, between pH 2.5 and 4.5. Solution pH is adjusted with HCl. The pH range was not found to be this restrictive, and sample pH was not adjusted as long as it was near neutral.

5. This procedure does not call for a blank determination before each analysis. Stray surfactant is usually not a problem if the water source is clean. However, it is sound practice to run blanks when changing water sources — for example, when switching to a new bottle of deionized water. Also, if using nondisposable beakers and other containers, be sure they are well rinsed after washing.

Interferences:

1. Rice University analysts did not investigate the full range of interferences. They did analyze samples containing different concentrations of monovalent and divalent cations and found them not to interfere. Rice University analysts also analyzed samples containing field contaminants and found no interferences. Results of these analyses appear in Table A1.9.

A1.4 DNAPL ANALYSIS

A1.4.1 Analysis of DNAPL in Soil Cores

The procedure is based on specific instruments, namely a gas chromatograph manufactured by Hewlett Packard, Model 6890; a purge-and-trap apparatus manufactured by OI Analytical, Model 4560, equipped with an automatic spiker; and an OI Analytical autosampler, Model 4551. Other instruments may serve as well. In fact, for toluene extracts, on-column injection should serve as well as purge-and-trap. In fact, on-column injection might produce lower detection limits than purge-and-trap in this application. Descriptions of instruments, along with applicable method parameters, follow the procedure.

Procedures:

1. Prepare a five-point set of standards by dissolving DNAPL components in toluene. Choose concentrations to correspond with DNAPL saturations of less than 1% to approximately 8%. To estimate appropriate concentrations in toluene, one must assume volumes of toluene extract, soil sample size, soil grain density, and soil porosity. Appropriate standard compositions can be estimated using the following equation:

$$\text{DNAPL}_{\text{toluene}} = \text{Wt}_{\text{soil}} \times \phi/((1-\phi) \times \rho) \times Sat/\text{V}_{\text{extract}} \qquad \text{(A1.4.1)}$$

where
$\text{DNAPL}_{\text{toluene}}$ = Volume of DNAPL per mL toluene in the standard
ϕ = Porosity
ρ = soil grain density
Wt_{soil} = Weight of the soil sample
$\text{V}_{\text{extract}}$ = Volume of toluene added to the soil sample to extract DNAPL
Sat = Fractional saturation

A porosity of 0.33, a soil grain density of 2.65 g/cm^3, a 50-g soil sample, and an extract volume of 80 mL were assumed.

2. Dissolve 5 μL of a standard in 40 mL deionized water contained in 40-mL sample vials.

3. Dissolve 5 μL of an unknown (i.e., toluene extracts containing DNAPL) in 40 mL deionized water contained in 40-mL sample vials.

4. Prepare diluted unknown samples containing known quantities of analyte, that is, matrix spikes in one of two ways:

   a. Mix aliquots of known solutions of analytes in toluene with aliquots of unknown samples, and then add 5 μL of the mixture to 40 mL deionized water.

   b. Add 1 μL of a known solution of analytes in toluene directly to 40 mL deionized water containing 5 μL unknown soil extract.

   The second technique is more convenient. An automatic spiker to deliver matrix spikes would be more convenient and more accurate.

5. Let diluted samples containing standards and unknowns age for at least 6 h at room temperature with frequent shaking. Adequate aging with shaking is essential. Five μL toluene in 40 mL deionized water produces a solution that is only about 25% saturated with toluene (cf. *Handbook of Physics and Chemistry*, 36th edition, 1954–55, Chemical Rubber Publishing Co., 2310 Superior Ave. N.E., Cleveland OH, p. 1178). However, toluene dissolves in water slowly. For accurate analyses, one must be certain that all components are dissolved.

6. Load the automatic spiker with a solution of the compound chosen for the required surrogate spike; this spiker delivered 10 μL spike into each sample. Carbon tetrachloride was chosen as the surrogate. Solutions of carbon tetrachloride in both methanol and water have been used successfully as surrogates.

7. Load standards, samples, spiked samples, and blanks into an autosampler that holds 40 vials. Usually, a five-point set of standards was included at the beginning of a run, and a three-point set of standards was placed either at the end of a run or distributed throughout the run. These extra standards serve as calibration checks. Runs usually included three or four blanks; these were positioned so as to minimize carry-over of analyte from anticipated high-concentration to low-concentration samples.

8. Run GC analysis.

9. Obtain GC-TCE peaks areas by integrating the chromatograms automatically, using Hewlett Packard Chem Station software. Analyte peaks were usually sharp and symmetrical.

10. Use peak areas and analyte concentrations to establish a calibration curve. Usually, a linear regression line forced through the origin is satisfactory, although occasionally a quadratic curve may better serve accuracy. An example of a standard curve appears in Figure A1.4. An example chromatogram appears in Figure A1.5.

11. Use the unknown peak areas and standard curves to calculate volume (mL) of DNAPL components (TCA, TCE, or PCE) per milliliter of toluene extract. In the following discussion, these components will be simply be called DNAPL. The equation for calculating the content of DNAPL component 'i' is:

$$\text{mL DNAPL}_i/\text{mL toluene} = \text{DNAPL}_{i-\text{sample}} = A \times S \qquad (A1.4.2)$$

where A = Area of the sample DNAPL peak and S is the slope of the standard curve in (mL DNAPL/mL toluene)/unit of peak area

12. Next, calculate the total TCE per gram of soil according to:

$$\text{DNAPL}_{i-\text{g soil}} = \text{DNAPL}_{i-\text{sample}} \times V_{\text{toluene}}/W_{\text{soil}} \qquad (A1.4.3)$$

where
$\text{DNAPL}_{i-\text{g soil}}$ = Volume of DNAPL component, i, per gram of soil
$\text{DNAPL}_{i-\text{sample}}$ = mL of DNAPL component, i, per mL of toluene extract
$V_{\text{toluene}}/W_{\text{soil}}$ = Volume of toluene preservative divided by the weight of the soil sample.

13. $$\% \text{ Saturation} = \sum_i \text{DNAPL}_{i-\text{g soil}}/Vp_{\text{g soil}} \qquad (A1.4.4)$$

$Vp_{\text{g soil}} = \phi/((1 - \phi)\rho_g)$
ϕ = Porosity

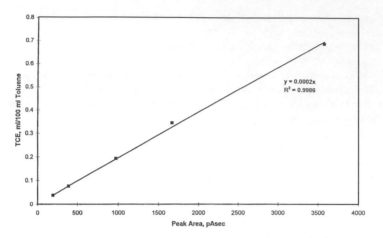

Figure A1.4 Calibration curve, TCE.

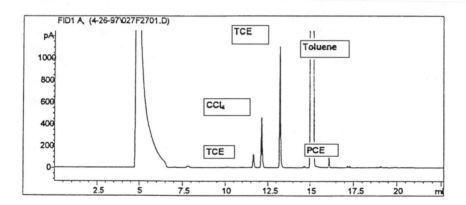

Figure A1.5 Chromatogram extract from sample CB-2, 46.1 ft bgs.

ρ_g = Soil grain density

Assuming 0.33 porosity and a soil grain density equal to that of quartz sand (2.65 g/cm³), $Vp_{g\,soil}$ = 0.186 cm³/g soil. These are the assumptions used to calculate DNAPL saturations. Sometimes, only TCE analyses were used to calculate total DNAPL saturation. Alternatively, INTERA calculated saturations using NAPLANAL, a program developed for that purpose.

Please note that the measured soil weight included the weight of whatever the pore space was filled with, either water or TCE, or both. These have been neglected. Therefore, Equation A1.4.4 underestimates DNAPL saturation. The true density of the soil sample is:

$$\rho_s = (1 - \phi)\rho_g + \phi S_w \rho_w + \phi S_{TCE} \rho_{TCE} \tag{A1.4.5}$$

where S = saturation

A1.4.1.1 Gas Chromatograph Set-up and Machine Parameters

Below is a description of the gas chromatograph set-up and the machine parameters used to analyze Hill AFB samples.

GC Apparatus and Parameters:

1. Gas Chromatograph Hewlett Packard Model 6890 with Hewlett Packard Chem Station Software.

Detector	FID
	Heater at 250°C
	Air flow = 400 mL/min
	Hydrogen flow = 30 mL/min
Column	Hewlett Packard VOCOL
	105 M × 0.53 mm × 3 mm
	He flow rate = 10 mL/min
Inlet temperature	250°C.
Temperature ramp	40° for 5 min, 10°/min to 150°C, 30°/min to 200°C,
	Hold for 5 min at 200°C.

2. Purge-and-trap unit and parameters

   OI Analytical Model 4560 equipped as follows:

   51-place autosampler, Model 4551 (this sampler accepts 40-mL vials).

   5 mL purge chamber

   10 µL spiker

   Infrared heater

   OI No. 9 trap silica gel, Tenax resin, activated charcoal

Purge rate	40 mL/min
Purge time	11 min
Sparge temp.	24°C
Desorb temp.	185°C
Desorb time	2 min
Transfer line temp.	100°C
Bake temp	185°C
Bake time	12 min

   These settings approximate EPA method 502.2

Discussion:

1. Determinations of detection limits (DL) and practical quantitation limits (PQL) appear in Table A1.11. The lowest concentrations investigated were approximately 0.002 mL analytes in 100 mL toluene. Analyses of a certified standard (I.E.-S3550, Lot A611080, exp 12-1-97, 0.7 vol% in toluene, Accustandard Inc., 10 replicates) containing TCE yielded an average error of 3% with a %RSD of 3%. For a three-component certified standard (I.D. S-3600, Lot A6120105, Accustandard Inc.), the average error and %RSD of 10 replicates analyses were: TCA, 7%, 7%; TCE 5%, 5%; and PCE, 3%, 2%.

2. For the method as presented, the saturation detection limit is approximately 0.01 saturation % to 0.02 saturation %. The practical quantitation limits are approximately 0.05 saturation % to 0.1 saturation %. In more recognizable units, the detection limit of this purge-and-trap method is about 2 to 4 µg/L in the diluted sample. Better method sensitivity can be accomplished using smaller sample dilutions. Sample dilution is limited by the solubility of toluene in water. Analyses of certified standards indicate that the average error for the method is about 3% for PCE, 5% for TCE, and 7% for TCA.

3. Another important consideration is over what range of analyte concentrations one can expect this purge-and-trap method to produce linear detector responses. This is important because, in a series of unknowns, concentrations are higher than expected and fall outside the range of calibration standards. If the range of linearity is limited, one must re-run samples at higher dilutions. GC detector responses to components in a sample of Hill DNAPL appear in Figure A1.6. These data show that this purge-and-trap method yields linear detector responses over the range 0 mg/L to at least 18 mg/L DNAPL in the diluted sample. For an 8000-fold dilution, as is proposed here, these values correspond to 0 mg/L to more than 100,000 mg/L in the undiluted unknown. Thus, the method produced linear responses over a wide range of concentrations.

Table A1.11 Detection Limits for DNAPL Components in Toluene

Sample	TCA (mL/100 mL toluene)	TCE (mL/100 mL toluene)	PCE (mL/100 mL toluene)
Standard 1	0.013	0.038	0.011
Standard 2	0.026	0.076	0.022
Standard 3	0.067	0.193	0.055
Standard 4	0.119	0.345	0.098
Standard 5	0.239	0.693	0.198
1	0.0023	0.0027	0.0008
2	0.0024	0.0029	0.0008
3	0.0023	0.0029	0.0008
4	0.0022	0.0029	0.0008
5	0.0020	0.0030	0.0008
6	0.0019	0.0021	0.0007
7	0.0021	0.0030	0.0008
8	0.0021	0.0029	0.0008
9	0.0022	0.0028	0.0008
Mean	0.0021	0.0028	0.0008
Standard Deviation	0.0001	0.0003	0.00003
Mean - $3 \times$ SD	0.0017	0.0019	0.0007
Detection limit	0.002	0.003	0.001
Saturation EL	0.02%	0.02%	0.01%
Saturation PQL	0.09%	0.12%	0.04%
Average recovery of surrogate	101%	100%	100%

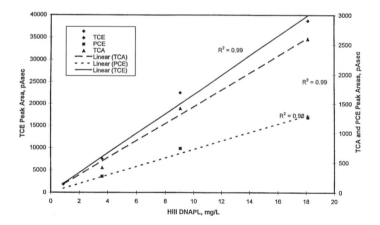

Figure A1.6 GC/FID response to components of Hill DNAPL.

A1.4.2 Hill DNAPL Composition

In the spring of 1996, Rice University requested and received a sample of Hill DNAPL from INTERA. Field personnel collected this sample from the product recovery tank of the SRS. It may have contained some recycled volatiles from the steam strippers.

To characterize this sample and confirm its overall composition, Rice analysts injected a sample directly onto a chromatograph column and ran it according to the schedule in Appendix A1.4.1. Three peak arrival times corresponded to those of 1,1,1-trichloroethane, trichloroethene, and tetra-chloroethene, as determined by injecting pure compounds and running using the same schedule. Analysts then analyzed Hill DNAPL as described below:

1. Prepare two standards by dissolving TCA, TCE, and PCE and an internal standard, CCl4, in toluene.
2. Dissolve an aliquot of Hill DNAPL, along with CCl_4, in toluene.
3. Dissolve 5 mL of each of the above in 40 mL deionized water contained in 40-mL vials.
4. Run all samples on the GC purge-and-trap apparatus described in Appendix A1.4.1, using the schedule in that section.
5. Calculate concentrations using the equation:

$$C_{i(\text{Hill DANPL})} = K_i \times PA_i \times (C_{CCl_4(\text{Hill DNAPL})}/PA_{CCl_4(\text{Hill DNAPL})}) \qquad (A1.4.6)$$

where PA = Peak area
$\quad C$ = Concentration
$\quad i$ = ith component of the DNAPL
K is obtained from:

$$C_{i(s\,\tan\,dard)} = K_i(C_{CCl_4(s\,\tan\,dard)}/PA_{CCl_4(s\,\tan\,dard)}) \qquad (A1.4.7)$$

The average value of K from the two standards served in final calculations.

DNAPL Produced from Well MW-2A:

Samples collected from the observation well, MW-2A, during the time period midday on 3/24/97 to near midnight of the same day, were single phase, dark colored, had densities of approximately that of Hill DNAPL (1.4 g/mL), and appeared to be mostly DNAPL. Sample 076M, collected 3/24/97 at 1800 h appeared to be entirely DNAPL. It had a density of approximately 1.4 g/mL, dissolved completely in toluene, and appeared to contain little, or no, water. Sample 076M contained 0.017 M surfactant and about an equivalent concentration of calcium, 0.006 M, suggesting that surfactant was present in the DNAPL as the calcium salt.

Discussion:

Hill DNAPL from the field separator was 6 wt% TCA, 72 wt% TCE, and 2 wt% PCE (Table A1.12). Produced DNAPL from Well MW-2A was 11 wt% TCA, 65 wt% TCE, and 6 wt% PCE (Table A1.13). The difference may be due to two factors. First, the sample of DNAPL from the SRS contained some effluent from the stripper and was shipped and stored unrefrigerated. The sample of DNAPL from MW-2A, on the other hand, was straight from the observation well and was shipped and stored at 4°C. Although the two compositions were similar, sample 076M from MW-2A probably more nearly represented reservoir DNAPL. Produced sample analyses contain yet another tie point, namely, the fraction of major components, TCA+TCE+PCE, attributable to each component. By way of comparison, average sample from well EX-2 was 14 wt% TCA, 76 wt% TCE, and 10 wt% PCE. This result is more in line with the DNAPL from observation well MW-2A.

A1.4.3 Measurement of DNAPL Constituent Concentration in Solutions

Before applying purge-and-trap analyses, surfactant solutions must be diluted enough to not foam. For 5% solutions of MA-80I in 11,500 mg/L NaCl, it was determined that 4000:1 dilutions with deionized water are enough to prevent foaming. A 5% solution of MA-80I is approximately 0.1 M in surfactant. Therefore, solutions of sodium dihexyl sulfosuccinate that are approximately 0.000025 M will not foam appreciably. This is much less that the CMC of this surfactant. If one is using a purge-and-trap analysis for surfactant solutions containing DNAPL, one must determine the foaming concentration of individual surfactants. Do not assume that the above dilution will be satisfactory for all surfactants. The CMC of sodium dihexyl sulfosuccinate is rather high. To be prudent, samples were diluted by factors of 8000.

Table A1.12 Composition of DNAPL from the Steam Stripper Using an Internal Standard

Sample	TCA (pk pAsec)	TCA (ppm)	CCl$_4$ (pk pAsec)	CCl$_4$ (ppm)	TCE (pk pAsec)	TCE (ppm)	PCE (pk pAsec)	PCE (ppm)
Standard 1	3022	4255	851	5126	1749	4707	3124	4255
Standard 2	2969	3923	831	4726	4193	8679	3006	3923
DNAPL	299	329	662	3163	2174	4160	125	134
K1	0.23					0.45		0.23
K2	0.23					0.36		0.23
Mean K	0.23					0.41		0.23
Wt% of Sample	6					72		2
Wt% of TCA+TCE+PCE	7					90		3

Note: Sample in toluene contained 5808 ppm DNAPL; measured density was 1.4 g/mL.

Table A1.13 Composition of DNAPL from MW-2A, Sample 076M

Component	TCA	TCE	PCE
Wt% of sample	11	65	6
Wt% of TCA + TCE + PCE	13	79	8

The procedure for analyzing toluene extracts for TCE and other DNAPL components follows. It is based on specific choices of instruments, namely a gas chromatograph manufactured by Hewlett Packard and a purge-and-trap apparatus manufactured by OI Analytical. Descriptions of these instruments, along with applicable method parameters appear in the preceding chapter sections.

Procedures:

1. Prepare a five-point set of standards by dissolving DNAPL components in surfactant solution — 5% MA-80I containing 11,500 ppm NaCl. Choose standard concentrations to span the expected range of contaminant concentrations. Usually, the range was selected to include 100 ppm to 2000 ppm for TCA and PCE, and 500 ppm to 10,000 ppm for TCE. Experience showed GC detector response to be linear over these ranges. Use the following equation to calculate the concentrations of DNAPL components in standards:

$$C_i = \rho_i V_i 10^6 / (\sum_i \rho_i V_i + \rho_{surf} V_{surf}) \tag{A1.4.8}$$

where

C_i = Concentration of component i, in ppm
V_i = Volume of component i, in mL
ρ_i = Density of component i, in g/mL

Always use at least 100 mL of solvent to prepare standards. Smaller volumes compromise accuracy. Standard concentrations should span the expected range of sample concentrations.

2. Dissolve 5 μL of the standards in 40 mL deionized water contained in 40-mL sample vials. Note that this is the 8000:1 dilution referred to earlier.

3. Dissolve 5 μL unknown samples in 40 mL deionized water contained in 40-mL sample vials. An optimum surfactant solution can contain hundreds of thousands of ppm analyte. The analyst must exercise judgment as to how much to dilute unknown samples before analysis. If surfactant is present, one must dilute to prevent foaming. Aside from preventing foaming, dilution is often required to maintain concentrations within the range of method linearity. Flame ionization detectors are linear over wide ranges. However, in purge-and-trap methods, detector linearity is less of an issue than the capacity of the trap. If one has an indication of unknown concentrations beforehand, it is prudent to dilute samples much more concentrated than 10,000 ppm with surfactant solution

before taking the 5-μL aliqout and adding to deionized water. A 10,000 ppm sample diluted 8000-fold provides about 1.25 ppm analyte, well below the solubility in water of TCA, TCE, or PCE and, certainly, well within the range of linearity of the method. However, this method provides considerable leeway in sample concentrations. Data showed that Rice University's purge-and-trap method produced linear responses for Hill DNAPL in water solutions up to at least 18 ppm. So, samples containing more than 100,000 ppm Hill DNAPL can be run safely with 8000-fold dilution. Also, a limited range of standard concentrations is sufficient to establish calibration parameters for a wide range of unknown concentrations. Do not assume, however, that one can analyze an undiluted water sample saturated with DNAPL; such a sample will completely overwhelm the trap.

4. Add known quantities of analytes to unknown samples. These are referred to as matrix spikes. The matrix spikes were prepared by adding 1 μL of a known solution of DNAPL components in surfactant solution to 40 mL deionized water containing 5 μL of unknown sample. (Because we began with an 8000:1 dilution, this extra surfactant solution is not enough to cause foaming.) More convenient, and more accurate, would be to use an automatic spiker to deliver matrix spikes.

5. Let diluted samples containing standards and unknowns age for at least 2 h at room temperature with frequent shaking. Adequate aging is not as critical as it is for samples in toluene, yet it is still sound practice to be certain all components are dissolved before proceeding with the analysis.

6. Load the automatic spiker with a solution of the compound chosen for the required surrogate spike. This spiker delivered 10 μL of spike into each sample. Carbon tetrachloride was selected as a surrogate. Solutions of carbon tetracholide in both methanol and water were successfully used as surrogates. The preferred surrogate spike evolved to be 150 μL CCl_4 in methanol.

7. Load standards, samples, spiked samples, and blanks into an autosampler. This autosampler held 51 samples. Place a blank at the beginning of each set. The first sample in a set is a dummy. That is, it is necessary to ensure no column carry-over from previous runs and to clear the valves of the automatic spiker. The five-point set of standards is at the beginning of a run. A three-point set of standards was placed either at the end of a run or distributed throughout the run. These extra standards served as calibration checks and allowed one to calculate precision. Runs usually included three or four blanks. It is sound practice to position blanks to minimize carry-over of analyte from anticipated high-concentration to low-concentration samples.

8. Run GC analysis as described in A1.4.1.

9. Obtain GC-TCE peak areas by integrating chromatograms automatically, using Hewlett Packard Chem Station software. Expect sharp and symmetrical analyte peaks when analyzing Hill DNAPL.

10. Use peak areas and analyte concentrations to establish a calibration curve. Linear regression was always satisfactory.

11. Use the unknown peak areas and standard curves to calculate concentrations of DNAPL components, TCA, TCE, or PCE in ppm. The equation for calculating the concentration of DNAPL component i is $DNAPL_i = A \times S$, where A is the area of the sample DNAPL peak and S is the slope of the standard curve in ppm/(unit of peak area). Occasionally, it is desirable to use regression equations. Always force regressions through the origin.

Discussion:

A determination of detection limits for DNAPL components dissolved in surfactant solutions appears in Table A1.14. Based on data in Table A1.14, detection limits for the three major DNAPL components were found to be less than 25 ppm. Taking 25 ppm as the detection limit, the Practical Quantitation Limit, PQL is $5 \times 25 = 125$ ppm. Please note that these limits apply only when one analyzes 5-mL aliquots of sample diluted with 40 mL water. An original sample, containing 25 ppm analyte, and diluted 8000:1, produces about 3 ppb in the diluted samples (i.e., in the diluent being analyzed by purge-and-trap). If better sensitivity than 25 ppm is desired, one only needs to use less dilution. This is not always possible with surfactant solutions, but it certainly is with water solutions. Lesser dilutions were used for samples near the beginning and end of the flood, which did not contain much surfactant. If circumstances require more sensitivity for concentrated surfactant solutions, consider using a larger purge chamber, say 25 mL, or heating the sample during purging. Analytical accuracy was approximately 5% for TCA and TCE, and 7% for PCE, based on analyses of a certified standard.

Table A1.14 Detection Limits of DNAPL Components in Surfactant

Sample	TCA (ppm)	TCE (ppm)	PCE (ppm)
Standard 1	77	244	77
Standard 2	349	1101	349
Standard 3	891	2811	891
Standard 4	1588	4999	1588
Standard 5	3172	9954	3172
1	21	23	27
2	22	25	28
3	22	24	27
4	23	24	28
5	23	24	27
6	21	23	27
7	23	24	27
8	21	22	27
9	21	23	28
10	22	24	30
Blank 1	0	0	0
Blank 2	0	0	0
Mean	22	24	28
Standard Deviation	1	1	1
Mean-3 × S.D.	21	23	27
PQL	109	119	138

Note: Detection limit is <25 ppm; PQL taken as 125 ppm, average recovery of surrogate = 100%.

A1.5 ANALYSIS OF TRACER ALCOHOLS

This method was used to determine the concentration of various volatile organic compounds in water containing potentially high concentrations of surfactants. The analytes to be analyzed by this method are various chlorinated ethenes and ethanes, and straight-chain and branched alcohols from ethanol to octanol.

Summary of Method:

This method, which is based on SW 846 Method 8015A, provides gas chromatographic conditions for the detection of various alcohols and chlorinated ethenes and ethanes in water samples containing potentially high concentrations of surfactants. Detection is achieved by a flame ionization detector. The method provides an optional GC column that may be helpful in resolving the analytes from coeluting nontarget compounds and for analyte confirmation.

SW846 Method 8015A Modifications:

1. Samples are introduced into the GC using direct injection. Although direct injection increases detection limits of the target analytes to around 1 to 10 mg/L, lower detection limits are unnecessary when high concentrations are expected.
2. A more rapid analysis is achieved by a shorter oven temperature program to separate target analytes.
3. A precolumn is placed between the injector and GC column. The precolumn is used to prevent surfactant from contaminating the GC column. The surfactant collects in the precolumn because it is nonvolatile. The precolumn has no effect on the chromatograms until a considerable amount of surfactant has been deposited. To avoid excessive surfactant buildup, the precolumn is replaced on a regular basis.
4. Surrogate spikes are not required because there is no sample extraction step.

5. Matrix spikes and matrix spike duplicates are used only in batches that contain samples composed of less than 99% water by weight.

Interferences:

1. Contamination by carry-over can occur whenever high-concentration and low-concentration samples are sequentially analyzed. To reduce carry-over, the sample syringe or purging device must be rinsed out between samples with water or solvent. Whenever an unusually concentrated sample is encountered, it should be followed by an analysis of a solvent blank or of water to check for cross-contamination. For volatile samples containing large amounts of water-soluble materials, suspended solids, high boiling compounds or high organohalide concentrations, it may be necessary to wash out the syringe or purging device with a detergent solution, rinse it with distilled water, and then dry it in a 105°C oven between analyses.
2. Samples can be contaminated by diffusion of volatile organics (particularly chlorofluorocarbons and methylene chloride) through the sample container septum during shipment and storage. However, because these compounds are highly volatile, they should not interfere with the analysis of the target compounds.

Apparatus and Materials:

1. Gas chromatograph: Analytical system complete with gas chromatograph suitable for on-column injections and all required accessories, including flame ionization detector, analytical columns, recorder, gases, and syringes. A data system for measuring peak heights and/or peak areas is recommended. Recommended column: 30 m × 0.53 mm ID capillary column with 1.0 μm film AT-WAX
2. Sample introduction apparatus
3. Syringes, 2, 10 mL
4. Volumetric flasks, Class A, appropriate sizes with ground glass stoppers
5. Microsyringes, 10μL with a 26-gauge needle (Hamilton 702N or equivalent) and a 100 μL or 250 μL
6. Analytical balance, 0.0001g sensitivity

Reagents:

1. Reagent-grade chemicals are to be used in all tests. Unless otherwise indicated, it is intended that all reagents conform to the specifications of the Committee on Analytical Reagents of the ACS, where specs are available. Other grades may be used if it is ascertained that the accuracy of the determination will not be diminished.
2. Reagent water is defined as a water in which an interferant is not observed at the method detection limit of the analytes of interest.
3. Carbon disulfide: pesticide quality or equivalent. Store away from other solvents.
4. Stock standards: stock solutions may be prepared from pure standard materials or purchased as certified solutions. Primary dilutions of organohalides should be prepared in a hood.
   a. Prepare a neat solution of the target analytes in a glass container with a foil cap. Record mass of each analyte added to container and compute final concentrations.
   b. Add reagent water to a tared glass flask until flask is about 80% to 90% full. Allow the flask to stand, unstoppered, for about 10 min until all wetted surfaces have dried. Weigh the flask to the nearest 0.0001 g.
   c. Add the assayed neat solution, then reweigh. The liquid must fall directly into the water without contacting the neck of the flask.
   d. Cap and then mix by inverting the flask several times. Calculate the concentration in milligrams per liter (mg/L) from the net gain in weight. When compound purity is assayed to be 96% or greater, the weight may be used without correction to calculate the concentration of the stock standard. Commercially prepared stock standards may be used at any concentration if they are certified by the manufacturer or by an independent source.

e. Place the stock standard solution into a vial or bottle with a foil-lined screw cap. Store, with minimal headspace, at 4°C, and protect from light.

f. Standards must be replaced after 6 months. Standards must be monitored closely by comparison with the initial calibration curve and by comparison with QC check standards. Replace the standards sooner if either check exceeds a 20% drift.

5. Secondary dilution standards (optional). Using stock standard solutions, prepare secondary dilution standards in reagent water, as needed, containing the compounds of interest, either singly or mixed together. Secondary dilution standards should be checked frequently for signs of degradation or evaporation, especially just prior to preparing calibration standards from them.

6. Calibration standards should be prepared in reagent water from the stock standards or secondary dilution standards at a minimum of five concentrations. One of the concentrations should correspond to the expected range of the concentrations found in real samples or should define the working range of the GC. Each standard should contain each analyte for detection by this method.

7. Optional internal standards are not used because external standards are used.

8. Surrogate standards are not used because there is no sample extraction step.

Sample Collection, Preservation, and Handling:

1. Containers: Water samples are collected and preserved in 40-mL glass VOA vials with aluminum foil-lined caps.

2. Sample collection: Liquid samples should be poured into the vial gently and without introducing air bubbles. The vial should be completely filled so that there is no headspace, sealed, and inverted. Vials should be labeled with sample name, collection time, preservative, and sampler's name. Chain-of-custody forms are kept with the samples at all times.

3. Sample storage and transportation: Samples are kept at 4°C in coolers with blue ice and then in refrigerators. The holding time for the samples is 7 d from the date of collection, or 6 weeks if preserved at pH < 2

Procedure:

1. Gas chromatographic conditions: helium flow rate, 7 mL/min; initial temperature, 35°C; hold for 8 minutes; ramp 10°C per min to 120°C, hold for 5 min.

2. Calibration: use the same sample introduction method as for sample analysis. Calibrate using the external standard technique.

External Standard Calibration Method:

1. Prepare calibration standards for each analyte of interest at a minimum of five concentrations. One should be slightly above the method detection limit; the other concentrations should be in the expected range of concentrations of the samples or define the working range of the detector.

2. Tabulate peak height or area responses against the mass injected. The results can be used to prepare a calibration curve. Alternatively, the ratio of the response to the amount injected, defined as the calibration factor (CF), can be calculated for each analyte at each standard concentration. If the percent relative standard deviation of the CF is less than 20% over the working range, the average CF can be used in place of a calibration curve.

3. The working calibration curve or CF must be verified on each working day by the injection of one or more calibration standards. The frequency of verification is dependent on the detector. If the response for any analyte varies from the predicted response by more than 20%, a new calibration curve must be prepared for that analyte.

4. The concentration of each analyte is determined as:

$$mg/L = (A \times D)/ (CF \times V)$$ (A1.5.1)

where
    A = Response for the analyte in either area counts or peak height
    D = Dilution factor
    CF = Area counts or peak height per nanogram
    V = Volume of sample injected, μL

5. For the recommended column, the order of retention times of several potential analytes is 1,1,1-TCA (3.7 min), isopropanol (4.7 min), TCE (6.6 min), PCE (7.6 min), 1-pentanol (14.1 min), 2-ethyl-1-butanol (15.2 min) and 1-heptanol (17.8 min).
6. If the response for a peak is off-scale, prepare a dilution from a second aliquot of the sample.

Quality Control:

1. Validate method with the QC check sample. The QC check sample concentrate should contain each analyte of interest at a concentration of 50 mg/L in reagent water.
2. Run continuing mid-range calibration check samples. Calculate the average recovery and standard deviation of each analyte of interest for four checks.
3. Ongoing QA/QC: For every batch of samples (20 or less), the following control samples are analyzed: reagent blank, continuing mid-range calibration check sample, matrix spike (msd if needed), and field duplicate. After the reagent blank and continuing mid-range calibration check, start the sequence. If the batch is not followed by another, run an additional reagent blank and calibration check to end the sequence.
4. The continuing calibration check sample analyses are plotted on a control chart. The control limit is the larger of ±20% of the average, or ±3 standard deviations of the percent recovery. Reagent blank results should be below quantitation limits.

Analytical Procedures Used in the SPME Field Demonstration

A2.1 STANDARD OPERATING PROCEDURE (SOP) FOR ANALYSIS OF ALCOHOLS USED AS PARTITIONING TRACERS

Scope and Application:

1. This SOP describes the analytical procedures utilized by the Soil and Water Science Department, University of Florida, IFAS, for analysis of alcohols used as partitioning tracers in both lab and field studies in order to quantify the amount and distribution of residual nonaqueous phase liquids (NAPLs) present in the saturated zone.
2. This SOP was written by R.D. Rhue, Soil and Water Science Department, University of Florida (UF), Gainesville. It is a modification of SOP-UF-Hill-95-07-0010-v.2, prepared by D.P. Dai, H.K. Kim, and P.S.C. Rao, Soil and Water Science Department, University of Florida. The SOP of Dai, Kim, and Rao was modified from a protocol provided to them by Professor Gary Pope at the University of Texas-Austin.
3. The alcohol tracers used in the UF lab and field studies are ethanol, n-butanol, n-pentanol, n-hexanol, n-heptanol, 2,2-dimethyl-3-pentanol, and 6-methyl-2-heptanol.
4. The method involves gas chromatography (GC) analysis for alcohol concentrations in aqueous samples. A flame ionization detector (FID) is used to quantify the analyte concentrations in the sample. The method has been found to provide reliable and reproducible quantitation of alcohols for concentrations >1 µg/mL. This value may be considered the method detection level (MDL). The standard calibration curve for FID response has been found to be linear up to 3000 µg/mL for ethanol.
5. Samples selected for GC/FID analysis may be chosen on the basis of preliminary screening that will provide approximate concentration ranges and appropriate sample injection volumes, standard concentrations, etc.

Purpose:

The purpose of this SOP is to ensure reliable and reproducible analytical results for alcohols in aqueous samples for laboratory-based or on-site (field-based) GC/FID analyses, and to permit tracing sources of error in analytical results.

Procedures:

1. Sample containers, collection, transportation and storage
   a. Sample containers: Field samples are collected in 5-mL glass sample vials (Fisher Catalog #06-406-19F) with teflon-faced septa caps. Glass vials and caps are not reused.

b. Sample collection: Each field sample vial is completely filled with liquid, so that no gas headspace exists, and capped. The vials are not opened until the time for analysis.

c. Transportation and storage: Field samples are stored in coolers containing "blue ice," and later stored in refrigerators in a trailer located on the site. Samples may be subjected to on-site GC analysis, and/or shipped back to UF labs; samples are packed in coolers and shipped via overnight air express (e.g., FedEx). The samples are stored in a cold storage room or refrigerator at 4°C until GC analysis. After subsampling, the samples are returned to cold storage. For lab studies, samples are collected directly in 2-mL GC vials whenever possible and stored in a refrigerator if analysis is expected to take more than a day.

2. Subsampling and dilution

   Field samples are subsampled into 2-mL vials for automated GC analysis. Disposable Pasteur glass pipets (Fisher Catalog #13-678-20B) are used to transfer samples from 5-mL sample vials to the 2-mL GC vials. For samples needing dilution prior to GC analysis, a dilution of 1:10 should be sufficient. Dilutions are made using double-distilled, deionized water.

3. Apparatus and materials

   a. Glassware

      • Disposable micropipets (100 μL; Fisher Catalog #21-175B; 21-175F) and Class A volumetric pipets (1 or 2 mL) are required for sample dilution.

      • Disposable Pasteur glass pipets (Fisher Catalog #13-678-20B) are required for subsampling.

      • GC vials (2-mL) with Teflon-faced caps (Fisher Catalog #03-375-16A) are required for GC analysis.

      • Volumetric Class A pipets and volumetric Class A flasks are required for preparations of the calibration standards.

   b. Gas chromatograph system: An analytical GC system with a temperature-programmable oven, an auto-injector capable of on-column injection, and either an integrator or a PC-based data acquisition/analysis software system are required. Also required are other accessories, including analytical columns and the gases required for GC/FID operation.

      A Perkin Elmer Autosystem with an FID and an integrated autosampler are used for analysis of field and laboratory samples. The Perkin Elmer system is linked to an IBM-compatible PC loaded with Turbochrom (version 4.01) software.

      A J&W Scientific DB-624 capillary column (30 m × 0.53 mm, 3-μm film thickness) is used. Zero-grade air and ultrahigh-purity hydrogen are used for the FID; ultrahigh-purity nitrogen or helium is used as the carrier gas.

4. Reagents

   a. Deionized, double-distilled water, which is prepared by double distillation of deionized water in a quartz still. This water will be referred to as reagent water.

   b. Alcohols: Certified ACS grade alcohols are purchased from Fisher Scientific and used as received.

5. Standard Solutions

   a. Stock standard solution: Analytical standards are prepared from reagent chemicals by the laboratory. Stock standards each contain a single alcohol dissolved in reagent water and stored in 20-mL glass vials (Fisher Catalog #03-393-D) with teflon-lined caps. These stock solutions are kept in a refrigerator at 4°C. Fresh stock standards are prepared every 6 months. The procedure for making stock standard solutions is essentially that given in the Federal Register, Rules and Regulations, Thursday, November 29, 1979, Part III, Appendix C, Section 5.10, "Standard Stock Solutions." The only modification of the procedure for the current study is that reagent water is used as the solvent in place of methanol.

   b. Calibration standards: Calibration standards are prepared by diluting the stock standards in reagent water. Each calibration standard will contain each of the alcohols listed above. Five concentrations are prepared that cover the approximate concentration range utilized in the partitioning tracer experiments.

6. QC blank spike/matrix spike

   Two 1-mL aliquots of the sample to be spiked are transferred to clean vials. To one vial, 1 mL reagent water is added. To the second vial, 1 mL of a calibration standard is added. The spike recovery is calculated using the difference between the two measured concentrations and the known spike concentration.

7. Quality control
   a. GC injector septa are changed every 80 to 100 injections, or sooner if any related problems occur.
   b. Injector liners are cleaned or changed every 80 to 100 injections or sooner if any related problems occur.
   c. A method blank is included in every 50 samples.
   d. A complete set of calibration standards (5) is run at the beginning of each day and after every 50th sample.
   e. One standard and a blank are included in every 25 samples.
   f. A sample spike and a blank spike are included in every 50 samples.

8. Instrumental procedures

Gas chromatography:	For J&W DB-624 column:
Injection port temperature	200°C
FID detector temperature	225°C
Temperature program	Isothermal at 60°C for 0 min; ramp to 120°C at 5°C/min

9. Sample preparation
   a. Subsampling: Field samples are transferred from the 5-mL sample vials to the 2-mL GC vials and capped with open-top, teflon-lined septa caps.
   b. Dilution: Samples are diluted if chromatographic peak areas for any of the alcohols exceed those of the highest calibration standard; 1 mL of sample is added to an appropriate amount of reagent water to make the dilution.

10. Sample analysis
    a. Analysis: The samples are allowed to reach ambient temperature prior to GC analysis.
    b. Sample vials (2 mL) are loaded onto the Perkin Elmer GC auto-injector. A 1-µL injection volume is used for both samples and standards.
    c. Analyte identification: Analyte identification is based on absolute retention times. The analytes of interest should elute at their characteristic retention times within ±0.1 min for the automated GC system.
    d. Analyte quantitation: When an analyte has been identified, the concentration is based on the peak area, which is converted to concentration using a standard calibration curve.

11. Interferences
    a. Contamination by carry-over can occur whenever high-level and low-level samples are sequentially analyzed. To reduce carry-over, the injector syringe should be rinsed with reagent water between samples.
    b. Potential carry-over is checked by running a highly concentrated sample, but one still within the standard concentration range, followed by a blank. A negligible reading for the blank will ensure that carry-over has been minimized.

12. Safety
    a. The main safety issue concerning the use of GC at a field site relates to the compressed gases. The FID gases (hydrogen and air) form explosive mixtures. It is important to keep this in mind at all times, and be aware of the hazard potential in the event of an undetected hydrogen leak. All gas connections are properly leak-tested at installation.
    b. High-pressure compressed-gas cylinders are secured to a firm mounting point, whether they are located internally or externally. Gas cylinders should preferably be located outside the trailer on a flat, level base, and the gas lines run inside through a duct or window opening. If the gases are located outside, then some form of weatherproofing for the gauges is necessary. As a temporary measure, heavy-duty polyethylene bags, secured with tie-wraps, have been used successfully; this may not be very elegant, but it is very effective for short-term use of the GC. A more permanent protective housing must be built if the GC is located outside the trailer for an extended time period.
    c. The main operating drawback to locating the gas cylinders externally is that it is not easy to monitor the cylinder contents from inside. The gas that could be used up most quickly is air for the FID, particularly if two instruments are hooked up to the same supply and they are running continuously. A reserve cylinder of air should be available at all times to prevent downtime. If it is not possible to arrange external siting easily, the gas cylinders should be secured to a wall inside the trailer.

    d. It is a good laboratory operating practice to make sure the flame is attended at all times.

    e. When it is necessary to change the injection liner on the GC, the detector gases should be shut off.

    f. The column must be connected to the detector before igniting the flame.

    g. The trailer should be kept well ventilated when using the GC.

    h. Reference is made to the Materials Safety Data Sheets (MSDS) for information on toxicity, flammability, and other hazard data.

A2.2 STANDARD OPERATING PROCEDURE FOR ANALYSIS OF PENTANOL IN AQUEOUS SURFACTANT SOLUTIONS USED WITH SPME FLUSHING TECHNOLOGY DEMONSTRATION

Scope and Application:

1. This SOP describes the analytical procedures to be utilized by the Soil and Water Science Department, University of Florida (UF), IFAS, for analysis of pentanol in lab and field studies of the SPME flushing technology.

2. This SOP was written by R.D. Rhue, Soil and Water Science Department, University of Florida, Gainesville.

3. This method involves gas chromatography (GC) analysis for pentanol concentrations in aqueous surfactant solutions used to solubilize OU1 LNAPL. A flame ionization detector is used to quantify the pentanol. Gas chromatography has been found to provide reliable and reproducible quantitation of alcohols in aqueous solutions for concentrations >1 µg/mL. This value may be considered the method detection level (MDL). Samples are diluted with methanol to give pentanol concentrations within the linear operating range.

4. Samples selected for GC/FID analysis may be chosen on the basis of preliminary screening, which will provide approximate concentration ranges and appropriate sample injection volumes, standard concentrations, etc.

Purpose:

The purpose of this SOP is to ensure reliable and reproducible analytical results for pentanol in aqueous surfactant samples for laboratory-based or on-site (field-based) GC/FID analyses, and to permit tracing sources of error in analytical results.

Procedures:

1. Sample containers, collection, transportation and storage

    a. Sample containers: Field samples are collected in 5-mL glass sample vials (Fisher Catalog #06-406-19F) with teflon-faced septa caps. Glass vials and caps are not reused.

    b. Sample collection: Each field sample vial is completely filled with liquid, so that no gas headspace exists, and capped. The vials will not be opened until time for analysis.

    c. Transportation and storage: For field studies, the samples are stored in coolers containing "blue ice," and later stored in refrigerators in a trailer located on the site. Samples may be subjected to on-site GC analysis, and/or shipped back to UF labs; samples are packed in coolers and shipped via overnight air express (e.g., FedEx). The samples are stored in the cold storage room or refrigerator at 4°C, until they are ready for GC analysis. After subsampling, the samples are returned to cold storage.

    For lab studies, samples are collected in 2-mL GC vials when possible and stored in a refrigerator if analysis is expected to take more than a day.

2. Dilution

    Fifty microliters of sample are diluted with 4 mL methanol prior to analysis. A portion of the diluted sample is transferred to 2-mL vials for automated GC analysis. Disposable Pasteur glass pipets (Fisher Catalog #13-678-20B) are used to transfer samples to the GC vials.

3. Apparatus and materials
   a. Glassware:

   Disposable micropipets (50 μL; Fisher Catalog #21-175B; 21-175F) and Class A volumetric pipets (1 or 2 mL) are required for sample dilution.

   Disposable Pasteur glass pipets (Fisher Catalog #13-678-20B) are required for subsampling. GC vials (2-mL) with Teflon-faced caps (Fisher Catalog #03-375-16A) are required for GC analysis.

   b. Gas chromatograph system:

   A Perkin Elmer Autosystem XL with an FID and an integrated autosampler is used for analysis of field and laboratory samples. The Perkin Elmer system is linked to an IBM-compatible PC loaded with Turbochrom (version 4.01) software.

   A J&W Scientific DB-624 capillary column (30 m × 0.53 mm, 3-μm film thickness) is used. Zero-grade air and ultrahigh-purity hydrogen is used for the FID. Ultrahigh-purity nitrogen or helium is used for carrier gas.

4. Reagents
   a. Deionized, double-distilled water: Deionized, double-distilled water is prepared by double distillation of deionized water in a quartz still. This water is referred to as reagent water.
   b. Alcohols: Certified ACS grade pentanol and methanol are purchased from Fisher Scientific and used as received.

5. Standard solutions
   a. Stock standard solution: A stock standard solution is prepared by dissolving a known amount of pentanol in methanol and storing at 4°C. Fresh stock standards are prepared every 6 months. The procedure for making stock standard solutions is essentially that given in the Federal Register, Rules and Regulations, Thursday, November 29, 1979, Part III, Appendix C, Section 5.10, "Standard Stock Solutions."
   b. Calibration standards: Calibration standards are prepared by diluting aliquots of the stock standard in methanol. Five concentrations are prepared that cover the approximate concentration range from 0 to 500 mg/L.
   c. Stock spiking solution: Since the matrix for the field and lab samples is an aqueous surfactant solution, the pentanol standards in methanol are not suitable for spiking samples. The stock spiking solution is an aqueous surfactant solution containing pentanol at a concentration similar to that used in field and lab experiments. The stock spiking solution is prepared by dissolving a known amount of pentanol in the aqueous surfactant solution.

6. QC blank spike/matrix spike

   Two 0.5-mL aliquots of the sample to be spiked are transferred to clean vials. To one, 0.5 mL aqueous surfactant solution containing no pentanol is added. To the second vial, 0.5 mL of the stock spiking solution is added. The spike recovery will be calculated using the difference between the two measured concentrations and the known spike concentration.

7. Quality control
   a. GC injector septa are changed every 80 to 100 injections, or sooner if any related problems occur.
   b. The injector liner is cleaned or changed every 80 to 100 injections or sooner if any related problems occur.
   c. A method blank is included in every 50 samples.
   d. A complete set of calibration standards (5) is run at the beginning of each day and after every 50th sample.
   e. One standard and a blank are included in every 25 samples.
   f. A sample spike and a blank spike are included in every 50 samples.

8. Instrumental procedures

Gas Chromatography:	For J&W DB-624 Column:
Injection port temperature	200°C
FID detector temperature	225°C
Temperature program:	Isothermal at 70°C for 8 min; ramp to 200°C at 3°C/min; hold at 200°C for 10 min.

9. Sample analysis
   a. Analysis: The samples are allowed to reach ambient temperature prior to GC analysis.

Sample vials (2 mL) are loaded on the Perkin Elmer GC auto-injector. A 1-μL injection volume is used for both samples and standards.

b. Analyte identification: Analyte identification is based on absolute retention times. Pentanol should elute at its characteristic retention times within ±0.1 min for the automated GC system.

c. Analyte quantitation: When an analyte has been identified, the concentration is based on the peak area, which is converted to concentration using a standard calibration curve.

10. Interferences

a. Contamination by carry-over can occur whenever high-level and low-level samples are sequentially analyzed. To reduce carry-over, the injector syringe should be rinsed with reagent water between samples.

b. Potential carry-over is checked by running a highly concentrated sample, but one still within the standard concentration range, followed by a blank. A negligible reading for the blank will ensure that carry-over has been minimized.

11. Safety

a. The main safety issue concerning the use of GC at a field site relates to the compressed gases. The FID gases (hydrogen and air) form explosive mixtures. It is important to keep this in mind at all times, and be aware of the hazard potential in the event of an undetected hydrogen leak. All gas connections are properly leak-tested at installation.

b. High-pressure compressed-gas cylinders are secured to a firm mounting point, whether they are located internally or externally. Gas cylinders should preferably be located outside the trailer on a flat, level base, and the gas lines run inside through a duct or window opening. If the gases are located outside, then some form of weatherproofing for the gauges is necessary. As a temporary measure, heavy-duty polyethylene bags, secured with tie-wraps, have been used successfully; this may not be very elegant, but it is very effective for short-term use of the GC. A more permanent protective housing must be built if the GC is located outside the trailer for an extended time period.

c. The main operating drawback to locating the gas cylinders externally is that it is not easy to monitor the cylinder contents from inside. The gas that could be used up most quickly is air for the FID, particularly if two instruments are hooked up to the same supply and they are running continuously. A reserve cylinder of air should be available at all times to prevent downtime. If it is not possible to arrange external siting easily, the gas cylinders should be secured to a wall inside the trailer.

d. It is a good laboratory operating practice to make sure the flame is attended at all times.

e. When it is necessary to change the injection liner on the GC, the detector gases should be shut off.

f. The column must be connected to the detector before igniting the flame.

g. The trailer should be kept well ventilated when using the GC.

h. Reference is made to the Materials Safety Data Sheets (MSDS) for information on toxicity, flammability, and other hazards.

A2.3 STANDARD OPERATING PROCEDURE FOR ANALYSIS OF SURFACTANT IN AQUEOUS SOLUTIONS USED WITH SPME FLUSHING TECHNOLOGY DEMONSTRATION

Scope and Application:

1. This SOP describes the analytical procedures to be utilized by the Soil and Water Science Department, University of Florida, IFAS, for nonionic, ethoxylated alcohol surfactant analysis in lab and field studies of the SPME flushing technology. However, since the microemulsion precursor has not yet been selected, a modification of this operating procedure may be necessary if a different type of surfactant is selected for the technology demonstration.

Commercial ethoxylated alcohol surfactants are polydisperse, meaning that they vary widely in both the degree of polymerization of the ethoxy group and length of the alcohol carbon chain. For example, the surfactant $C_{12}E_{3.8}$ may contain chromatographically detectable amounts of C_mE_n with n ranging from 0 to 14 and m ranging from 8 to 16 (Stancher and Favretto, 1978). Since the identification of every component of a commercial surfactant formulation would be impractical for this study, we propose to measure the concentration of one or two components that are representative of the mixture (i.e., $C_{12}E_3$ and $C_{12}E_4$ for the above example).

Cross (1987) reviewed GC methods for analyzing nonionic surfactants. Measurement of ethoxylated alcohol surfactants by GC requires columns that can be temperature programmed to 350°C or higher since the elution temperature increases almost linearly with increasing number of oxyethylene units per molecule. There appears to be little or no benefit from derivitization of the ethoxylated alcohols to less polar compounds (Stancher and Favretto, 1978). It is proposed to analyze nonionic surfactants by direct injection onto capillary columns designed for high-temperature analysis. Giger et al. (1981) successfully separated nonylphenol ethoxylates on glass capillary columns with split injection and temperature programming from 50°C to 270°C.

2. This SOP was written by R.D. Rhue, Soil and Water Science Department, University of Florida, Gainesville, FL.

3. This method involves gas chromatography (GC) analysis of surfactant concentrations in aqueous solutions used to solubilize OU1 LNAPL. A flame ionization detector (FID) is used to quantify the surfactant. A method detection limit is established using standard procedures. However, the MDL should be on the order of 0.1 g/L.

4. Samples selected for GC/FID analysis may be chosen on the basis of preliminary screening that will provide approximate concentration ranges and appropriate sample injection volumes, standard concentrations, etc.

Purpose:

The purpose of this SOP is to ensure reliable and reproducible analytical results for surfactant in aqueous solutions for laboratory-based or on-site (field-based) GC/FID analyses, and to permit tracing sources of error in analytical results.

Procedure:

1. Sample containers, collection, transportation, and storage
   a. Sample containers: Field samples are collected in 5-mL glass sample vials (Fisher Catalog #06-406-19F) with teflon-faced septa caps. Glass vials and caps are not reused.
   b. Sample collection: Each field sample vial is completely filled with liquid, such that no gas headspace exists, and capped. The vials will not be opened until the time for analysis.
   c. Transportation and storage: Field samples are stored in coolers containing "blue ice," and later stored in refrigerators in a trailer located on the site. Samples may be subjected to on-site GC analysis, and/or shipped back to UF labs; samples are packed in coolers and shipped via overnight air express (e.g., FedEx). The samples are stored in the cold storage room or refrigerator at 4°C until GC analysis. After subsampling, the samples are returned to cold storage.

   For lab studies, samples will be collected directly in 2-mL vials whenever possible and stored in a refrigerator if analysis is expected to take more than a day.

2. Subsampling and dilution
   a. Samples from field experiments are subsampled into 2-mL vials for automated GC analysis. Disposable Pasteur glass pipets (Fisher Catalog #13-678-20B) are used to transfer samples from 5-mL sample vials to the 2-mL GC vials.
   b. For samples needing dilution prior to GC analysis, a dilution of 1:10 should be sufficient. Dilutions are made using double-distilled, deionized water.

3. Apparatus and materials
   a. Glassware:
      Disposable micropipets (50 µL; Fisher Catalog #21-175B; 21-175F) and Class A volumetric pipets (1 or 2 mL) are required for sample dilution.
      Disposable Pasteur glass pipets (Fisher Catalog #13-678-20B) are required for subsampling. GC vials (2-mL) with Teflon-faced caps (Fisher Catalog #03-375-16A) are required for GC analysis.

   b. Gas chromatograph system: A Perkin Elmer Autosystem XL with an FID and an integrated
      autosampler is used for analysis of field and laboratory samples. The Perkin Elmer system is
      linked to an IBM-compatible PC loaded with Turbochrom (version 4.01) software.
         A Heliflex AT-1 column (10 m × 0.53 mm, 0.25-μm film thickness) is used. Zero-grade air
      and ultrahigh-purity hydrogen are used for the FID. Ultrahigh-purity nitrogen or helium is used
      for carrier gas.

4. Reagents
   a. Deionized, double-distilled water: Deionized, double-distilled water is prepared by double
      distillation of deionized water in a quartz still. This water is referred to as reagent water.
   b. Surfactants: Commercial-grade surfactants are used in the field studies. Therefore, samples of
      these same commercial-grade materials are obtained from the manufacturers and used in labo-
      ratory and field samples. These surfactants are identified by the manufacturer's trade name and
      are used as received without purification. Monodisperse polyoxyethylene mono-n-alkyl ethers
      (C_mE_n) are obtained and used to identify peaks of target components in the commercial surfactant
      formulations.

5. Standard solutions
   a. Stock standard solution: The surfactant concentration will depend on the particular surfactant
      chosen for the SPME experiments, but will probably be 3 wt%. The stock standard solution is
      prepared by dissolving the commercial-grade surfactant in reagent water. The amount of sur-
      factant added to the water is based on the manufacturer's lot analysis for purity.
   b. Calibration standards: Calibration standards are prepared by diluting the stock standard with
      reagent water. Five concentrations are prepared that cover the approximate concentration range
      from 0 to 3 wt%.

6. QC blank spike/matrix spike
      Two 0.5-mL aliquots of the sample to be spiked are transferred to clean vials. To one vial, 0.5
      mL reagent water is added. To the second, 0.5 mL of a calibration standard is added. The spike
      recovery is calculated using the difference between the two measured concentrations and the
      known spike concentration.

7. Quality control
   a. GC injector septa are changed every 80 to 100 injections, or sooner if any related problems occur.
   b. Injector liner is cleaned or changed every 80 to 100 injections or sooner if any related problems
      occur.
   c. A method blank is included in every 50 samples.
   d. A complete set of calibration standards (5) is run at the beginning of each day and after every
      50th sample.
   e. One standard and a blank are included in every 25 samples.
   f. A sample spike and a blank spike are included in every 50 samples.

8. Instrumental procedures
Gas chromatography:	For Heliflex AT-1 column
Injection port temperature	320°C
FID detector temperature	345°C
Temperature program:	Initial temperature at 100°C for 0 min; ramp to 320°C at 8°C/min; hold at 320°C for 10 min.

9. Sample analysis
   a. Analysis: The samples are allowed to reach ambient temperature prior to GC analysis.
      Sample vials (2 mL) are loaded on the Perkin Elmer GC autoinjector. A 1-μL injection volume
      is used for both samples and standards.
   b. Analyte identification: Analyte identification is based on absolute retention times. The surfactant
      components should elute at their characteristic retention times within ±0.1 min for the automated
      GC system.
   c. Analyte quantitation: When an analyte has been identified, the concentration is based on the
      peak area, which is converted to concentration using a standard calibration curve.

10. Interferences
   a. Contamination by carry-over can occur whenever high-level and low-level samples are sequen-
      tially analyzed. To reduce carry-over, the injector syringe should be rinsed with reagent water
      between samples.

b. Potential carry-over is checked by running a highly concentrated sample, but one still within the standard concentration range, followed by a blank. A negligible reading for the blank will ensure that carry-over has been minimized.

11. Safety

a. The main safety issue concerning the use of GC at a field site relates to the compressed gases. The FID gases (hydrogen and air) form explosive mixtures. It is important to keep this in mind at all times, and be aware of the hazard potential in the event of an undetected hydrogen leak. All gas connections are properly leak-tested at installation.

b. High-pressure compressed-gas cylinders are secured to a firm mounting point, whether they are located internally or externally. Gas cylinders should preferably be located outside the trailer on a flat, level base, and the gas lines run inside through a duct or window opening. If the gases are located outside, then some form of weatherproofing for the gauges is necessary. As a temporary measure, heavy-duty polyethylene bags, secured with tie-wraps, have been used successfully; this may not be very elegant, but it is very effective for short-term use of the GC. A more permanent protective housing must be built if the GC is located outside the trailer for an extended time period.

c. The main operating drawback to locating the gas cylinders externally is that it is not easy to monitor the cylinder contents from inside. The gas that could be used up most quickly is air for the FID, particularly if two instruments are hooked up to the same supply and they are running continuously. A reserve cylinder of air should be available at all times to prevent downtime. If it is not possible to arrange external siting easily, the gas cylinders should be secured to a wall inside the trailer.

d. It is good laboratory operating practice to make sure the flame is attended at all times.

e. When it is necessary to change the injection liner on the GC, the detector gases should be shut off.

f. The column must be connected to the detector before igniting the flame.

g. The trailer should be kept well ventilated when using the GC.

h. Reference is made to the Materials Safety Data Sheets (MSDS) for information on toxicity, flammability, and other hazard data.

References

Cross, J. (1987). Gas-liquid Chromatography of Nonionic Surfactants and Their Derivatives. In *Nonionic Surfactants: Chemical Analysis*, Surfactant Science Series, Vol. 19. John Cross, Ed., Marcel Dekker, New York.

Giger, W., Stephanou, E., and Schaffner, C. (1981). Persistent organic chemicals in sewage effluents. I. Identifications of nonylphenols and nonylphenolethoxylates by glass capillary gas chromatography/mass spectrometry. *Chemosphere* 10:1253–1263.

Stancher, B. and Favretto, L. (1978). Gas-liquid chromatographic fractionation of polyoxyethylene non-ionic surfactants. Polyoxyethylene mono-n-alkyl ethers. *J. Chromatography* 150:447–453.

A2.4 STANDARD OPERATING PROCEDURE FOR ANALYSIS OF ANALYTES WITH SPME FLUSHING

Scope and Application:

1. This SOP describes the analytical procedures to be utilized by the Soil and Water Science Department, University of Florida, IFAS, for analysis of select constituents of nonaqueous phase liquids (NAPLs) in both lab and field SPME flushing studies for the purpose of quantifying the amount and distribution of residual nonaqueous phase liquids (NAPLs) present in the saturated zone. The matrix for these samples is an aqueous solution containing a surfactant and an alcohol. The maximum surfactant concentration should be around 3 weight percent (wt%); that of the alcohol should be between 1 and 3 wt%, depending on the surfactant.

2. This SOP was written by R.D. Rhue, Soil and Water Science Department, University of Florida, Gainesville, FL. It is a modification of SOP-UF-Hill-95-07-0012-v.2, prepared by D.P. Dai and P.S.C. Rao, Soil and Water Science Department, University of Florida.

3. The selected constituents are n-decane, n-undecane, n-dodecane, n-tridecane, 1,2,4,-trimethylbenzene, 1,2-dichlorobenzene, and 1,2,4-trichlorobenzene.

4. The method involves gas chromatography (GC) analysis for concentrations of above constituents in aqueous surfactant samples. A flame ionization detector (FID) is used to quantify the analyte concentrations in the sample. Gas chromatography has been found to provide reliable and reproducible quantitation of the above constituents in organic solvents for concentrations >0.10 µg/mL. The standard calibration curve for FID response is linear to greater than 200 µg/mL. For concentrations beyond this value, sample dilution is generally required.

5. Samples selected for GC/FID analysis may be chosen on the basis of preliminary screening, which will provide approximate concentration ranges and appropriate sample injection volumes, standard concentrations, etc.

Purpose:

The purpose of this SOP is to insure reliable and reproducible analytical results of NAPL constituents in aqueous surfactant and alcohol flushing solutions for laboratory-based or on-site (field-based) GC/FID analyses, and to permit tracing sources of error in analytical results.

Procedures:

1. Sample containers, collection, transportation and storage
   a. Sample containers: Field samples are collected in 5-mL glass sample vials (Fisher Catalog #06-406-19F) with teflon-faced septa caps. Glass vials and caps are not reused.
   b. Sample collection: Each field sample vial is completely filled with liquid, such that no gas headspace exists, and capped. The vials will not be opened until time for analysis.
   c. Transportation and storage: Field samples are stored in coolers containing "blue ice," and later stored in refrigerators in a trailer located on the site. Samples may be subjected to on-site GC analysis, and/or shipped back to UF labs; samples are packed in coolers and shipped via overnight air express (e.g., FedEx). The samples are stored in the cold storage room or refrigerator at 4°C until GC analysis. After subsampling, the samples are returned to cold storage.
      For lab studies, samples are collected directly in 2-mL GC vials when possible and stored in a refrigerator if analysis is expected to take more than a day.

2. Subsampling and dilution
   SPME samples from field experiments are subsampled into 2-mL vials for automated GC analysis. Disposable Pasteur glass pipets (Fisher Catalog #13-678-20B) are used to transfer samples from 5-mL sample vials to the 2-mL GC vials.
   For samples needing dilution prior to GC analysis, dilution is usually 1:10 in aqueous surfactant/alcohol solution, but the best dilution ratio can be determined following a preliminary screening analysis.

3. Apparatus and materials
   a. Glassware:
      • Disposable micropipets (100 µL; Fisher Catalog #21-175B; 21-175F) and Class A volumetric pipets (1 or 2 mL) are required for sample dilution.
      • Disposable Pasteur glass pipets (Fisher Catalog #13-678-20B) are required for subsampling.
      • GC vials (2-mL) with Teflon-faced caps (Fisher Catalog #03-375-16A) are required for GC analysis.
      • Volumetric Class A pipets and volumetric Class A flasks are required for preparations of the calibration standards.
   b. Gas chromatograph system: A Perkin Elmer Autosystem XL with an FID and an integrated autosampler are used for analysis of field and laboratory samples. The Perkin Elmer system is linked to an IBM-compatible PC loaded with Turbochrom (version 4.01) software.

A J&W Scientific DB-624 capillary column (30 m × 0.53 mm, 3-μm film thickness) is used. Zero-grade air and ultrahigh-purity hydrogen are used for the FID. Ultrahigh-purity nitrogen or helium is used for carrier gas.

4. Reagents
   a. Deionized, double-distilled water: Deionized, double-distilled water is prepared by double distillation of deionized water in a quartz still. This water is referred to as reagent water.
   b. *n*-Hexane: High-purity, *Pro Analysis* grade hexane is purchased from Fisher Scientific and used as received.
   c. Alcohols: Certified ACS grade *n*-pentanol and *n*-butanol are purchased from Fisher Scientific and used as received.
   d. Surfactants: Commercial-grade surfactants are used in the field studies. Therefore, samples of these same commercial-grade materials are obtained from the manufacturers and used in laboratory and field samples. These surfactants are identified by the manufacturer's trade name and are used as received without purification.

5. Standard solutions
   a. Stock standard solution: Analytical standards are prepared from reagent chemicals by the laboratory. Stock standards will each contain a single analyte dissolved in hexane and stored in 20-mL glass vials (Fisher Catalog #03-393-D) with teflon-lined caps. These stock solutions are kept in a refrigerator at 4°C. Fresh stock standards are prepared every 6 months. The procedure for making stock standard solutions is essentially that given in the Federal Register, Rules and Regulations, Thursday, November 29, 1979, Part III, Appendix C, Section 5.10, "Standard Stock Solutions." The only modification of the procedure for the current study is that hexane is used as the solvent in place of methanol.
   b. Calibration standards: Calibration standards are prepared by diluting the stock standards in hexane. Each calibration standard will contain each of the eight analytes listed above. Five concentrations are prepared that cover the approximate concentration range from 0 to 200 mg/L.
   c. Stock spiking solution: Since the matrix for the field and lab samples is an aqueous surfactant/alcohol solution, the standards in hexane are not suitable for spiking samples. The matrix for the stock spiking solution is an aqueous surfactant/alcohol solution containing surfactant and alcohol at the same concentrations as those used in the field and lab experiments. The stock spiking solution is prepared by dissolving a known amount of the eight target analytes in the aqueous surfactant/alcohol solution. The analyte concentrations will reflect the concentrations expected in the samples.

6. QC blank spike/matrix spike
   Two 0.5-mL aliquots of the sample to be spiked are transferred to clean vials. To one, 0.5 mL aqueous surfactant/alcohol solution containing no analytes is added. To the second vial, 0.5 mL stock spiking solution is added. The spike recovery is calculated using the difference between the two measured concentrations and the known spike concentration.

7. Quality control
   a. GC injector septa are changed every 80 to 100 injections, or sooner if any related problems occur.
   b. Injector liner is cleaned or changed every 80 to 100 injections or sooner if any related problems occur.
   c. A method blank is included in every 50 samples.
   d. A complete set of calibration standards (5) is run at the beginning of each day and after every 50th sample.
   e. One standard and a blank are included in every 25 samples.
   f. A sample spike and a blank spike are included in every 50 samples.

8. Instrumental procedures

Gas chromatography:	For J&W DB-624 Column:
Injection port temperature	200°C
FID detector temperature	225°C
Temperature program:	Isothermal at 100°C for 10 min; ramp to 200°C at 10°C/min.

9. Sample preparation
   a. Subsampling: Field samples are transferred from the 5-mL sample vials to the 2-mL GC vials and capped with open-top, teflon-lined septa caps.

    b. Dilution: Samples are diluted if chromatographic peak areas for any of the eight target analytes exceed those of the calibration standards; 1 mL of sample is added to an appropriate amount of stock surfactant solution to make the dilution.

10. Sample analysis

    a. Analysis: The samples are allowed to reach ambient temperature prior to GC analysis. Sample vials (2 mL) are loaded on the Perkin Elmer GC auto-injector. A 1-μL injection volume is used for both samples and standards.

    b. Analyte identification: Analyte identification is based on absolute retention times. The analytes of interest should elute at their characteristic retention times within ±0.1 min for the automated GC system.

    c. Analyte quantitation: When an analyte has been identified, the concentration is based on the peak area, which is converted to concentration using a standard calibration curve.

11. Interferences

    a. Contamination by carry-over can occur whenever high-level and low-level samples are sequentially analyzed. To reduce carry over, the injector syringe should be rinsed with a blank surfactant/alcohol solution between samples.

    b. Potential carry-over is checked by running a highly concentrated sample, but one still within the standard concentration range, followed by a blank. A negligible reading for the blank will ensure that carry-over has been minimized.

12. Safety

    a. The main safety issue concerning the use of the GC at a field site relates to the compressed gases. The FID gases (hydrogen and air) form explosive mixtures. It is important to keep this in mind at all times, and be aware of the hazard potential in the event of an undetected hydrogen leak. All gas connections are properly leak-tested at installation.

    b. High-pressure compressed-gas cylinders are secured to a firm mounting point, whether they are located internally or externally. Gas cylinders should preferably be located outside the trailer on a flat, level base, and the gas lines run inside through a duct or window opening. If the gases are located outside, then some form of weatherproofing for the gauges is necessary. As a temporary measure, heavy-duty polyethylene bags, secured with tie-wraps, have been used successfully; this may not be very elegant, but it is very effective for short-term use of the GC. A more permanent protective housing must be built if the GC is located outside the trailer for an extended time period.

    c. The main operating drawback to locating the gas cylinders externally is that it is not easy to monitor the cylinder contents from inside. The gas that could be used up most quickly is air for the FID, particularly if two instruments are hooked up to the same supply and they are running continuously. A reserve cylinder of air should be available at all times to prevent downtime. If it is not possible to arrange external siting easily, the gas cylinders should be secured to a wall inside the trailer.

    d. It is good laboratory operating practice to make sure the flame is attended at all times.

    e. When it is necessary to change the injection liner on the GC, the detector gases should be shut off.

    f. The column must be connected to the detector before igniting the flame.

    g. The trailer should be kept well ventilated when using the GC.

    h. Reference is made to the Materials Safety Data Sheets (MSDS) for information on toxicity, flammability, and other hazard data.

A2.5 STANDARD OPERATING PROCEDURE (SOP) FOR ANALYSIS OF TARGET ANALYTES IN GROUNDWATER SAMPLES

Scope and Application:

1. This SOP describes the analytical procedures utilized by the Department of Environmental Engineering Sciences, University of Florida, for analysis of target analytes in groundwater samples from both lab and field studies. This analysis provides characterization of existing site and lab column aqueous contamination both before and following flushing technology applications.

2. This SOP was written by M.D. Annable, Department of Environmental Engineering Sciences, University of Florida, Gainesville. It is a modification of SOP-UF-Hill-95-07-0012-v.2, prepared by D.P. Dai and P.S.C. Rao, Soil and Water Science Department, University of Florida.

3. The selected constituents are benzene, toluene, o-xylene, 1,1,1-trichloroethane, 1,3,5,-trimethylbenzene, 1,2-dichlorobenzene, decane, and naphthalene.

4. The method involves gas chromatography (GC) analysis for target analyte concentrations in aqueous samples. Headspace analysis with a flame ionization detector (FID) is used to quantify the analyte concentrations in the sample. The method has been found to provide reliable and reproducible quantitation of the above constituents for concentrations >5 µg/L. This value may be considered the method detection level (MDL).

5. Samples selected for GC/FID analysis may be chosen on the basis of preliminary screening, which will provide approximate concentration ranges and appropriate sample injection times, and standard concentrations, etc.

Purpose:

The purpose of this SOP is to ensure reliable and reproducible analytical results for soluble NAPL constituents in aqueous samples for laboratory-based GC/FID analyses, and to permit tracing sources of error in analytical results.

Procedures:

1. Sample containers, collection, transportation and storage
   a. Sample containers: Field samples are collected in 20-mL glass sample vials (Fisher Catalog #03-340-121) with teflon-faced rubber-backed caps. Glass vials and caps are not reused.
   b. Sample collection: Each field sample vial will be completely filled with liquid, such that no gas headspace exists, and capped. The vials will not be opened until the time for analysis.
   c. Transportation and storage: Field samples are stored in coolers containing "blue ice," and later stored in refrigerators in a trailer located on the site. Samples are sent to UF labs packed in coolers and shipped via overnight air express (e.g., FedEx). The samples are stored in the cold storage room or refrigerator at 4°C until GC analysis. After subsampling, the samples are returned to cold storage.

   For lab studies, samples are collected directly in 20-mL headspace vials whenever possible and stored in a refrigerator if analysis is expected to take more than a day.

2. Subsampling and dilution

   Field samples are subsampled by placing 10 mL into 20-mL headspace vials containing 2 g sodium chloride for automated GC analysis. Pipets are used to transfer samples from 20-mL sample vials to the 20-mL GC headspace vials.

3. Apparatus and materials
   a. Glassware: Glass pipets are required for subsampling. GC headspace vials (20-mL) with Teflon-faced caps are required for GC analysis. Volumetric Class A pipets and volumetric Class A flasks are required for preparations of the calibration standards.
   b. Gas chromatograph system: An analytical GC system with a temperature-programmable oven, headspace sample injection system, and either an integrator or a PC-based data acquisition/analysis software system are required. Also required are other accessories, including analytical columns and the gases required for GC/FID operation.

   A Perkin Elmer Autosystems with an HS40 Autoheadspace sampler and an FID is used for analysis of field and laboratory samples. The Perkin Elmer system is linked to an IBM-compatible PC loaded with Turbochrom (version 4.01) software.

   A J&W Scientific DB-624 capillary column (50 m × 0.53 mm, 3-mm film thickness) is used. Zero-grade air and high-purity hydrogen are used for the FID. Ultrahigh-purity nitrogen or helium is the carrier gas.

4. Reagents
   a. Deionized, double-distilled water which is prepared by double distillation of deionized water in a quartz still. This water is referred to as reagent water.

5. Standard solutions

   a. Stock standard solution: Analytical standards are prepared from reagent chemicals by the laboratory. Stock standards each contain a single analyte dissolved in methanol and stored in 20-mL glass vials (Fisher Catalog #03-393-D) with teflon-lined caps. These stock solutions will be kept in a refrigerator at 4°C. Fresh stock standards are prepared every 6 months. The procedure for making stock standard solutions is essentially that given in the Federal Register, Rules and Regulations, Thursday, November 29, 1979, Part III, Appendix C, Section 5.10, "Standard Stock Solutions."

   b. Calibration standards: Calibration standards are prepared by diluting the stock standards in water. Each calibration standard contains each of the eight analytes listed above. Five concentrations are prepared that cover the approximate concentration range from 0 to 20 mg/L.

6. QC blank spike/matrix spike

   Two 1-mL aliquots of the sample to be spiked are transferred to clean vials. To one, 1 mL reagent water is added. To the second vial, 1 mL of a calibration standard is added. The spike recovery is calculated using the difference between the two measured concentrations and the known spike concentration.

7. Quality Control

   a. A method blank is included in every 50 samples.

   b. A complete set of calibration standards (5) is run at the beginning of each day and after every 50th sample.

   c. One standard and a blank are included in every 25 samples.

   d. A sample spike and a blank spike are included in every 50 samples.

8. Instrumental Procedures

Gas Chromatography:	For J&W DB-624 Column:
Headspace sample temperature	90°C
Injection needle temperature	120°C
Transfer line Temperature	150°C
FID detector temperature	225°C
Carrier gas pressure	12 psi
Temperature program:	Isothermal at 50°C for 0 min; ramp to 200°C at 5°C/min; hold for 10 min

9. Sample preparation

   a. Subsampling: Field samples are transferred from the 20-mL sample vials to the 20-mL GC headspace vials and capped with open-top, teflon-lined septa caps.

   b. Dilution: Samples are diluted if chromatographic peak areas for any of the analytes exceed those of the highest calibration standard; 1 mL sample is added to an appropriate amount of reagent water to make the dilution.

10. Sample analysis

   a. Analysis: Sample headspace vials (20 mL) are loaded onto the Perkin Elmer HS40 autosampler. Samples are pressurized for 1 min followed by a 0.1-min injection time; the injection needle is then flushed for 1 min.

   b. Analyte identification: Analyte identification is based on absolute retention times. The analytes of interest should elute at their characteristic retention times within ±0.1 minute for the automated GC system.

   c. Analyte quantitation: When an analyte has been identified, the concentration is based on the peak area, which is converted to concentration using a standard calibration curve.

11. Interferences

   a. Contamination by carry-over can occur whenever high-level and low-level samples are sequentially analyzed. To reduce carry over, the injector needle should be purged with carrier gas between samples.

   b. Potential carry-over will be checked by running a highly concentrated sample, but one still within the standard concentration range, followed by a blank. A negligible reading for the blank will insure that carry-over has been minimized.

12. Safety
   a. The main safety issue concerning the use of the GC relates to the compressed gases. The FID gases (hydrogen and air) form explosive mixtures. It is important to keep this in mind at all times, and be aware of the hazard potential in the event of an undetected hydrogen leak. All gas connections are properly leak-tested at installation.
   b. High-pressure compressed-gas cylinders are secured to a firm mounting point, whether they are located internally or externally.
   c. When it is necessary to change the injection liner on the GC, the detector gases should be shut off. The column must be connected to the detector before igniting the flame.
   d. Reference is made to the Materials Safety Data Sheets (MSDS) for information on toxicity, flammability, and other hazard data.

APPENDIX 3

Multilevel Sampler Data

This appendix contains example data collected from the multilevel sampling points during the three stages of SPME field demonstration — the pre-flushing partitioning tracer test, the SPME flood, and the post-flushing partitioning tracer test. For reference as to sampling point locations, Figure 14.2 is included. As can be seen, the first digit of the MLS location refers to the injection well from which it is downgradient. The second digit increases with distance downgradient of the injection well. The color refers to the depth of the sampling point below ground surface, as shown on the key.

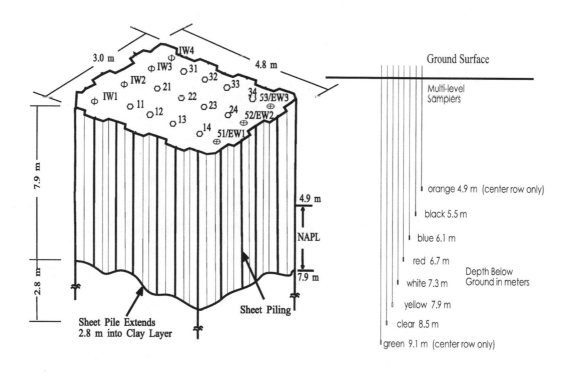

Figure 14.2 Schematic of the SPME test cell.

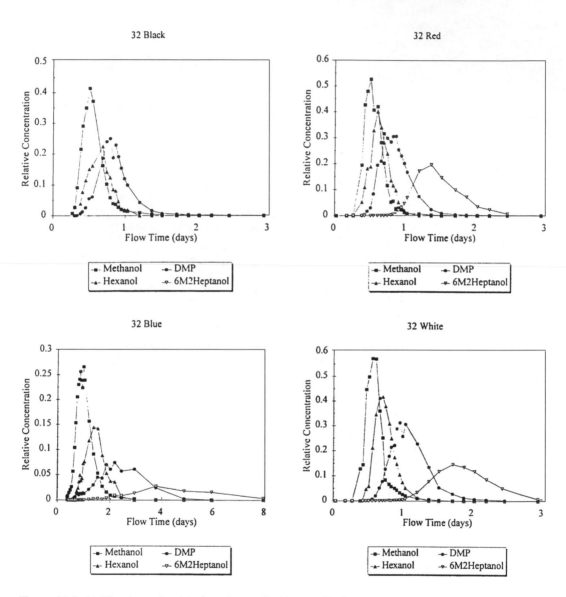

Figure A3.1 Multilevel sampler data from the pre-flushing partitioning tracer test.

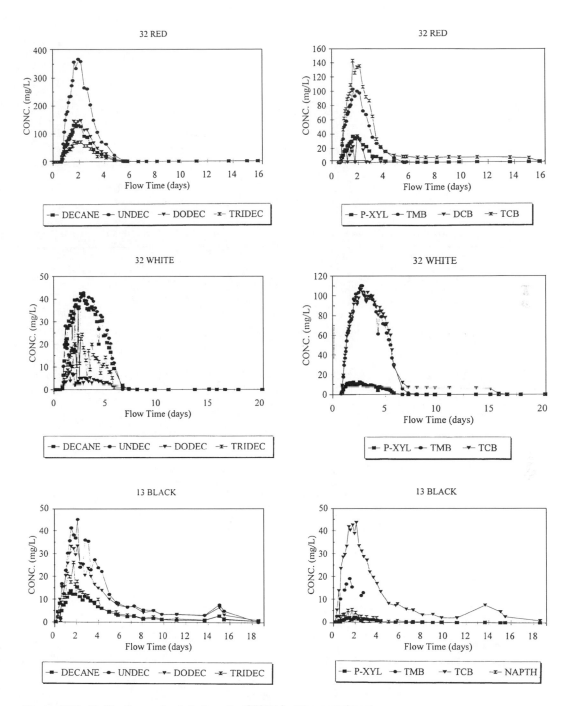

Figure A3.2 Multilevel sampler data from the SPME flushing experiment.

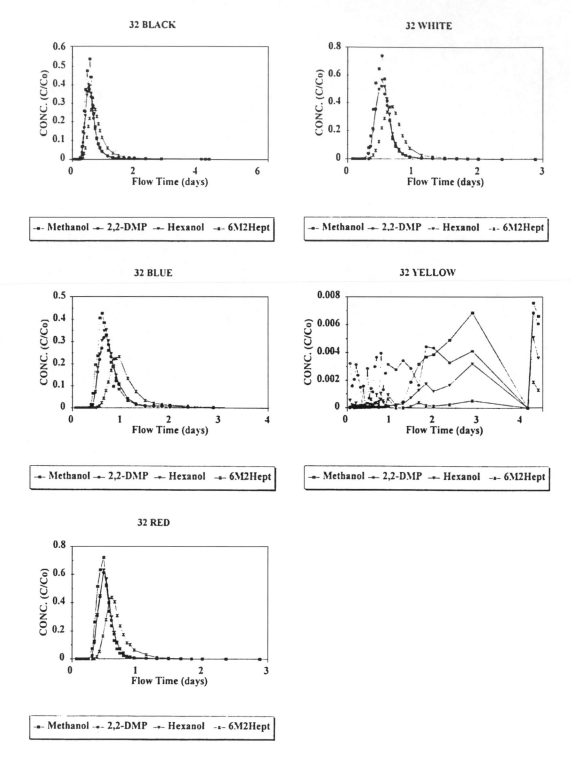

Figure A3.3 Multilevel sampler data from the post-flushing partitioning tracer test.

Drawings for Surfactant/foam Scale-up Design

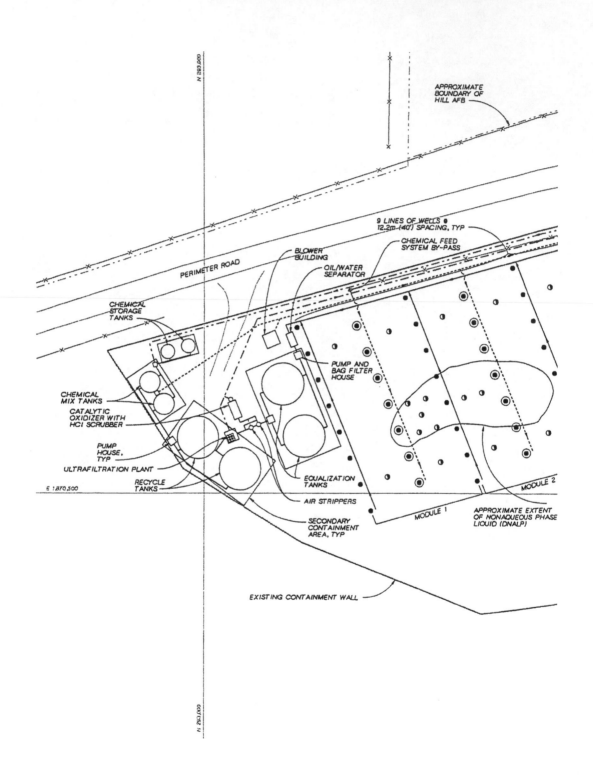

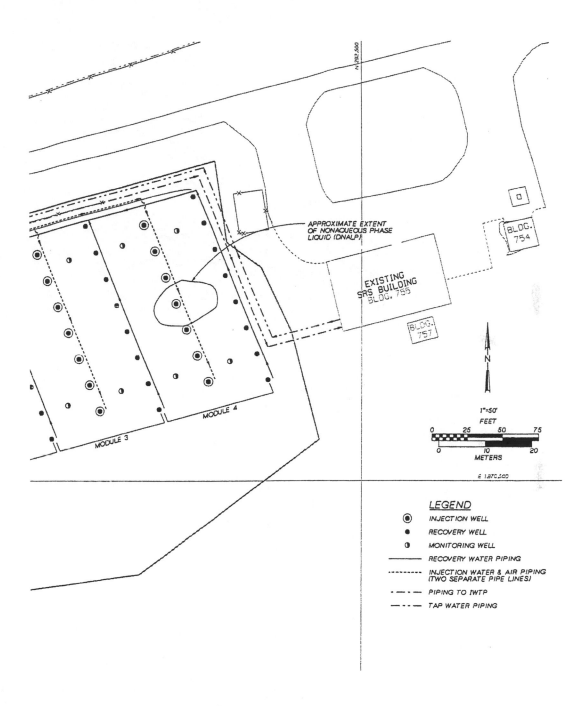

Figure A4.1 Surfactant/foam site plan.

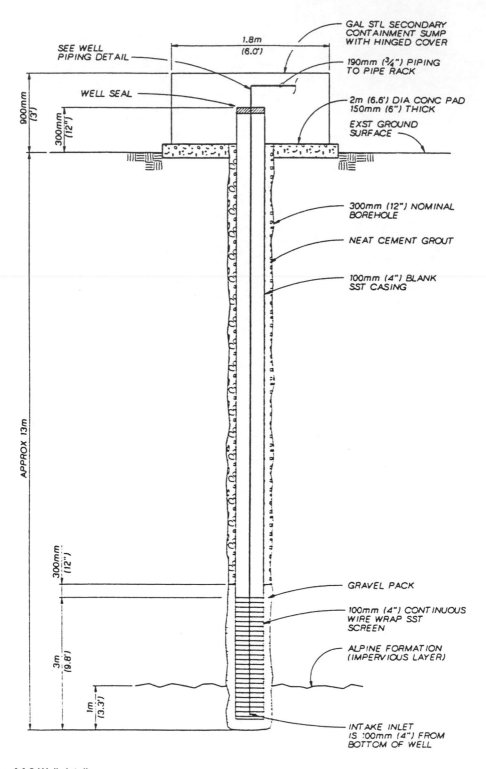

Figure A4.2 Well detail.

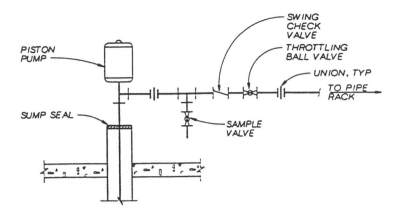

RECOVERY WELL PIPING DETAIL
NTS

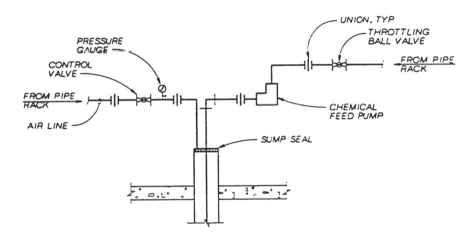

Figure A4.3 Well piping details.

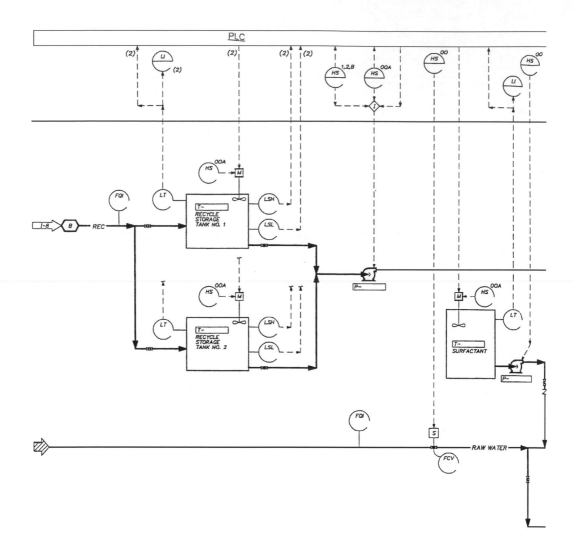

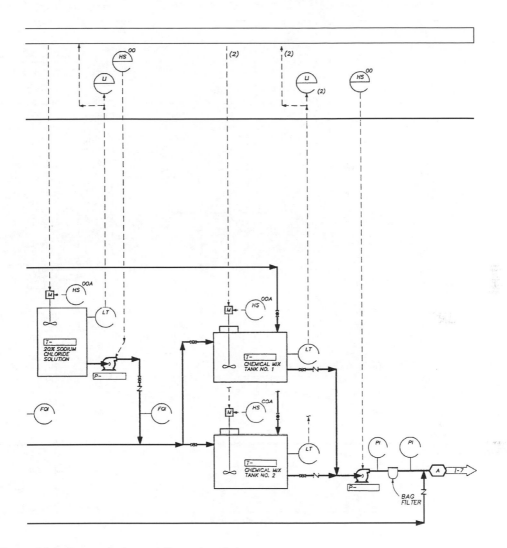

Figure A4.4a Process instrumentation and controls.

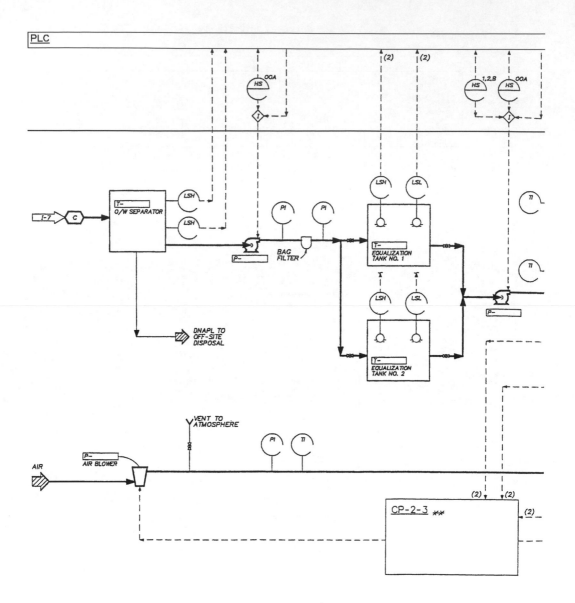

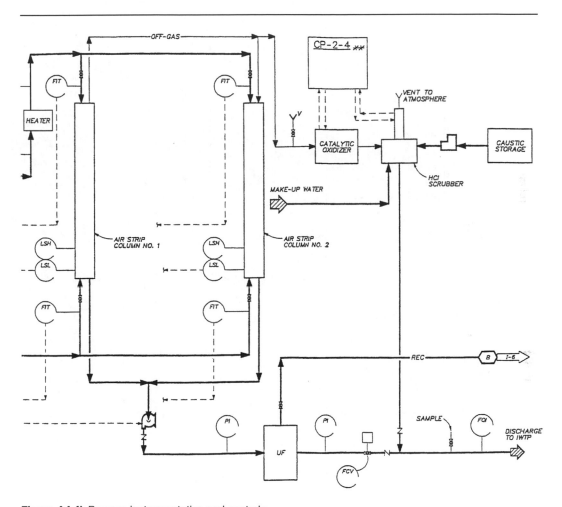

Figure A4.4b Process instrumentation and controls.

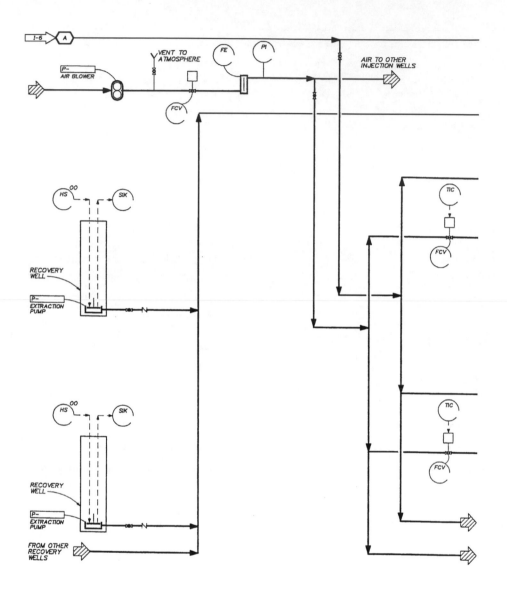

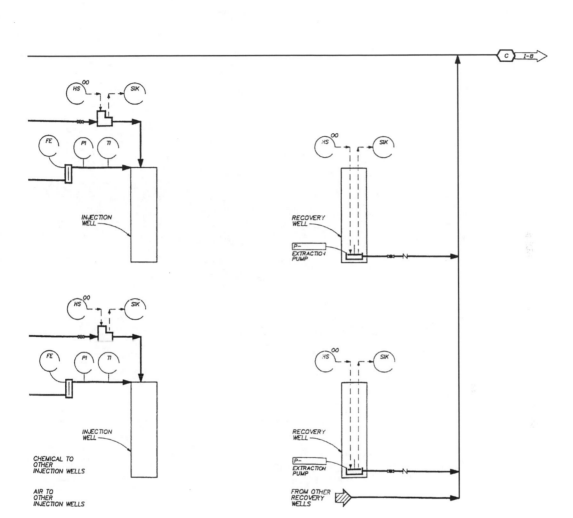

Figure A4.4c Process instrumentation and controls.

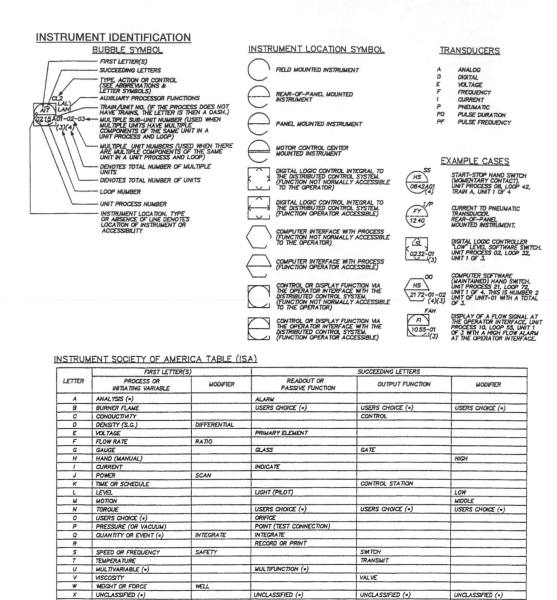

Figure A4.5a Legend sheet 1.

FLOW STREAM IDENTIFICATION

BYP	BYPASS
OF	OVERFLOW
REC	RECYCLE

ABBREVIATION & LETTER SYMBOLS

AC	ALTERNATING CURRENT
ACK	ACKNOWLEDGE
AFD	ADJUSTABLE FREQUENCY DRIVE
AM	AUTO-MANUAL
AS	ADJUSTABLE SPEED
CAM	COMPUTER-AUTO-MANUAL
CCS	CENTRAL CONTROL SYSTEM
CIP	CHEMICAL INTERFACE PANEL
CL2	CHLORINE (TYPICAL: USE STANDARD CHEMICAL ELEMENT ABBREVIATION)
CL RESD	CHLORINE RESIDUAL
COND	CONDUCTIVITY
CP-WWXX	COMPUTER-MANUAL
	CONTROL PANEL (CONTROL PANELS ASSOCIATED WITH A PACKAGED SYSTEM) WW = UNIT PROCESS NUMBER XX = LOOP NUMBER
CS	CONSTANT SPEED
DC	DIRECT CURRENT
DO	DISSOLVED OXYGEN
E STOP	EMERGENCY STOP
FCC-XX-AA	FILTER CONTROL CONSOLE (SPECIFIED UNDER PROCESS INSTRUMENTATION AND CONTROLS) WW = UNIT PROCESS NUMBER AA = PANEL UNIT NUMBER
FP-XX-AA	FIELD PANEL (SPECIFIED UNER PROCESS INSTRUMENTATION AND CONTROLS) WW = UNIT PROCESS NUMBER AA = PANEL UNIT NUMBER
F/R	FORWARD/REVERSE
HOA	HAND-OFF-AUTO
HOR	HAND-OFF-REMOTE
KK	TIMER CONTROL STATION
LCP-XX-AA	LOCAL CONTROL PANEL (SPECIFIED UNDER PROCESSS INSTRUMENTATION AND CONTROLS) WW = UNIT PROCESS NUMBER AA = PANEL UNIT NUMBER
L/L/S/A	LEAD/LAG/STANDBY/AUTO
LR	LOCAL-REMOTE
M	MANUAL
MA	MANUAL-AUTO
MCC-X	MOTOR CONTROL CENTER NUMBER - X
MCP	MAIN CONTROL PANEL (SPECIFIED UNDER PROCESS INSTRUMENTATION AND CONTROLS)
MMI	MAN-MACHINE INTERFACE
NIC	NOT IN CONTRACT
OC	OPEN-CLOSE
OCA	OPEN-CLOSE-AUTO
OO	ON-OFF (MAINTAINED CONTROL)
OOA	ON-OFF-AUTO
OOC	ON-OFF-COMPUTER
OOR	ON-OFF-REMOTE
OSC	OPEN-STOP-CLOSE
PERM	PERMISSIVE
PLC	PROGRAMMABLE LOGIC CONTROLLER
PMP	PLANT MONITORING PANEL (SPECIFIED PROCESS INSTRUMENTATION AND CONTROLS)
POT	POTENTIOMETER
R/T/A	RECYCLE TRANSFER AUTOMATIC
SCADA	SUPERVISORY CONTROL AND DATA ACQUISITION
SCU	SPEED CONTROL UNIT
SS	START-STOP (MOMENTARY CONTROL)
TURB	TURBIDITY
VHC	VOLATILE HYCROCARBON
1/2/B	1/2/BOTH
<	LESS THAN
>	GREATER THAN
X	MULTIPLIER
1/2	ONE TWO SELECTOR
+	SUMMATION
SqR	SQUARE ROOT

GENERAL NOTES

1. COMPONENTS AND PANELS SHOWN WITH A DIAMOND (◆) ARE TO BE PROVIDED UNDER SECTION "PROCESS INSTRUMENTATION AND CONTROLS."

2. COMPONENTS AND PANELS SHOWN WITH A DOUBLE ASTERISK (**) ARE TO BE PROVIDED AS PART OF A PACKAGED OR MECHANICAL SYSTEM.

3. COMPONENTS AND PANELS SHOWN WITH A TRIANGLE (▲) ARE TO BE PROVIDED BY OTHERS (NIC).

4. THIS IS A STANDARD LEGEND SHEET. THEREFORE, NOT ALL OF THE INFORMATION SHOWN MAY BE USED ON THIS PROJECT.

Figure A4.5a (continued) Legend sheet 1.

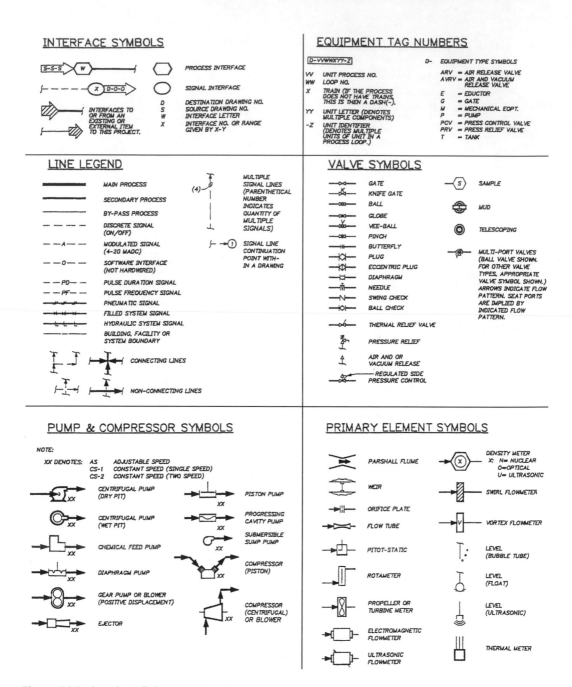

Figure A4.5a (continued) Legend sheet 1.

MISCELLANEOUS SYMBOLS

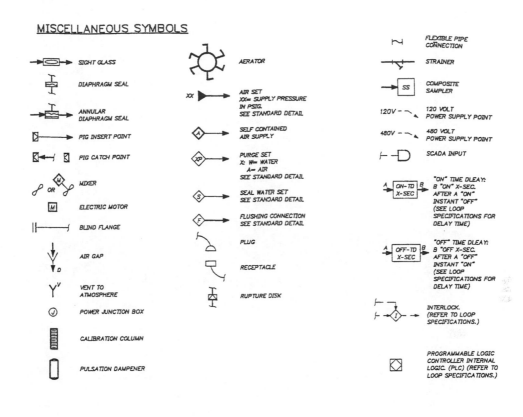

ACTUATOR SYMBOLS

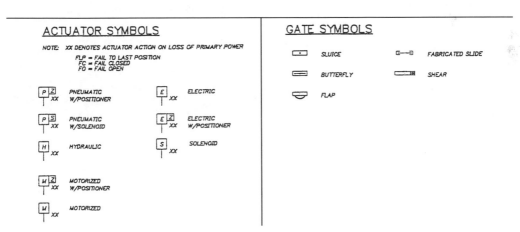

Figure A4.5b Legend sheet 2.

APPENDIX 5

Surfactant/foam Scale-up Mass Balance and Design Spreadsheets

BASE CASE DESIGN WORKSHEET

Surfacant/Foam Design, Base Case

Technology Evaluation Report for Surfactant-Foam Aquifer Remediation

Design Worksheet
Base Case

General Site Assumptions:

Total Area =	1.03 ac =	0.42 hectare
H, Thickness of target DNAPL zone =	2 m =	6.56 ft
W, Width per module =	24.4 m =	80 ft
L, Length per module =	42.7 m =	140 ft Spacing = 0.5 * W

Number of Injection Wells per module   7 wells per module

Phase I of remediation will utilize   1 module
The maximum no. of modules to be operated   1 module(s) at a time.

porosity =	0.27
K (water or surfactant) =	0.011 cm/s
Viscosity (water) =	0.01 poise
Viscosity (surfactant/alcohol) =	0.02 poise
Hydraulic Head, Δh =	1 m

These parameters were adjusted to give the desired flow rate (based on a reasonable time frame. A maximum K of 0.1 cm/s is possible at this site.

No. of Pore Volumes Of Flushing Required For Remediation
Initial Waterflood	=	10
Surfactant Flood	=	2.5
Post Waterflood	=	10

Assumptions Used to Calculate Present Worth
Number of Years of   1   (This is the calculated treatment time, rounded up)
Interest Rate =   5.00%

Area and Volume Calculation for the Site

Number of Surfactant modules =	4.00 =	4 rounded
Area per module=	0.26 ac	
Volume of Alluvium with DNAPL per module =	2,726 cy per module =	2,084 cu m per module

Pore Volume Calculation:

PV = 2 * (H+Δh/2) * L * (W*0.5) * porosity =   703.3 cu m per module

Volume of DNAPL Calculation

Assume DNAPL is 4.3% of the total volume (16% DNAPL saturation at a porosity of 0.27) over 1/4 of the site

and 1.1% over 1/2 of the site (4% DNAPL saturation, assumed based on variability at the site)

and 0.3% over the other 1/4 of the site (1% DNAPL saturation, assumed based on variability at the site)

Thus 42.9 cu m = 11,329 gal

Specific Gravity of DNAPL is 1.4

Flow Rate Calculations per Module Q=KAi (in 2 directions from the source)

$Q_{water} = 2 \cdot K \cdot (H + \Delta h/2) \cdot L \cdot \Delta h / (.5 \cdot W) =$ 0.0020 cu m/s = 170.1 cu m/d = 31 gpm

$Q_{surfactant} = 2 \cdot K \cdot (\text{viscosity}_{water}/\text{viscosity}_{surfactant}) \cdot (H + \Delta h/2) \cdot L \cdot \Delta h / (.5 \cdot W) =$ 0.00098 cu m/d = 85.05 cu m/d = 16 gpm

For a module containing 7 wells, the water flow rate is 4.5 gpm per well

the surfactant flow rate is 2.2 gpm per well

Time & Volume Calculations for Waterflood, Surfactant/Foam Flood and Post Waterflood

Time per PV = PV/Q =	Water	Surfactant/Alcohol
	4.1 days	8.3 days

Sequence of Operations

Initial Waterflood	10 PV:	=	7,033 cu m per module for	41.3 days
		=	1,858,037 gal per module	
Surfactant Flood	2.5 PV:	=	1,758 cu m per module for	20.7 days
		=	464,509 gal per module	
Post Waterflood	10 PV:	=	7,033 cu m per module for	41.3 days
		=	1,858,037 gal per module	

Total time : 103.4 days = 0.28 years

Assume that Initial Waterflood is conducted on all modules at once.

Surfactant Flood and Post Waterflood will be conducted individually during each of 4 phases.

Total time for remediation 289 days = 0.79 years

Aerosol MA-80 Storage Requirement

Assume:

Recycling Rate: Volume of surfactant needed = initial pore volume + 22% % loss for each subsequent pore volume, (in ground and losses to IWTP)

Injected surfactant is 4% wt active sodium dihexyl sulfosuccinate

Aerosol MA-80 is supplied as a concentrate at 80% wt active

Specific gravity of supplied Aerosol MA-80 is 1.1

Specific gravity of surfactant solution is 1.

Fiberglass tanks shall be used for storing the concentrated chemicals. A convenient size of chemical storage tank is 5,000 gallons (8 foot diameter, 13.3 feet high)

Total quantity of Aerosol MA-80 needed:

83,323	kg active surfactant	
104,154	kg total =	229,139 lb total for remediation
104	mtons	115 tons
94.7	cu m =	25,016 gal

The maximum usage rate = 3.09 cu m/d = 817.1 gal/d

Use 1 tank(s), @ 5,000 gal ea = 18.9 cu m each

The actual specified storage volume represents the useage requirements 4.1 days

Sodium Chloride Storage

Assume:

Recycling Rate: Volume of NuCl needed = Initial pore volume + 20% % loss for each subsequent pore volume

Inject solution that is 1.1% wt active NaCl

Make stock solution of 20% wt NaCl

Specific gravity of stock NaCl solution is 1.1

Fiberglass tanks shall be used for storing the concentrated chemicals. A convenient size of chemical storage tank is 5,000 gallons (8 foot diameter, 13.3 feet high)

Quantity of NaCl Needed

40,227	kg active NaCl =	
88,499	lb active NaCl	
201,135	kg total weight of stock solution	201 mtons
442,497	lb total weight of stock solution	221 tons
182.8	cu m	
48,309	gal	

The maximum usage rate = 4.25 cu m/d = 1,124 gal/d

Use 1 tank(s), @ 5,000 gal ea = 18.9 cu m ea

The actual specified storage volume represents the useage requirements 2.1 days

Solution Mix/Storage Tanks

Assume that two (2) chemical mix/storage tanks are needed. The chemical mixture will be made in batches. One tank can be mixing while the other is dispensing solution to the injection wells.

Volume of Solution Mix Storage Tank

Assume operate a maximum of 1 module(s) at one time

If we assume that we make up a surfactant solution every 2 days

thus need solution tank 170.1 cu m = 44,940 gal

Use 2 tank(s), @ 20,000 gal ea = 75.7 cu m ea

The actual specified storage volume represents the useage requirements for 1.8 days

Air Injection System Requirements

Assume:

Air will be injected at a maximum of   2 scfm per well   2 hours on, 4 hours off
20 psi

The maximum number of wells to be operated at a time is   7 wells which correlates to   1 module(s)
The maximum required blower air flow rate is   14 scfm at   20 psi

Produced Fluids Handling

Assume:

A maximum of   1 module(s) will be operated at one time.

Recovery at a rate   1.1 times the injection rate

Water produced during the initial waterflood will be recycled through an oil water separator and does not need to pass through the water treatment plant.
Water produced during the surfactant flood and post waterflood will require pretreatment before being sent to the industrial waste water treatment plant (IWTP).
Pretreatment will consist of passing through an oil/water separator, equalization tank, air stripper, and ultrafiltration unit. The equalization tank will have a holding holding   1.5 days
The excess water will be discharged to the IWTP.

Injection Flow rate:

During Initial Waterflood -   170.1 cu m/d =   31 gpm   for   41 days
During Surfactant Flood -   85.1 cu m/d =   16 gpm   for   21 days
During Post Waterflood -   170.1 cu m/d =   31 gpm   for   41 days

Recovery Flow Rates and Volumes

During Initial Waterflood -   187.1 cu m/d =   7,736 cu m   34 gpm = 2,043,840 gal
During Surfactant Flood -   93.6 cu m/d =   1,934 cu m   17 gpm = 510,960 gal
During Post Waterflood -   187.1 cu m/d =   7,736 cu m   34 gpm = 2,043,840 gal

Equalization Tank Volume for Air Stripping

Assume we need one (1) tank before the air stripper

Volume based on   1 days of holding time for the recovered fluids
The flow rate governing design is   187.1 cu m/d which occurs during   Post Waterflood
Required Volume =   187 cu m =   49,434 gal
Use   3 each   20,000 gal ea =   75.7 cu m each
The actual storage volume represents the useage requirements for   1.2 days

Volume Requiring Air Stripping and Ultrafiltration (Recovered volumes during surfactant flood and post waterflood)

Volume =   9,670 cu m =   2,554,800 gal per module
Total volume =   38,680 cu m =   10,219,202 gal

Volume of Water Requiring Treatment Through the IWTP

29% of surfactant flood and post waterflood water

Process Volume =   2,761 cu m =   729,464 gal per module
Final Pore Volume =   352 cu m =   92,902 gal, fraction not sent off-site
Total Volume =   11,396 cu m =   3,010,758 gal

Recycle Storage Tank Volume

Provide storage for   40% of a pore volume to allow flexibility in operations

Volume required =   141 cu m =   37,161 gal

Use   2 each of   20,000 gal or   76 cu m tank(s)

Volume Requiring Off-Site Incineration

Assume the last pore volume will be re-concentrated by   50%

In the ultrafilter unit, and then sent off for incineration

Thus, volume requiring off-site incineration =   352 cu m =   92,902 gal

Volume from oil/water separator =   39 cu m =   10,196 gal

total volume for incineration =   390 cu m =   103,098 gal

BASE CASE MASS BALANCE

Surfactant/Foam MB, Optimized Case

Technology Evaluation Report for Surfactant-Foam Aquifer Remediation

Flow Diagram and Mass Balance
Optimized Case

Chemical Feed System
<u>Parameters During Surfactant Flood</u>
Parameters are specified for operation of   1 module at a time
The required surfactant/alcohol injection rate is   85.1 cu m/d for the above module(s) of operation
The injected solution will be   4.0% by weight surfactant
The injected solution will be   1.1% by weight NaCl
The specific gravity of the supplied surfactant is   1.1
The specific gravity of the premixed 20% NaCl is   1.1

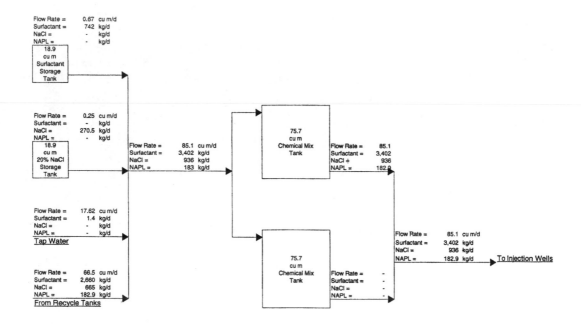

Surfactant/Foam MB, Base Case
(continued)

Technology Evaluation Report for Surfactant-Foam Aquifer Remediation

Flow Diagram and Mass Balance
Base Case

Parameters During Surfactant Flood

Assumptions:

Air to Water ratio for air stripper =	40
Ratio of COD/surfactant or LNAPL =	2 based on Surfactant/Foam report
Concentrations are constant for	21 days during surfactant injection for each module.
The recovery rate is	1.1 times the injection rate
Surfactant recovery from the subsurface is	85% by weight of injected surfactant
NaCl is conserved in the subsurface.	
Total DNAPL recovery from the subsurface is	90% by weight of DNAPL present in the subsurface
The oil/water separator recovers	25% by weight of the total recovered DNAPL
The oil/water separator recovers	0% by weight of the surfactant
The air stripper recovers	90% by weight of the NAPL from the oil/water separator effluent
The ultrafiltration plant recovers	92% by weight of the surfactant in micelles from the air stripper
The ultrafiltration plant recovers	90% by weight of the NAPL in micelles from the air stripper
The ultrafiltration plant concentrate waste steam is	4.0% by weight surfactant
Assume an average carbon pickup of	10% by weight (lb COD/lb carbon)

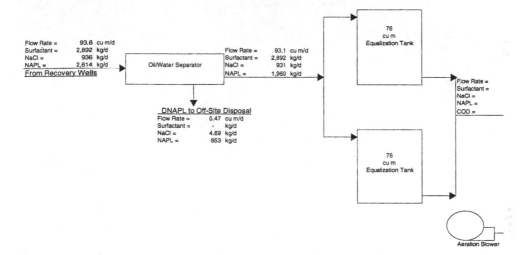

Surfactant/Foam MB, Base Case
(continued)

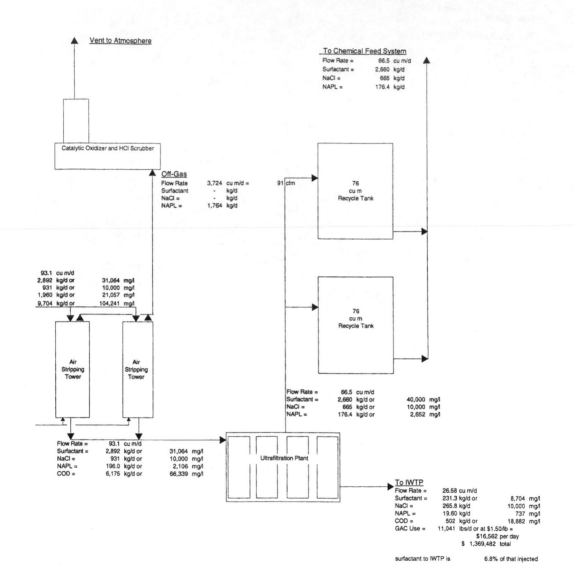

Vent to Atmosphere

<u>To Chemical Feed System</u>

Flow Rate =	66.5	cu m/d
Surfactant =	2,660	kg/d
NaCl =	665	kg/d
NAPL =	176.4	kg/d

Catalytic Oxidizer and HCl Scrubber

<u>Off-Gas</u>

Flow Rate	3,724	cu m/d =	91 cfm
Surfactant	-	kg/d	
NaCl =	-	kg/d	
NAPL =	1,764	kg/d	

76
cu m
Recycle Tank

93.1	cu m/d
2,892 kg/d or	31,064 mg/l
931 kg/d or	10,000 mg/l
1,960 kg/d or	21,057 mg/l
9,704 kg/d or	104,241 mg/l

76
cu m
Recycle Tank

Air
Stripping
Tower

Air
Stripping
Tower

Flow Rate =	66.5	cu m/d
Surfactant =	2,660 kg/d or	40,000 mg/l
NaCl =	665 kg/d or	10,000 mg/l
NAPL =	176.4 kg/d or	2,652 mg/l

Flow Rate =	93.1	cu m/d
Surfactant =	2,892 kg/d or	31,064 mg/l
NaCl =	931 kg/d or	10,000 mg/l
NAPL =	196.0 kg/d or	2,106 mg/l
COD =	6,175 kg/d or	66,339 mg/l

Ultrafiltration Plant

<u>To IWTP</u>

Flow Rate =	26.58 cu m/d	
Surfactant =	231.3 kg/d or	8,704 mg/l
NaCl =	265.8 kg/d	10,000 mg/l
NAPL =	19.60 kg/d	737 mg/l
COD =	502 kg/d or	18,882 mg/l
GAC Use =	11,041 lbs/d or at $1.50/lb =	
		$16,562 per day
	$ 1,369,482	total

surfactant to IWTP is   6.8% of that injected

OPTIMIZED CASE DESIGN WORKSHEET

Surfacant/Foam Design, Optimized Case

Technology Evaluation Report for Surfactant-Foam Aquifer Remediation

Design Worksheet
Optimized Case

General Site Assumptions:

Total Area =	1.03 ac =	0.42 hectare	
H, Thickness of target DNAPL zone =	2 m =	6.56 ft	
W, Width per module =	24.4 m =	80 ft	
L, Length per module =	42.7 m =	140 ft	Spacing = 0.5 * W

Number of Injection Wells per module   7 wells per module

Phase I of remediation will utilize   1 module
The maximum no. of modules to be operated   1 module(s) at a time.

porosity =   0.27

K (water or surfactant) =   0.011 cm/s   These parameters were adjusted to give the desired flow rate (based on a reasonable time frame).
Viscosity (water) =   0.01 poise   A maximum K of 0.1 cm/s is possible at this site.
Viscosity (surfactant/alcohol) =   0.02 poise
Hydraulic Head, Δh =   1 m

No. of Pore Volumes Of Flushing Required For Remediation
Initial Waterflood   =   5
Surfactant Flood   =   1.5
Post Waterflood   =   5

Assumptions Used to Calculate Present Worth
Number of Years of   1   (This is the calculated treatment time, rounded up)
Interest Rate =   5.00%

Area and Volume Calculation for the Site

Number of Surfactant modules =	4.00 =	4 rounded
Area per module=	0.26 ac	
Volume of Alluvium with DNAPL per module =	2,726 cy per module =	2,084 cu m per module

Pore Volume Calculation:
PV = 2 * (H+Δh/2) * L * (W*0.5) * porosity =   703.3 cu m per module

Volume of DNAPL Calculation

Assume DNAPL is   4.3% of the total volume (16% DNAPL saturation at a porosity of 0.27) over 1/4 of the site
and   1.1% over 1/2 of the site (4% DNAPL saturation, assumed based on variability at the site)
and   0.3% over the other 1/4 of the site (1% DNAPL saturation, assumed based on variability at the site)

Thus   42.9 cu m =   11,329 gal
Specific Gravity of DNAPL is   1.4

Flow Rate Calculations per Module Q=KAI (in 2 directions from the source)

$Q_{water} = 2 * K * (H+\Delta h/2) * L * \Delta h / (.5*W) =$ 0.0020 cu m/s =   170.1 cu m/d =   31 gpm

$Q_{surfactant} = 2 * K * (viscosity_{water}/viscosity_{surfactant}) * (H+\Delta h/2) * L * \Delta h / (.5*W) :$ 0.00098 cu m/s =   85.05 cu m/d =   16 gpm

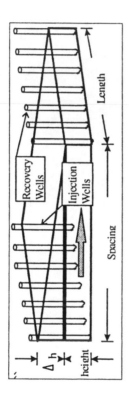

For a module containing   7 wells, the water flow rate is   4.5 gpm per well

the surfactant flow rate is   2.2 gpm per well

Time & Volume Calculations for Waterflood, Surfactant/Foam Flood and Post Waterflood

Time per PV = PV/Q =

	Water	Surfactant/Alcohol
	4.1 days	8.3 days

Sequence of Operations

Initial Waterflood   5 PV: =   3,516 cu m per module for   20.7 days

    =   929,018 gal per module

Surfactant Flood   1.5 PV: =   1,055 cu m per module for   12.4 days

    =   278,706 gal per module

Post Waterflood   5 PV: =   3,516 cu m per module for   20.7 days

    =   929,018 gal per module

Total time =   53.7 days =   0.15 years

Assume that Initial Waterflood is conducted on all modules at once.

Surfactant Flood and Post Waterflood will be conducted individually during each of   4 phases.

Total time for remediation   153 days =   0.42 years

Aerosol MA-80 Storage Requirement

Assume:

Recycling Rate: Volume of surfactant needed = initial pore volume +   22% % loss for each subsequent pore volume, (in ground and losses to IWTP)

Injected surfactant is   4% wt active sodium dihexyl sulfosuccinate

Aerosol MA-80 is supplied as a concentrate at   80% wt active

Specific gravity of supplied Aerosol MA-80 is   1.1

Specific gravity of surfactant solution is 1.

Fiberglass tanks shall be used for storing the concentrated chemicals. A convenient size of chemical storage tank is 5,000 gallons (8 foot diameter, 13.3 foot high)

Total quantity of Aerosol MA-80 needed: 124,788 kg active surfactant

                                           155,985 kg total =   343,167 lb total for remediation

                                               156 mtons   172 tons

                                             141.8 cu m =   37,465 gal

The maximum usage rate =   3.09 cu m/d =   817.1 gal/d

    Use   1 tank(s), @   5,000 gal ea =   18.9 cu m each
    The actual specified storage volume represents the useage requirements   1.7 days

Sodium Chloride Storage

Assume:

Recycling Rate: Volume of NaCl needed = initial pore volume +   20% % loss for each subsequent pore volume

Inject solution that is   1.1% wt active NaCl

Make stock solution of   20% wt NaCl

Specific gravity of stock NaCl solution is   1.1

Fiberglass tanks shall be used for storing the concentrated chemicals. A convenient size of chemical storage tank is 5,000 gallons (8 foot diameter, 13.3 foot high)

Quantity of NaCl Needed   34,038 kg active NaCl =

                                            74,884 lb active NaCl

                                           170,191 kg total weight of stock solution   170 mtons

                                           374,420 lb total weight of stock solution   187 tons

                                             154.7 cu m

                                            40,877 gal

The maximum usage rate =   4.25 cu m/d =   1,124 gal/d

    Use   1 tank(s), @   5,000 gal ea =   18.9 cu m ea
    The actual specified storage volume represents the useage requirements   1.5 days

---

Solution Mix/Storage Tanks

Assume that two (2) chemical mix/storage tanks are needed. The chemical mixture will be made in batches. One tank can be mixing while the other is dispensing solution to the injection wells.

Volume of Solution Mix Storage Tank

    Assume operate a maximum of   1 module(s) at one time

    If we assume that we make up a surfactant solution every   2 days

    thus need solution tank   170.1 cu m   =   44,940 gal

    Use   2 tank(s), @   20,000 gal ea =   75.7 cu m ea
    The actual specified storage volume represents the useage requirements for   1.8 days

Surfacant/Foam Design, Optimized Case

Air Injection System Requirements

Assume:

Air will be injected at a maximum of   2 scfm per well   2 hours on, 4 hours off
   20 psi

The maximum number of wells to be operated at a time is   7 wells which correlates to   1 module(s)
The maximum required blower air flow rate is   14 scfm at   20 psi

Produced Fluids Handling

Assume:

A maximum of   1 module(s) will be operated at one time.
Recovery at a rate   1.1 times the injection rate
Water produced during the initial waterflood will be recycled through an oil water separator and does not need to pass through the water treatment plant.
Water produced during the surfactant flood and post waterflood will require pretreatment before being sent to the industrial waste water treatment plant (IWTP).
Pretreatment will consist of passing through an oil/water separator, equalization tank, air stripper, and ultrafiltration unit. The equalization tank will have a holding holding   1.5 days
The excess water will be discharged to the IWTP.

Injection Flow rate:

During Initial Waterflood -	170.1 cu m/d =	31 gpm	for	21 days
During Surfactant Flood -	85.1 cu m/d =	16 gpm	for	12 days
During Post Waterflood -	170.1 cu m/d =	31 gpm	for	21 days

Recovery Flow Rates and Volumes

During Initial Waterflood -	187.1 cu m/d =	3,868 cu m	34 gpm =	1,021,920 gal
During Surfactant Flood -	93.6 cu m/d =	1,160 cu m	17 gpm =	306,576 gal
During Post Waterflood -	187.1 cu m/d =	3,868 cu m	34 gpm =	1,021,920 gal

Equalization Tank Volume for Air Stripping

Assume we need one (1) tank before the air stripper

Volume based on   1 days of holding time for the recovered fluids

The flow rate governing design is   187.1 cu m/d which occurs during   Post Waterflood

Required Volume =   187 cu m =   49,434 gal

Use   3 each   20,000 gal ea =   75.7 cu m each

The actual storage volume represents the useage requirements for   1.2 days

Volume Requiring Air Stripping and Ultrafiltration (Recovered volumes during surfactant flood and post waterflood)

Volume =   5,028 cu m =   1,328,496 gal per module
Total volume =   20,113 cu m =   5,313,985 gal

Volume of Water Requiring Treatment Through the IWTP

28% of surfactant flood and post waterflood water

Process Volume =   1,416 cu m =   374,182 gal per module
Final Pore Volume =   352 cu m =   92,902 gal, fraction not sent off-site
Total Volume =   6,017 cu m =   1,589,631 gal

Recycle Storage Tank Volume

Provide storage for   40% of a pore volume to allow flexibility in operations

Volume required =   141 cu m =   37,161 gal

Use   2 each of   20,000 gal or   76 cu m tank(s)

Volume Requiring Off-Site Incineration

Assume the last pore volume will be re-concentrated by   50%

In the ultrafilter unit. and then sent off for incineration

Thus, volume requiring off-site incineration =   352 cu m =   92,902 gal

Volume from oil/water separator =   48 cu m =   12,689 gal

total volume for incineration =   400 cu m =   105,591 gal

OPTIMIZED CASE MASS BALANCE

Technology Evaluation Report for Surfactant-Foam Aquifer Remediation

Flow Diagram and Mass Balance
Base Case

Chemical Feed System
Parameters During Surfactant Flood
Parameters are specified for operation of   1 module at a time
The required surfactant/alcohol injection rate is   85.1 cu m/d for the above module(s) of operation
The injected solution will be   4.0% by weight surfactant
The injected solution will be   1.1% by weight NaCl
The specific gravity of the supplied surfactant is   1.1
The specific gravity of the premixed 20% NaCl is   1.1

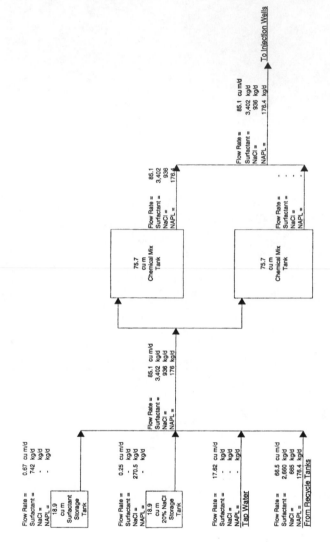

Surfactant/Foam MB, Optimized Case
(continued)

Technology Evaluation Report for Surfactant-Foam Aquifer Remediation

Flow Diagram and Mass Balance
Optimized Case

Parameters During Surfactant Flood

Assumptions:

Air to Water ratio for air stripper =	20	
Ratio of COD/surfactant or LNAPL =	2	based on Surfactant/Foam report
Concentrations are constant for	12	days during surfactant injection for each module.
The recovery rate is	1.1	times the injection rate
Surfactant recovery from the subsurface is	85%	by weight of injected surfactant
NaCl is conserved in the subsurface.		
Total DNAPL recovery from the subsurface is	70%	by weight of DNAPL present in the subsurface
The oil/water separator recovers	40%	by weight of the total recovered DNAPL
The oil/water separator recovers	0%	by weight of the surfactant
The air stripper recovers	90%	by weight of the NAPL from the oil/water separator effluent
The ultrafiltration plant recovers	92%	by weight of the surfactant in micelles from the air stripper
The ultrafiltration plant recovers	90%	by weight of the NAPL in micelles from the air stripper
The ultrafiltration plant concentrate waste steam is	4.0%	by weight surfactant
Assume an average carbon pickup of	20%	#DIV/0!

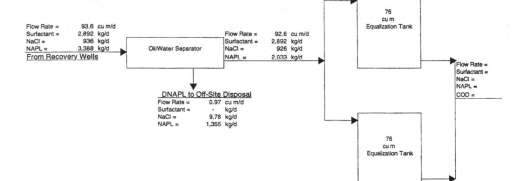

Flow Rate = 93.6 cu m/d
Surfactant = 2,892 kg/d
NaCl = 936 kg/d
NAPL = 3,388 kg/d
From Recovery Wells

Oil/Water Separator

Flow Rate = 92.6 cu m/d
Surfactant = 2,892 kg/d
NaCl = 926 kg/d
NAPL = 2,033 kg/d

76 cu m Equalization Tank

Flow Rate =
Surfactant =
NaCl =
NAPL =
COD =

DNAPL to Off-Site Disposal
Flow Rate = 0.97 cu m/d
Surfactant = - kg/d
NaCl = 9.78 kg/d
NAPL = 1,355 kg/d

76 cu m Equalization Tank

Aeration Blower

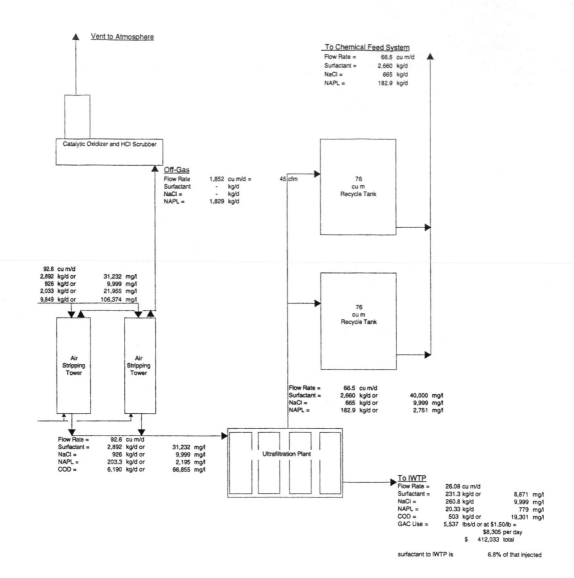

Vent to Atmosphere

To Chemical Feed System
Flow Rate =   66.5 cu m/d
Surfactant =   2,660 kg/d
NaCl =   665 kg/d
NAPL =   182.9 kg/d

Catalytic Oxidizer and HCl Scrubber

Off-Gas
Flow Rate   1,852 cu m/d =   45 cfm
Surfactant   -   kg/d
NaCl =   -   kg/d
NAPL =   1,829 kg/d

76
cu m
Recycle Tank

92.6 cu m/d
2,892 kg/d or   31,232 mg/l
926 kg/d or   9,999 mg/l
2,033 kg/d or   21,955 mg/l
9,849 kg/d or   106,374 mg/l

76
cu m
Recycle Tank

Air
Stripping
Tower

Air
Stripping
Tower

Flow Rate =   66.5 cu m/d
Surfactant =   2,660 kg/d or   40,000 mg/l
NaCl =   665 kg/d or   9,999 mg/l
NAPL =   182.9 kg/d or   2,751 mg/l

Flow Rate =   92.6 cu m/d
Surfactant =   2,892 kg/d or   31,232 mg/l
NaCl =   926 kg/d or   9,999 mg/l
NAPL =   203.3 kg/d or   2,195 mg/l
COD =   6,190 kg/d or   66,855 mg/l

Ultrafiltration Plant

To IWTP
Flow Rate =   26.08 cu m/d
Surfactant =   231.3 kg/d or   8,871 mg/l
NaCl =   260.8 kg/d   9,999 mg/l
NAPL =   20.33 kg/d   779 mg/l
COD =   503 kg/d or   19,301 mg/l
GAC Use =   5,537 lbs/d or at $1.50/lb =
                                      $8,305 per day
                      $ 412,033 total

surfactant to IWTP is   6.8% of that injected

Surfactant/Foam Scale-up Cost Estimates

Details of the cost estimates are presented in this appendix. Two general methods were used to prepare these cost estimates. Excel spreadsheets were used to prepare estimates for "additional capital costs" and operations and maintenance costs. Spreadsheets were also used to compile all of the costs into summaries. Baseline capital costs were prepared using Hard Dollar Estimating Office System (EOS). EOS, recognized as the most comprehensive, commercially available construction estimating software system available, is designed for use by contractors to develop detailed costs estimates for the remediation and construction industries. EOS is formatted to develop estimates "from the ground up," whether employing a total cost buildup method, or generating a more general unit price estimate.

This estimate was developed by entering subcontractor and equipment vendor quotes, combined with detailed estimated costs for labor to install the system.

The following list is a brief description of the cost estimate sheets that are contained in this appendix. Cost summaries are provided in Chapter 20 of the text. The base case sheets are provided first, followed by the optimized case.

- Additional Capital Costs: Includes additional items not included in the capital costs. Items such as field demonstrations are included here.
- Operations and Maintenance Costs: Operating costs per module.
- Cost Summary: Pulls all the costs together. At the bottom of the summary table are calculations of the cost on a unit basis (dollar per area or volume).
- WBS Outline from EOS: Includes where the markups need to be included.
- Work Plan Summary from EOS: Summarizes the cost for each category. It should be noted that these costs do not inlude the markup. The markup was added on the summary sheet.
- Activity Detail Forms from EOS: Includes the details for each category. Note that there are forms for subcategories in some cases.

BASE CASE ADDITIONAL CAPITAL COST

Adtnl Cptl Cost, Base Case

Technology Evaluation Report for Surfactant-Foam Aquifer Remediation

Additional Capital Cost
Base Case

Description	Qty	Unit	Unit Cost ($)	Total Cost ($)	Comments
Additional Site Characterization and Laboratory Testing					
Soil Borings	20	ea	$1,200	$24,000	Assumes 20 additional borings that will extend to 33 feet, cost based on vendor quote and RACER
Soil Analysis	20	sample	$500	$10,000	Assumes samples will be analyzed for 3 constituents, costs based on prior experience
Bench-Scale Testing	1	ls	$250,000	$250,000	Based on prior experience. Includes surfactant IFT, phase behavior, solubilization potential, linear and radial coreflood tests, two dimensional sandpacks, testing of a bench-scale UF unit and air stripper, work plans, test labor, and reports.
Subtotal				$284,000	
Numerical Simulations					
Model Simulations	240	hrs	$100	$24,000	Cost includes operator and computer time. It does not include cost of software
Field Demonstrations					
Installation of Sheet Pile Wall	2,153	sf	$22	$47,366	Assumes 5m x 5m x 10m deep sheet pile wall for containment, cost based on prior experience
Installation of Injection Wells	4	ea	$4,600	$18,400	Wells will be 33 feet deep and stainless steel, cost based on vendor quote and RACER
Installation of Extraction Well	3	ea	$4,600	$13,800	Wells will be 33 feet deep and stainless steel, cost based on vendor quote and RACER
Monitoring Piezometer - 30 foot well	4	ea	$4,200	$16,800	Cost based on vendor quote and RACER. Nested with 3 MP.
Soil Borings	6	ea	$1,200	$7,200	Assumes borings will extend to 33 feet, cost based on vendor quote and RACER
Soil Analysis	28	sample	$500	$14,000	Assumes 2 soil samples per well will be analyzed for 3 constituents, costs based on prior experience
Analysis of UF and air stripper effluent	30	sample	$125	$3,750	Once per day analysis of VOC for approximately 1 month
Analysis of treated produced fluids	30	sample	$125	$3,750	Once per day analysis of VOC for approximately 1 month.
Analysis of monitoring well fluids	132	sample	$125	$16,500	Once per day analysis of VOC for approximately 4 weeks.
Tanks for temporary storage	2	ea	$9,500	$19,000	10,000 gal single-walled steel tanks with containment berm. Cost from RACER
Pilot testing of ultrafiltration	1	ls	$75,000	$75,000	Tests will be conducted on 25% of the fluids. Cost based on prior experience
Pilot testing of steam stripping	1	ls	$75,000	$75,000	Tests will be conducted on 25% of the fluids. Cost based on prior experience
Pilot testing of biological treatment	1	ls	$75,000	$75,000	Tests will be conducted on 25% of the fluids. Cost based on prior experience
Fluid Incineration	481,913	lb	$0.17	$81,925	Assumes that 75% of the produced fluids will be incinerated. Assumes 22.5 pore volumes of water will be generated. No recycling will be conducted. Quote from Laidlaw Environmental (SLC, UT), <3,000 BTU/lb, >50% water, <10% Cl.
Discharge to IWTP	21	1,000 gal	$11.00	$235	Balance of water not incinerated will be sent to the IWTP.
Field Demonstration Labor	1	ls	$62,400	$62,400	Assumes demonstration will require 30 days, 2 technicians and 2 engineers per day. Cost based on prior experience.
Field Demonstration Work Plan and Report	2	ea	$25,000	$50,000	Cost based on prior experience
Subtotal				$580,126	
Facility Design	10%		$2,243,988	$224,399	
Engineering During Construction	8%		$2,243,988	$179,519	
Post Flushing Soil Coring, Analysis, and Tracer Tests					
Soil Borings	10	ea	$1,200	$12,000	Assumes 1 boring every 20 m of drainline in Phase I (boring will extend to 33 feet), cost based on vendor quote and RACER
Soil Analysis	10	sample	$500	$5,000	Assumes samples will be analyzed for 3 constituents, costs based on prior experience
Tracer Test	1	ls	$40,000	$40,000	Assumes the test will require 40 days of an engineer's time and $25,000 for analysis of tracers
Subtotal				$57,000	
Present Worth of Post Flushing Soil Coring and Analysis				$56,218	Assumes costs are incurred after Phase 1 of remediation, 5% interest rate
Demolition and Site Restoration After Remediation	2.5%		$2,243,988	$56,100	Percentage is based on total construction costs
Present Worth of Dem & Rest. Costs				$53,971	Assume costs are incurred at the end of remediation, 5% interest rate
Salvage Value of Equipment	25%		$1,893,022	$473,255	Based on physical treatment equipment
Present Worth of Salvage Value				$455,297	

BASE CASE O&M COST

O&M Cost, Base Case

Technology Evaluation Report for Surfactant-Foam Aquifer Remediation

Operations and Maintenance Cost
Base Case

Description	Qty	Unit	Unit Cost ($)	Total Cost ($)	Comments
OPERATIONS/MAINTENANCE AND MONITORING COSTS - INITIAL WATER FLOOD					
Labor					
Operators	1,323	hrs	$40	$52,921	4 operators will be required for the duration of the project (56 hrs/wk)
Supervisor	331	hrs	$50	$16,538	1 supervisor will be required for the duration of the project (56 hrs/wk)
Maintenance Technician	118	hrs	$50	$5,906	1 maintenance technician will be required for the duration of the project (20 hrs/wk)
Engineer	354	hrs	$80	$28,350	1.5 engineers will be required for the duration of the project (40 hrs/wk)
Laboratory Technician	331	hrs	$25	$8,269	1 laboratory technician will be required for the duration of the project (56 hrs/wk)
Temporary Site Facilities					
Office Trailer	1.4	month	$250	$339	32' x 8' Temporary Office Trailer, cost from RACER, $250/month
Portable Toilet	1.4	month	$100	$136	Quote from vendor, $100/month
Move process equipment	1	ls	$10,000	$10,000	Cost for moving the pumping equipment to the next phase wells
Maintenance Materials	0.34%		$2,000,000	$6,789	Engineers estimate actual fraction of 3% per year of construction cost.
Utilities					
Electrical for pumps, blower, etc.	19,035	kwh	$0.08	$1,523	Electrical requirements for pumps, blower, etc. (25.75 hp assumed for initial water flood)
Off-Site Laboratory Analysis					
Analysis of produced fluids	0	sample	$125	$760	Once per week analysis of VOC (1 analysis per day during initial water flood)
Tracer Test	1	ls	$40,000	$40,000	Assumes the test will require 40 days of an engineer's time and $25,000 for analysis of tracers, cost
Discharge to IWTP	1,850	1,000 gal	$11.00	$20,438	IWTP costs were obtained from Hill Air Force Base personnel. (all water from initial water flood)
Subtotal				$191,960	
Contingency	20%		$191,960	$38,392	
Subtotal - Operations/Maintenance and Monitoring Costs - Initial WF				$230,352	

O&M Cost, Base Case

Technology Evaluation Report for Surfactant-Foam Aquifer Remediation

Operations and Maintenance Cost

Base Case

Description	Qty	Unit	Unit Cost ($)	Total Cost ($)	Comments
OPERATIONS/MAINTENANCE AND MONITORING COSTS PER MODULE - SURFACTANT AND POST WATER FLOOD					
Labor					
Operators	1,985	hrs	$40	$79,381	4 operators will be required for the duration of the project (56 hrs/wk)
Supervisor	496	hrs	$50	$24,807	1 supervisor will be required for the duration of the project (56 hrs/wk)
Maintenance Technician	177	hrs	$50	$8,860	1 maintenance technician will be required for the duration of the project (20 hrs/wk)
Engineer	532	hrs	$80	$42,526	1.5 engineers will be required for the duration of the project (40 hrs/wk)
Laboratory Technician	496	hrs	$25	$12,403	1 laboratory technician will be required for the duration of the project (56 hrs/wk)
Temporary Site Facilities					
Office Trailer	2.0	month	$250	$500	32' x 8' Temporary Office Trailer, cost from RACER, $250/month
Portable Toilet	2.0	month	$100	$204	Quote from vendor, $100/month
Move process equipment	1	ls	$10,000	$10,000	Cost for moving the pumping equipment to the next phase wells
Chemicals					
Aerosol MA-80 (surfactant)	57,285	lbs	$1.28	$73,325	Quote from Cytec Industries, bulk rate
NaCl	22,125	lbs	$0.10	$2,212	Quote from vendor
Maintenance materials	0.51%		$2,000,000	$10,183	Engineers estimate actual fraction of 3% per year of construction cost.
Utilities					
Electrical for pumps, blower, etc.	66,531	kwh	$0.08	$5,323	Electrical requirements for pumps, blower, etc. (60 hp assumed per module)
Fuel for air stripper off-gas treatment system	1,488	hr	$1.84	$2,739	Fuel cost for catox unit from Global, 24 hr per day operation
Off-Site Laboratory Analysis					
Analysis of recovered fluids	63	sample	$125	$7,875	Once per day analysis of VOC (1 analysis per day during surfactant and post water flood)
Analysis of air stripper effluent	63	sample	$125	$7,875	Once per day analysis of VOC (1 analysis per day during surfactant and post water flood)
Analysis of treated produced fluids	63	sample	$125	$7,875	Once per day analysis of VOC (1 analysis per day during surfactant and post water flood)
Analysis of monitoring well fluids	63	sample	$125	$7,875	Once per day analysis of VOC (1 analysis per day during surfactant and post water flood), 1 well per day per module
Disposal					
Incineration of NAPL from O/W separator	29,715	lb	$0.16	$4,754	Quote from Laidlaw Environmental (SLC, UT), 3,000-8,000 BTU/lb, <50% water, <10% CI (fluids produced during surfactant flood period only)
Surcharge for GAC Costs at IWTP	228,247	lb carbon	$1.50	$342,370	Assume 10% by weight pickup of organic contaminants on carbon (lb organic/lb carbon)
Discharge to IWTP	2,323	1,000 gal	$11.00	$25,548	IWTP costs were obtained from Hill Air Force Base personnel. (all water from surfactant and post water flood)
Subtotal				$676,644	
Contingency	20%		$676,644	$135,329	
Total - Operations/Maintenance and Monitoring Costs - Surf. and Post- WF				**$811,973**	
Total No. of Modules	4 modules				
Total Time for Operation of all Modules after IWF	0.68 years				
Total Operations and Maintence Cost for all Modules				$3,247,890	
Interest Rate for Present Worth Calculation	5%				
Present Worth - Operations/Maintenance and Monitoring of All Modules				**$3,478,242**	

COST SUMMARY

Cost Summary, Base Case

Technology Evaluation Report for Single Phase Micro-Emulsion

Cost Summary
Base Case

WBS No.	Standard Description	Cost	Subtotal	% of Total		Raw Capital Costs	Markup
321	Pre-Construction and Project Management Activities		$1,112,525		14.8%		
32190	Additional Site Characterization and Laboratory Testing	$ 284,000		3.8%			
32191	Numerical Simulations	$ 24,000		0.3%			
32192	Field Demonstration	$ 580,126		7.7%			
32193	Design	$ 224,399		3.0%			
331	HTRW Remedial Action (Construction)		$2,243,988		29.8%		
33101	Mobilization and Preparatory Work	$ 96,991		1.3%		$ 84,340	15%
33103	Site Work	$ 25,300		0.3%		$ 22,000	15%
33106	Groundwater Collection and Control	$ 668,829		8.9%		$ 581,590	15%
33113	Physical Treatment	$1,893,022		25.2%		$1,646,106	15%
33121	Demobilization	$ 15,143		0.2%		$ 13,168	15%
33190	Salvage Value of Equipment	$ (455,297)		-0.1%			
332	Engineering During Construction	$ 179,519	$ 179,519	2.4%	2.4%		
333	Construction Management	$ 398,674	$ 398,674	5.3%	5.3%	$ 346,673	15%
342	HTRW Operation and Maintenance (Post Construction)		$3,588,431		47.7%		
34202	Monitoring, Sampling, Testing, and Analysis	$ 325,777		4.3%			
34203	Site Work	$ 121,019		1.6%			
34206	Groundwater Collection and Control	$ 455,866		6.1%			
34213	Physical Treatment	$ 455,866		6.1%			
34218	Disposal (other than commercial)	$1,813,356		24.1%			
34290	Chemicals	$ 362,578		4.8%			
33120	Site Restoration	$ 53,971		0.7%			
	Total Project Cost		$7,523,137	100.0%	100.0%		

Cost Analysis:

Treatment Area		Volume of Impacted Soil (1)			Volume of DNAPL Removed		
hectares	$ / hectare	cu m	$/cu m		l DNAPL	$ / l DNAPL	
acres	$ / acre	cy	$/cy		gal DNAPL	$ / gal DNAPL	
0.42	$18,047,976	16,675	$ 451		42,879	$175	
1.03	$ 7,304,016	21,809	$ 345		11,329	$664	

(1) Volume of impacted soil = Volume of soil likely to be excavated = 2 times the target NAPL volume.

Adtnl Cptl Cost, Optimized Case

Technology Evaluation Report for Surfactant-Foam Aquifer Remediation

Additional Capital Cost
Optimized Case

Description	Qty	Unit	Unit Cost ($)	Total Cost ($)
Additional Equipment/Facilities for Optimized Case				
Automation and Monitoring System	1	ls	$200,000	$200,000
Subtotal				$200,000
Engineering During Construction	8%		$2,376,029	$190,082
Post Flushing Soil Coring, Analysis, and Tracer Tests				
Soil Borings	10	ea	$1,200	$12,000
Soil Analysis	10	sample	$500	$5,000
Tracer Test	1	ls	$40,000	$40,000
Subtotal				$57,000
Present Worth of Post Flushing Soil Coring and Analysis				$57,000
Demolition and Site Restoration After Remediation				
Demolition and Site Restoration after Remediation	2.5%		$2,376,029	$59,401
Present Worth of Dem & Rest. Costs				$59,401
Salvage Value of Equipment	25%		$2,093,022	$523,255
Present Worth of Salvage Value				$523,255

O&M Cost, Optimized Case

Technology Evaluation Report for Surfactant-Foam Aquifer Remediation

Operations and Maintenance Cost
Optimized Case

Description	Qty	Unit	Unit Cost ($)	Total Cost ($)	Comments
OPERATIONS/MAINTENANCE AND MONITORING COSTS - INITIAL WATER FLOOD					
Labor					
Operators	662	hrs	$40	$26,460	4 operators will be required for the duration of the project (56 hrs/wk)
Supervisor	165	hrs	$50	$8,269	1 supervisor will be required for the duration of the project (56 hrs/wk)
Maintenance Technician	59	hrs	$50	$2,953	1 maintenance technician will be required for the duration of the project (20 hrs/wk)
Engineer	177	hrs	$80	$14,175	1.5 engineers will be required for the duration of the project (40 hrs/wk)
Laboratory Technician	165	hrs	$25	$4,134	1 laboratory technician will be required for the duration of the project (56 hrs/wk)
Temporary Site Facilities					
Office Trailer	0.7	month	$250	$170	32' x 8' Temporary Office Trailer, cost from RACER, $250/month
Portable Toilet	0.7	month	$100	$68	Quote from vendor, $100/month
Move process equipment	1	ls	$10,000	$10,000	Cost for moving the pumping equipment to the next phase wells
Maintenance Materials	0.17%	ls	$2,000,000	$3,394	Engineers estimate actual fraction of 3% per year of construction cost.
Utilities					
Electrical for pumps, blower, etc.	9,518	kwh	$0.08	$761	Electrical requirements for pumps, blower, etc. (25.75 hp assumed for initial water flood)
Off-Site Laboratory Analysis					
Analysis of produced fluids	3	sample	$125	$375	Once per week analysis of VOC (1 analysis per day during initial water flood)
Tracer Test	1	ls	$40,000	$40,000	Assumes the test will require 40 days of an engineer's time and $25,000 for analysis of tracers, cost based on prior experience
Discharge to IWTP	929	1,000 gal	$11.00	$10,219	IWTP costs were obtained from Hill Air Force Base personnel. (all water from initial water flood)
Subtotal				$120,980	
Contingency	20%			$24,196	
Subtotal - Operations/Maintenance and Monitoring Costs - IWF			$120,980	$145,176	

O&M Cost, Optimized Case

Technology Evaluation Report for Surfactant-Foam Aquifer Remediation

Operations and Maintenance Cost
Optimized Case

Description	Qty	Unit	Unit Cost ($)	Total Cost ($)	Comments
OPERATIONS/MAINTENANCE AND MONITORING COSTS PER MODULE					
Labor					
Operators	1,058	hrs	$40	$42,337	4 operators will be required for the duration of the project (56 hrs/wk)
Supervisor	265	hrs	$50	$13,230	1 supervisor will be required for the duration of the project (56 hrs/wk)
Maintenance Technician	95	hrs	$50	$4,725	1 maintenance technician will be required for the duration of the project (20 hrs/wk)
Engineer	284	hrs	$80	$22,680	1.5 engineers will be required for the duration of the project (40 hrs/wk)
Laboratory Technician	265	hrs	$25	$6,615	1 laboratory technician will be required for the duration of the project (56 hrs/wk)
Temporary Site Facilities					
Office Trailer	1.1	month	$250	$272	32' x 8' Temporary Office Trailer, cost from RACER, $250/month
Portable Toilet	1.1	month	$100	$109	Quote from vendor, $100/month
Move process equipment	1	ls	$10,000	$10,000	Cost for moving the pumping equipment to the next phase wells
Chemicals					
Aerosol MA-80 (surfactant)	85,792	lbs	$1.28	$109,813	Quote from Cytec Industries, bulk rate
NaCl	18,721	1.4	$0.10	$1,872	Quote from vendor
Maintenance Materials	0.27%		$2,000,000	$5,431	Engineers estimate actual fraction of 3% per year of construction cost.
Utilities					
Electrical for pumps, blower, etc.	35,483	kwh	$0.08	$2,839	Electrical requirements for pumps, blower, etc. (60 hp assumed per module)
Fuel for air stripper off-gas treatment system	794	hr	$1.84	$1,461	Fuel cost for catox unit from Global, 24 hr per day operation
Off-Site Laboratory Analysis					
Analysis of injected fluid quality	34	sample	$125	$4,250	Once per day analysis of VOC (1 analysis per day during surfactant and post water flood)
Analysis of recovered fluids	34	sample	$125	$4,250	Once per day analysis of VOC (1 analysis per day during surfactant and post water flood)
Analysis of treated produced fluids	34	sample	$125	$4,250	Once per day analysis of VOC (1 analysis per day during surfactant and post water flood)
Analysis of monitoring well fluids	34	sample	$125	$4,250	Once per day analysis of VOC (1 analysis per 2 days during surfactant and post water flood), sample 1 well per day per module
Disposal					
Incineration of NAPL from O/W separator	36,978	lb	$0.16	$5,917	Quote from Laidlaw Environmental (SLC, UT), 3,000-8,000 BTU/lb, <50% water, <10% Cl (fluids produced during surfactant flood period only)
Surcharge for GAC Costs at IWTP	68,672	lb carbon	$1.50	$103,008	Assume 10% by weight pickup of organic contaminants on carbon (lb organic/lb carbon)
Discharge to IWTP	1,208	1,000 gal	$11.00	$13,285	IWTP costs were obtained from Hill Air Force Base personnel. (all water from surfactant and post water flood)
Subtotal				$360,593	
Contingency	20%		$360,593	$72,119	
Total - Annual Operations/Maintenance and Monitoring Costs				**$432,712**	

Total No. of Modules	4 modules	
Total Time for Operation of all Modules after IWF	0.36 years	
Total Operations and Maintenance Cost for all Modules	$1,730,849	
Interest Rate for Present Worth Calculation	5%	
Present Worth - Operations/Maintenance and Monitoring of All Modules	$1,876,025	

Cost Summary, Optimized Case

Technology Evaluation Report for Single Phase Micro-Emulsion

Cost Summary
Optimized Case

WBS No.	Standard Description	Cost	Subtotal	% of Total	% of Total	Raw Capital Costs	Markup
321	Pre-Construction and Project Management Activities	$ -			0.0%		
32190	Additional Site Characterization and Laboratory Testing	$ -		0.0%			
32191	Numerical Simulations	$ -		0.0%			
32192	Field Demonstration	$ -		0.0%			
32193	Design	$ -		0.0%			
331	HTRW Remedial Action (Construction)		$2,376,029		47.9%		
33101	Mobilization and Preparatory Work	$ 96,991		2.0%		$ 84,340	15%
33103	Site Work	$ 25,300		0.5%		$ 22,000	15%
33106	Groundwater Collection and Control	$ 668,829		13.5%		$ 581,590	15%
33113	Physical Treatment	$2,093,022		42.2%		$ 1,646,106	15%
33121	Demobilization	$ 15,143		0.3%		$ 13,168	15%
33190	Salvage Value of Equipment	$ (523,255)		-10.6%			
332	Engineering During Construction	$ 190,082	$ 190,082	3.8%	3.8%		
333	Construction Management	$ 398,074	$ 398,074	8.0%	8.0%	$ 346,073	15%
342	HTRW Operation and Maintenance (Post Construction)		$1,992,425		40.2%		
34202	Monitoring, Sampling, Testing, and Analysis	$ 223,764		4.5%			
34203	Site Work	$ 92,253		1.9%			
34206	Groundwater Collection and Control	$ 241,024		4.9%			
34213	Physical Treatment	$ 241,024		4.9%			
34218	Disposal (other than commercial)	$ 598,870		12.1%			
34290	Chemicals	$ 536,081		10.8%			
33120	Site Restoration	$ 59,401		1.2%			
	Total Project Cost		$4,957,211	100.0%	100.0%		

Cost Analysis:

Treatment Area			Volume of Impacted Soil (1)		I DNAPL	$ / I DNAPL
hectares		$ / hectare	cu m	$/cu m	gal DNAPL	$ / gal DNAPL
acres		$ / acre	cy	$/cy		
0.42		$ 11,892,330	16,675	$ 297	42,879	$ 116
1.03		$ 4,812,826	21,809	$ 227	11,329	$ 438
1.4						

(1) Volume of impacted soil = Volume of soil likely to be excavated = 2 times the target NAPL volume.

EOS OUTPUT

```
                              CH2M HILL
                         (SFSRBASE) SFSR Base Bid
                          W B S   O U T L I N E


  Code                         Description              Quantity  UM  Source
  -------------------------------------------------------------------------
  33                           CONSTRUCTION ACTIVITIES    1.0000  LS  Detail
  33 10                        Remedial Action            1.0000  LS  Detail
  33 10  1                     Mobe and Preparation Work  1.0000  LS  Detail
  33 10  1  1                  Install Access Roads     300.0000  LF  Plug
  33 10  1  2                  Install Secondary Contain 6100.000 SF  Plug
  33 10  1  3                  Pre-fab Buildings          2.0000  EA  Plug
  33 10  1  4                  Site Utilities             1.0000  LS  Plug
  33 10  1  5                  Site Temporary Facilities  1.0000  MO  Plug
  33 10  3                     Site Work                  1.0000  LS  Plug
  33 10  4                     G'water Collection/Contrl  1.0000  LS  Detail
  33 10  4  1                  Monitoring Wells          16.0000  EA  Quote
  33 10  4  2                  Recovery Wells            40.0000  EA  Quote
  33 10  4  3                  Injection Wells           28.0000  EA  Quote
  33 10  4  4                  Soil Borings              20.0000  EA  Detail
  33 10  4  5                  Soil Samples              20.0000  EA  Plug
  33 10  5                     Physical Treatment         1.0000  LS  Detail
  33 10  5  1                  Electrical Package         1.0000  LS  Detail
  33 10  5  2                  Plumbing Package           1.0000  LS  Detail
  33 10  8                     Demobilization             1.0000  LS  Detail
  33 10 11                     Construction Management  104.0000  DA  Detail
  33 10 11  1                  Mobe/Demobe               30.0000  DA  Detail
  33 10 11  2                  Operations                74.0000  DA  Detail
```

(SFSRBASE) SFSR Base Bid
W O R K   P L A N   S U M M A R Y

```
                              ------------Activity's immediate cost (excluding subordinates' cost)------------  S
                              ----Labor----  ----Equipment----                                                  r
                              Base  Burden  Ownershp  Operatn  Rentals  Material  Supply  Subcon  Other  Taxes  Total  c
```

WBS	Description	Qty	UM		Material	Subcon	Other	Taxes	Total	Src
33	CONSTRUCTION ACTIVITIES	1.0000	LS	U					0.0000	D
				T					0.00	
33 10	Remedial Action	1.0000	LS	U					0.0000	D
				T					0.00	
33 10 1	Mobe and Preparation Work	1.0000	LS	U					0.0000	D
				T					0.00	
33 10 1 1	Install Access Roads	300.0000	LF	U		84.0000			84.0000	P
				T		25200.00			25200.00	
33 10 1 2	Install Secondary Contain	6100.000	SF	U		2.4000			2.4000	P
				T		14640.00			14640.00	
33 10 1 3	Pre-fab Buildings	2.0000	EA	U	10000.00				10000.0000	P
				T	20000.00				20000.00	
33 10 1 4	Site Utilities	1.0000	LS	U		15000.00			15000.0000	P
				T		15000.00			15000.00	
33 10 1 5	Site Temporary Facilities	1.0000	MO	U			9500.000		9500.0000	P
				T			9500.00		9500.00	
33 10 3	Site Work	1.0000	LS	U	10000.00	12000.00			22000.0000	P
				T	10000.00	12000.00			22000.00	
33 10 4	G'water Collection/Contrl	1.0000	LS	U					0.0000	D
				T					0.00	
33 10 4 1	Monitoring Wells	16.0000	EA	U		2587.500			2587.5000	Q
				T		41400.00			41400.00	
33 10 4 2	Recovery Wells	40.0000	EA	U		7762.500			7762.5000	Q
				T		310500.0			310500.00	
33 10 4 3	Injection Wells	28.0000	EA	U		7417.500			7417.5000	Q
				T		207690.0			207690.00	
33 10 4 4	Soil Borings	20.0000	EA	U		600.0000			600.0000	D
				T		12000.00			12000.00	
33 10 4 5	Soil Samples	20.0000	EA	U		500.0000			500.0000	P
				T		10000.00			10000.00	

| 33 10 5 | Physical Treatment | 1.0000 | LS | U 29752.50 | Base 25500.00 | Burden 4250.000 | Material 1275550 | Supply 2500.000 | Subcon 41888.00 | Taxes 63902.50 | Total 1443343.0000 |

(SFSRBASE) SFSR Base Bid

W O R K   P L A N   S U M M A R Y

------------Activity's immediate cost (excluding subordinates' cost)------------

	Description	Qty	UM	Base (Labor)	Burden (Labor)	Ownershp (Equip)	Operatn (Equip)	Rentals	Material	Supply	Subcon	Other	Taxes	Total	Src
33 10 5 1	Electrical Package	1.0000	LS U								129004.2			129004.2000	D
			T	29752.50		25500.00	4250.00		1275550	2500.00	41888.00 / 129004.2		63902.50	1443343.00 / 129004.20	D
33 10 5 2	Plumbing Package	1.0000	LS U						7080.000		66325.00		354.0000	73759.0000	D
			T						7080.00		66325.00		354.00	73759.00	D
33 10 8	Demobilization	1.0000	LS U	5008.000	4887.440			1150.000		2000.000			123.0000	13168.4400	D
			T	5008.00	4887.44			1150.00		2000.00			123.00	13168.44	D
33 10 11	Construction Management	104.0000	DA U											0.0000	D
			T											0.00	D
33 10 11 1	Mobe/Demobe	30.0000	DA U	1177.553	977.4880			240.0000		20.0000		861.3333	48.0250	3324.3997	D
			T	35326.60	29324.64			7200.00		600.00		25840.00	1440.75	99731.99	D
33 10 11 2	Operations	74.0000	DA U	1177.644	977.4880			240.1351		20.2703		872.9054	48.6000	3337.0423	D
			T	87145.62	72334.11			17770.00		1500.00		64595.00	3596.40	24694.13	D
	Total			157232.7	106546.2	25500.00	4250.00	26120.00	1312630	6600.00	885647.2	99935.00	69416.65	2693877.76	

```
                        (SFSRBASE) SFSR Base Bid
                                M A R K U P

Currency: US Dollars

Summary
-------
                          --------Markup------------
                          Average
Escalated Cost   Amount   Rate          Amount   Escalated Cost + MU
-------------------------------------------------------------------------
Direct        2693877.76   15.00%     404081.66         3097959.43
Overhead                    0.00%
-------------------------------------------------------------------------
Total         2693877.76   15.00      404081.66         3097959.43


Direct (DC) Breakdown
--------------------------------
                          --------Markup------------
                          Average
Escalated Cost   Amount   Rate          Amount   Escalated Cost + MU
-------------------------------------------------------------------------
Labor Base     157232.72   15.00%      23584.91          180817.63
Labor Brdn     106546.19   15.00%      15981.93          122528.12
Equip Ownr      25500.00   15.00%       3825.00           29325.00
Equip Oprn       4250.00   15.00%        637.50            4887.50
Rentals         26120.00   15.00%       3918.00           30038.00
Materials     1312630.00   15.00%     196894.50         1509524.50
Supplies         6600.00   15.00%        990.00            7590.00
Subcontrct     885647.20   15.00%     132847.08         1018494.28
Other Cost      99935.00   15.00%      14990.25          114925.25
Tax-RSMO        69416.65   15.00%      10412.50           79829.15
-------------------------------------------------------------------------
DC Total      2693877.76   15.00%     404081.66         3097959.43


Overhead (OH) Breakdown
------------------------------------
                          --------Markup------------
                          Average
Escalated Cost   Amount   Rate          Amount   Escalated Cost + MU
-------------------------------------------------------------------------
Labor Base                 15.00%                         23584.91
Labor Brdn                 15.00%                         15981.93
Equip Ownr                 15.00%                          3825.00
Equip Oprn                 15.00%                           637.50
Rentals                    15.00%                          3918.00
Materials                  15.00%                        196894.50
Supplies                   15.00%                           990.00
Subcontrct                 15.00%                        132847.08
Other Cost                 15.00%                         14990.25
Tax-RSMO                   15.00%                         10412.50
-------------------------------------------------------------------------
OH Total                    0.00%

Subcontractor Markup Breakdown
------------------------------
                                    -----Markup---------
Code     Subcontractor   Award Amount  Rate    Amount   Award + Markup
-------------------------------------------------------------------------
                  Total    885647.20   0.00%              885647.20
-------------------------------------------------------------------------
00000000  Plugged Amount w  326057.20  0.00%             326057.20
WELLS                       559590.00  0.00%             559590.00
```

(SFSRBASE) SFSR Base Bid
A C T I V I T Y - D E T A I L   F O R M S

```
+33----------------------CONSTRUCTION ACTIVITIES---Qty_UM--  1.00-LS- 2693878--+
|
|A s s i g n m n t s    S c h e d u l e          C o s t  Unit     Total
|-------------------    ---------------          ----------------------------
|Bid Item Line        1 Duration                 Source    Details
|Quote Grp    NONE      Bar Start
|Qry Lbl 1              Bar Finsh                 Lab Base  157232.7   157233
|Qry Lbl 2              CPM ES                     Lab Brdn  106546.2   106546
|Qry Lbl 3              CPM EF                     Eqp Own   25500.00    25500
|Acnt                   CPM LS                     Eqp Op    4250.000     4250
|                       CPM LF                     Rental    26120.00    26120
|S h i f t s            Float                      Material  1312630   1312630
|-----------------                                 Supply    6600.000     6600
|Work Hrs/Shift    10   P r o d u c t i o n        Subconts  885647.2   885647
|Pay  Hrs/Shift    10   -------------------        Other     99935.00    99935
|Shifts/Day       1.0   Days       175.6000        Tax RMSO  69416.65    69417
|Days/Week        5.0   Shifts     175.6000       +----------------------------+
|                       Hours      1756.000       |Total      2693878   2693878|
|R a t e s              Man-Hrs    6227.000       +----------------------------+
|---------              Eqp-Hrs    5610.000        > Job      2693878
|Rates Unique           UM/Da        0.0057
|                       UM/Sh        0.0057           >DBE Goal
|                       UM/Hr        0.0006           >MBE Goal
|                       UM/MH        0.0002           >WBE Goal
|                       MH/UM      6227.000           >O-1 Goal
|                                                     >O-2 Goal
+------------------------------------------------------------------------------+


+33 10--------------------Remedial Action-----------Qty_UM--  1.00-LS- 2693878--+
|
|A s s i g n m n t s    S c h e d u l e          C o s t  Unit     Total
|-------------------    ---------------          ----------------------------
|Bid Item Line        1 Duration                 Source    Details
|Quote Grp    NONE      Bar Start
|Qry Lbl 1              Bar Finsh                 Lab Base  157232.7   157233
|Qry Lbl 2              CPM ES                     Lab Brdn  106546.2   106546
|Qry Lbl 3              CPM EF                     Eqp Own   25500.00    25500
|Acnt                   CPM LS                     Eqp Op    4250.000     4250
|                       CPM LF                     Rental    26120.00    26120
|S h i f t s            Float                      Material  1312630   1312630
|-----------------                                 Supply    6600.000     6600
|Work Hrs/Shift    10   P r o d u c t i o n        Subconts  885647.2   885647
|Pay  Hrs/Shift    10   -------------------        Other     99935.00    99935
|Shifts/Day       1.0   Days       175.6000        Tax RMSO  69416.65    69417
|Days/Week        5.0   Shifts     175.6000       +----------------------------+
|                       Hours      1756.000       |Total      2693878   2693878|
|R a t e s              Man-Hrs    6227.000       +----------------------------+
|---------              Eqp-Hrs    5610.000        > Level 1  2693878
|Rates Unique           UM/Da        0.0057
|                       UM/Sh        0.0057           >DBE Goal
|                       UM/Hr        0.0006           >MBE Goal
|                       UM/MH        0.0002           >WBE Goal
|                       MH/UM      6227.000           >O-1 Goal
|                                                     >O-2 Goal
+------------------------------------------------------------------------------+
```

(SFSRBASE) SFSR Base Bid
A C T I V I T Y - D E T A I L   F O R M S

```
+33 10 1-----------------Mobe and Preparation Work-Qty_UM--  1.00-LS-84340.00--+
|                                                                              |
|A s s i g n m n t s   S c h e d u l e          C o s t  Unit     Total        |
|-------------------   ----------------         ----------------------------   |
|Bid Item Line      1 Duration                  Source    Details              |
|Quote Grp   NONE     Bar Start                                                |
|Qry Lbl 1            Bar Finsh                 Lab Base                        |
|Qry Lbl 2            CPM ES                     Lab Brdn                       |
|Qry Lbl 3            CPM EF                     Eqp Own                        |
|Acnt                 CPM LS                     Eqp Op                         |
|                     CPM LF                     Rental                         |
|S h i f t s          Float                      Material  20000.00    20000    |
|----------------                                Supply                         |
|Work Hrs/Shift   10  P r o d u c t i o n        Subconts  54840.00    54840    |
|Pay Hrs/Shift    10  -------------------        Other     9500.000     9500    |
|Shifts/Day      1.0  Days                       Tax RMSO                       |
|Days/Week       5.0  Shifts                    +---------------------------+   |
|                     Hours                     |Total     84340.00    84340|   |
|R a t e s            Man-Hrs                   +---------------------------+   |
|---------            Eqp-Hrs                    > Level 2 84340.00             |
|Rates Unique         UM/Da                                                     |
|                     UM/Sh                                 >DBE Goal           |
|                     UM/Hr                                 >MBE Goal           |
|                     UM/MH                                 >WBE Goal           |
|                     MH/UM                                 >O-1 Goal           |
|                                                           >O-2 Goal           |
+------------------------------------------------------------------------------+
```

```
+33 10 1 1----------------Install Access Roads------Qty_UM--300.00-LF-  84.00--+
|                                                                              |
|A s s i g n m n t s  S c h e d u l e            C o s t  Unit     Total        |
|-------------------  ----------------           ----------------------------   |
|Bid Iten Line      1 Duration                   Source    Plug                 |
|Quote Grp   NONE     Bar Start                                                 |
|Qry Lbl 1            Bar Finsh                   Lab Base                       |
|Qry Lbl 2            CPM ES                      Lab Brdn                       |
|Qry Lbl 3            CPM EF                       Eqp Own                       |
|                     CPM LS                       Eqp Op                        |
|                     CPM LF                       Rental                        |
|                     Float                        Material                      |
|                                                  Supply                        |
|                                                  Subconts  84.000    25200     |
|                                                  Other                         |
|                                                  Tax RMSO                      |
|                                                 +--------------------------+   |
|                                                 |Total     84.000    25200 |   |
|                                                 +--------------------------+   |
|                                                  > Level 3 25200.00            |
|                                                                                |
|                                                                                |
|                                                                                |
+------------------------------------------------------------------------------+
```

```
                              (SFSRBASE) SFSR Base Bid
                    A C T I V I T Y - D E T A I L   F O R M S


+33 10 1 2----------------Install Secondary Contain-Qty_UM--6100.0-SF-   2.40--+
|
|Assignmnts     Schedule            Cost  Unit    Total
|-----------    -----------         --------------------------
|Bid Iten Line     1 Duration       Source   Plug
|Quote Grp  NONE     Bar Start
|Qry Lbl 1           Bar Finsh       Lab Base
|Qry Lbl 2           CPM ES          Lab Brdn
|Qry Lbl 3           CPM EF          Eqp Own
|                    CPM LS          Eqp Op
|                    CPM LF          Rental
|                    Float           Material
|                                    Supply
|                                    Subconts   2.400    14640
|                                    Other
|                                    Tax RMSO
|                                    +------------------------
|                                    |Total     2.400    14640
|                                    +------------------------
|                                    > Level 3 14640.00
|
|
|
|
|
|
+-----------------------------------------------------------------+


+33 10 1 3----------------Pre-fab Buildings---------Qty_UM--  2.00-EA-10000.00--+
|
|Assignmnts     Schedule            Cost  Unit    Total
|-----------    -----------         --------------------------
|Bid Iten Line     1 Duration       Source   Plug
|Quote Grp  NONE     Bar Start
|Qry Lbl 1           Bar Finsh       Lab Base
|Qry Lbl 2           CPM ES          Lab Brdn
|Qry Lbl 3           CPM EF          Eqp Own
|                    CPM LS          Eqp Op
|                    CPM LF          Rental
|                    Float           Material  10000.00   20000
|                                    Supply
|                                    Subconts
|                                    Other
|                                    Tax RMSO
|                                    +------------------------
|                                    |Total    10000.00   20000
|                                    +------------------------
|                                    > Level 3 20000.00
|
|
|
|
+-----------------------------------------------------------------+
```

(SFSRBASE) SFSR Base Bid
A C T I V I T Y - D E T A I L   F O R M S

```
+33 10 1 4----------------Site Utilities-----------Qty_UM--  1.00-LS-15000.00--+
|                                                                             |
|A s s i g n m n t s    S c h e d u l e          C o s t  Unit    Total       |
|-------------------    ---------------          -------------------------    |
|Bid Iten Line         1 Duration                Source   Plug                |
|Quote Grp    NONE       Bar Start                                            |
|Qry Lbl 1               Bar Finsh               Lab Base                     |
|Qry Lbl 2               CPM ES                  Lab Brdn                     |
|Qry Lbl 3               CPM EF                  Eqp Own                      |
|                        CPM LS                  Eqp Op                       |
|                        CPM LF                  Rental                       |
|                        Float                   Material                     |
|                                                Supply                       |
|                                                Subconts  15000.00    15000  |
|                                                Other                        |
|                                                Tax RMSO                     |
|                                                +----------------------------|
|                                                |Total    15000.00    15000  |
|                                                +----------------------------|
|                                                > Level 3 15000.00           |
|                                                                             |
|                                                                             |
|                                                                             |
|                                                                             |
|                                                                             |
+-----------------------------------------------------------------------------+
```

```
+33 10 1 5----------------Site Temporary Facilities-Qty_UM--  1.00-MO- 9500.00--+
|                                                                             |
|A s s i g n m n t s    S c h e d u l e          C o s t  Unit    Total       |
|-------------------    ---------------          -------------------------    |
|Bid Iten Line         1 Duration                Source   Plug                |
|Quote Grp    NONE       Bar Start                                            |
|Qry Lbl 1               Bar Finsh               Lab Base                     |
|Qry Lbl 2               CPM ES                  Lab Brdn                     |
|Qry Lbl 3               CPM EF                  Eqp Own                      |
|                        CPM LS                  Eqp Op                       |
|                        CPM LF                  Rental                       |
|                        Float                   Material                     |
|                                                Supply                       |
|                                                Subconts                     |
|                                                Other     9500.000     9500  |
|                                                Tax RMSO                     |
|                                                +----------------------------|
|                                                |Total    9500.000     9500  |
|                                                +----------------------------|
|                                                > Level 3 9500.000           |
|                                                                             |
|                                                                             |
|                                                                             |
|                                                                             |
|                                                                             |
+-----------------------------------------------------------------------------+
```

```
                              (SFSRBASE) SFSR Base Bid
                       A C T I V I T Y · D E T A I L   F O R M S


+33 10 3-----------------Site Work----------------Qty_UM--  1.00-LS-22000.00--+
|
| A s s i g n m n t s   S c h e d u l e          C o s t  Unit    Total
| --------------------   ------------------       ----------------------------
| Bid Iten Line      1 Duration                   Source    Plug
| Quote Grp    NONE    Bar Start
| Qry Lbl 1            Bar Finsh                   Lab Base
| Qry Lbl 2            CPM ES                      Lab Brdn
| Qry Lbl 3            CPM EF                       Eqp Own
|                      CPM LS                      Eqp Op
|                      CPM LF                      Rental
|                      Float                       Material  10000.00    10000
|                                                  Supply
|                                                  Subconts  12000.00    12000
|                                                  Other
|                                                  Tax RMSO
|                                                  +-------------------------
|                                                  |Total     22000.00    22000
|                                                  +-------------------------
|
|                                                  > Level 2 22000.00
|
|
|
|
+-----------------------------------------------------------------------------+


+33 10 4-----------------G'water Collection/Contrl-Qty_UM--  1.00-LS-581590.0--+
|
| A s s i g n m n t s   S c h e d u l e          C o s t  Unit    Total
| --------------------   ------------------       ----------------------------
| Bid Item Line      1 Duration                   Source    Details
| Quote Grp    NONE    Bar Start
| Qry Lbl 1            Bar Finsh                   Lab Base
| Qry Lbl 2            CPM ES                      Lab Brdn
| Qry Lbl 3            CPM EF                       Eqp Own
| Acnt                 CPM LS                      Eqp Op
|                      CPM LF                      Rental
| S h i f t s          Float                       Material
| ------------------                               Supply
| Work Hrs/Shift  10   P r o d u c t i o n         Subconts  581590.0    581590
| Pay Hrs/Shift   10   -------------------          Other
| Shifts/Day     1.0   Days                        Tax RMSO
| Days/Week      5.0   Shifts                      +-------------------------+
|                      Hours                       |Total     581590.0   581590|
| R a t e s            Man-Hrs                     +-------------------------+
| ---------            Eqp-Hrs                     > Level 2 581590.0
| Rates Unique         UM/Da
|                      UM/Sh                                 >DBE Goal
|                      UM/Hr                                 >MBE Goal
|                      UM/MH                                 >WBE Goal
|                      MH/UM                                 >O-1 Goal
|                                                            >O-2 Goal
+------------------------------------------------------------------------------+
```

```
                         (SFSRBASE) SFSR Base Bid
                      A C T I V I T Y - D E T A I L   F O R M S


 +33 10 4 1----------------Monitoring Wells----------Qty_UM-- 16.00-EA- 2587.50--+
 |
 |A s s i g n m n t s    S c h e d u l e        C o s t  Unit    Total
 |--------------------    -----------------      -------------------------
 |Bid Item Line      1 Duration                 Source    Quote
 |Query Grp    NONE    Bar Start
 |Qry Lbl 1            Bar Finsh                 Lab Base
 |Qry Lbl 2            CPM ES                    Lab Brdn
 |Qry Lbl 3            CPM EF                    Eqp Own
 |                     CPM LS                    Eqp Op
 |                     CPM LF                    Rental
 |                     Float                     Material
 |                                               Supply
 |                                               Subconts  2587.500    41400
 |                                               Other
 |                                               Tax RMSO
 |                                              +-------------------------
 |                                              |Total     2587.500    41400
 |                                              +-------------------------
 |                                               > Level 3 41400.00
 |Awardee
 |        WELLS                                          >DBE Goal
 |                                                       >MBE Goal
 |DBE Allow      1.00%                                   >WBE Goal
 |                                                       >O-1 Goal
 |                                                       >O-2 Goal
 +-----------------------------------------------------------------------+


 +33 10 4 2----------------Recovery Wells-----------Qty_UM-- 40.00-EA- 7762.50--+
 |
 |A s s i g n m n t s    S c h e d u l e        C o s t  Unit    Total
 |--------------------    -----------------      -------------------------
 |Bid Item Line      1 Duration                 Source    Quote
 |Query Grp    NONE    Bar Start
 |Qry Lbl 1            Bar Finsh                 Lab Base
 |Qry Lbl 2            CPM ES                    Lab Brdn
 |Qry Lbl 3            CPM EF                    Eqp Own
 |                     CPM LS                    Eqp Op
 |                     CPM LF                    Rental
 |                     Float                     Material
 |                                               Supply
 |                                               Subconts  7762.500   310500
 |                                               Other
 |                                               Tax RMSO
 |                                              +-------------------------
 |                                              |Total     7762.500   310500
 |                                              +-------------------------
 |                                               > Level 3 310500.0
 |Awardee
 |        WELLS                                          >DBE Goal
 |                                                       >MBE Goal
 |DBE Allow      1.00%                                   >WBE Goal
 |                                                       >O-1 Goal
 |                                                       >O-2 Goal
 +-----------------------------------------------------------------------+
```

```
                         (SFSRBASE) SFSR Base Bid
                    A C T I V I T Y - D E T A I L   F O R M S


+33 10 4 3----------------Injection Wells-----------Qty_UM-- 28.00-EA- 7417.50--+

A s s i g n m n t s    S c h e d u l e         C o s t   Unit    Total
-------------------    -----------------       --------------------------
Bid Item Line       1 Duration                 Source   Quote
Query Grp    NONE     Bar Start
Qry Lbl 1             Bar Finsh                 Lab Base
Qry Lbl 2             CPM ES                     Lab Brdn
Qry Lbl 3             CPM EF                     Eqp Own
                      CPM LS                     Eqp Op
                      CPM LF                     Rental
                      Float                      Material
                                                 Supply
                                                 Subconts  7417.500   207690
                                                 Other
                                                 Tax RMSO
                                                +------------------------
                                                |Total     7417.500   207690
                                                +------------------------
                                                 > Level 3 207690.0
Awardee
        WELLS                                            >DBE Goal
                                                         >MBE Goal
DBE Allow       1.00%                                    >WBE Goal
                                                         >O-1 Goal
                                                         >O-2 Goal
+------------------------------------------------------------------------+



+33 10 4 4----------------Soil Borings-------------Qty_UM-- 20.00-EA-  600.00--+

A s s i g n m n t s    S c h e d u l e         C o s t   Unit    Total
-------------------    -----------------       --------------------------
Bid Item Line       1 Duration                 Source   Details  L
Quote Grp    NONE     Bar Start
Qry Lbl 1             Bar Finsh                 Lab Base
Qry Lbl 2             CPM ES                     Lab Brdn
Qry Lbl 3             CPM EF                     Eqp Own
Acnt                  CPM LS                     Eqp Op
                      CPM LF                     Rental
S h i f t s           Float                      Material
-------------------                             Supply
Work Hrs/Shift  10  P r o d u c t i o n          Subconts   600.000    12000
Pay  Hrs/Shift  10  -------------------          Other
Shifts/Day     1.0  Days                         Tax RMSO
Days/Week      5.0  Shifts                      +------------------------
                    Hours                       |Total      600.000    12000|
R a t e s           Man-Hrs                     +------------------------
---------           Eqp-Hrs                      > Level 3 12000.00
Rates Unique        UM/Da
                    UM/Sh                                >DBE Goal
                    UM/Hr                                >MBE Goal
                    UM/MH                                >WBE Goal
                    MH/UM                                >O-1 Goal
                                                         >O-2 Goal
+------------------------------------------------------------------------+
```

(SFSRBASE) SFSR Base Bid
A C T I V I T Y - D E T A I L   F O R M S

```
+-33 10 4 4-------------------Soil Borings-------------------------------+
|RESOURCE LIST Cost Detail                        Source Worksheet:      |
+=======================================================================+
                                        +----------------------+
       Man-Hrs             Eqp-Hrs       |>Unit     600.00     |
       UM/Man-Hr           UM/Eqp-Hr     |>TOTAL    12000       |
       Man-Hr/UM           Eqp-Hr/UM     +----------------------+



   Resource                   Quantity UM Count  Unit Cost       TOTAL
   --------------------------------------------------------------------
                                                 Total:    12000
                                                 Unit:      600.00
   USOILBOR soil borings        20.00 EA 1.00     600.00    12000.00 |
+-----------------------------------------------------------------------+


+=======================================================================+
+-33 10 4 5----------------Soil Samples--------------Qty_UM-- 20.00-EA- 500.00--+

   A s s i g n m n t s     S c h e d u l e        C o s t  Unit     Total
   -------------------     -------------------     -------------------------
   Bid Iten Line       1 Duration                  Source    Plug
   Quote Grp    NONE     Bar Start
   Qry Lbl 1             Bar Finsh                  Lab Base
   Qry Lbl 2             CPM ES                     Lab Brdn
   Qry Lbl 3             CPM EF                     Eqp Own
                         CPM LS                     Eqp Op
                         CPM LF                     Rental
                         Float                      Material
                                                    Supply
                                                    Subconts   500.000     10000
                                                    Other
                                                    Tax RMSO
                                                   +------------------------
                                                   |Total      500.000     10000
                                                   +------------------------
                                                    > Level 3 10000.00




+-----------------------------------------------------------------------+
```

(SFSRBASE) SFSR Base Bid
A C T I V I T Y - D E T A I L   F O R M S

```
+33 10 5-------------------Physical Treatment--------Qty_UM--  1.00-LS- 1646106--+
|                                                                                |
| A s s i g n m n t s   S c h e d u l e          C o s t  U n i t     T o t a l  |
| --------------------   -----------------        ----------------------------   |
| Bid Item Line       1 Duration                  Source   Details  LC          |
| Quote Grp    NONE     Bar Start                                                |
| Qry Lbl 1             Bar Finsh                 Lab Base 29752.50    29753     |
| Qry Lbl 2             CPM ES                    Lab Brdn                       |
| Qry Lbl 3             CPM EF                     Eqp Own  25500.00    25500     |
| Acnt                  CPM LS                     Eqp Op   4250.000     4250     |
|                       CPM LF                     Rental                        |
| S h i f t s           Float                      Material 1282630  1282630     |
| -----------------                                Supply   2500.000    2500     |
| Work Hrs/Shift   10   P r o d u c t i o n        Subconts 237217.2  237217     |
| Pay Hrs/Shift    10   -----------------          Other                         |
| Shifts/Day      1.0   Days      25.0000          Tax RMSO 64256.50    64257    |
| Days/Week       5.0   Shifts    25.0000          +-----------------------------+
|                       Hours    250.0000          |Total     1646106  1646106|  |
| R a t e s             Man-Hrs  1250.000          +-----------------------------+
| ---------             Eqp-Hrs  1250.000          > Level 2  1646106            |
| Rates Unique          UM/Da      0.0400                                        |
|                       UM/Sh      0.0400               >DBE Goal                 |
|                       UM/Hr      0.0040               >MBE Goal                 |
|                       UM/MH      0.0008               >WBE Goal                 |
|                       MH/UM    1250.000               >O-1 Goal                 |
|                                                       >O-2 Goal                 |
+--------------------------------------------------------------------------------+
```

```
+-33 10 5---------------------Physical Treatment---------------------------------+
|RESOURCE LIST Cost Detail                    Source Worksheet:                   |
+================================================================================+
|                                             +----------------------+          |
|   Man-Hrs           Eqp-Hrs                 |>Unit     1383e3      |          |
|   UM/Man-Hr         UM/Eqp-Hr               |>TOTAL    1383e3      |          |
|   Man-Hr/UM         Eqp-Hr/UM               +----------------------+          |
|                                                                                |
|                                                                                |
|                                                                                |
| Resource                     Quantity UM Count  Unit Cost       TOTAL         |
| -----------------------------------------------------------------------------  |
|                                                     Total:   1383e3           |
|                                                     Unit:    1383e3           |
|                                                                                |
| MFILTER3 BAG FILTER (44 GPM)       2.00 EA 1.00    535.50     1071.00          |
| MMIXER1  CHEM STORAGE TANK MIXER   2.00 EA 1.00    5250.0    10500.00          |
| MMIXER2  CHEMICAL MIX TANKS MIXER  2.00 EA.1.00    5250.0    10500.00          |
| MPUMP1   CHEMICAL FEED PUMP (25G)  2.00 EA 1.00    5250.0    10500.00          |
| MPUMP2   CHEMICAL FEED PUMP (55G)  2.00 EA 1.00    8925.0    17850.00          |
| MPUMP4   CHEM. FEED INJ. PUMP (5G)28.00 EA 1.00    6825.0   191100.00          |
| MPUMP6   VAR.SPEED RECOV.PMP (5G) 40.00 EA 1.00    4305.0   172200.00          |
| MPUMP8   TREAT.PLNT.FEED PUMP (65) 8.00 EA 1.00    4200.0    33600.00          |
| MSUMP1   GALV. CMP W/ COVER       68.00 EA 1.00    525.00    35700.00          |
| MCHEMTNK ChemStorageFracTank-10k   2.00 EA 1.00    7875.0    15750.00          |
| MCHMTNK2 ChemStorTnk.20K           1.00 EA 1.00    11550     11550.00          |
| MTANK3   EQUALIZATION TANK (20K)   3.00 EA 1.00    36750    110250.00          |
| MTANK4   CONCENTRATE STOR.TNK.(20K 2.00 EA 1.00    36750     73500.00          |
| MTWRSTRP PAcked Tower Strip.30gpm  2.00 EA 1.00    47013     94025.40          |
+--------------------------------------------------------------------------------+
```

```
                            (SFSRBASE) SFSR Base Bid
                        A C T I V I T Y - D E T A I L   F O R M S


+-33 10 5--------------------Physical Treatment----------------------------+
|RESOURCE LIST Cost Detail                        Source Worksheet:         |
+==========================================================================+
                                        +--------------------+
   Man-Hrs               Eqp-Hrs        |>Unit      1383e3   |
   UM/Man-Hr             UM/Eqp-Hr      |>TOTAL     1383e3   |
   Man-Hr/UM             Eqp-Hr/UM      +--------------------+



Resource                    Quantity UM Count  Unit Cost       TOTAL
--------------------------------------------------------------------------
                                                   Total:   1383e3
                                                   Unit:    1383e3
MOXYSYS  Catalytic Oxidation SYS.      1.00 EA 1.00   470400   470400.00
SMISC100 MISC. SUPPLIES               25.00 EA 1.00   105.00     2625.00
USUBPRDM SUBCONTRACTOR PER DIEM      175.00 MD 1.00    65.00    11375.00
MBLOWER1 CENTRIFUGAL BLOWER(15PSI)     1.00 EA 1.00    15750    15750.00
UPIPE15T 1.5" BLACK PIPE (TREAT.)    100.00 LF 1.00    12.00     1200.00
UMISCP&F MISC. PIPE & FITTINGS         0.33 LS 1.00    75000    24750.00
MBLWRSHD Blower Shed(BUILDING)         1.00 EA 1.00    10500    10500.00
MO/WSEPR OIL WATER SEPARATOR 30GPM     1.00 EA 1.00   6300.0     6300.00
UPIPE15T 1.5" BLACK PIPE (TREAT.)    295.00 LF 1.00    12.00     3540.00
M1.5"REC 1.5"RECY.Tnk. Discharge     164.00 LF 1.00    12.60     2066.40
UPIPE4   4" BLACK PIPE                33.00 LF 1.00    31.00     1023.00
M.5SCRUB .5"Steel Scrubber Dischar    33.00 LF 1.00     8.40      277.20
MFLUMISC MISC.Pipe F&E FLUID HANDL     0.05 LS 1.00    78750     3937.50
MULTRAFL Ultrafiltration Unit17gpm     1.00 EA 1.00    42000    42000.00
+--------------------------------------------------------------------------+

+-33 10 5--------------------Physical Treatment----------------------------+
|CREW CONFIGURATION Cost Detail                   Source Worksheet:         |
+==========================================================================+
                                          +--------------------+
Days       25.000  UM/Da  0.0400  Da/UM  25.000 |>Unit      59503   |
Shifts     25.000  UM/Sh  0.0400  Sh/UM  25.000 |>TOTAL     59503   |
Hours      250.00  UM/Hr  0.0040  Hr/UM  250.00 +--------------------+
Man-hours  1250.0  UM/MH  0.0008  MH/UM  1250.0



Resource   Qty                              Unit Cost  UM       TOTAL
--------------------------------------------------------------------------
Man-Count:  5.00                                    Total:     59503
Eqp-Count:  5.00                                    Unit:      59503
LTCCI5     1.00   CCI TECHNICIAN 5           27.60     HR     6900.00
LCM5       1.00   CONSTRUCTION MGR. 5        34.62     HR     8655.00
LTCCI3     3.00   CCI TECHNICIAN 3           18.93     HR    14197.50
ECAT960    1.00   CAT 960 WHEEL LOADER       34.00     HR     8500.00
ECRN60     1.00   TRUCK CRANE - 60 TN        35.00     HR     8750.00
EPUTRK     2.00   PICKUP TRUCK                6.00     HR     3000.00
ECAT215    1.00   CAT 215 EXCAVATOR          38.00     HR     9500.00
+--------------------------------------------------------------------------+


+==========================================================================+
```

(SFSRBASE) SFSR Base Bid
A C T I V I T Y - D E T A I L   F O R M S

```
+-33 10 5 1----------------Electrical Package--------Qty_UM--  1.00-LS-129004.2--+
|                                                                               |
| A s s i g n m n t s   S c h e d u l e        C o s t  Unit     Total          |
| --------------------  ------------------     -------------------------------- |
| Bid Item Line     1 Duration                 Source     Details  L            |
| Quote Grp    NONE   Bar Start                                                 |
| Qry Lbl 1           Bar Finsh                Lab Base                          |
| Qry Lbl 2           CPM ES                   Lab Brdn                          |
| Qry Lbl 3           CPM EF                   Eqp Own                           |
| Acnt                CPM LS                   Eqp Op                            |
|                     CPM LF                   Rental                           |
| S h i f t s         Float                    Material                         |
| ------------------                           Supply                           |
| Work Hrs/Shift   10 P r o d u c t i o n      Subconts  129004.2    129004     |
| Pay  Hrs/Shift   10 -------------------      Other                            |
| Shifts/Day      1.0 Days                     Tax RMSO                         |
| Days/Week       5.0 Shifts                   +----------------------------+   |
|                     Hours                    |Total      129004.2    129004|  |
| R a t e s           Man-Hrs                  +----------------------------+   |
| ---------           Eqp-Hrs                  > Level 3 129004.2               |
| Rates Unique        UM/Da                                                     |
|                     UM/Sh                          >DBE Goal                  |
|                     UM/Hr                          >MBE Goal                  |
|                     UM/MH                          >WBE Goal                  |
|                     MH/UM                          >O-1 Goal                  |
|                                                    >O-2 Goal                  |
+-------------------------------------------------------------------------------+
```

ALL PRICES ARE FURNISH AND INSTALLED.

```
+-33 10 5 1-------------------Electrical Package--------------------------------+
|RESOURCE LIST Cost Detail                     Source Worksheet:                |
+==============================================================================+|
|                                              +--------------------+           |
|   Man-Hrs           Eqp-Hrs                  |>Unit    129004     |           |
|   UM/Man-Hr         UM/Eqp-Hr                |>TOTAL   129004     |           |
|   Man-Hr/UM         Eqp-Hr/UM                +--------------------+           |
|                                                                               |
|                                                                               |
| Resource                       Quantity UM Count  Unit Cost      TOTAL        |
| ---------------------------------------------------------------------------   |
|                                                     Total:   129004           |
|                                                     Unit:    129004           |
| U2C4WIRE 2" CONDUIT W/ 3-#1 WIRE   624.00 LF 1.00     20.20    12604.80        |
| U480VPNL 480 VOLT PANEL BOARD        1.00 EA 1.00   10200      10200.00        |
| UMPC     MINI POWER CENTER           4.00 EA 1.00    2100.0     8400.00        |
| U1C3WIRE 1" CONDUIT W/ 2-#8 WIRE  1444.00 LF 1.00     11.40    16461.60        |
| UMISCW&E MISC. WIRING AND ELECTRIC   0.50 LS 1.00   32830      16415.00        |
| U2C4WIRE 2" CONDUIT W/ 3-#1 WIRE  3214.00 LF 1.00     20.20    64922.80        |
+-------------------------------------------------------------------------------+


+==============================================================================+
```

(SFSRBASE) SFSR Base Bid
A C T I V I T Y - D E T A I L   F O R M S

```
+33 10 5 2----------------Plumbing Package----------Qty_UM--  1.00-LS-73759.00--+
|                                                                              |
| A s s i g n m n t s   S c h e d u l e        C o s t  Unit    Total          |
| -------------------   ----------------       ----------------------------    |
| Bid Item Line    1 Duration                  Source   Details  L             |
| Quote Grp    NONE  Bar Start                                                 |
| Qry Lbl 1          Bar Finsh                 Lab Base                         |
| Qry Lbl 2          CPM ES                    Lab Brdn                         |
| Qry Lbl 3          CPM EF                    Eqp Own                          |
| Acnt               CPM LS                    Eqp Op                           |
|                    CPM LF                    Rental                           |
| S h i f t s        Float                     Material  7080.000      7080     |
| ---------------                              Supply                           |
| Work Hrs/Shift   10  P r o d u c t i o n     Subconts 66325.00      66325     |
| Pay Hrs/Shift    10  -------------------     Other                            |
| Shifts/Day      1.0  Days                    Tax RMSO   354.000       354|    |
| Days/Week       5.0  Shifts                 +---------------------------+     |
|                      Hours                  |Total     73759.00      73759|   |
| R a t e s            Man-Hrs                +---------------------------+     |
| ---------            Eqp-Hrs                 > Level 3 73759.00               |
| Rates Unique         UM/Da                                                    |
|                      UM/Sh                             >DBE Goal              |
|                      UM/Hr                             >MBE Goal              |
|                      UM/MH                             >WBE Goal              |
|                      MH/UM                             >O-1 Goal              |
|                                                        >O-2 Goal              |
+------------------------------------------------------------------------------+


+-33 10 5 2-----------------Plumbing Package----------------------------------+
|RESOURCE LIST Cost Detail                     Source Worksheet:              |
+============================================================================+
|                                             +-----------------------+      |
|   Man-Hrs             Eqp-Hrs               |>Unit     73759        |      |
|   UM/Man-Hr           UM/Eqp-Hr             |>TOTAL    73759        |      |
|   Man-Hr/UM           Eqp-Hr/UM             +-----------------------+      |
|                                                                            |
|                                                                            |
|                                                                            |
| Resource                 Quantity UM Count  Unit Cost         TOTAL        |
| ---------------------------------------------------------------------------|
|                                             Total:     73759              |
|                                             Unit:      73759              |
|UPIPE2   2" BLACK PIPE             656.00 LF 1.00      16.00    10496.00    |
|UPIPE2IN 2" BLACK PIPE (INJECTION) 427.00 LF 1.00      16.00     6832.00    |
|UPIP2DB1 2" BLACK PIPE (DBL. WALL) 361.00 LF 1.00      19.00     6859.00    |
|U1"BSPIN 1"BSPInjection Well Pipe  590.00 LF 1.00      12.60     7434.00    |
|USFSRMSC SFSR Misc.F&E Piping        1.00 LS 1.00    18300.00   18300.00    |
|U1.5PRIR 1.5" Primary air Piping   330.00 LF 1.00      12.00     3960.00    |
|U1.5SEC  1.5" Secondary air piping 590.00 LF 1.00      12.00     7080.00    |
|U1.ORECW 1"Recovery Well Sec.pipe  755.00 LF 1.00      10.00     7550.00    |
|U.5IWTP  .5"BSP IWTP Discharge     656.00 LF 1.00       8.00     5248.00    |
+----------------------------------------------------------------------------+


+============================================================================+
```

```
                              (SFSRBASE) SFSR Base Bid
                         A C T I V I T Y - D E T A I L   F O R M S


+33 10 8------------------Demobilization-----------Qty_UM--  1.00-LS-13168.44--+

 A s s i g n m n t s   S c h e d u l e           C o s t  Unit     Total
 -------------------   ------------------         -----------------------------
 Bid Item Line    1 Duration                      Source     Details  LC
 Quote Grp   NONE   Bar Start
 Qry Lbl 1          Bar Finsh                      Lab Base  5008.000     5008
 Qry Lbl 2          CPM ES                         Lab Brdn  4887.440     4887
 Qry Lbl 3          CPM EF                         Eqp Own
 Acnt               CPM LS                         Eqp Op
                    CPM LF                         Rental    1150.000     1150
 S h i f t s        Float                          Material
 ------------------                               Supply    2000.000     2000
 Work Hrs/Shift  10 P r o d u c t i o n           Subconts
 Pay Hrs/Shift   10 -------------------           Other
 Shifts/Day     1.0 Days        5.0000            Tax RMSO   123.000      123
 Days/Week      5.0 Shifts      5.0000           +----------------------------+
                    Hours      50.0000           |Total       13168.44    13168|
 R a t e s          Man-Hrs   200.0000           +----------------------------+
 ---------          Eqp-Hrs   200.0000            > Level 2 13168.44
 Rates Unique       UM/Da       0.2000
                    UM/Sh       0.2000                     >DBE Goal
                    UM/Hr       0.0200                     >MBE Goal
                    UM/MH       0.0050                     >WBE Goal
                    MH/UM     200.0000                     >O-1 Goal
                                                          >O-2 Goal
+-----------------------------------------------------------------------------+


+-33 10 8--------------------Demobilization------------------------------------+
|RESOURCE LIST Cost Detail                        Source Worksheet:           |
+=============================================================================+
                                                 +----------------------+
    Man-Hrs              Eqp-Hrs                  |>Unit     2100.0      |
    UM/Man-Hr            UM/Eqp-Hr                |>TOTAL    2100.0      |
    Man-Hr/UM            Eqp-Hr/UM                +----------------------+



 Resource                       Quantity UM Count  Unit Cost      TOTAL
 --------------------------------------------------------------------
                                                 Total:   2100.0
                                                 Unit:    2100.0
|SMISC200 MISC. SUPPLIES          10.00 EA 1.00    210.00         2100.00 |
+-----------------------------------------------------------------------------+

+-33 10 8--------------------Demobilization------------------------------------+
|CREW CONFIGURATION Cost Detail                   Source Worksheet:           |
+=============================================================================+
                                                 +----------------------+
|Days       5.0000  UM/Da 0.2000   Da/UM 5.0000   |>Unit     11068      |
|Shifts     5.0000  UM/Sh 0.2000   Sh/UM 5.0000   |>TOTAL    11068      |
|Hours     50.000   UM/Hr 0.0200   Hr/UM 50.000   +----------------------+
|Man-hours 200.00   UM/MH 0.0050   MH/UM 200.00


|Resource   Qty                                  Unit Cost  UM     TOTAL
 --------------------------------------------------------------------
Man-Count:   4.00                                        TOTAL:  11068
Eqp-Count:   4.00                                        Unit:   11068
|ONSITE     1.00  ONSITE MGMT. CREW                221.37     HR  11068.44 |
+-----------------------------------------------------------------------------+
```

```
                        (SFSRBASE) SFSR Base Bid
                   A C T I V I T Y - D E T A I L   F O R M S


+33 10 11----------------Construction Management---Qty_UM--104.00-DA- 3333.40--+
|                                                                              |
|A s s i g n m n t s    S c h e d u l e         C o s t   Unit     Total       |
|-------------------    ---------------         -----------------------------  |
|Bid Item Line       1 Duration                 Source    Details              |
|Quote Grp    NONE     Bar Start                                               |
|Qry Lbl 1             Bar Finsh                Lab Base  1177.618   122472     |
|Qry Lbl 2             CPM ES                    Lab Brdn   977.488   101659     |
|Qry Lbl 3             CPM EF                    Eqp Own                        |
|Acnt                  CPM LS                    Eqp Op                         |
|                      CPM LF                    Rental    240.096    24970     |
|S h i f t s           Float                     Material                       |
|-----------------                              Supply     20.192     2100     |
|Work Hrs/Shift   10   P r o d u c t i o n      Subconts                       |
|Pay  Hrs/Shift   10   -------------------       Other     869.567    90435     |
|Shifts/Day      1.0   Days       145.6000       Tax RMSO   48.434     5037     |
|Days/Week       5.0   Shifts     145.6000      +----------------------------  |
|                      Hours     1456.000       |Total    3333.395   346673|    |
|R a t e s             Man-Hrs   4777.000       +----------------------------  |
|---------             Eqp-Hrs   4160.000        > Level 2 346673.1            |
|Rates Unique          UM/Da        0.7143                                      |
|                      UM/Sh        0.7143                >DBE Goal             |
|                      UM/Hr        0.0714                >MBE Goal             |
|                      UM/MH        0.0218                >WBE Goal             |
|                      MH/UM       45.9327                >O-1 Goal             |
|                                                         >O-2 Goal             |
+------------------------------------------------------------------------------+


+33 10 11 1--------------Mobe/Demobe---------------Qty_UM-- 30.00-DA- 3324.40--+
|                                                                              |
|A s s i g n m n t s    S c h e d u l e         C o s t   Unit     Total       |
|-------------------    ---------------         -----------------------------  |
|Bid Item Line       1 Duration                 Source    Details  LC          |
|Quote Grp    NONE     Bar Start                                               |
|Qry Lbl 1             Bar Finsh                Lab Base  1177.553    35327     |
|Qry Lbl 2             CPM ES                    Lab Brdn   977.488    29325     |
|Qry Lbl 3             CPM EF                    Eqp Own                        |
|Acnt                  CPM LS                    Eqp Op                         |
|                      CPM LF                    Rental    240.000     7200     |
|S h i f t s           Float                     Material                       |
|-----------------                              Supply     20.000      600     |
|Work Hrs/Shift   10   P r o d u c t i o n      Subconts                       |
|Pay  Hrs/Shift   10   -------------------       Other     861.333    25840     |
|Shifts/Day      1.0   Days        42.0000       Tax RMSO   48.025     1441     |
|Days/Week       5.0   Shifts      42.0000      +----------------------------  |
|                      Hours      420.0000      |Total    3324.400    99732|    |
|R a t e s             Man-Hrs   1378.000       +----------------------------  |
|---------             Eqp-Hrs   1200.000        > Level 3  958.961            |
|Rates Unique          UM/Da        0.7143                                      |
|                      UM/Sh        0.7143                >DBE Goal             |
|                      UM/Hr        0.0714                >MBE Goal             |
|                      UM/MH        0.0218                >WBE Goal             |
|                      MH/UM       45.9333                >O-1 Goal             |
|                                                         >O-2 Goal             |
+------------------------------------------------------------------------------+
```

```
                           (SFSRBASE) SFSR Base Bid
                     A C T I V I T Y - D E T A I L   F O R M S


+--33 10 11 1------------------Mobe/Demobe-------------------------------------+
|RESOURCE LIST Cost Detail                        Source Worksheet:           |
+=============================================================================+
|                                                                             |
|                                          +----------------------+           |
|    Man-Hrs   173.00    Eqp-Hrs           |>Unit     764.21      |           |
|    UM/Man-Hr 0.1685    UM/Eqp-Hr         |>TOTAL    22926       |           |
|    Man-Hr/UM 5.9333    Eqp-Hr/UM         +----------------------+           |
|                                                                             |
|                                                                             |
|                                                                             |
| Resource                        Quantity UM Count  Unit Cost        TOTAL   |
| --------------------------------------------------------------------------- |
|                                                               Total:  22926 |
|                                                               Unit:  764.21 |
| LCM5      CONSTRUCTION MGR. 5    120.00 HR 1.00       34.62       4154.40    |
| LCCIOFF   CCI OFFICE/ADMIN.       48.00 HR 1.00       15.40        739.20    |
| LCM6      CONSTRUCTION MGR. 6     10.00 HR 1.00       38.50        385.00    |
| SMISC100 MISC. SUPPLIES            6.00 EA 1.00      105.00        630.00    |
| UAIRFAR1 Not Found                         1.00                             |
| UCAR1     RENTAL CAR               6.00 DA 1.00       50.00        300.00    |
| UPRDIEM1 PER DIEM W/ HOTEL         3.00 DA 1.00       89.25        267.75    |
| UPRDIEM2 PER DIEM W/OUT HOTEL      3.00 DA 1.00       25.00         75.00    |
| UAIRFAR1 Not Found                         1.00                             |
| UPRDIEM1 PER DIEM W/ HOTEL       180.00 DA 1.00       89.25      16065.00    |
| UMILEAGE POV MILEAGE            1000.00 MI 1.00        0.31        310.00    |
+-----------------------------------------------------------------------------+


+--33 10 11 1------------------Mobe/Demobe-------------------------------------+
|CREW CONFIGURATION Cost Detail                   Source Worksheet:           |
+=============================================================================+
|                                                                             |
|                                          +----------------------+           |
| Days      30.000  UM/Da  1.0000  Da/UM  1.0000 |>Unit   2213.7   |          |
| Shifts    30.000  UM/Sh  1.0000  Sh/UM  1.0000 |>TOTAL  66411    |          |
| Hours     300.00  UM/Hr  0.1000  Hr/UM  10.000 +-----------------+          |
| Man-hours 1200.0  UM/MH  0.0250  MH/UM  40.000                              |
|                                                                             |
|                                                                             |
|                                                                             |
| Resource    Qty                              Unit Cost  UM       TOTAL      |
| --------------------------------------------------------------------------- |
| Man-Count:   4.00                                        Total:   66411     |
| Eqp-Count:   4.00                                        Unit:   2213.7     |
| ONSITE     1.00  ONSITE MGMT. CREW            221.37     HR  66410.64       |
+-----------------------------------------------------------------------------+


+=============================================================================+
```

```
                        (SFSRBASE) SFSR Base Bid
                    A C T I V I T Y - D E T A I L   F O R M S


 +33 10 11 2--------------Operations----------------Qty_UM-- 74.00-DA- 3337.04--+

 |A s s i g n m n t s    S c h e d u l e         C o s t  Unit     Total
 |------------------     -----------------        -------------------------
 |Bid Item Line        1 Duration                 Source    Details  LC
 |Quote Grp    NONE      Bar Start
 |Qry Lbl 1              Bar Finsh                 Lab Base  1177.644   87146
 |Qry Lbl 2              CPM ES                    Lab Brdn   977.488   72334
 |Qry Lbl 3              CPM EF                    Eqp Own
 |Acnt                   CPM LS                    Eqp Op
 |                       CPM LF                    Rental     240.135   17770
 |                       Float                     Material
 |S h i f t s                                      Supply      20.270    1500
 |------------------                               Subconts
 |Work Hrs/Shift    10   P r o d u c t i o n       Other      872.905   64595
 |Pay Hrs/Shift     10   -------------------       Tax RMSO    48.600    3596
 |Shifts/Day       1.0   Days         103.6000     +-------------------------+
 |Days/Week        5.0   Shifts       103.6000     |Total     3337.042  246941|
 |                       Hours        1036.000     +-------------------------+
 |R a t e s              Man-Hrs      3399.000      > Level 3 2374.434
 |---------              Eqp-Hrs      2960.000
 |Rates Unique           UM/Da          0.7143          >DBE Goal
 |                       UM/Sh          0.7143          >MBE Goal
 |                       UM/Hr          0.0714          >WBE Goal
 |                       UM/MH          0.0218          >O-1 Goal
 |                       MH/UM         45.9324          >O-2 Goal
 +------------------------------------------------------------------------+


 +-33 10 11 2----------------Operations-----------------------------------+
 |RESOURCE LIST Cost Detail                    Source Worksheet:          |
 +========================================================================+

    |Man-Hrs    439.00   Eqp-Hrs            |>Unit    765.79  |
     UM/Man-Hr  0.1686   UM/Eqp-Hr           >TOTAL   56668
     Man-Hr/UM  5.9324   Eqp-Hr/UM          +-----------------+



 Resource                   Quantity UM Count  Unit Cost       TOTAL
 -------------------------------------------------------------------
                                                   Total:      56668
                                                   Unit:      765.79
 LCM5      CONSTRUCTION MGR. 5    296.00 HR 1.00     34.62    10247.52
 LCCIOFF   CCI OFFICE/ADMIN.      118.00 HR 1.00     15.40     1817.20
 LCM6      CONSTRUCTION MGR. 6     25.00 HR 1.00     38.50      962.50
 SMISC100  MISC. SUPPLIES          15.00 EA 1.00    105.00     1575.00
 UAIRFAR1  Not Found                       1.00
 UCAR1     RENTAL CAR              15.00 DA 1.00     50.00      750.00
 UPRDIEM1  PER DIEM W/ HOTEL        8.00 DA 1.00     89.25      714.00
 UPRDIEM2  PER DIEM W/OUT HOTEL     8.00 DA 1.00     25.00      200.00
 UAIRFAR1  Not Found                       1.00
 UPRDIEM1  PER DIEM W/ HOTEL      444.00 DA 1.00     89.25    39627.00
 UMILEAGE  POV MILEAGE           2500.00 MI 1.00      0.31      775.00
 +------------------------------------------------------------------------+
```

(SFSRBASE) SFSR Base Bid
A C T I V I T Y - D E T A I L   F O R M S

```
+-33 10 11 2------------------Operations---------------------------------------+
|CREW CONFIGURATION Cost Detail                      Source Worksheet:         |
+==============================================================================+
                                                 +---------------------+
|Days        74.000  UM/Da  1.0000   Da/UM  1.0000  |>Unit     2213.7     |
|Shifts      74.000  UM/Sh  1.0000   Sh/UM  1.0000  |>TOTAL    163813     |
|Hours       740.00  UM/Hr  0.1000   Hr/UM  10.000  +---------------------+
|Man-hours   2960.0  UM/MH  0.0250   MH/UM  40.000

|Resource    Qty                                  Unit Cost  UM      TOTAL
 ----------------------------------------------------------------------------
|Man-Count:   4.00                                        Total:   163813
|Eqp-Count:   4.00                                        Unit:    2213.7
|ONSITE      1.00  ONSITE MGMT. CREW                221.37    HR 163812.91 |
+------------------------------------------------------------------------------+
```

APPENDIX 7

SPME Scale-up Drawings

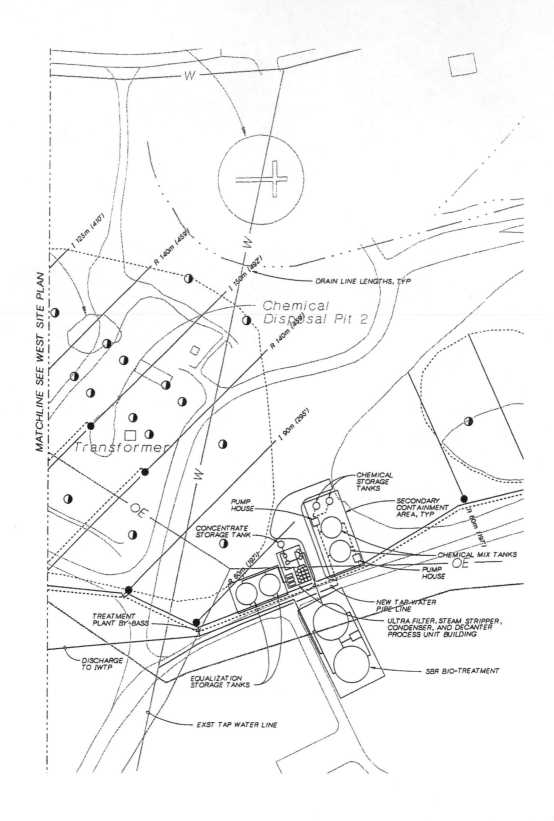

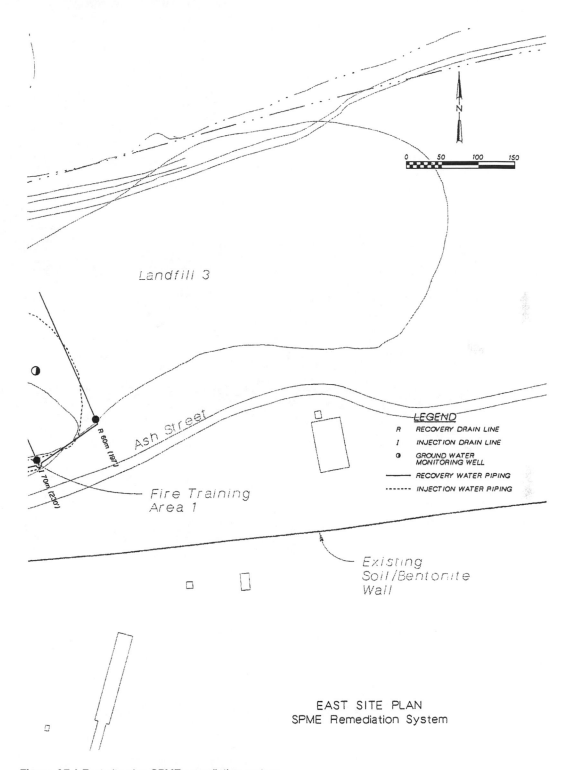

Figure A7.1 East site plan SPME remediation system.

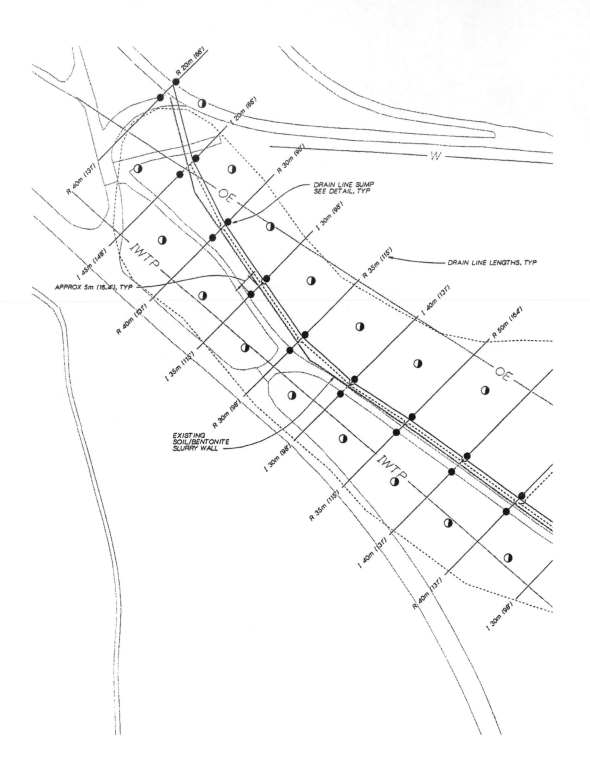

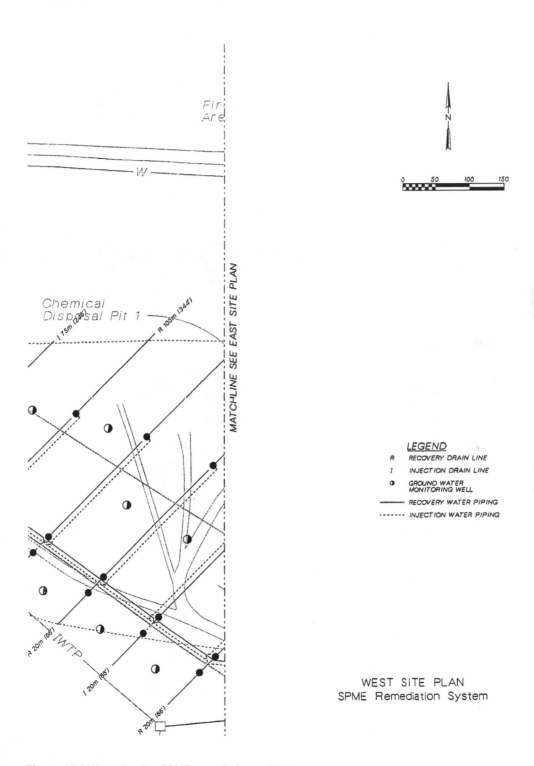

Figure A7.2 West site plan SPME remediation system.

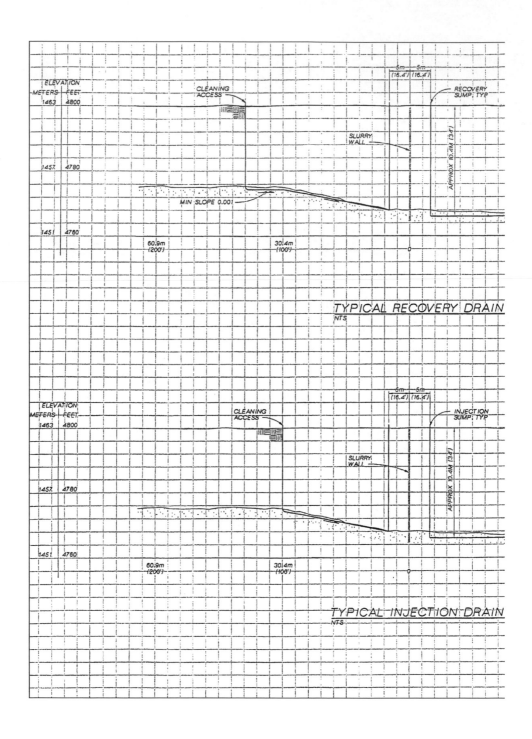

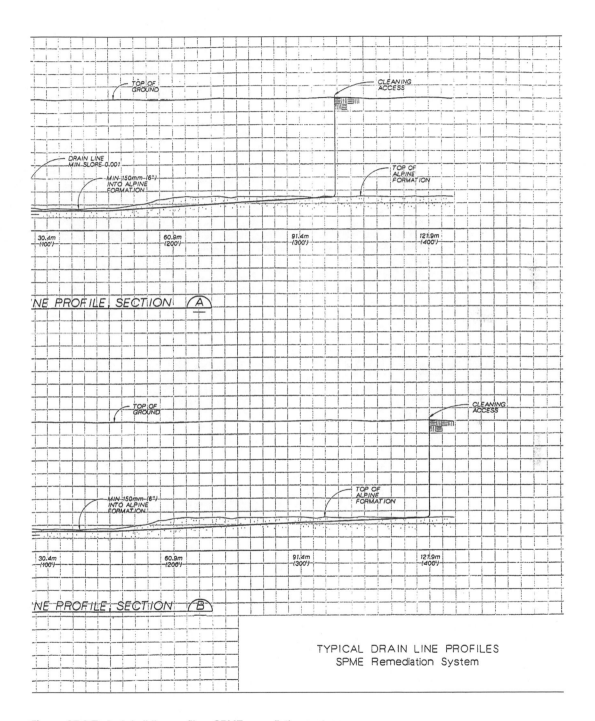

Figure A7.3 Typical drail line profiles, SPME remediation system.

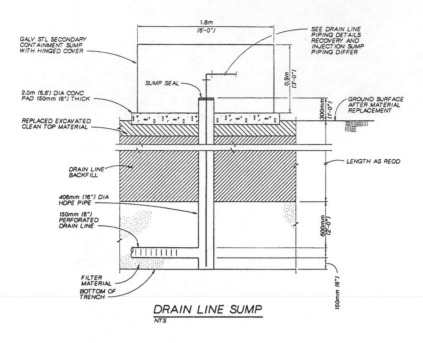

DRAIN LINE SUMP
NTS

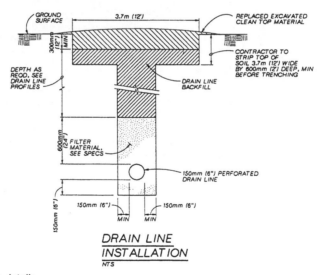

DRAIN LINE
INSTALLATION
NTS

Figure A7.4a Drain line details.

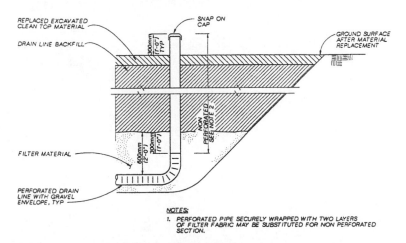

NOTES:
1. PERFORATED PIPE SECURELY WRAPPED WITH TWO LAYERS OF FILTER FABRIC MAY BE SUBSTITUTED FOR NON PERFORATED SECTION.

DRAIN LINE CLEANING ACCESS
NTS

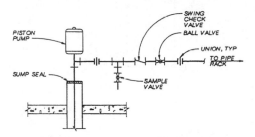

RECOVERY DRAIN LINE PIPING DETAIL
NTS

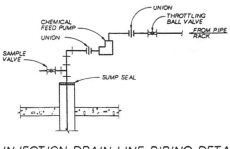

INJECTION DRAIN LINE PIPING DETAIL
NTS

DRAIN LINE DETAILS
SPME Remediation System

Figure A7.4b Drain line details.

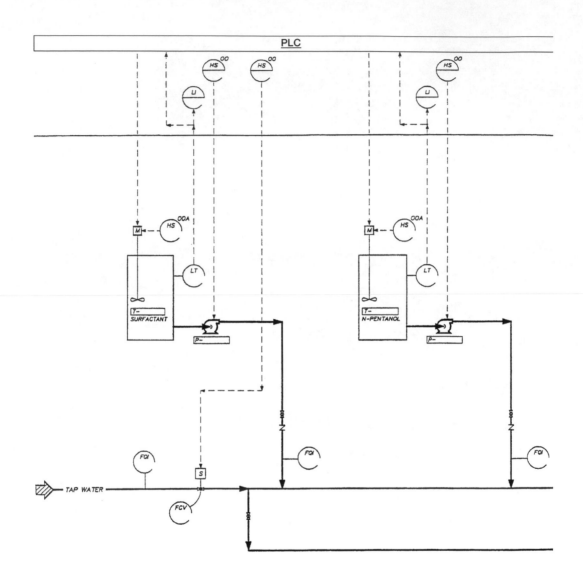

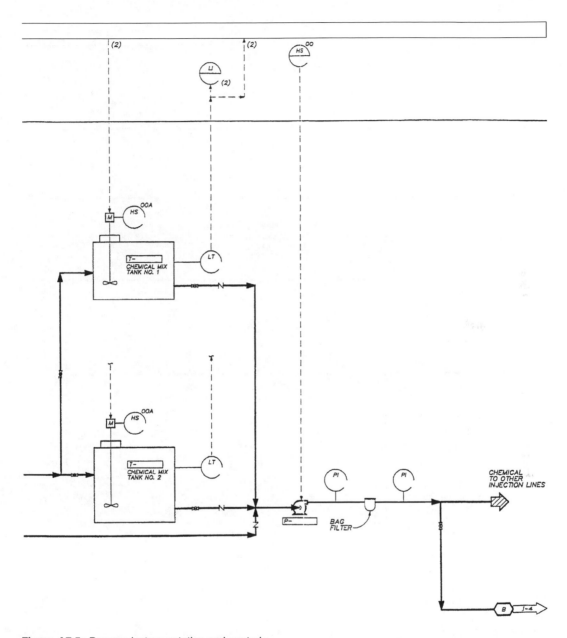

Figure A7.5a Process instrumentation and controls.

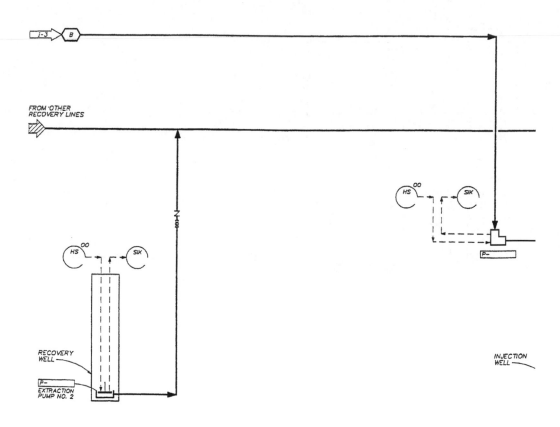

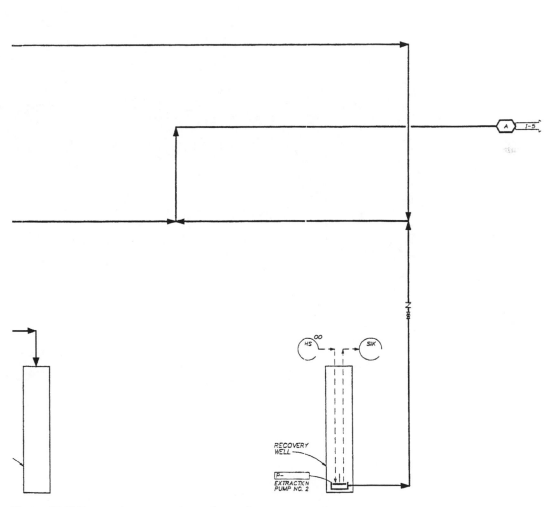

Figure A7.5b Process instrumentation and controls.

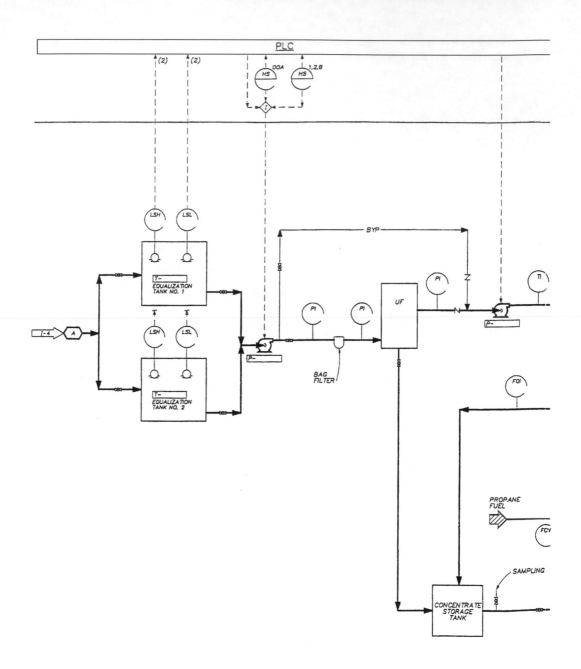

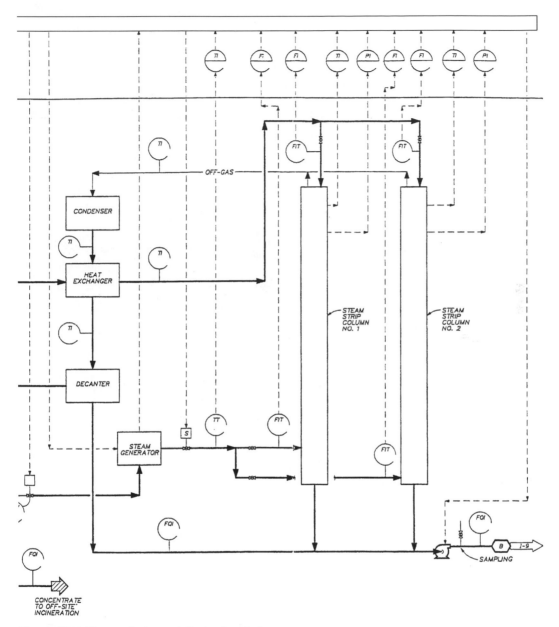

Figure A7.5c Process instrumentation and controls.

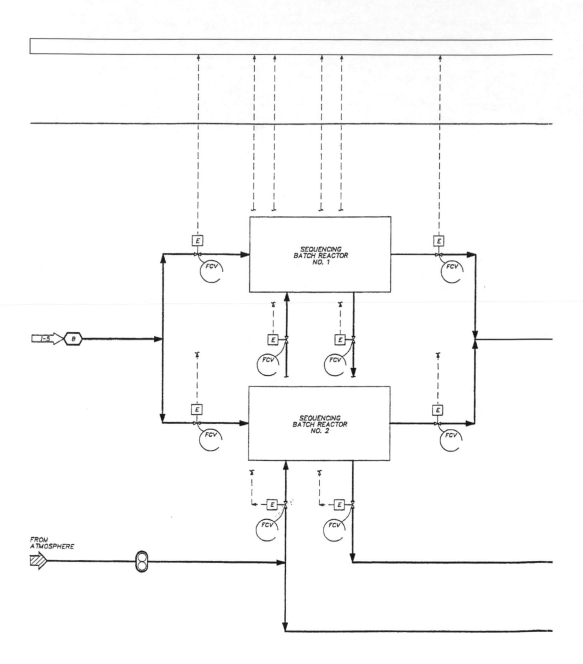

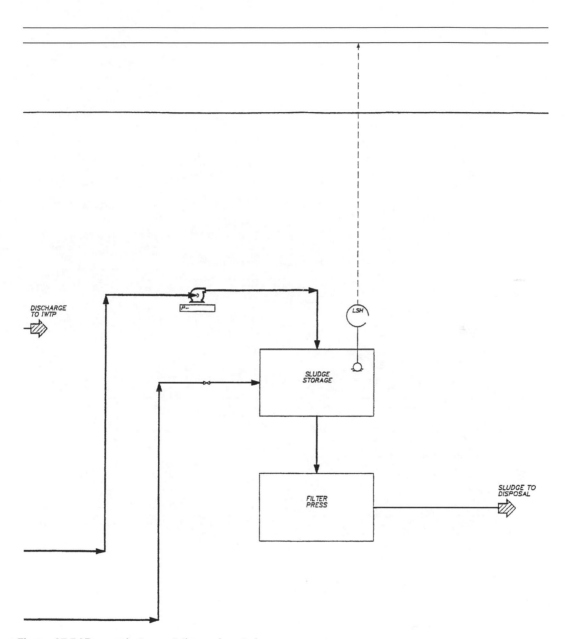

Figure A7.5d Process instrumentation and controls.

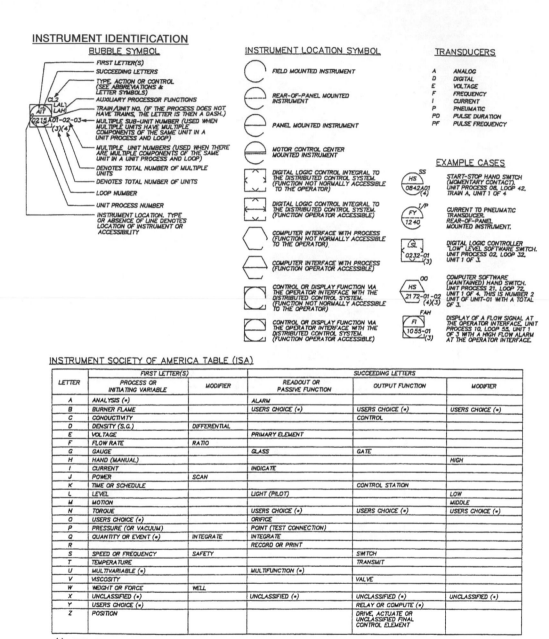

Figure A7.6a Legend sheet 1.

FLOW STREAM IDENTIFICATION

BYP	BYPASS
OF	OVERFLOW
REC	RECYCLE

ABBREVIATION & LETTER SYMBOLS

AC	ALTERNATING CURRENT
ACK	ACKNOWLEDGE
VFD	ADJUSTABLE FREQUENCY DRIVE
AM	AUTO-MANUAL
AS	ADJUSTABLE SPEED
CAM	COMPUTER-AUTO-MANUAL
CCS	CENTRAL CONTROL SYSTEM
CIF	CHEMICAL INTERFACE PANEL
CL2	CHLORINE (TYPICAL: USE STANDARD CHEMICAL ELEMENT ABBREVIATION)
CL RESD	CHLORINE RESIDUAL
COND	CONDUCTIVITY
CP-WWXX	COMPUTER-MANUAL CONTROL PANEL (CONTROL PANELS ASSOCIATED WITH A PACKAGED SYSTEM) WW = UNIT PROCESS NUMBER XX = LOOP NUMBER
CS	CONSTANT SPEED
DC	DIRECT CURRENT
DO	DISSOLVED OXYGEN
E STOP	EMERGENCY STOP
FCC-XX-AA	FILTER CONTROL CONSOLE (SPECIFIED UNDER PROCESS INSTRUMENTATION AND CONTROLS) WW = UNIT PROCESS NUMBER AA = PANEL UNIT NUMBER
FP-XX-AA	FIELD PANEL (SPECIFIED UNER PROCESS INSTRUMENTATION AND CONTROLS) WW = UNIT PROCESS NUMBER AA = PANEL UNIT NUMBER
F/R	FORWARD/REVERSE
HOA	HAND-OFF-AUTO
HOR	HAND-OFF-REMOTE
KK	TIMER CONTROL STATION
LCP-XX-AA	LOCAL CONTROL PANEL (SPECIFIED UNDER PROCESSS INSTRUMENTATION AND CONTROLS) WW = UNIT PROCESS NUMBER AA = PANEL UNIT NUMBER
L/L/S/A	LEAD/LAG/STANDBY/AUTO
LR	LOCAL-REMOTE
M	MANUAL
MA	MANUAL-AUTO
MCC-X	MOTOR CONTROL CENTER NUMBER - X
MCP	MAIN CONTROL PANEL (SPECIFIED UNDER PROCESS INSTRUMENTATION AND CONTROLS)
MMI	MAN-MACHINE INTERFACE
NIC	NOT IN CONTRACT
OC	OPEN-CLOSE
OCA	OPEN-CLOSE-AUTO
OO	ON-OFF (MAINTAINED CONTROL)
OOA	ON-OFF-AUTO
OOC	ON-OFF-COMPUTER
OOR	ON-OFF-REMOTE
OSC	OPEN-STOP-CLOSE
PERM	PERMISSIVE
PLC	PROGRAMMABLE LOGIC CONTROLLER
PMP	PLANT MONITORING PANEL (SPECIFIED PROCESS INSTRUMENTATION AND CONTROLS)
POT	POTENTIOMETER
R/T/A	RECYCLE TRANSFER AUTOMATIC
SCADA	SUPERVISORY CONTROL AND DATA ACQUISITION
SCU	SPEED CONTROL UNIT
SS	START-STOP (MOMENTARY CONTROL)
TURB	TURBIDITY
VHC	VOLATILE HYCROCARBON
1/2/B	1/2/BOTH
<	LESS THAN
>	GREATER THAN
X	MULTIPLIER
1/2	ONE TWO SELECTOR
+	SUMMATION
SqR	SQUARE ROOT

GENERAL NOTES

1. COMPONENTS AND PANELS SHOWN WITH A DIAMOND (◆) ARE TO BE PROVIDED UNDER SECTION "PROCESS INSTRUMENTATION AND CONTROLS."

2. COMPONENTS AND PANELS SHOWN WITH A DOUBLE ASTERISK (✱✱) ARE TO BE PROVIDED AS PART OF A PACKAGED OR MECHANICAL SYSTEM.

3. COMPONENTS AND PANELS SHOWN WITH A TRIANGLE (▲) ARE TO BE PROVIDED BY OTHERS (NIC).

4. THIS IS A STANDARD LEGEND SHEET. THEREFORE, NOT ALL OF THE INFORMATION SHOWN MAY BE USED ON THIS PROJECT.

Figure A7.6a (continued) Legend sheet 1.

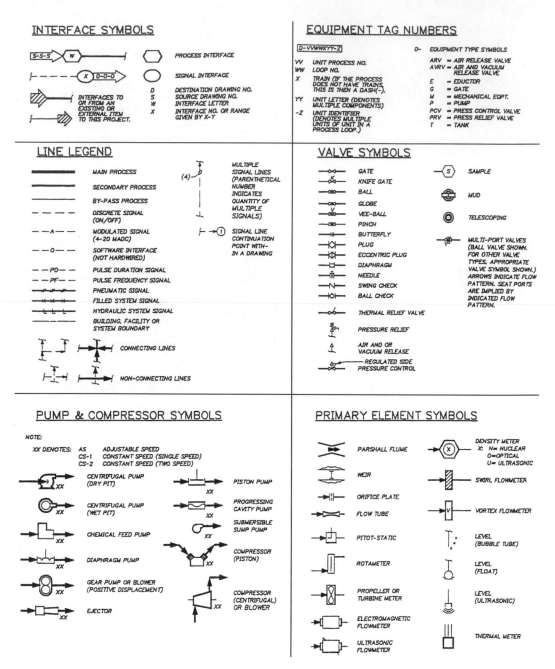

Figure A7.6a (continued) Legend sheet 1.

MISCELLANEOUS SYMBOLS

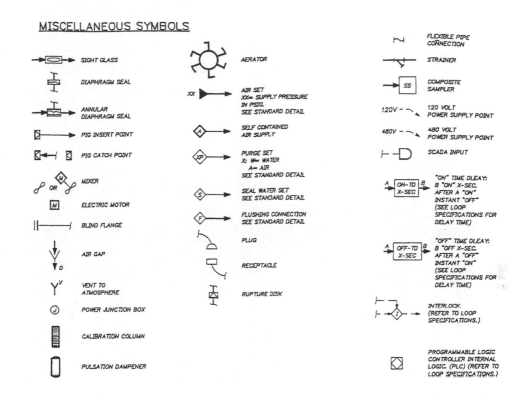

SIGHT GLASS

DIAPHRAGM SEAL

ANNULAR DIAPHRAGM SEAL

PIG INSERT POINT

PIG CATCH POINT

MIXER

ELECTRIC MOTOR

BLIND FLANGE

AIR GAP

VENT TO ATMOSPHERE

POWER JUNCTION BOX

CALIBRATION COLUMN

PULSATION DAMPENER

AERATOR

AIR SET
XX= SUPPLY PRESSURE IN PSIG.
SEE STANDARD DETAIL

SELF CONTAINED AIR SUPPLY

PURGE SET
X: W= WATER
A= AIR
SEE STANDARD DETAIL

SEAL WATER SET
SEE STANDARD DETAIL

FLUSHING CONNECTION
SEE STANDARD DETAIL

PLUG

RECEPTACLE

RUPTURE DISK

FLEXIBLE PIPE CONNECTION

STRAINER

COMPOSITE SAMPLER

120 VOLT POWER SUPPLY POINT

480 VOLT POWER SUPPLY POINT

SCADA INPUT

"ON" TIME DLEAY:
B "ON" X-SEC.
AFTER A "ON"
INSTANT "OFF"
(SEE LOOP
SPECIFICATIONS FOR
DELAY TIME)

"OFF" TIME DLEAY:
B "OFF X-SEC.
AFTER A "OFF"
INSTANT "ON"
(SEE LOOP
SPECIFICATIONS FOR
DELAY TIME)

INTERLOCK.
(REFER TO LOOP SPECIFICATIONS.)

PROGRAMMABLE LOGIC CONTROLLER INTERNAL LOGIC. (PLC) (REFER TO LOOP SPECIFICATIONS.)

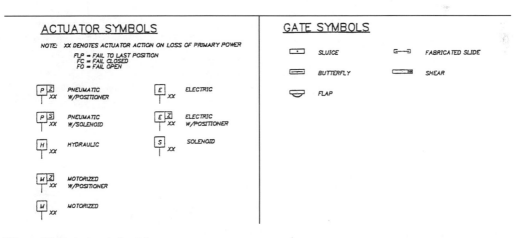

ACTUATOR SYMBOLS

NOTE: XX DENOTES ACTUATOR ACTION ON LOSS OF PRIMARY POWER
FLP = FAIL TO LAST POSITION
FC = FAIL CLOSED
FO = FAIL OPEN

PNEUMATIC W/POSITIONER

PNEUMATIC W/SOLENOID

HYDRAULIC

MOTORIZED W/POSITIONER

MOTORIZED

ELECTRIC

ELECTRIC W/POSITIONER

SOLENOID

GATE SYMBOLS

SLUICE

BUTTERFLY

FLAP

FABRICATED SLIDE

SHEAR

Figure A7.6b Legend sheet 2.

APPENDIX **8**

SPME Scale-up Mass Balance and Design Spreadsheets

BASE CASE DESIGN WORKSHEET

Technology Evaluation Report for Single Phase Micro-Emulsion

Design Worksheet
Base Case

General Site Assumptions:

Total Area =	11.87 ac =	4.80 hectare
Thickness of target LNAPL zone =	1.5 m =	4.92 ft
Spacing Between Drainlines=	31 m =	101.7 ft
Estimated Length of Injection Line	800 m =	2,624.0 ft
Estimated Length of Recovery Line	925 m =	3,034.0 ft

Assume perform surfactant flushing remediation in   3 phases

Phase I of remediation will inject surfactant through	260 m of injection drain line	852.8 ft of injection line
Phase II of remediation will inject surfactant through	270 m of injection drain line	885.6 ft of injection line
Phase III of remediation will inject surfactant through	270 m of injection drain line	885.6 ft of injection line

Assume the maximum flow rates for design conditions will occur during   Phase II and III

porosity =	0.16	
K, Soil Permeability (water) =	0.016 cm/s	for the granular alluvium (Provo Formation)
Viscosity (water) =	0.010 poise	
Viscosity (surfactant/alcohol) =	0.020 poise	
Hydraulic Head, Δh =	1 m	

Is Recycling of Chemicals to be used?   N   (Input "Y" for yes, and "N" or any other character for no.)

Recycling Rate all Chemicals: Volume of Chemical needed = Initial pore volume +   20% loss for each subsequent pore volume

The chemicals to be recycled at the above rate are: none

No. of Pore Volumes Of Flushing Required For Remediation

Initial Waterflood	=	2
Surfactant/Alcohol Flood	=	10
Post Waterflood	=	6.5

Assumptions Used to Calculate Present Worth

Number of Years of	3.00	(This is calculated. See Time and Volume Calcs For the Operations)
Interest Rate =	5.00%	

Volume of LNAPL Calculation:
Assume the LNAPL occurs in two concentrations
  One half of the site contains   8% of the pore space as LNAPL
                                                                         1.3% of the total volume as LNAPL
  and the other half contains   0.64% of the total volume as LNAPL
Volume of Soil with LNAPL =   94,219   cy =   72,040   cu m
Thus   691.6 cu m =   182,717   gal =   691,583   liters
Specific Gravity of LNAPL is   0.85

Pore Volume Calculation (in two directions from the injection drainline on a per meter basis):
PV_{module} = 2 * (thickness+Δh/2) * length * spacing * porosity =   19.84 cu m per meter of drain line

Flow Rate Calculation Q=KAI (in two directions from the injection drainline on a per meter basis)
Q_{water} = 2 * K * (thickness+Δh/2) * length * Δh / spacing =
$Q_{surfactant/alcohol}$ = 2 * K * (viscosity$_{water}$/viscosity$_{surfactant/alcohol}$) * (thickness+Δh/2) * length * Δh / spacing =

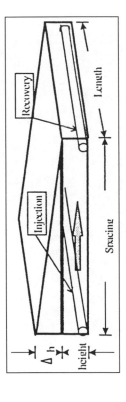

Water -   0.0000206 cu m/s =   1.78 cu m/d =   0.33 gpm per meter of drain line
Surfactant/Alcohol -   0.0000103 cu m/s =   0.89 cu m/d =   0.16 gpm per meter of drain line

Total System Max Flow Rate Calc for   270 m maximum operation of injection drain line
Water -   0.00557 cu m/s =   481.6 cu m/d =   88.4 gpm
Surfactant/Alcohol -   0.00279 cu m/s =   240.8 cu m/d =   44.2 gpm

Time & Volume Calculations for Initial Waterflood, Surfactant Flood and Post Waterflood Operations
Time per PV = PV/Q =

	Water	Surfactant/Alcohol
	11.1 days	22.2 days

Sequence of Operations

Initial Waterflood	2 PV: =	40 cu m for	=	10,483 gal	22 days
Surfactant Flood	10 PV: =	198 cu m for	=	52,417 gal	222 days
Post Waterflood	6.5 PV: =	129 cu m for	=	34,071 gal	72 days

total time = 317 days = 0.87 years

Assume total time to treat one (1) phase of the site is   0.96 years. (This allows 10% additional time to move the process equipment.)
Operate for   3 phases, so total time req'd to treat the site is: 2.87 years (used to calculate the quantities of chemicals req'd)

3 years rounded up for Present Value Calculations

Surfactant/Alcohol Solution Handling and Storage - General
Assume:

Make up surfactant solutions every   1 days during surfactant/alcohol flood
Use   2 surfactant/alcohol solution tanks, (one can be in use, one can be mixing)

Size of Surfactant/Alcohol Storage Tank req'd   240.8 cu m =   63,621 gal Total

Use   3 tanks, @   20,000 gal ea =   75.7 cu m

Brij 97 - Handling and Storage

Assume:

Injected surfactant concentration is	3% wt, active Brij 97
Brij 97 is supplied as a concentrate at	100% wt active
Specific gravity of supplied Brij 97 is	1.1
Specific gravity of surfactant solution is 1.	
Provide minimum	7 days of Brij 97 storage at the site

Total quantity of Brij 97 needed:

4,761,600 kg active surfactant	
4,761,600 kg total =	4,762 mtons
10,475,520 lb =	5,238 tons
4,329 cu m =	1,143,650 gal

The maximum usage rate =   6.57 cu m/d =   1,735 gal/d

Brij 97 - Req'd. Storage Volume: 46.0 cu m =   12,146 gal for

Use   1 tanks, @   10,000 gal ea =   37.85 cu m ea

The actual specified storage volume represents the usage requirements for   5.0 days of operations

7.0 days of Brij 97 storage at the site

n-Pentanol - Handling and Storage

Assume

Injected n-pentanol concentration is	2.5% wt active
n-Pentanol is supplied as concentrate at	100% wt
Specific gravity of supplied n-Pentanol is	0.82
Provide	7 days of n-pentanol storage at the site

Total quantity of n-pentanol needed:

3,968,000 kg pure n-pentanol	
3,968,000 kg total =	3,968 mtons
8,729,600 lb total =	4,365 tons
3,254 cu m =	859,643 gal

The maximum usage rate =   7.34 cu m/d =   1,940 gal/d

n-Pentanol - Req'd. Storage Volume: 51.4 cu m =   13,578 gal for

Use   1 tanks, @   10,000 gal ea =   37.85 cu m ea

The actual specified storage volume represents the usage requirements for   5.2 days of operation

7.0 days of n-pentanol storage at the site

Produced Fluids Handling System

Provide fluids handling system for   Phase II and III   of remediation which is the maximum design flow condition.

Assumptions:

Recovery at a rate   1.1 times the injection rate so that the system is contained.

Water produced during the initial waterflood will be sent directly to the industrial wastewater treatment plant (IWTP) and does not need to pass through the water treatment plant.

Water produced during the surfactant flood and post waterflood will require pretreatment before being sent to the IWTP.

Pretreatment will consist of passing through an equalization tank with a holding time of   1.5 days based on the surfactant flow rate.

The equalization tank will be followed by an ultrafiltration, steam stripping, and biological treatment prior to discharge to the IWTP.

Assume   0% of treated water will be recycled

The following recoveries will be achieved by the ultrafiltration

     90% by weight of the total LNAPL

     90% by weight of the surfactant will be recovered

     30% by weight of the n-pentanol will be recovered

Maimum Injection Flow rate:

During Initial Waterflood -	481.6	cu m/d	=	88	gpm for	22 days
During Surfactant Flood	240.8	cu m/d	=	44	gpm for	222 days
During Post Waterflood	481.6	cu m/d	=	88	gpm for	72 days

Maximum Recovery Flow Rates and Volumes

During Initial Waterflood -	529.8	cu m/d	or	11,765	cu m	97.2 gpm	=	3,113,500 gal
During Surfactant Flood	264.9	cu m/d	or	58,925	cu m	48.6 gpm	=	15,567,932 gal
During Post Waterflood	529.8	cu m/d	or	38,301	cu m	97.2 gpm	=	10,119,156 gal

Equalization tank volume =   **397** cu m =

Use   4 tanks, @   20,000 gal ea =   75.7 cu m

The actual specified storage volume represents the useage requirements for   1.1 days

104,974 gal based on surfactant flow rate.

Storage Tank For Concentrated Fluids Requiring Off-Site Incineration From the Ultrafiltration Unit

Assume

   Haul concentrate in trucks w/ a capacity of: 5,000 gal =   18.9 cu m

   Number of truckloads req'd to be stored   4

Volume of Ultrafiltration Concentrate Storage Tank   20,000 gal =   75.7 cu m

The actual specified concentrate storage volume represents approximately   1.9 days of storage

Storage Tank For Fluids Requiring Off-Site Incineration From the Decanter

Assume

   Haul concentrate in trucks w/ a capacity of: 5,000 gal =   18.9 cu m

   Number of truckloads req'd to be stored   4

Volume of Decant Storage Tank   20,000 gal =   75.7 cu m

The actual specified concentrate storage volume represents approximately   21.0 days of storage

<u>DESIGN OF BIOLOGICAL FACILITIES: SBRs</u>
Based on treating steam stripper effluent

ASSUMPTIONS

<u>INFLUENT CHARACTERISTICS</u>   (sized based on characteristics during surfactant flood)

MAX. FLOW TO SBR SYSTEM =	0.073 mgd	51	GPM =	73,488 GPD
AVG COD OF INFLUENT =		12,588 MG/L =	7,715	lb COD/d
BOD/COD =		0.5		
AVG BOD OF INFLUENT =		6,294 MG/L =	3857	LB/D
Influent TSS =		50 MG/L =	31	LB/D
Avg. TKN OF INFLUENT =		1 MG/L =	0	LB/D

<u>EFFLUENT CHARACTERISTICS</u>

ASSUMED % COD REMOVAL ACHIEVABLE	90%	
COD =	1,259	MG/L
TSS =	20	MG/L
Nitrogen =	1 MG/L	1 LB/D

<u>OPERATING CHARACTERISTICS</u>

TSS YIELD =	0.25 G TSS/G COD	
OPERATING MIXED LIQUOR (MLSS) =	3,500 MG/L	
Desired SRT =	8 DAYS	ASSUMED based on previous experience

<u>VARIOUS POSSIBLE OPERATING CONDITIONS</u>   WITH VOLUME =   300,000   GAL

CONDITION	No. OF TANKS	No. OF CYCLES	Q (GPM)	COD (MG/L)	NET Y	MLSS (MG/L)	SRT * (DAYS)	HRT (DAYS)	% DECANT REQUIRED
1	1	1	51	12,588	0.25	3500	5	4.1	24
2	2	2	51	12,588	0.25	3500	8	8.2	6

* SRT REDUCED TO TAKE INTO COUNT NON-REACTIVE PERIODS (2 HOURS PER CYCLE).
Select 2   300,000 gal tanks

<u>AERATION REQUIREMENTS</u>

LB COD REMOVE PER DAY =	6,943 LB COD/D	
ASSUME /	0.7	LB O2/LB COD REMOVED
OXYGEN REQUIRED FOR COD REMOVAL =	4,860 LB O2/D =	2202 KG/D

TKN NITRIFIED = INFLUENT - NUTRIENT REQUIREMENT =
    OXYGEN DEMAND OF   4.63 O2/LB TKN
    THUS, NEED   (1) LB O2/D FOR NITRIFICATION

O2 FOR CARBONACEOUS AND N =	4,859	LB/D
ASSUME 75% IS IS CONSUMED IN		6 HOURS
THUS, NEED TO SUPPLY A DEMAND OF	14,577	LB O2/D = 6603 KG/D

THUS, TOTAL O2 DEMAND =	14,577 LB O2/D =	6603 KG/D

ASSUME AN AOTE OF   10%

SCFM REQUIRED =	7,335	SCFM
=	207.74	CU M/MIN
APPROX. hp =	291	hp
	217.18	KW

FOR MIXING:
ASSUME NEED   40 HP/MG   WITH A VOLUME OF =   300,000   GAL

THUS,   12 HP =   8.95   KW

SLUDGE PRODUCTION

ASSUME YIELD =   0.25
ASSUME WASTE SLUDGE IS   10,000 MG/L TSS

COD REMOVED =   6,943 LB/D

SLUDGE PRODUCTION   1,736 LB/D =   317   T/YR
           =   786   KG/D   287,013   kg/yr
VOLUME =   20,813 GAL/D =   634,811   GAL/MO =   7,617,728   GAL/YR
           =   79   cu m/d   28795   cu m/yr

liquid storage
provide   7 days
           145,694 gal   use   40,000 gal
concentrate to   150,000 mg/l   in filter press
volume =   1388 gal/d
sp of sludge =   1.2 =   13,887 lb/d
total number of days =   667 days thus, total of   4,634 tons
                                   =   4,585 cy

Alternative GAC cost without SBRs
wgt pickup of carbon   10%
thus, need   69,434 lb/d
                       46,337,365 lb total
at $ 1.50 $ 104,151 per day
                       $ 69,506,048 total

BASE CASE MASS BALANCE

Technology Evaluation Report for Single Phase Micro-Emulsion

Flow Diagram and Mass Balance
Base Case

Chemical Feed System
Parameters During Surfactant/Alcohol Flood

Assumptions:

Parameters are specified for operation of	270 m of injection drain line, which corresponds to	Phase II and III	of the remediation alone.

The required surfactant/alcohol injection rate is   241 cu m/d for the above phase(s) of operation
The injected solution will be   3.0% by weight surfactant
The injected solution will be   2.5% by weight n-pentanol
The specific gravity of the supplied surfactant is   1.1
The specific gravity of the supplied n-pentanol is   0.82

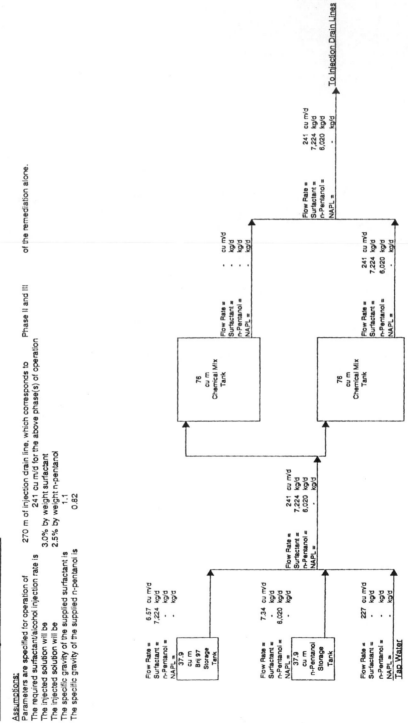

Chemical Recovery and Treatment System
Parameters During Surfactant/Alcohol Flood

Assumptions:

Ratio of COD/surfactant or pentanol or LNAPL =	1.8 with in the typical range of ethers	
Concentrations are constant for	222 days during surfactant injection.	
Parameters are specified for operation of	270 m of injection drain line, which corresponds to	Phase II and III of the remediation.
The recovery rate is	1.1 times the injection rate	
Surfactant recovery from the subsurface is	90% by weight of mass injected	
Total LNAPL recovery from the subsurface is	85% of total volume present in the subsurface	
n-pentanol is conserved in the subsurface.		
n-pentanol recovery from the ultrafilter is	30% by weight in the micelles	
The ultrafiltration plant recovers	90% by weight of the total recovered LNAPL in micelles	
The ultrafiltration plant recovers	90% by weight of the surfactant in micelles	
The ultrafiltration plant concentrate waste steam is	15.0% by volume of the total influent	
The steam stripper utilizes a steam to feed ratio of	25% by volume	
The quantity of steam that condenses into the liquid phase in the stripper is	20% by volume	
The steam stripper recovers	99.9% by weight of the NAPL from the ultrafilter effluent	
The steam stripper recovers	99% by weight of the n-pentanol from the ultrafilter effluent	
The organic phase decant flow rate coming off the decanter is	8% of the influent flow rate	
Assume the bottoms discharge of the decanter is saturated with n-pentanol and NAPL.		
The solubility limit of n-pentanol is approximately	27,000 mg/l in water	
The solubility limit of NAPL is approximately	1,000 mg/l in water	
The recycle ratio is	0.00 (flow rate of recycle over flow rate of stream to IWTP)	
Assume an average carbon pickup of	10% by weight (lb NAPL/lb carbon)	

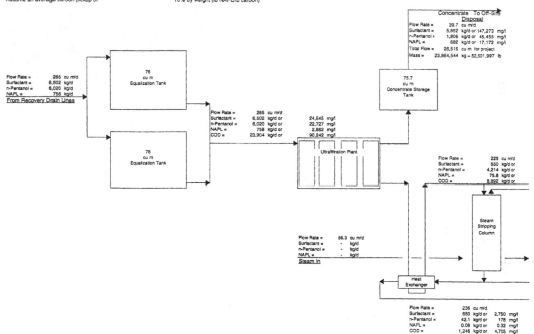

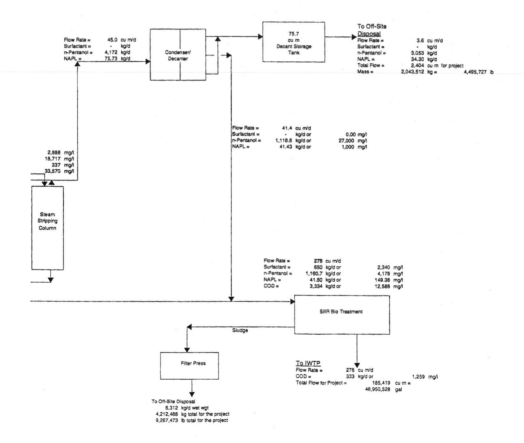

OPTIMIZED CASE DESIGN WORKSHEET

Technology Evaluation Report for Single Phase Micro-Emulsion

Design Worksheet
Optimized Case

General Site Assumptions:

Total Area =	11.87 ac =	4.80 hectare
Thickness of target LNAPL zone =	1.5 m =	4.92 ft
Spacing Between Drainlines=	31 m =	101.7 ft
Estimated Length of Injection Line	800 m =	2,624.0 ft
Estimated Length of Recovery Line	925 m =	3,034.0 ft

Assume perform surfactant flushing remediation in   3 phases

Phase I of remediation will inject surfactant through   200 m of injection drain line   852.0 ft of injection line

Phase II of remediation will inject surfactant through   270 m of injection drain line   885.0 ft of injection line

Phase III of remediation will inject surfactant through   270 m of injection drain line   885.0 ft of injection line

Assume the maximum flow rates for design conditions will occur during   Phase II and III

porosity =	0.10	
K, Soil Permeability (water) =	0.016 cm/s	for the granular alluvium (Provo Formation)
Viscosity (water) =	0.010 poise	
Viscosity (surfactant/alcohol) =	0.020 poise	
Hydraulic Head, Δh =	1 m	

Is Recycling of Chemicals to be used?   N   (Input "Y" for yes, and "N" or any other character for no.)

Recycling Rate all Chemicals: Volume of Chemical needed = Initial pore volume +   20% loss for each subsequent pore volume

The chemicals to be recycled at the above rate are: none

No. of Pore Volumes Of Flushing Required For Remediation

Initial Waterflood =	2
Surfactant/Alcohol Flood =	4
Post Waterflood =	3

Assumptions Used to Calculate Present Worth

Number of Years of	2.00	(This is calculated. See Time and Volume Calcs For the Operations)
Interest Rate =	5.00%	

Volume of LNAPL Calculation:

Assume the LNAPL occurs in two concentrations

One half of the site contains   8% of the pore space as LNAPL
                                        1.3% of the total volume as LNAPL
and the other half contains   0.64% of the total volume as LNAPL

Volume of Soil with LNAPL =   691.6 cu m =   94,219   cy =   72,040   cu m
Thus   182,717   gal =   691,583   liters

Specific Gravity of LNAPL is   0.85

Pore Volume Calculation (in two directions from the injection drainline on a per meter basis):

PV_{module} = 2 * (thickness+Δh/2) * length * spacing * porosity =   19.84 cu m per meter of drain line

Flow Rate Calculation Q=KAi (in two directions from the injection drainline on a per meter basis)

Q_{water} = 2 * K * (thickness+Δh/2) * length * Δh / spacing =

$Q_{surfactant/alcohol}$ = 2 * K * ($viscosity_{water}$/$viscosity_{surfactant/alcohol}$) * (thickness+$\Delta$h/2) * length * Δh / spacing =

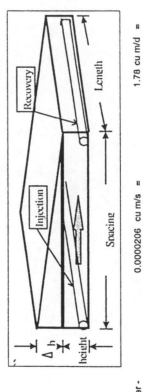

Water -   0.0000206 cu m/s =   1.78 cu m/d =   0.33 gpm per meter of drain line
Surfactant/Alcohol -   0.0000103 cu m/s =   0.89 cu m/d =   0.16 gpm per meter of drain line

Total System Max Flow Rate Calc for   270 m maximum operation of injection drain line

Water -   0.00557 cu m/s =   481.6 cu m/d =   88.4 gpm
Surfactant/Alcohol -   0.00279 cu m/s =   240.8 cu m/d =   44.2 gpm

Time & Volume Calculations for Initial Waterflood, Surfactant Flood and Post Waterflood Operations

	Water	Surfactant/Alcohol
Time per PV = PV/Q =	11.1 days	22.2 days

Sequence of Operations

Initial Waterflood	2 PV: =	40 cu m for	22 days	
		= 10,483 gal		
Surfactant Flood	4 PV: =	79 cu m for	89 days	
		= 20,967 gal		
Post Waterflood	3 PV: =	60 cu m for	33 days	
		= 15,725 gal		
		total time =	145 days =	0.40 years

Assume total time to treat one (1) phase of the site is　0.44 years.　(This allows 10% additional time to move the process equipment.)

Operate for　3 phases, so total time req'd to treat the site is:　1.31 years (used to calculate the quantities of chemicals req'd)

2 years rounded up for Present Value Calculations

Surfactant/Alcohol Solution Handling and Storage - General

Assume:

Make up surfactant solutions every　1 days during surfactant/alcohol flood

Use　2 surfactant/alcohol solution tanks, (one can be in use, one can be mixing)

Size of Surfactant/Alcohol Storage Tank req'd　240.8 cu m =　63,621 gal Total

Use　3 tanks, @　20,000 gal ea =　75.7 cu m

Brij 97 - Handling and Storage

Assume:

Injected surfactant concentration is	3% wt, active Brij 97	
Brij 97 is supplied as a concentrate at	100% wt active	
Specific gravity of supplied Brij 97 is	1.1	
Specific gravity of surfactant solution is 1.		
Provide minimum	7 days of Brij 97 storage at the site	

Total quantity of Brij 97 needed:

1,904,640	kg active surfactant	
1,904,640	kg total =	1,905 mtons
4,190,208	lb =	2,095 tons
1,731	cu m =	457,460 gal

The maximum usage rate = 6.57 cu m/d = 1,735 gal/d

Brij 97 - Req'd. Storage Volume: 46.0 cu m = 12,146 gal for

Use 1 tanks, @ 10,000 gal ea = 37.85 cu m ea

The actual specified storage volume represents the usage requirements for 5.8 days of operations

7.0 days of Brij 97 storage at the site

n-Pentanol - Handling and Storage

Assume:

Injected n-pentanol concentration is	2.5% wt active	
n-Pentanol is supplied as concentrate at	100% wt	
Specific gravity of supplied n-Pentanol is	0.82	
Provide	7 days of n-pentanol storage at the site	

Total quantity of n-pentanol needed:

1,587,200	kg pure n-pentanol	
1,587,200	kg total =	1,587 mtons
3,491,840	lb total =	1,746 tons
1,302	cu m =	343,857 gal

The maximum usage rate = 7.34 cu m/d = 1,940 gal/d

n-Pentanol - Req'd. Storage Volume: 51.4 cu m = 13,578 gal for

Use 1 tanks, @ 10,000 gal ea = 37.85 cu m ea

The actual specified storage volume represents the usage requirements for 5.2 days of operation

7.0 days of n-pentanol storage at the site

Produced Fluids Handling System

Provide fluids handling system for   Phase II and III   of remediation which is the maximum design flow condition.

Assumptions:

Recovery at a rate   1.1 times the injection rate so that the system is contained.

Water produced during the initial waterflood will be sent directly to the industrial wastewater treatment plant (IWTP) and does not need to pass through the water treatment plant.

Water produced during the surfactant flood and post waterflood will require pretreatment before being sent to the IWTP.

Pretreatment will consist of passing through an equalization tank with a holding time of   1.5 days based on the surfactant flow rate.

The equalization tank will be followed by salt addition, oil/water separation, and steam stripping prior to discharge to the IWTP.

Assume   0% of treated water will be recycled

The following recoveries will be achieved by the ultrafiltration

    99% by weight of the total LNAPL
    99% by weight of the surfactant will be recovered
    30% by weight of the n-pentanol will be recovered

Maximum Injection Flow rate:

During Initial Waterflood -	481.6	cu m/d	=	88 gpm for	22 days
During Surfactant Flood	240.8	cu m/d	=	44 gpm for	89 days
During Post Waterflood	481.6	cu m/d	=	88 gpm for	33 days

Maximum Recovery Flow Rates and Volumes

During Initial Waterflood -	529.8	cu m/d	or	11,785	cu m	97.2 gpm =	3,113,586 gal
During Surfactant Flood	264.9	cu m/d	or	23,570	cu m	48.6 gpm =	6,227,173 gal
During Post Waterflood	529.8	cu m/d	or	17,677	cu m	97.2 gpm	4,670,380 gal

Equalization tank volume = 397 cu m = 104,974 gal based on surfactant flow rate.

Use   4 tanks, @   20,000 gal ea =   75.7 cu m ea

The actual specified storage volume represents the useage requirements for   1.1 days

Salt Addition

Add to   11% by wgt

Flow during surfactant flushing =	264.9	cu m/d
=	264,886	kg/d
Thus, mass of salt needed =	29,137	kg/d
Total for project =	7,778,074 kg =	17,111,762 lb

Storage Tank For Concentrated Fluids Requiring Incineration From the Oil/Water Separator Unit

Assume

Haul concentrate in trucks w/ a capacity of: 5,000 gal   =   18.9 cu m
Number of truckloads req'd to be stored   4

Volume of the Oil/Water Separator Fluids Storage Tank   20,000 gal   =   75.7 cu m

The actual specified concentrate storage volume represents approximately   4.8 days of storage

Storage Tank For Concentrated Fluids Requiring Incineration From the Decanter

Assume

Haul concentrate in trucks w/ a capacity of: 5,000 gal   =   18.9 cu m
Number of truckloads req'd to be stored   4

Volume of the Decant Storage Tank   20,000 gal   =   75.7 cu m

The actual specified concentrate storage volume represents approximately   380 days of storage

OPTIMIZED CASE MASS BALANCE

Technology Evaluation Report for Single Phase Micro-Emulsion

Flow Diagram and Mass Balance
Optimizes Case

Chemical Feed System
Parameters During Surfactant/Alcohol Flood

<u>Assumptions:</u>

Parameters are specified for operation of 270 m of injection drain line, which corresponds to Phase II and III
The required surfactant/alcohol injection rate is 241 cu m/d for the above phase(s) of operation
The injected solution will be 3.0% by weight surfactant
The injected solution will be 2.5% by weight n-pentanol
The specific gravity of the supplied surfactant is 1.1
The specific gravity of the supplied n-pentanol is 0.82

of the remediation alone.

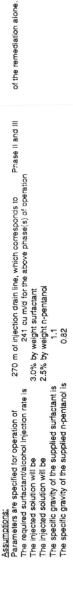
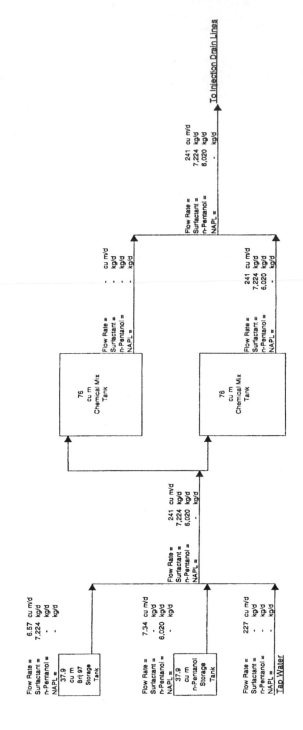

Technology Evaluation Report for Single Phase Micro-Emulsion

Flow Diagram and Mass Balance
Optimized Case

Chemical Recovery and Treatment System
Parameters During Surfactant/Alcohol Flood

Assumptions:

Ratio of COD/surfactant or pentanol or LNAPL =	1.8	with in the typical range of ethers
Concentrations are constant for	89 days during surfactant injection.	
Parameters are specified for operation of	270 m of injection drain line, which corresponds to	Phase II and III of the remediation.
The recovery rate is	1.1 times the injection rate	
Surfactant recovery from the subsurface is	99% by weight of mass injected	
Total LNAPL recovery from the subsurface is	70% of total volume present in the subsurface	
n-pentanol is conserved in the subsurface.		
n-pentanol recovery from the ultrafilter is	75% by weight in the micelles	
The salt addition with o/w sep. recovers	99.3% by weight of the total recovered LNAPL in micelles	
The salt addition with o/w sep. recovers	99.3% by weight of the surfactant in micelles	
The o/w sep. concentrate waste steam is	6.0% by volume of the total influent	
The steam stripper utilizes a steam to feed ratio of	25% by volume	
The quantity of steam that condenses into the liquid phase in the stripper is	20% by volume	
The steam stripper recovers	99.9% by weight of the NAPL from the ultrafilter effluent	
The steam stripper recovers	99% by weight of the n-pentanol from the ultrafilter effluent	
The condenser/decanter recovers	99% be volume of the NAPL, n-pentanol, and surfactant	
The organic phase decant flow rate coming off the decanter is	0.4% of the influent flow rate	
Assume the bottoms discharge of the decanter is saturated with n-pentanol and NAPL.		
The solubility limit of n-pentanol is approximately	27,000 mg/l in water	
Assume all NAPL will be solubilized in the water phase off the condenser/decanter.		
The recycle ratio is	0.00 (flow rate of recycle over flow rate of stream to IWTP)	
Assume an average carbon pickup of	10% by weight (lb NAPL/lb carbon)	

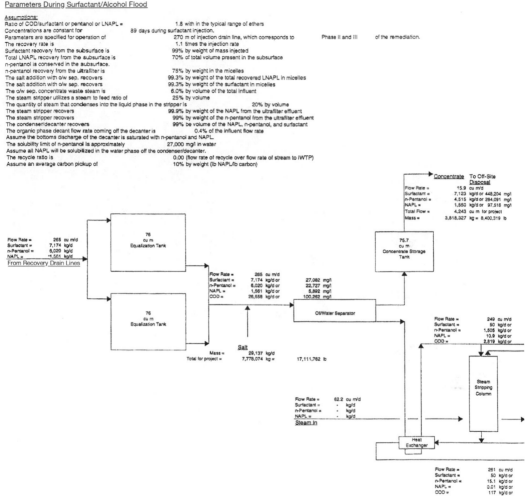

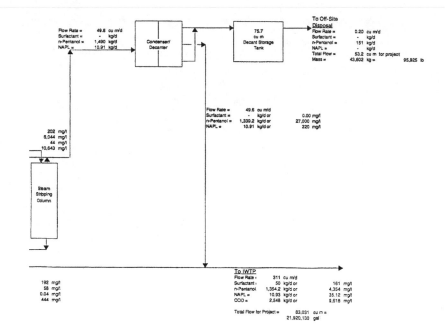

Cost Estimates

Details of the cost estimates are presented in this appendix. Two general methods were used to prepare these cost estimates. Excel spreadsheets were used to prepare estimates for "additional capital costs" and operations and maintenance costs. Spreadsheets were also used to compile all of the costs into summaries. Baseline capital costs were prepared using Hard Dollar Estimating Office System (EOS). EOS, recognized as the most comprehensive, commercially available construction estimating software system available, is designed for use by contractors to develop detailed costs estimates for the remediation and construction industries. EOS is formatted to develop estimates "from the ground up," whether employing a total cost buildup method, or generating a more general unit price estimate.

This estimate was developed by entering subcontractor and equipment vendor quotes, combined with detailed estimated costs for labor to install the system.

The following list is a brief description of the cost estimate sheets that are contained in this appendix. Cost summaries are provided in Chapter 26 of the text. The base case sheets are provided first, followed by the optimized case.

- Additional Capital Costs: Includes additional items not included in the capital costs. Items such as field demonstrations are included here.
- Operations and Maintenance Costs: Operating costs per module.
- Cost Summary: Pulls all the costs together. At the bottom of the summary table are calculations of the cost on a unit basis (dollar per area or volume).
- WBS Outline from EOS: Includes where the markups need to be included.
- Work Plan Summary from EOS: Summarizes the cost for each category. It should be noted that these costs do not inlude the markup. The markup was added on the summary sheet.
- Activity Detail Forms from EOS: Includes the details for each category. Note that there are forms for subcategories in some cases.

BASE CASE ADDITIONAL CAPITAL COST

Adtnl Cptl Cost, Base Case

Technology Evaluation Report for Single Phase Micro-Emulsion

Additional Capital Cost

Base Case

Description	Qty	Unit	Unit Cost ($)	Total Cost ($)	Comments
Additional Site Characterization and Laboratory Testing					
Soil Borings	108	ea	$1,200	$129,600	Assumes 3 borings per drainline (36) that will extend to 33 feet, cost based on vendor quote and RACER
Soil Analysis	108	sample	$500	$54,000	Assumes samples will be analyzed for 3 constituents, costs based on prior experience
Bench-Scale Testing	1	ls	$250,000	$250,000	Based on prior experience. Includes surfactant IFT, phase behavior, solubilization potential, linear and radial coreflood tests, two dimensional sandpacks, testing of a bench-scale UF unit, steam stripper, and SBR, work plans, test labor, and reports.
Subtotal				$433,600	
Numerical Simulations					
Model Simulations	240	hrs	$100	$24,000	Cost includes operator and computer time. It does not include cost of software
Field Demonstrations					
Installation of Sheet Pile Wall	2,153	sf	$22	$47,366	Assumes 5m x 5m x 10m deep sheet pile wall for containment, cost based on prior experience
Installation of Injection Wells	4	ea	$4,600	$18,400	Wells will be 33 feet deep and stainless steel, cost based on vendor quote and RACER
Installation of Extraction Well	3	ea	$4,600	$13,800	Wells will be 33 feet deep and stainless steel, cost based on vendor quote and RACER
Monitoring Piezometer - 30 foot well	4	ea	$4,200	$16,800	Cost based on vendor quote and RACER. Nested at 3 depths.
Soil Borings	6	ea	$1,200	$7,200	Assumes borings will extend to 33 feet, cost based on vendor quote and RACER
Soil Analysis	28	sample	$500	$14,000	Assumes 2 soil samples per well will be analyzed for 3 constituents, costs based on prior experience
Analysis of injected fluid quality	6	sample	$100	$600	Thrice per week analysis of alcohol for approximately 2 weeks of surfactant injection.
Analysis of UF and steam stripper effluent	2	sample	$450	$1,013	Thrice per month analysis of alcohol, TPH, and VOC/SVOC for approximately 3 weeks.
Analysis of treated produced fluids	2	sample	$450	$1,013	Thrice per month analysis of alcohol, TPH, and VOC/SVOC for approximately 3 weeks.
Analysis of monitoring well fluids	33	sample	$450	$14,850	Once per week analysis of alcohol, TPH, and VOC/SVOC for approximately 3 weeks.
Tanks for temporary storage	2	ea	$9,500	$19,000	10,000 gal single-walled steel tanks with containment berm. Cost from RACER
Pilot testing of ultrafiltration	1	ls	$75,000	$75,000	Tests will be conducted on 25% of the fluids. Cost based on prior experience
Pilot testing of steam stripping	1	ls	$75,000	$75,000	Tests will be conducted on 25% of the fluids. Cost based on prior experience
Pilot testing of biological treatment	1	ls	$75,000	$75,000	Tests will be conducted on 25% of the fluids. Cost based on prior experience
Fluid Incineration	167,720	lb	$0.17	$29,512	Assumes that 75% of the produced fluids will be incinerated. Assumes 18.5 pore volumes of water will be generated. No recycling will be conducted. Quote from Laidlaw Environmental (SLC, UT), <3,000 BTU/lb, >50% water, <10% Cl.
Discharge to IWTP	7	1,000 gal	$11.00	$82	Balance of water not incinerated will be sent to the IWTP.
Field Demonstration Labor	1	ls	$41,600	$41,600	Assumes demonstration will require 20 days, 2 technicians and 2 engineers per day. Cost based on prior experience.
Field Demonstration Work Plan and Report	2	ea	$25,000	$50,000	Cost based on prior experience
Subtotal				$499,235	
Facility Design					
Design	10%		$5,831,806	$583,181	Percentage is based on total construction costs
Engineering During Construction	8%		$5,831,806	$466,545	
Post Flushing Soil Coring and Analysis					
Soil Borings	10	ea	$1,200	$12,000	Assumes 1 boring every 20 m of drainline in Phase I (boring will extend to 33 feet), cost based on vendor quote and RACER
Soil Analysis	10	sample	$500	$5,000	Assumes samples will be analyzed for 3 constituents, costs based on prior experience
Subtotal				$17,000	
Present Worth of Post Flushing Soil Coring and Analysis				$16,295	Assumes costs are incurred after Phase 1 of remediation, 5% interest rate
Demolition and Site Restoration After Remediation	2.5%		$5,831,806	$145,795	Percentage is based on total construction costs
Present Worth of Dem & Rest. Costs				$126,769	Assume costs are incurred at the end of remediation, 5% interest rate
Salvage Value of Equipment	10%		$3,835,562	$383,556	Based on physical treatment equipment
Present Worth of Salvage Value				$333,504	

BASE CASE O&M COST

O&M Cost, Base Case

Technology Evaluation Report for Single Phase Micro-Emulsion

Operations and Maintenance Cost
Base Case

Description	Qty	Unit	Unit Cost ($)	Total Cost ($)	Comments
OPERATIONS/MAINTENANCE AND MONITORING COSTS PER MODULE					
Labor					
Operators	10,144	hrs	$40	$405,756	4 operators will be required for the duration of the project (56 hrs/wk)
Supervisor	2,536	hrs	$50	$126,799	1 supervisor will be required for the duration of the project (56 hrs/wk)
Maintenance Technician	906	hrs	$50	$45,285	1 maintenance technician will be required for the duration of the project (20 hrs/wk)
Engineer	1,811	hrs	$80	$144,913	1 engineer will be required for the duration of the project (40 hrs/wk)
Laboratory Technician	2,536	hrs	$25	$63,399	1 laboratory technician will be required for the duration of the project (56 hrs/wk)
Temporary Site Facilities					
Office Trailer	10.4	month	$250	$2,603	32' x 8' Temporary Office Trailer, cost from RACER, $250/month
Portable Toilet	10.4	month	$100	$1,041	Quote from vendor, $100/month
Move process equipment	1	ls	$10,000	$10,000	Cost for moving the sump pumping equipment to the next phase drainline sumps after each module.
Chemicals					
Brij 97	3,491,840	lbs	$2.07	$7,228,109	Quote from ICI Americas, bulk rate
n-pentanol	2,909,867	lbs	$0.59	$1,702,272	Quote from Union Carbide, bulk rate
Maintenance Materials	3%		$4,000,000	$120,000	Engineers estimate 3% of construction cost.
Utilities					
Electric - Initial Water Flood	20,882	kwh	$0.08	$1,671	Electrical requirements for pumps, blowers, etc. (52.5 hp assumed per module)
Electric - Surfactant and Post Water Flood	1,449,291	kwh	$0.08	$115,943	Electrical requirements for pumps, blowers, etc. (275 hp assumed per module, 150 hp electric boiler)
Off-Site Laboratory Analysis					
Analysis of injected fluid quality	32	sample	$100	$3,200	Once per week analysis of alcohol (1 analysis per week of surfactant flood)
Analysis of UF and steam stripper effluent	10	sample	$450	$4,500	Once per month analysis of alcohol, TPH, and VOC/SVOC (4 analyses per month of surfactant and post water flood)
Analysis of treated produced fluids	10	sample	$450	$4,500	Once per month analysis of alcohol, TPH, and VOC/SVOC (4 analyses per month of surfactant and post water flood)
Analysis of monitoring well fluids	92	sample	$450	$41,400	Once per week analysis of alcohol, TPH, and VOC/SVOC (2 analyses per week of the project), sample 2 wells per week per module
Disposal					
Incineration of sludge from SBR	3,089,158	lb	$0.57	$1,760,820	Quote from Laidlaw Environmental (SLC, UT), <3,000 BTU/lb, <50% water, <10% CI (sludge produced during surfactant flood period only)
Incineration of concentrate from UF	17,500,666	lb	$0.17	$2,975,113	Quote from Laidlaw Environmental (SLC, UT), <3,000 BTU/lb, >50% water, <10% CI (fluids produced during surfactant flood period only)
Incineration of decant from steam stripper condenser	1,499,576	lb	$0.16	$239,772	Quote from Laidlaw Environmental (SLC, UT), 3,000-8,000 BTU/lb, <50% water, <10% CI (fluids produced during surfactant flood period only)
Discharge to IWTP	28,347	1,000 gal	$11.00	$311,813	IWTP costs were obtained from Hill Air Force Base personnel. (all water from pre- and post-water flood plus residual from treatment during surfactant flood)
Subtotal			$15,308,907	$15,308,907	
Contingency	20%			$3,061,781	
Total - Operations/Maintenance and Monitoring Costs Per Module				$18,370,689	
Total No. of Modules	3 modules				
Total Time for Operation of all Modules	2.87 years				
Annual Operations/Maintenance and Monitoring Costs				$19,229,638	
Interest Rate for Present Worth Calculation	5%				
Present Worth - Operations/Maintenance and Monitoring of All Modules				$50,187,842	

BASE CASE CAPITAL COST DETAIL

(SPME) spme base
W B S   O U T L I N E

Code				Description	Quantity	UM	Source
33				CONSTRUCTION ACTIVITIES	1.0000	LS	Detail
33	10			Remedial Action	1.0000	LS	Detail
33	10	1		Mobe and Preparation Work	1.0000	LS	Detail
33	10	1	1	Install Access Roads	300.0000	LF	Plug
33	10	1	2	Install Secondary Contain	5000.000	SF	Plug
33	10	1	3	Pre-fab Buildings	2.0000	EA	Plug
33	10	1	4	Site Utilities	1.0000	LS	Plug
33	10	1	5	Site Temporary Facilities	1.0000	MO	Plug
33	10	2		Monitoring/Testing/Analys	1.0000	LS	Detail
33	10	3		Site Work	1.0000	LS	Plug
33	10	4		G'water Collection/Contrl	1.0000	LS	Detail
33	10	4	1	Monitoring Wells	40.0000	EA	Quote
33	10	4	2	Recovery Line System	5800.000	LF	Quote
33	10	4	3	Injection Line System	5200.000	LF	Quote
33	10	4	4	Soil Borings	75.0000	EA	Plug
33	10	4	5	Soil Samples	25.0000	EA	Plug
33	10	5		Physical Treatment	1.0000	LS	Detail
33	10	5	1	Electrical Package	1.0000	LS	Detail
33	10	5	2	Plumbing Package	1.0000	LS	Detail
33	10	8		Demobilization	1.0000	LS	Detail
33	10	11		Construction Management	104.0000	DA	Detail
33	10	11	1	Mobe/Demobe	30.0000	DA	Detail
33	10	11	2	Operations	74.0000	DA	Detail

(SPME) spme base
M A R K U P

Summary

```
                        --------Markup------------
                        Average
Escalated Cost    Amount Rate              Amount   Escalated Cost + MU
------------------------------------------------------------------------
Direct          5707812.66   15.00%       856171.90          6563984.56
Overhead                      0.00%
------------------------------------------------------------------------
Total           5707812.66   15.00        856171.90          6563984.56
```

Direct (DC) Breakdown

```
                        --------Markup------------
                        Average
Escalated Cost    Amount Rate              Amount   Escalated Cost + MU
------------------------------------------------------------------------
Labor Base       157232.72   15.00%        23584.91           180817.63
Labor Brdn       106546.19   15.00%        15981.93           122528.12
Equip Ownr        25500.00   15.00%         3825.00            29325.00
Equip Oprn         4250.00   15.00%          637.50             4887.50
Rentals           26120.00   15.00%         3918.00            30038.00
Materials       2667484.00   15.00%       400122.60          3067606.60
Supplies           6600.00   15.00%          990.00             7590.00
Subcontrct      2476985.40   15.00%       371547.81          2848533.21
Other Cost        99935.00   15.00%        14990.25           114925.25
Tax-RSMO         137159.35   15.00%        20573.90           157733.25
------------------------------------------------------------------------
DC Total        5707812.66   15.00%       856171.90          6563984.56
```

Overhead (OH) Breakdown

```
                        --------Markup------------
                        Average
Escalated Cost    Amount Rate              Amount   Escalated Cost + MU
------------------------------------------------------------------------
Labor Base                   15.00%                            23584.91
Labor Brdn                   15.00%                            15981.93
Equip Ownr                   15.00%                             3825.00
Equip Oprn                   15.00%                              637.50
Rentals                      15.00%                             3918.00
Materials                    15.00%                           400122.60
Supplies                     15.00%                              990.00
Subcontrct                   15.00%                           371547.81
Other Cost                   15.00%                            14990.25
Tax-RSMO                     15.00%                            20573.90
------------------------------------------------------------------------
OH Total                      0.00%
```

Subcontractor Markup Breakdown

```
                                  -----Markup--------
Code     Subcontractor   Award Amount  Rate      Amount    Award + Markup
-------------------------------------------------------------------------
                 Total    2476985.40  0.00%                    2476985.40
-------------------------------------------------------------------------
00000000  Plugged Amount w  625485.40  0.00%                    625485.40
UGMECH                     1771000.00  0.00%                   1771000.00
WELLS                        80500.00  0.00%                     80500.00
```

WORK PLAN SUMMARY
(SPME) spme base

------Activity's immediate cost (excluding subordinates' cost)------

	Description	Qty	UM	Labor Base	Burden	Equipment Ownershp	Operatn	Rentals	Material	Supply	Subcon	Other	Taxes	Total	s r c
33	CONSTRUCTION ACTIVITIES	1.0000	LS U T											0.0000 0.00	D
33 10	Remedial Action	1.0000	LS U T											0.0000 0.00	D
33 10 1	Mobe and Preparation Work	1.0000	LS U T											0.0000 0.00	D
33 10 1 1	Install Access Roads	300.0000	LF U T								84.0000 25200.00			84.0000 25200.00	P
33 10 1 2	Install Secondary Contain	5000.000	SF U T								2.4000 12000.00			2.4000 12000.00	P
33 10 1 3	Pre-fab Buildings	2.0000	EA U T						10000.00 20000.00					10000.0000 20000.00	P
33 10 1 4	Site Utilities	1.0000	LS U T								15000.00 15000.00			15000.0000 15000.00	P
33 10 1 5	Site Temporary Facilities	1.0000	MO U T									9500.000 9500.00		9500.0000 9500.00	P
33 10 2	Monitoring/Testing/Analys	1.0000	LS U T											0.0000 0.00	D
33 10 3	Site Work	1.0000	LS U T						10000.00 10000.00		12000.00 12000.00			22000.0000 22000.00	P
33 10 4	G'water Collection/Contrl	1.0000	LS U T											0.0000 0.00	D
33 10 4 1	Monitoring Wells	40.0000	EA U T								2012.500 80500.00			2012.5000 80500.00	Q
33 10 4 2	Recovery Line System	5800.000	LF U T								161.0000 933800.0			161.0000 933800.00	Q
33 10 4 3	Injection Line System	5200.000	LF U T								161.0000 837200.0			161.0000 837200.00	Q
33 10 4 4	Soil Borings	75.0000	EA U T								600.0000 45000.00			600.0000 45000.00	P
33 10 4 5	Soil Samples	25.0000	EA U T								500.0000			500.0000	

(SPME) spme base

WORK PLAN SUMMARY

	Description	Qty	UM	Labor		Equipment		Rentals	Material	Supply	Subcon	Other	Taxes	Total	s r c
				Base	Burden	Ownership	Operatn								
			T								12500.00			12500.00	P
33 10 5	Physical Treatment	1.0000	LS U	29752.50		25500.00	4250.000		2486220	2500.000	11375.00		124436.0	2684033.5000	P
			T	29752.50		25500.00	4250.00		2486220	2500.00	11375.00		124436.0	2684033.50	D
33 10 5 1	Electrical Package	1.0000	LS U								204820.4			204820.4000	
			T								204820.4			204820.40	D
33 10 5 2	Plumbing Package	1.0000	LS U						151264.0		287590.0		7563.200	446417.2000	
			T						151264.0		287590.0		7563.20	446417.20	D
33 10 8	Demobilization	1.0000	LS U	5008.000	4887.440			1150.000		2000.000			123.0000	13168.4400	
			T	5008.00	4887.44			1150.00		2000.00			123.00	13168.44	D
33 10 11	Construction Management	104.0000	DA U											0.0000	
			T											0.00	D
33 10 11 1	Mobe/Demobe	30.0000	DA U	1177.553	977.4880			240.0000		20.0000		861.3333	48.0250	3324.3997	
			T	35326.60	29324.64			7200.00		600.00		25840.00	1440.75	99731.99	D
33 10 11 2	Operations	74.0000	DA U	1177.644	977.4880			240.1351		20.2703		872.9054	48.6000	3337.0423	
			T	87145.62	72334.11			17770.00		1500.00		64595.00	3596.40	246941.13	D
	Total			157232.7	106546.2	25500.00	4250.00	26120.00	2667484	6600.00	2476985	99935.00	137159.4	5707812.66	

----- Activity's immediate cost (excluding subordinates' cost) -----

(SPME) spme base
A C T I V I T Y - D E T A I L   F O R M S

```
+33----------------------CONSTRUCTION ACTIVITIES---Qty_UM--  1.00-LS- 5707813--+

Assignmnts    Schedule            Cost  Unit     Total
--------------------  ------------------    ----------------------------
Bid Item Line    1 Duration          Source    Details
Quote Grp   NONE    Bar Start
Qry Lbl 1           Bar Finsh         Lab Base   157232.7   157233
Qry Lbl 2           CPM ES            Lab Brdn   106546.2   106546
Qry Lbl 3           CPM EF            Eqp Own    25500.00   25500
Acnt                CPM LS            Eqp Op     4250.000    4250
                    CPM LF            Rental     26120.00   26120
S h i f t s         Float             Material   2667484   2667484
-----------------                     Supply     6600.000    6600
Work Hrs/Shift  10  P r o d u c t i o n Subconts  2476985   2476985
Pay  Hrs/Shift  10  -----------------  Other      99935.00   99935
Shifts/Day     1.0  Days      175.6000 Tax RMSO   137159.4   137159
Days/Week      5.0  Shifts    175.6000 +----------------------------+
                    Hours     1756.000 |Total      5707813   5707813|
R a t e s           Man-Hrs   6227.000 +----------------------------+
---------           Eqp-Hrs   5610.000 > Job       5707813
Rates Unique        UM/Da       0.0057
                    UM/Sh       0.0057      >DBE Goal
                    UM/Hr       0.0006      >MBE Goal
                    UM/MH       0.0002      >WBE Goal
                    MH/UM     6227.000      >0-1 Goal
                                            >0-2 Goal
+------------------------------------------------------------------+


+33 10---------------------Remedial Action-----------Qty_UM--  1.00-LS- 5707813--+

Assignmnts    Schedule            Cost  Unit     Total
--------------------  ------------------    ----------------------------
Bid Item Line    1 Duration          Source    Details
Quote Grp   NONE    Bar Start
Qry Lbl 1           Bar Finsh         Lab Base   157232.7   157233
Qry Lbl 2           CPM ES            Lab Brdn   106546.2   106546
Qry Lbl 3           CPM EF            Eqp Own    25500.00   25500
Acnt                CPM LS            Eqp Op     4250.000    4250
                    CPM LF            Rental     26120.00   26120
S h i f t s         Float             Material   2667484   2667484
-----------------                     Supply     6600.000    6600
Work Hrs/Shift  10  P r o d u c t i o n Subconts  2476985   2476985
Pay  Hrs/Shift  10  -----------------  Other      99935.00   99935
Shifts/Day     1.0  Days      175.6000 Tax RMSO   137159.4   137159
Days/Week      5.0  Shifts    175.6000 +----------------------------+
                    Hours     1756.000 |Total      5707813   5707813|
R a t e s           Man-Hrs   6227.000 +----------------------------+
---------           Eqp-Hrs   5610.000 > Level 1   5707813
Rates Unique        UM/Da       0.0057
                    UM/Sh       0.0057      >DBE Goal
                    UM/Hr       0.0006      >MBE Goal
                    UM/MH       0.0002      >WBE Goal
                    MH/UM     6227.000      >0-1 Goal
                                            >0-2 Goal
+------------------------------------------------------------------+
```

(SPME) spme base
A C T I V I T Y - D E T A I L   F O R M S

```
+33 10 1-----------------Mobe and Preparation Work-Qty_UM--  1.00-LS-81700.00--+
|                                                                              |
| A s s i g n m n t s    S c h e d u l e         C o s t  Unit     Total       |
| -------------------    -----------------       --------------------------    |
| Bid Item Line       1 Duration                 Source    Details            |
| Quote Grp     NONE    Bar Start                                              |
| Qry Lbl 1             Bar Finsh                 Lab Base                     |
| Qry Lbl 2             CPM ES                     Lab Brdn                     |
| Qry Lbl 3             CPM EF                     Eqp Own                      |
| Acnt                  CPM LS                     Eqp Op                       |
|                       CPM LF                     Rental                       |
| S h i f t s           Float                      Material  20000.00   20000  |
| -----------------                               Supply                       |
| Work Hrs/Shift   10   P r o d u c t i o n        Subconts  52200.00   52200  |
| Pay Hrs/Shift    10   -------------------        Other      9500.000    9500  |
| Shifts/Day      1.0   Days                       Tax RMSO                     |
| Days/Week       5.0   Shifts                     +--------------------------+ |
|                       Hours                      |Total     81700.00   81700| |
| R a t e s             Man-Hrs                    +--------------------------+ |
| ---------             Eqp-Hrs                    > Level 2 81700.00          |
| Rates Unique          UM/Da                                                  |
|                       UM/Sh                                 >DBE Goal        |
|                       UM/Hr                                 >MBE Goal        |
|                       UM/MH                                 >WBE Goal        |
|                       MH/UM                                 >O-1 Goal        |
|                                                             >O-2 Goal        |
+------------------------------------------------------------------------------+
```

```
+33 10 1 1----------------Install Access Roads------Qty_UM--300.00-LF-   84.00--+
|                                                                              |
| A s s i g n m n t s    S c h e d u l e         C o s t  Unit     Total       |
| -------------------    -----------------       --------------------------    |
| Bid Item Line       1 Duration                 Source    Plug                |
| Quote Grp     NONE    Bar Start                                              |
| Qry Lbl 1             Bar Finsh                 Lab Base                     |
| Qry Lbl 2             CPM ES                     Lab Brdn                     |
| Qry Lbl 3             CPM EF                     Eqp Own                      |
|                       CPM LS                     Eqp Op                       |
|                       CPM LF                     Rental                       |
|                       Float                      Material                     |
|                                                  Supply                       |
|                                                  Subconts  84.000    25200   |
|                                                  Other                        |
|                                                  Tax RMSO                     |
|                                                  +-------------------------+  |
|                                                  |Total     84.000    25200|  |
|                                                  +-------------------------+  |
|                                                  > Level 3 25200.00          |
|                                                                              |
|                                                                              |
+------------------------------------------------------------------------------+
```

(SPME) spme base
A C T I V I T Y - D E T A I L   F O R M S

+33 10 1 2----------------Install Secondary Contain-Qty_UM--5000.0-SF-   2.40--+

```
A s s i g n m n t s   S c h e d u l e                C o s t  Unit    Total
-------------------   -----------------               --------------------------
Bid Iten Line       1 Duration                        Source    Plug
Quote Grp    NONE     Bar Start
Qry Lbl 1             Bar Finsh                        Lab Base
Qry Lbl 2             CPM ES                           Lab Brdn
Qry Lbl 3             CPM EF                            Eqp Own
                      CPM LS                            Eqp Op
                      CPM LF                            Rental
                      Float                             Material
                                                        Supply
                                                        Subconts   2.400     12000
                                                        Other
                                                        Tax RMSO
                                                       +--------------------------
                                                       |Total      2.400     12000
                                                       +--------------------------
                                                        . Level 3 12000.00
```

+--+

+33 10 1 3----------------Pre-fab Buildings---------Qty_UM-- 2.00-EA-10000.00--+

```
A s s i g n m n t s   S c h e d u l e                C o s t  Unit    Total.
-------------------   -----------------               --------------------------
Bid Iten Line       1 Duration                        Source    Plug
Quote Grp    NONE     Bar Start
Qry Lbl 1             Bar Finsh                        Lab Base
Qry Lbl 2             CPM ES                           Lab Brdn
Qry Lbl 3             CPM EF                            Eqp Own
                      CPM LS                            Eqp Op
                      CPM LF                            Rental
                      Float                             Material 10000.00   20000
                                                        Supply
                                                        Subconts
                                                        Other
                                                        Tax RMSO
                                                       +--------------------------
                                                       |Total     10000.00   20000
                                                       +--------------------------
                                                        > Level 3 20000.00
```

+--+

A C T I V I T Y - D E T A I L   F O R M S

```
+33 10 1 4--------------Site Utilities-----------Qty_UM--  1.00-LS-15000.00--+
|                                                                            |
| A s s i g n m n t s   S c h e d u l e           C o s t  Unit     Total    |
| -------------------   -----------------         -----------------------    |
| Bid Iten Line      1 Duration                   Source    Plug             |
| Quote Grp    NONE    Bar Start                                             |
| Qry Lbl 1            Bar Finsh                   Lab Base                   |
| Qry Lbl 2            CPM ES                      Lab Brdn                   |
| Qry Lbl 3            CPM EF                      Eqp Own                    |
|                      CPM LS                      Eqp Op                     |
|                      CPM LF                      Rental                     |
|                      Float                       Material                   |
|                                                  Supply                     |
|                                                  Subconts  15000.00   15000 |
|                                                  Other                      |
|                                                  Tax RMSO                   |
|                                                 +------------------------   |
|                                                 |Total     15000.00   15000 |
|                                                 +------------------------   |
|                                                 > Level 3 15000.00          |
|                                                                            |
|                                                                            |
|                                                                            |
+----------------------------------------------------------------------------+
```

```
+33 10 1 5---------------Site Temporary Facilities-Qty_UM--  1.00-MO- 9500.00--+
|                                                                             |
| A s s i g n m n t s   S c h e d u l e            C o s t  Unit     Total     |
| -------------------   -----------------          -----------------------    |
| Bid Iten Line      1 Duration                    Source    Plug             |
| Quote Grp    NONE    Bar Start                                              |
| Qry Lbl 1            Bar Finsh                    Lab Base                   |
| Qry Lbl 2            CPM ES                       Lab Brdn                   |
| Qry Lbl 3            CPM EF                       Eqp Own                    |
|                      CPM LS                       Eqp Op                     |
|                      CPM LF                       Rental                     |
|                      Float                        Material                  |
|                                                   Supply                    |
|                                                   Subconts                  |
|                                                   Other    9500.000   9500  |
|                                                   Tax RMSO                  |
|                                                  +------------------------  |
|                                                  |Total    9500.000   9500  |
|                                                  +------------------------  |
|                                                  > Level 3 9500.000         |
|                                                                             |
|                                                                             |
|                                                                             |
+-----------------------------------------------------------------------------+
```

```
                          (SPME) spme base
                  A C T I V I T Y - D E T A I L   F O R M S


 +33 10 2-----------------Monitoring/Testing/Analys-Qty_UM--  1.00-LS-        --+
 |                                                                              |
 | A s s i g n m n t s    S c h e d u l e      C o s t   Unit      Total        |
 | -------------------    ---------------      --------------------------       |
 | Bid Item Line      1 Duration               Source    Details               |
 | Quote Grp    NONE    Bar Start                                               |
 | Qry Lbl 1            Bar Finsh              Lab Base                          |
 | Qry Lbl 2            CPM ES                 Lab Brdn                          |
 | Qry Lbl 3            CPM EF                 Eqp Own                           |
 | Acnt                 CPM LS                 Eqp Op                            |
 |                      CPM LF                 Rental                           |
 | S h i f t s          Float                  Material                         |
 | -----------------                           Supply                           |
 | Work Hrs/Shift   10  P r o d u c t i o n    Subconts                         |
 | Pay  Hrs/Shift   10  -------------------     Other                           |
 | Shifts/Day      1.0  Days                   Tax RMSO                         |
 | Days/Week       5.0  Shifts                +----------------------------+    |
 |                      Hours                 |Total                       |    |
 | R a t e s            Man-Hrs               +----------------------------+    |
 | ---------            Eqp-Hrs               > Level 2                        |
 | Rates Unique         UM/Da                                                   |
 |                      UM/Sh                          >DBE Goal                |
 |                      UM/Hr                          >MBE Goal                |
 |                      UM/MH                          >WBE Goal                |
 |                      MH/UM                          >O-1 Goal                |
 |                                                     >O-2 Goal                |
 |                                                                              |
 +------------------------------------------------------------------------------+



 +33 10 3-----------------Site Work----------------Qty_UM--  1.00-LS-22000.00--+
 |                                                                              |
 | A s s i g n m n t s    S c h e d u l e      C o s t   Unit      Total        |
 | -------------------    ---------------      --------------------------       |
 | Bid Item Line      1 Duration               Source    Plug                   |
 | Quote Grp    NONE    Bar Start                                               |
 | Qry Lbl 1            Bar Finsh              Lab Base                          |
 | Qry Lbl 2            CPM ES                 Lab Brdn                          |
 | Qry Lbl 3            CPM EF                 Eqp Own                           |
 |                      CPM LS                 Eqp Op                            |
 |                      CPM LF                 Rental                           |
 |                      Float                  Material  10000.00     10000      |
 |                                             Supply                           |
 |                                             Subconts  12000.00     12000      |
 |                                             Other                            |
 |                                             Tax RMSO                         |
 |                                            +----------------------------+    |
 |                                            |Total     22000.00     22000|    |
 |                                            +----------------------------+    |
 |                                             > Level 2 22000.00               |
 |                                                                              |
 +------------------------------------------------------------------------------+
```

(SPME) spme base
A C T I V I T Y - D E T A I L   F O R M S

```
+33 10 4----------------G'water Collection/Contrl-Qty_UM--  1.00-LS- 1909000--+
|                                                                             |
| A s s i g n m n t s    S c h e d u l e         C o s t  Unit     Total      |
| --------------------    ------------------      ---------------------------  |
| Bid Item Line      1 Duration                  Source    Details            |
| Quote Grp    NONE    Bar Start                                              |
| Qry Lbl 1            Bar Finsh                 Lab Base                      |
| Qry Lbl 2            CPM ES                    Lab Brdn                      |
| Qry Lbl 3            CPM EF                    Eqp Own                       |
| Acnt                 CPM LS                    Eqp Op                        |
|                      CPM LF                    Rental                        |
| S h i f t s          Float                     Material                      |
| ------------------                             Supply                        |
| Work Hrs/Shift   10  P r o d u c t i o n       Subconts  1909000  1909000   |
| Pay Hrs/Shift    10  -------------------       Other                         |
| Shifts/Day     1.0   Days                      Tax RMSO                      |
| Days/Week      5.0   Shifts                    +------------------------------+
|                      Hours                     |Total     1909000  1909000| |
| R a t e s            Man-Hrs                   +------------------------------+
| ---------            Eqp-Hrs                   > Level 2  1909000            |
| Rates Unique         UM/Da                                                   |
|                      UM/Sh                              >DBE Goal            |
|                      UM/Hr                              >MBE Goal            |
|                      UM/MH                              >WBE Goal            |
|                      MH/UM                              >O-1 Goal            |
|                                                        >O-2 Goal            |
|                                                                             |
+-----------------------------------------------------------------------------+
```

```
+33 10 4 1----------------Monitoring Wells----------Qty_UM-- 40.00-EA- 2012.50--+
|                                                                              |
| A s s i g n m n t s    S c h e d u l e         C o s t  Unit     Total       |
| --------------------    ------------------      ---------------------------   |
| Bid Item Line      1 Duration                  Source    Quote               |
| Query Grp    NONE    Bar Start                                               |
| Qry Lbl 1            Bar Finsh                 Lab Base                       |
| Qry Lbl 2            CPM ES                    Lab Brdn                       |
| Qry Lbl 3            CPM EF                    Eqp Own                        |
|                      CPM LS                    Eqp Op                         |
|                      CPM LF                    Rental                         |
|                      Float                     Material                       |
|                                                Supply                         |
|                                                Subconts  2012.500   80500    |
|                                                Other                          |
|                                                Tax RMSO                       |
|                                                +------------------------------+
|                                                |Total     2012.500   80500| |
|                                                +------------------------------+
|                                                > Level 3  80500.00           |
| wardee                                                                       |
|         WELLS                                          >DBE Goal             |
|                                                        >MBE Goal             |
|  E Allow      1.00%                                    >WBE Goal             |
|                                                        >O-1 Goal             |
|                                                        >O-2 Goal             |
|                                                                              |
+------------------------------------------------------------------------------+
```

```
                         (SPME) spme base
                 A C T I V I T Y - D E T A I L   F O R M S


+33 10 4 2----------------Recovery Line System------Qty_UM--5800.0-LF-  161.00--+

A s s i g n m n t s   S c h e d u l e          C o s t  Unit     Total
--------------------   ------------------        -----------------------------
Bid Item Line      1 Duration                    Source    Quote
Query Grp    NONE    Bar Start
Qry Lbl 1            Bar Finsh                    Lab Base
Qry Lbl 2            CPM ES                       Lab Brdn
Qry Lbl 3            CPM EF                        Eqp Own
                     CPM LS                        Eqp Op
                     CPM LF                        Rental
                     Float                         Material
                                                   Supply
                                                   Subconts   161.000    933800
                                                   Other
                                                   Tax RMSO
                                                  +----------------------------
                                                  |Total      161.000    933800
                                                  +----------------------------
                                                   > Level 3 933800.0
Awardee
        UGMECH                                            >DBE Goal
                                                          >MBE Goal
DBE Allow      1.00%                                      >WBE Goal
                                                          >O-1 Goal
                                                          >O-2 Goal
+------------------------------------------------------------------------------+


+33 10 4 3----------------Injection Line System-----Qty_UM--5200.0-LF-  161.00--+

A s s i g n m n t s   S c h e d u l e          C o s t  Unit     Total
--------------------   ------------------        -----------------------------
Bid Item Line      1 Duration                    Source    Quote
Query Grp    NONE    Bar Start
Qry Lbl 1            Bar Finsh                    Lab Base
Qry Lbl 2            CPM ES                       Lab Brdn
Qry Lbl 3            CPM EF                        Eqp Own
                     CPM LS                        Eqp Op
                     CPM LF                        Rentai
                     Float                         Material
                                                   Supply
                                                   Subconts   161.000    837200
                                                   Other
                                                   Tax RMSO
                                                  +----------------------------
                                                  |Total      161.000    837200
                                                  +----------------------------
                                                   > Level 3 837200.0
Awardee
        UGMECH                                            >DBE Goal
                                                          >MBE Goal
DBE Allow      1.00%                                      >WBE Goal
                                                          >O-1 Goal
                                                          >O-2 Goal
+------------------------------------------------------------------------------+
```

```
                            (SPME) spme base
                     A C T I V I T Y - D E T A I L   F O R M S


  +33 10 4 4----------------Soil Borings------------Qty_UM-- 75.00-EA-  600.00--+
  |
  | A s s i g n m n t s   S c h e d u l e               C o s t  Unit    Total
  | -------------------   -----------------             ---------------------------
  | Bid Iten Line     1 Duration                        Source    Plug
  | Quote Grp    NONE   Bar Start
  | Qry Lbl 1           Bar Finsh                        Lab Base
  | Qry Lbl 2           CPM ES                           Lab Brdn
  | Qry Lbl 3           CPM EF                            Eqp Own
  |                     CPM LS                            Eqp Op
  |                     CPM LF                            Rental
  |                     Float                             Material
  |                                                       Supply
  |                                                       Subconts   600.000    45000
  |                                                       Other
  |                                                       Tax RMSO
  |                                                       +-----------------------
  |                                                       |Total      600.000    45000
  |                                                       +-----------------------
  |                                                       > Level 3 45000.00
  |
  |
  +--------------------------------------------------------------------------------+


  +33 10 4 5----------------Soil Samples------------Qty_UM-- 25.00-EA-  500.00--+
  |
  | A s s i g n m n t s   S c h e d u l e               C o s t  Unit    Total
  | -------------------   -----------------             ---------------------------
  | Bid Iten Line     1 Duration                        Source    Plug
  | Quote Grp    NONE   Bar Start
  | Qry Lbl 1           Bar Finsh                        Lab Base
  | Qry Lbl 2           CPM ES                           Lab Brdn
  | Qry Lbl 3           CPM EF                            Eqp Own
  |                     CPM LS                            Eqp Op
  |                     CPM LF                            Rental
  |                     Float                             Material
  |                                                       Supply
  |                                                       Subconts   500.000    12500
  |                                                       Other
  |                                                       Tax RMSO
  |                                                       +-----------------------
  |                                                       |Total      500.000    12500
  |                                                       +-----------------------
  |                                                       > Level 3 12500.00
  |
  |
  +--------------------------------------------------------------------------------+
```

```
                              CH2M HILL
                           (SPME) spme base
                    A C T I V I T Y - D E T A I L   F O R M S


  +33 10 5------------------Physical Treatment--------Qty_UM--  1.00-LS- 3335271---+

  A s s i g n m n t s     S c h e d u l e          C o s t   Unit      Total
  -------------------     ------------------        ---------------------------
  Bid Item Line       1 Duration                   Source    Details  LC
  Quote Grp    NONE     Bar Start
  Qry Lbl 1             Bar Finsh                   Lab Base  29752.50    29753
  Qry Lbl 2             CPM ES                      Lab Brdn
  Qry Lbl 3             CPM EF                       Eqp Own  25500.00    25500
  Acnt                  CPM LS                       Eqp Op   4250.000     4250
                        CPM LF                       Rental
  S h i f t s           Float                        Material 2637484   2637484
  -----------------                                  Supply   2500.000     2500
  Work Hrs/Shift    10  P r o d u c t i o n          Subconts 503785.4   503785
  Pay  Hrs/Shift    10  -------------------          Other
  Shifts/Day       1.0  Days         25.0000         Tax RMSO 131999.2   131999
  Days/Week        5.0  Shifts       25.0000        +----------------------------+
                        Hours       250.0000        |Total       3335271  3335271|
  R a t e s             Man-Hrs    1250.000         +----------------------------+
  ---------             Eqp-Hrs    1250.000          > Level 2 3335271
  Rates Unique          UM/Da         0.0400
                        UM/Sh         0.0400                  >DBE Goal
                        UM/Hr         0.0040                  >MBE Goal
                        UM/MH         0.0008                  >WBE Goal
                        MH/UM      1250.000                   >O-1 Goal
                                                              >O-2 Goal


  +-33 10 5---------------------Physical Treatment-----------------------------------+
  |RESOURCE LIST Cost Detail                      Source Worksheet:                   |
  +=================================================================================+
                                                  +----------------------+
     Man-Hrs             Eqp-Hrs                  |>Unit    2624e3        |
     UM/Man-Hr           UM/Eqp-Hr                |>TOTAL   2624e3        |
     Man-Hr/UM           Eqp-Hr/UM                +----------------------+



  Resource                   Quantity UM Count  Unit Cost      TOTAL
  ----------------------------------------------------------------------
                                                 Total:    2624e3
                                                 Unit:     2624e3
  MDECANT1 DECANTER (8GPM)           1.00 EA 1.00    21000     21000.00
  MFILTER1 BAG FILTER (55GPM)        2.00 EA 1.00    1155.0     2310.00
  MFILTER4 ULTRAFILTRATION UNIT(50G) 1.00 EA 1.00   115500    115500.00
  MFILTER3 BAG FILTER (44 GPM)       2.00 EA 1.00   535.50      1071.00
  MGENR'R1 STEAM GENERATOR (150HP)   1.00 EA 1.00    48300     48300.00
  MMIXER1  CHEM STORAGE TANK MIXER   2.00 EA 1.00   5250.0     10500.00
  MMIXER2  CHEMICAL MIX TANKS MIXER  3.00 EA 1.00   5250.0     15750.00
  MPUMP1   CHEMICAL FEED PUMP (25G)  2.00 EA 1.00   5250.0     10500.00
  MPUMP2   CHEMICAL FEED PUMP (55G)  2.00 EA 1.00   8925.0     17850.00
  MPUMP3   CHEM. FEED INJ.PUMP (15G) 9.00 EA 1.00   7875.0     70875.00
  MPUMP4   CHEM. FEED INJ. PUMP (5G) 13.00 EA 1.00   6825.0     88725.00
  MPUMP5   VAR.SPEED RECOV.PMP (15G) 9.00 EA 1.00   4305.0     38745.00
  MPUMP6   VAR.SPEED RECOV.PMP (5G)  16.00 EA 1.00   4305.0     68880.00
  MPUMP8   TREAT.PLNT.FEED PUMP (65) 4.00 EA 1.00   4200.0     16800.00
```

```
                              (SPME) spme base
                      A C T I V I T Y - D E T A I L    F O R M S


+-33 10 5--------------------Physical Treatment-----------------------------+
|RESOURCE LIST Cost Detail                        Source Worksheet:         |
+========================================================================== +
|                                       +----------------------+            |
|   Man-Hrs              Eqp-Hrs         |>Unit    2624e3       |            |
|   UM/Man-Hr            UM/Eqp-Hr       |>TOTAL   2624e3       |            |
|   Man-Hr/UM            Eqp-Hr/UM       +----------------------+            |
|                                                                           |
|                                                                           |
|                                                                           |
| Resource                      Quantity UM Count  Unit Cost        TOTAL   |
| ------------------------------------------------------------------------  |
|                                                  Total:   2624e3          |
|                                                  Unit:    2624e3          |
| MSTRIP2  STEAM STRIP. COLM. (42G)    1.00 EA 1.00   446250      446250.00 |
| MSUMP1   GALV. CMP W/ COVER         51.00 EA 1.00   525.00       26775.00 |
| MTANK1   CHEM. STORAGE TANK (10K)    2.00 EA 1.00   18375        36750.00 |
| MTANK2   CHEM. MIX TANK (20K)        3.00 EA 1.00   36750       110250.00 |
| MTANK3   EQUALIZATION TANK (20K)     4.00 EA 1.00   36750       147000.00 |
| MTANK4   CONCENTRATE STOR.TNK.(20K   1.00 EA 1.00   36750        36750.00 |
| MVPORIZR PROPANE VAPORIZER           1.00 EA 1.00   68250        68250.00 |
| MCONDENS CONDENSOR (10GPM)           1.00 EA 1.00   36750        36750.00 |
| MTANK5   SBR TANKS (300K)            2.00 EA 1.00   551250     1102500.0  |
| MBLOWER1 CENTRIFUGAL BLOWER(15PSI)   2.00 EA 1.00   15750        31500.00 |
| MFLTRPRS FILTER PRESS (SMALL)        1.00 EA 1.00   36750        36750.00 |
| MFEEDSYS CHEMICAL FEED SYSTEM        2.00 EA 1.00   2100.0        4200.00 |
| SMISC100 MISC. SUPPLIES             25.00 EA 1.00   105.00        2625.00 |
| USUBPRDM SUBCONTRACTOR PER DIEM    175.00 MD 1.00   65.00        11375.00 |
+---------------------------------------------------------------------------+


+-33 10 5--------------------Physical Treatment-----------------------------+
|CREW CONFIGURATION Cost Detail                   Source Worksheet:         |
+========================================================================== +
|                                       +----------------------+            |
| Days     25.000 UM/Da 0.0400 Da/UM 25.000 |>Unit    59503      |          |
| Shifts   25.000 UM/Sh 0.0400 Sh/UM 25.000 |>TOTAL   59503      |          |
| Hours    250.00 UM/Hr 0.0040 Hr/UM 250.00 +--------------------+          |
| Man-hours 1250.0 UM/MH 0.0008 MH/UM 1250.0                                |
|                                                                           |
|                                                                           |
| Resource   Qty                            Unit Cost  UM         TOTAL     |
| ------------------------------------------------------------------------  |
| Man-Count:   5.00                              Total:          59503      |
| Eqp-Count:   5.00                              Unit:           59503      |
| LTCCI5    1.00  CCI TECHNICIAN 5               27.60    HR       6900.00   |
| LCM5      1.00  CONSTRUCTION MGR. 5            34.62    HR       8655.00   |
| LTCCI3    3.00  CCI TECHNICIAN 3               18.93    HR      14197.50   |
| ECAT960   1.00  CAT 960 WHEEL LOADER           34.00    HR       8500.00   |
| ECRN60    1.00  TRUCK CRANE - 60 TN            35.00    HR       8750.00   |
| EPUTRK    2.00  PICKUP TRUCK                    6.00    HR       3000.00   |
| ECAT215   1.00  CAT 215 EXCAVATOR              38.00    HR       9500.00   |
+---------------------------------------------------------------------------+


+========================================================================== +
```

```
                              (SPME) spme base
                       A C T I V I T Y - D E T A I L   F O R M S


   +33 10 5 1----------------Electrical Package--------Qty_UM--  1.00-LS-204820.4--+
   |                                                                               |
   | A s s i g n m n t s   S c h e d u l e      C o s t  Unit     Total            |
   | -------------------   -----------------    -----------------------            |
   | Bid Item Line      1 Duration              Source    Details  L               |
   | Quote Grp    NONE    Bar Start                                                |
   | Qry Lbl 1            Bar Finsh             Lab Base                            |
   | Qry Lbl 2            CPM ES                Lab Brdn                            |
   | Qry Lbl 3            CPM EF                Eqp Own                             |
   | Acnt                 CPM LS                Eqp Op                             |
   |                      CPM LF                Rental                            |
   |                      Float                 Material                           |
   | S h i f t s                                Supply                             |
   | -----------------                          Subconts  204820.4    204820       |
   | Work Hrs/Shift   10  P r o d u c t i o n   Other                              |
   | Pay  Hrs/Shift   10  -------------------   Tax RMSO                           |
   | Shifts/Day      1.0  Days                  +-------------------------+        |
   | Days/Week       5.0  Shifts                |Total     204820.4   204820|      |
   |                      Hours                 +-------------------------+        |
   | R a t e s            Man-Hrs                                                  |
   | ---------            Eqp-Hrs              > Level 3 204820.4                   |
   | Rates Unique         UM/Da                                                    |
   |                      UM/Sh                          >DBE Goal                 |
   |                      UM/Hr                          >MBE Goal                 |
   |                      UM/MH                          >WBE Goal                 |
   |                      MH/UM                          >O-1 Goal                 |
   |                                                     >O-2 Goal                 |
   +-------------------------------------------------------------------------------+


   ALL PRICES ARE FURNISH AND INSTALLED.




   +-33 10 5 1-------------------Electrical Package----------------------------+
   |RESOURCE LIST Cost Detail                       Source Worksheet:          |
   +=========================================================================+
   |                                          +--------------------+          |
   |  Man-Hrs             Eqp-Hrs             |>Unit     132325    |          |
   |  UM/Man-Hr           UM/Eqp-Hr           |>TOTAL    132325    |          |
   |  Man-Hr/UM           Eqp-Hr/UM           +--------------------+          |
   |                                                                          |
   |                                                                          |
   |                                                                          |
   |Resource                      Quantity UM Count  Unit Cost        TOTAL   |
   | ------------------------------------------------------------------------ |
   |                                                 Total:  132325           |
   |                                                 Unit:   132325           |
   |U2C4WIRE 2" CONDUIT W/ 3-#1 WIRE    262.00 LF 1.00    20.20     5292.40    |
   |U480VPNL 480 VOLT PANEL BOARD         1.00 EA 1.00    10200    10200.00    |
   |UMPC     MINI POWER CENTER            8.00 EA 1.00    2100.0   16800.00    |
   |U1C3WIRE 1" CONDUIT W/ 2-#8 WIRE    200.00 LF 1.00    11.40     2280.00    |
   |UMISCW&E MISC. WIRING AND ELECTRIC    1.00 LS 1.00    32830    32830.00    |
   |U2C4WIRE 2" CONDUIT W/ 3-#1 WIRE   3214.00 LF 1.00    20.20    64922.80    |
   +--------------------------------------------------------------------------+


   +=========================================================================+
```

(SPME) spme base
A C T I V I T Y - D E T A I L   F O R M S

```
+33 10 5 2----------------Plumbing Package---------Qty_UM--  1.00-LS-446417.2--+
|
| A s s i g n m n t s   S c h e d u l e          C o s t  Unit    Total
| --------------------   -----------------        --------------------------
| Bid Item Line    1 Duration                     Source    Details  L
| Quote Grp    NONE   Bar Start
| Qry Lbl 1           Bar Finsh                    Lab Base
| Qry Lbl 2           CPM ES                       Lab Brdn
| Qry Lbl 3           CPM EF                       Eqp Own
| Acnt                CPM LS                       Eqp Op
|                     CPM LF                       Rental
| S h i f t s         Float                        Material  151264.0    151264
| ------------------                               Supply
| Work Hrs/Shift   10 P r o d u c t i o n          Subconts  287590.0    287590
| Pay  Hrs/Shift   10 -------------------          Other
| Shifts/Day      1.0 Days                         Tax RMSO  7563.200      7563
| Days/Week       5.0 Shifts                       +--------------------------+
|                     Hours                        |Total     446417.2    446417|
| R a t e s           Man-Hrs                      +--------------------------+
| ---------           Eqp-Hrs                      > Level 3 446417.2
| Rates Unique        UM/Da
|                     UM/Sh                                >DBE Goal
|                     UM/Hr                                >MBE Goal
|                     UM/MH                                >WBE Goal
|                     MH/UM                                >O-1 Goal
|                                                          >O-2 Goal
+------------------------------------------------------------------------------+


+-33 10 5 2-------------------Plumbing Package---------------------------------+
|RESOURCE LIST Cost Detail                        Source Worksheet:           |
+=============================================================================+
|                                                 +-----------------------+
|  Man-Hrs              Eqp-Hrs                    |>Unit    287590        |
|  UM/Man-Hr            UM/Eqp-Hr                  |>TOTAL   287590        |
|  Man-Hr/UM            Eqp-Hr/UM                  +-----------------------+
|
|
|
|Resource                        Quantity UM Count  Unit Cost      TOTAL
|---------------------------------------------------------------------------
|                                                    Total:    287590
|                                                    Unit:     287590
|
|UPIPE2   2" BLACK PIPE            3936.00 LF 1.00      16.00   62976.00
|UPIPE2IN 2" BLACK PIPE (INJECTION) 3542.00 LF 1.00    16.00   56672.00
|UPIP2DB1 2" BLACK PIPE (DBL. WALL) 3838.00 LF 1.00    19.00   72922.00
|UPIP2DB2 2" BLACK PIPE (IWTP)      524.00 LF 1.00     19.00    9956.00
|UMISCP&F MISC. PIPE & FITTINGS       1.00 LS 1.00     75000    75000.00
|UPIPE4   4" BLACK PIPE              32.00 LF 1.00     31.00     992.00
|UPIPE15S 1.5" BLACK PIPE (STRIPPER  100.00 LF 1.00    12.00    1200.00
|UPIPE15T 1.5" BLACK PIPE (TREAT.)   656.00 LF 1.00    12.00    7872.00
+---------------------------------------------------------------------------+


+=============================================================================+
```

```
                        (SPME) spme base
                  A C T I V I T Y - D E T A I L   F O R M S


  +33 10 8------------------Demobilization------------Qty_UM--  1.00-LS-13168.44--+

  |A s s i g n m n t s    S c h e d u l e        C o s t  Unit    Total
  |--------------------   ------------------      ----------------------------
  |Bid Item Line      1 Duration                 Source     Details LC
  |Quote Grp    NONE    Bar Start
  |Qry Lbl 1            Bar Finsh                 Lab Base  5008.000    5008
  |Qry Lbl 2            CPM ES                    Lab Brdn  4887.440    4887
  |Qry Lbl 3            CPM EF                    Eqp Own
  |Acnt                 CPM LS                    Eqp Op
  |                     CPM LF                    Rental    1150.000    1150
  |S h i f t s          Float                     Material
  |------------------                             Supply    2000.000    2000
  |Work Hrs/Shift   10  P r o d u c t i o n       Subconts
  |Pay  Hrs/Shift   10  -------------------       Other
  |Shifts/Day      1.0  Days        5.0000        Tax RMSO   123.000     123
  |Days/Week       5.0  Shifts      5.0000        +--------------------------+
  |                     Hours      50.0000        |Total     13168.44   13168|
  |R a t e s            Man-Hrs   200.0000        +--------------------------+
  |---------            Eqp-Hrs   200.0000        > Level 2 13168.44
  |Rates Unique         UM/Da       0.2000
  |                     UM/Sh       0.2000             >DBE Goal
  |                     UM/Hr       0.0200             >MBE Goal
  |                     UM/MH       0.0050             >WBE Goal
  |                     MH/UM     200.0000             >O-1 Goal
  |                                                    >O-2 Goal
  +------------------------------------------------------------------------+


  +-33 10 8--------------------Demobilization--------------------------------+
  |RESOURCE LIST Cost Detail                     Source Worksheet:          |
  +==========================================================================+
  |                                              +--------------------+
  |  Man-Hrs              Eqp-Hrs                 |>Unit    2100.0     |
  |  UM/Man-Hr            UM/Eqp-Hr               |>TOTAL   2100.0     |
  |  Man-Hr/UM            Eqp-Hr/UM               +--------------------+
  |
  |
  |
  |
  |Resource                    Quantity UM Count  Unit Cost     TOTAL
  |-------------------------------------------------------------------------
  |                                              Total:  2100.0
  |                                              Unit:   2100.0
  |SMISC200 MISC. SUPPLIES          10.00 EA 1.00   210.00       2100.00 |
  +-------------------------------------------------------------------------+

  +-33 10 8--------------------Demobilization--------------------------------+
  |CREW CONFIGURATION Cost Detail                Source Worksheet:          |
  +==========================================================================+
  |                                              +--------------------+
  |Days      5.0000  UM/Da 0.2000  Da/UM 5.0000  |>Unit    11068     |
  |Shifts    5.0000  UM/Sh 0.2000  Sh/UM 5.0000  |>TOTAL   11068     |
  |Hours    50.000   UM/Hr 0.0200  Hr/UM 50.000  +--------------------+
  |Man-hours 200.00  UM/MH 0.0050  MH/UM 200.00
  |
  |
  |Resource   Qty                             Unit Cost  UM     TOTAL
  |-------------------------------------------------------------------------
  |Man-Count:  4.00                              Total:  11068
  |Eqp-Count:  4.00                              Unit:   11068
  |ONSITE    1.00  ONSITE MGMT. CREW          221.37     HR  11068.44 |
  +-------------------------------------------------------------------------+
```

(SPME) spme base
A C T I V I T Y - D E T A I L   F O R M S

```
+33 10 11----------------Construction Management---Qty_UM--104.00-DA- 3333.40--+
|                                                                              |
| A s s i g n m n t s    S c h e d u l e          C o s t  Unit    Total       |
| -------------------    -----------------        ---------------------------  |
| Bid Item Line     1 Duration                    Source    Details            |
| Quote Grp    NONE    Bar Start                                                |
| Qry Lbl 1            Bar Finsh                   Lab Base  1177.618   122472  |
| Qry Lbl 2            CPM ES                      Lab Brdn   977.488   101659  |
| Qry Lbl 3            CPM EF                      Eqp Own                      |
| Acnt                 CPM LS                      Eqp Op                       |
|                      CPM LF                      Rental     240.096    24970  |
| S h i f t s          Float                       Material                     |
| -----------------                               Supply      20.192     2100  |
| Work Hrs/Shift   10  P r o d u c t i o n         Subconts                     |
| Pay Hrs/Shift    10  -------------------         Other      869.567    90435  |
| Shifts/Day      1.0  Days        145.6000        Tax RMSO    48.434     5037  |
| Days/Week       5.0  Shifts      145.6000        +--------------------------+ |
|                      Hours      1456.000         |Total     3333.395   346673| |
| R a t e s            Man-Hrs    4777.000         +--------------------------+ |
| ---------            Eqp-Hrs    4160.000         > Level 2 346673.1          |
| Rates Unique         UM/Da        0.7143                                      |
|                      UM/Sh        0.7143                   >DBE Goal          |
|                      UM/Hr        0.0714                   >MBE Goal          |
|                      UM/MH        0.0218                   >WBE Goal          |
|                      MH/UM       45.9327                   >O-1 Goal          |
|                                                           >O-2 Goal          |
+------------------------------------------------------------------------------+
```

```
+33 10 11 1---------------Mobe/Demobe---------------Qty_UM-- 30.00-DA- 3324.40--+
|                                                                              |
| A s s i g n m n t s    S c h e d u l e          C o s t  Unit    Total       |
| -------------------    -----------------        ---------------------------  |
| Bid Item Line     1 Duration                    Source    Details LC         |
| Quote Grp    NONE    Bar Start                                                |
| Qry Lbl 1            Bar Finsh                   Lab Base  1177.553    35327  |
| Qry Lbl 2            CPM ES                      Lab Brdn   977.488    29325  |
| Qry Lbl 3            CPM EF                      Eqp Own                      |
| Acnt                 CPM LS                      Eqp Op                       |
|                      CPM LF                      Rental     240.000     7200  |
| S h i f t s          Float                       Material                     |
| -----------------                               Supply      20.000      600  |
| Work Hrs/Shift   10  P r o d u c t i o n         Subconts                     |
| Pay Hrs/Shift    10  -------------------         Other      861.333    25840  |
| Shifts/Day      1.0  Days         42.0000        Tax RMSO    48.025     1441  |
| Days/Week       5.0  Shifts       42.0000        +--------------------------+ |
|                      Hours       420.0000        |Total     3324.400    99732| |
| R a t e s            Man-Hrs    1378.000         +--------------------------+ |
| ---------            Eqp-Hrs    1200.000         > Level 3  958.961          |
| Rates Unique         UM/Da        0.7143                                      |
|                      UM/Sh        0.7143                   >DBE Goal          |
|                      UM/Hr        0.0714                   >MBE Goal          |
|                      UM/MH        0.0218                   >WBE Goal          |
|                      MH/UM       45.9333                   >O-1 Goal          |
|                                                           >O-2 Goal          |
+------------------------------------------------------------------------------+
```

```
                              (SPME) spme base
                  A C T I V I T Y - D E T A I L   F O R M S


+-33 10 11 1------------------Mobe/Demobe----------------------------------+
|RESOURCE LIST Cost Detail                      Source Worksheet:          |
+==========================================================================+
|                                    +--------------------+                |
|    Man-Hrs    178.00    Eqp-Hrs    |>Unit      764.21   |                |
|    UM/Man-Hr  0.1685    UM/Eqp-Hr  |>TOTAL     22926    |                |
|    Man-Hr/UM  5.9333    Eqp-Hr/UM  +--------------------+                |
|                                                                          |
|                                                                          |
|   Resource                     Quantity UM Count  Unit Cost     TOTAL    |
|   -----------------------------------------------------------------------|
|                                                   Total:     22926       |
|                                                    Unit:     764.21      |
|                                                                          |
|   LCM5     CONSTRUCTION MGR. 5   120.00 HR 1.00    34.62      4154.40     |
|   LCCIOFF  CCI OFFICE/ADMIN.      48.00 HR 1.00    15.40       739.20     |
|   LCM6     CONSTRUCTION MGR. 6    10.00 HR 1.00    38.50       385.00     |
|   SMISC100 MISC. SUPPLIES          6.00 EA 1.00   105.00       630.00     |
|   UAIRFAR1 Not Found                       1.00                          |
|   UCAR1    RENTAL CAR              6.00 DA 1.00    50.00       300.00     |
|   UPRDIEM1 PER DIEM W/ HOTEL       3.00 DA 1.00    89.25       267.75     |
|   UPRDIEM2 PER DIEM W/OUT HOTEL    3.00 DA 1.00    25.00        75.00     |
|   UAIRFAR1 Not Found                       1.00                          |
|   UPRDIEM1 PER DIEM W/ HOTEL     180.00 DA 1.00    89.25     16065.00     |
|   UMILEAGE POV MILEAGE          1000.00 MI 1.00     0.31       310.00     |
+--------------------------------------------------------------------------+


+-33 10 11 1------------------Mobe/Demobe----------------------------------+
|CREW CONFIGURATION Cost Detail                 Source Worksheet:          |
+==========================================================================+
|                                              +--------------------+      |
|Days      30.000  UM/Da 1.0000  Da/UM 1.0000  |>Unit     2213.7    |      |
|Shifts    30.000  UM/Sh 1.0000  Sh/UM 1.0000  |>TOTAL    66411     |      |
|Hours     300.00  UM/Hr 0.1000  Hr/UM 10.000  +--------------------+      |
|Man-hours 1200.0  UM/MH 0.0250  MH/UM 40.000                              |
|                                                                          |
|                                                                          |
|Resource   Qty                              Unit Cost  UM      TOTAL      |
|--------------------------------------------------------------------------|
|Man-Count:   4.00                                Total:    66411          |
|Eqp-Count:   4.00                                 Unit:    2213.7         |
|ONSITE     1.00  ONSITE MGMT. CREW            221.37    HR  66410.64       |
+--------------------------------------------------------------------------+


+==========================================================================+
```

```
                          (SPME) spme base
                 A C T I V I T Y - D E T A I L   F O R M S


  +33 10 11 2---------------Operations----------------Qty_UM-- 74.00-DA- 3337.04--+
  |                                                                               |
  | A s s i g n m n t s     S c h e d u l e         C o s t  Unit     Total       |
  | -------------------     ----------------         ----------------------------  |
  | Bid Item Line     1 Duration                    Source    Details  LC         |
  | Quote Grp    NONE   Bar Start                                                  |
  | Qry Lbl 1           Bar Finsh                   Lab Base 1177.644    87146     |
  | Qry Lbl 2           CPM ES                      Lab Brdn  977.488    72334     |
  | Qry Lbl 3           CPM EF                       Eqp Own                       |
  | Acnt                CPM LS                       Eqp Op                        |
  |                     CPM LF                       Rental   240.135    17770     |
  | S h i f t s         Float                        Material                      |
  | -----------------                               Supply    20.270     1500     |
  | Work Hrs/Shift   10  P r o d u c t i o n         Subconts                      |
  | Pay  Hrs/Shift   10  -----------------           Other    872.905    64595     |
  | Shifts/Day     1.0  Days       103.6000          Tax RMSO  48.600     3596     |
  | Days/Week      5.0  Shifts     103.6000         +--------------------------+   |
  |                     Hours      1036.000         |Total     3337.042   246941|   |
  | R a t e s           Man-Hrs    3399.000         +--------------------------+   |
  | ---------           Eqp-Hrs    2960.000          > Level 3 2374.434           |
  | Rates Unique        UM/Da        0.7143                                        |
  |                     UM/Sh        0.7143                   >DBE Goal            |
  |                     UM/Hr        0.0714                   >MBE Goal            |
  |                     UM/MH        0.0218                   >WBE Goal            |
  |                     MH/UM       45.9324                   >O-1 Goal            |
  |                                                          >O-2 Goal            |
  +-------------------------------------------------------------------------------+


  +-33 10 11 2------------------Operations----------------------------------------+
  |RESOURCE LIST Cost Detail                        Source Worksheet:             |
  +===============================================================================+
  |                                                 +--------------------+        |
  |  Man-Hrs    439.00    Eqp-Hrs                   |>Unit     765.79    |        |
  |  UM/Man-Hr  0.1686    UM/Eqp-Hr                 |>TOTAL    56668     |        |
  |  Man-Hr/UM  5.9324    Eqp-Hr/UM                 +--------------------+        |
  |                                                                               |
  |                                                                               |
  |                                                                               |
  | Resource                     Quantity UM Count  Unit Cost       TOTAL         |
  | ---------------------------------------------------------------------------   |
  |                                                   Total:     56668            |
  |                                                   Unit:      765.79           |
  | LCM5     CONSTRUCTION MGR. 5  296.00 HR 1.00      34.62      10247.52          |
  | LCCIOFF  CCI OFFICE/ADMIN.    118.00 HR 1.00      15.40       1817.20          |
  | LCM6     CONSTRUCTION MGR. 6   25.00 HR 1.00      38.50        962.50          |
  | SMISC100 MISC. SUPPLIES        15.00 EA 1.00     105.00       1575.00          |
  | UAIRFAR1 Not Found                      1.00                                  |
  | UCAR1    RENTAL CAR            15.00 DA 1.00      50.00        750.00          |
  | UPRDIEM1 PER DIEM W/ HOTEL      8.00 DA 1.00      89.25        714.00          |
  | UPRDIEM2 PER DIEM W/OUT HOTEL   8.00 DA 1.00      25.00        200.00          |
  | UAIRFAR1 Not Found                      1.00                                  |
  | UPRDIEM1 PER DIEM W/ HOTEL    444.00 DA 1.00      89.25      39627.00          |
  | UMILEAGE POV MILEAGE         2500.00 MI 1.00       0.31        775.00          |
  +-------------------------------------------------------------------------------+
```

(SPME) spme base
A C T I V I T Y - D E T A I L   F O R M S

```
+-33 10 11 2-----------------Operations-------------------------------------+
|CREW CONFIGURATION Cost Detail                    Source Worksheet:        |
+==========================================================================+
                                          +---------------------+
Days         74.000  UM/Da  1.0000  Da/UM  1.0000  |>Unit     2213.7    |
Shifts       74.000  UM/Sh  1.0000  Sh/UM  1.0000  |>TOTAL    163813    |
Hours        740.00  UM/Hr  0.1000  Hr/UM  10.000  +---------------------+
Man-hours    2960.0  UM/MH  0.0250  MH/UM  40.000


Resource     Qty                                 Unit Cost  UM     TOTAL
-----------------------------------------------------------------------------
Man-Count:   4.00                                         Total:  163813
Eqp-Count:   4.00                                          Unit:  2213.7
ONSITE       1.00  ONSITE MGMT. CREW              221.37    HR 163812.91
+---------------------------------------------------------------------------+
```

OPTIMIZED CASE ADDITIONAL CAPITAL COST

Adtnl Cptl Cost, Optimized Case

Technology Evaluation Report for Single Phase Micro-Emulsion

Additional Capital Cost
Optimized Case

Description	Qty	Unit	Unit Cost ($)	Total Cost ($)	Comments
Additional Equipment/Facilities for Optimized Case					
Automation and Monitoring System	1	ls	$200,000	$200,000	Assumes automation system will replace night shift labor, cost based on experience
Subtotal				$200,000	
Facility Design					
Design	10%		$4,667,049	$466,705	Percentage is based on total construction costs
Engineering During Construction	8%		$4,667,049	$373,364	
Post Flushing Soil Coring and Analysis					
Soil Borings	10	ea	$1,200	$12,000	Assumes 1 boring every 20 m of drainline in Phase I (boring will extend to 33 feet), cost based on vendor quote and RACER
Soil Analysis	10	sample	$500	$5,000	Assumes samples will be analyzed for 3 constituents, costs based on prior experience
Subtotal				$17,000	
Present Worth of Post Flushing Soil Coring and Analysis				$16,675	Assumes costs are incurred after Phase 1 of remediation, 5% interest rate
Demolition and Site Restoration After Remediation					
Demolition and Site Restoration after Remediation	2.5%		$4,667,049	$116,676	Percentage is based on total construction costs.
Present Worth of Dem & Rest. Costs				$109,467	Assume costs are incurred at the end of remediation, 5% interest rate
Salvage Value of Equipment	10%		$2,579,293	$257,929	Based on physical treatment equipment
Present Worth of Salvage Value				$241,991	

OPTIMIZED CASE O&M COST

O&M Cost, Optimized Case

Technology Evaluation Report for Single Phase Micro-Emulsion

Operations and Maintenance Cost
Optimized Case

Description	Qty	Unit	Unit Cost ($)	Total Cost ($)	Comments
OPERATIONS/MAINTENANCE AND MONITORING COSTS PER MODULE					
Labor					
Operators	3,470	hrs	$40	$138,811	3 operators will be required for the duration of the project (56 hrs/wk), automated night operation
Supervisor	1,157	hrs	$50	$57,838	1 supervisor will be required for the duration of the project (56 hrs/wk)
Maintenance Technician	413	hrs	$50	$20,658	1 maintenance technician will be required for the duration of the project (20 hrs/wk)
Engineer	826	hrs	$80	$66,101	1 engineer will be required for the duration of the project (40 hrs/wk)
Laboratory Technician	1,157	hrs	$25	$28,919	1 laboratory technician will be required for the duration of the project (56 hrs/wk)
Temporary Site Facilities					
Office Trailer	5	month	$250	$1,187	32' x 8' Temporary Office Trailer, cost from RACER, $250/month
Portable Toilet	5	month	$100	$475	Quote from vendor, $100/month
Move process equipment	1	ls	$10,000	$10,000	Cost for moving the sump pumping equipment to the next phase drainline sumps after each
Chemicals					
Brij 97	1,396,736	lbs	$2.07	$2,891,244	Quote from ICI Americas, bulk rate
n-pentanol	1,163,947	lbs	$0.59	$680,909	Quote from Union Carbide, bulk quanity
salt	5,703,921	lbs	$0.10	$570,392	Quote from vendor.
Maintenance Materials	3%		$4,000,000	$120,000	Engineers estimate 3% of construction cost.
Utilities					
Electric - Initial Water Flood	20,882	kwh	$0.08	$1,671	Electrical requirements for pumps, blowers, etc. (52.5 hp assumed per module)
Electric - Surfactant and Post Water Flood	503,150	kwh	$0.08	$40,252	Electrical requirements for pumps, blowers, etc. (230 hp assumed per module, 150 hp electric boi
Off-Site Laboratory Analysis					
Analysis of injected fluid quality	13	sample	$100	$1,300	Once every week analysis of alcohol (1 analysis per week of surfactant flood)
Analysis of UF and steam stripper effluent	5	sample	$450	$2,250	Once every month analysis of alcohol, TPH, and VOC/SVOC (4 analyses per month of surfactant and post water flood)
Analysis of treated produced fluids	5	sample	$450	$2,250	Once every month analysis of alcohol, TPH, and VOC/SVOC (4 analyses per month of surfactant and post water flood)
Analysis of monitoring well fluids	42	sample	$450	$18,900	Once every week analysis of alcohol, TPH, and VOC/SVOC (2 analyses per week of the project), sample 2 wells per 10 days per module
Disposal					
Incineration of sludge from SBR	0	lb	$0.57	$0	Quote from Laidlaw Environmental (SLC, UT), <3,000 BTU/lb, <50% water, <10% Cl (sludge produced during surfactant flood period only)
Incineration of NAPL from O/W separator	2,800,106	lb	$0.17	$476,018	Quote from Laidlaw Environmental (SLC, UT), <3,000 BTU/lb, >50% water, <10% Cl (fluids produced during surfactant flood period only)
Incineration of decant from steam stripper condenser	31,975	lb	$0.06	$1,919	Quote from Laidlaw Environmental (SLC, UT), >12,000 BTU/lb, <50% water, <10% Cl (fluids produced during surfactant flood period only)
Operation of Ultrafiltration	0	1,000 gal	$0.45	$0	Journal AWWA, May 1996 (UF operated during surfactant flood period only)
Discharge to IWTP	14,383	1,000 gal	$11.00	$158,213	IWTP costs were obtained from Hill Air Force Base personnel. (all water from pre- and post-water flood plus residual from treatment during surfactant flood)
Subtotal				$5,289,304	
Contingency	20%		$5,289,304	$1,057,861	
Total - Annual Operations/Maintenance and Monitoring Costs				**$6,347,165**	
Total No. of Modules	3 modules				
Total Time for Operation of all Modules	1.31 years				
Annual Operations/Maintenance and Monitoring Costs				$14,565,552	
Interest Rate for Present Worth Calculation	5%				
Present Worth - Operations/Maintenance and Monitoring of All Modules				**$18,000,583**	

```
                      (SPMEFALT) spme final Alternate
                           W B S   O U T L I N E


  Code                      Description              Quantity  UM  Source
  --------------------------------------------------------------------------

       33                   CONSTRUCTION ACTIVITIES    1.0000  LS  Detail
       33 10               Remedial Action            1.0000  LS  Detail
       33 10  1            Mobe and Preparation Work   1.0000  LS  Detail
       33 10  1  1         Install Access Roads      300.0000  LF  Plug
       33 10  1  2         Install Secondary Contain 5000.000  SF  Plug
       33 10  1  3         Pre-fab Buildings           2.0000  EA  Plug
       33 10  1  4         Site Utilities              1.0000  LS  Plug
       33 10  1  5         Site Temporary Facilities   1.0000  MO  Plug
       33 10  2            Monitoring/Testing/Analys   1.0000  LS  Detail
       33 10  3            Site Work                   1.0000  LS  Detail
       33 10  4            G'water Collection/Contrl   1.0000  LS  Detail
       33 10  4  1         Monitoring Wells           40.0000  EA  Quote
       33 10  4  2         Recovery Line System      5800.000  LF  Quote
       33 10  4  3         Injection Line System     5200.000  LF  Quote
       33 10  4  4         Soil Borings               75.0000  EA  Plug
       33 10  4  5         Soil Samples               25.0000  EA  Plug
       33 10  5            Physical Treatment          1.0000  LS  Detail
       33 10  5  1         Electrical Package          1.0000  LS  Detail
       33 10  5  2         Plumbing Package            1.0000  LS  Detail
       33 10  8            Demobilization              1.0000  LS  Detail
       33 10 11            Construction Management   104.0000  DA  Detail
       33 10 11  1         Mobe/Demobe                30.0000  DA  Detail
       33 10 11  2         Operations                 74.0000  DA  Detail
```

```
                    (SPMEFALT) spme final Alternate
                            M A R K U P

   Currency: US Dollars

   Summary
   -------

                       --------Markup------------
                       Average
   Escalated Cost    Amount Rate            Amount   Escalated Cost + MU
   ----------------------------------------------------------------------
   Direct         4441491.66    15.00%    666223.75         5107715.41
   Overhead                      0.00%
   ----------------------------------------------------------------------
   Total          4441491.66    15.00     666223.75         5107715.41


   Direct (DC) Breakdown
   ---------------------------------

                       --------Markup------------
                       Average
   Escalated Cost    Amount Rate            Amount   Escalated Cost + MU
   ----------------------------------------------------------------------
   Labor Base      157232.72    15.00%     23584.91          180817.63
   Labor Brdn      106546.19    15.00%     15981.93          122528.12
   Equip Ownr       25500.00    15.00%      3825.00           29325.00
   Equip Oprn        4250.00    15.00%       637.50            4887.50
   Rentals          26120.00    15.00%      3918.00           30038.00
   Materials      1461464.00    15.00%    219219.60         1680683.60
   Supplies          6600.00    15.00%       990.00            7590.00
   Subcontrct     2476985.40    15.00%    371547.81         2848533.21
   Other Cost       99935.00    15.00%     14990.25          114925.25
   Tax-RSMO         76858.35    15.00%     11528.75           88387.10
   ----------------------------------------------------------------------
   DC Total       4441491.66    15.00%    666223.75         5107715.41


   Overhead (OH) Breakdown
   -----------------------------------

                       --------Markup------------
                       Average
   Escalated Cost    Amount Rate            Amount   Escalated Cost + MU
   ----------------------------------------------------------------------
   Labor Base                   15.00%                        23584.91
   Labor Brdn                   15.00%                        15981.93
   Equip Ownr                   15.00%                         3825.00
   Equip Oprn                   15.00%                          637.50
   Rentals                      15.00%                         3918.00
   Materials                    15.00%                       219219.60
   Supplies                     15.00%                          990.00
   Subcontrct                   15.00%                       371547.81
   Other Cost                   15.00%                        14990.25
   Tax-RSMO                     15.00%                        11528.75
   ----------------------------------------------------------------------
   OH Total                      0.00%

   Subcontractor Markup Breakdown
   ------------------------------

                                      -----Markup---------
   Code     Subcontractor   Award Amount  Rate      Amount    Award + Markup
   ----------------------------------------------------------------------
                     Total    2476985.40  0.00%                   2476985.40
   ----------------------------------------------------------------------
   00000000  Plugged Amount w   625485.40  0.00%                    625485.40
   UGMECH                      1771000.00  0.00%                   1771000.00
   WELLS                         80500.00  0.00%                     80500.00
```

(SPMEFALT) spme final Alternate
WORK PLAN SUMMARY

| | | | Labor | | Equipment | | | | | | | | | |
Description	Qty	UM	Base	Burden	Ownershp	Operatn	Rentals	Material	Supply	Subcon	Other	Taxes	Total	s/r/c
33 CONSTRUCTION ACTIVITIES	1.0000	LS	U T										0.0000 / 0.00	D
33 10 Remedial Action	1.0000	LS	U T										0.0000 / 0.00	D
33 10 1 Mobe and Preparation Work	1.0000	LS	U T										0.0000 / 0.00	D
33 10 1 1 Install Access Roads	300.0000	LF	U T							84.0000 / 25200.00			84.0000 / 25200.00	P
33 10 1 2 Install Secondary Contain	5000.000	SF	U T							2.4000 / 12000.00			2.4000 / 12000.00	P
33 10 1 3 Pre-fab Buildings	2.0000	EA	U T					10000.00 / 20000.00					10000.0000 / 20000.00	P
33 10 1 4 Site Utilities	1.0000	LS	U T							15000.00 / 15000.00			15000.0000 / 15000.00	P
33 10 1 5 Site Temporary Facilities	1.0000	MO	U T								9500.000 / 9500.00		9500.0000 / 9500.00	F
33 10 2 Monitoring/Testing/Analys	1.0000	LS	U T										0.0000 / 0.00	D
33 10 3 Site Work	1.0000	LS	U T					10000.00 / 10000.00		12000.00 / 12000.00			22000.0000 / 22000.00	P
33 10 4 G'water Collection/Contrl	1.0000	LS	U T										0.0000 / 0.00	D
33 10 4 1 Monitoring Wells	40.0000	EA	U T							2012.500 / 80500.00			2012.5000 / 80500.00	Q
33 10 4 2 Recovery Line System	5800.000	LF	U T							161.0000 / 933800.0			161.0000 / 933800.00	Q
33 10 4 3 Injection Line System	5200.000	LF	U T							161.0000 / 837200.0			161.0000 / 837200.00	Q
33 10 4 4 Soil Borings	75.0000	EA	U T							600.0000 / 45000.00			600.0000 / 45000.00	P
33 10 4 5 Soil Samples	25.0000	EA	U							500.0000			500.0000	

Page: 1

(SPMEFALT) spme final Alternate
W O R K   P L A N   S U M M A R Y

------Activity's immediate cost (excluding subordinates' cost)------

				Labor		Equipment									Src
Code	Description	Qty	UM		Base	Burden	Ownershp	Operatn	Rentals	Material	Supply	Subcon	Other	Taxes	Total	
												12500.00			12500.00	P
33 10 5	Physical Treatment	1.0000	LS	U	29752.50		25500.00	4250.000		1280200	2500.000	11375.00		64135.00	1417712.5000	P
				T	29752.50		25500.00	4250.00		1280200	2500.00	11375.00		64135.00	1417712.50	D
33 10 5 1	Electrical Package	1.0000	LS	U								204820.4			204820.4000	
				T								204820.4			204820.40	D
33 10 5 2	Plumbing Package	1.0000	LS	U						151264.0		287590.0		7563.200	446417.2000	
				T						151264.0		287590.0		7563.20	446417.20	D
33 10 8	Demobilization	1.0000	LS	U	5008.000	4887.440			1150.000		2000.000			123.0000	13168.4400	
				T	5008.00	4887.44			1150.00		2000.00			123.00	13168.44	D
33 10 11	Construction Management	104.0000	DA	U											0.0000	
				T											0.00	D
33 10 11 1	Mobe/Demobe	30.0000	DA	U	1177.553	977.4880			240.0000		20.0000		861.3333	48.0250	3324.3997	
				T	35326.60	29324.64			7200.00		600.00		25840.00	1440.75	99731.99	D
33 10 11 2	Operations	74.0000	DA	U	1177.644	977.4880			240.1351		20.2703		872.9054	48.6000	3337.0423	
				T	87145.62	72334.11			17770.00		1500.00		64595.00	3596.40	246941.13	D
	Total				157232.7	106546.2	25500.00	4250.00	26120.00	1461464	6600.00	2476985	99935.00	76858.35	4441491.66	

```
                        (SPMEFALT) spme final Alternate
                     A C T I V I T Y - D E T A I L   F O R M S


 +33-------------------------CONSTRUCTION ACTIVITIES---Qty_UM--  1.00-LS- 4441492--+

 | A s s i g n m n t s    S c h e d u l e          C o s t  Unit      Total
 | --------------------    -----------------        -----------------------------
 | Bid Item Line        1 Duration                  Source     Details
 | Quote Grp    NONE      Bar Start
 | Qry Lbl 1              Bar Finsh                  Lab Base  157232.7   157233
 | Qry Lbl 2              CPM ES                     Lab Brdn  106546.2   106546
 | Qry Lbl 3              CPM EF                     Eqp Own   25500.00    25500
 | Acnt                   CPM LS                     Eqp Op    4250.000     4250
 |                        CPM LF                     Rental    26120.00    26120
 | S h i f t s            Float                      Material  1461464   1461464
 | ------------------                               Supply    6600.000     6600
 | Work Hrs/Shift    10   P r o d u c t i o n        Subconts  2476985   2476985
 | Pay Hrs/Shift     10   ------------------         Other     99935.00    99935
 | Shifts/Day       1.0   Days         175.6000      Tax RMSO  76858.35    76858|
 | Days/Week        5.0   Shifts       175.6000     +-------------------------+
 |                        Hours        1756.000     |Total     4441492  4441492|
 | R a t e s             Man-Hrs      6227.000     +-------------------------+
 | ---------             Eqp-Hrs      5610.000      > Job      4441492
 | Rates Unique          UM/Da          0.0057
 |                        UM/Sh          0.0057           >DBE Goal
 |                        UM/Hr          0.0006           >MBE Goal
 |                        UM/MH          0.0002           >WBE Goal
 |                        MH/UM        6227.000           >O-1 Goal
 |                                                        >O-2 Goal
 +--------------------------------------------------------------------------------+



 +33 10-------------------Remedial Action-----------Qty_UM--  1.00-LS- 4441492--+

 | A s s i g n m n t s    S c h e d u l e          C o s t  Unit      Total
 | --------------------    -----------------        -----------------------------
 | Bid Item Line        1 Duration                  Source     Details
 | Quote Grp    NONE      Bar Start
 | Qry Lbl 1              Bar Finsh                  Lab Base  157232.7   157233
 | Qry Lbl 2              CPM ES                     Lab Brdn  106546.2   106546
 | Qry Lbl 3              CPM EF                     Eqp Own   25500.00    25500
 | Acnt                   CPM LS                     Eqp Op    4250.000     4250
 |                        CPM LF                     Rental    26120.00    26120
 | S h i f t s            Float                      Material  1461464   1461464
 | ------------------                               Supply    6600.000     6600
 | Work Hrs/Shift    10   P r o d u c t i o n        Subconts  2476985   2476985
 | Pay Hrs/Shift     10   ------------------         Other     99935.00    99935
 | Shifts/Day       1.0   Days         175.6000      Tax RMSO  76858.35    76858|
 | Days/Week        5.0   Shifts       175.6000     +-------------------------+
 |                        Hours        1756.000     |Total     4441492  4441492|
 | R a t e s             Man-Hrs      6227.000     +-------------------------+
 | ---------             Eqp-Hrs      5610.000      > Level 1  4441492
 | Rates Unique          UM/Da          0.0057
 |                        UM/Sh          0.0057           >DBE Goal
 |                        UM/Hr          0.0006           >MBE Goal
 |                        UM/MH          0.0002           >WBE Goal
 |                        MH/UM        6227.000           >O-1 Goal
 |                                                        >O-2 Goal
 +--------------------------------------------------------------------------------+
```

```
                    (SPMEFALT) spme final Alternate
                    A C T I V I T Y - D E T A I L   F O R M S


   +33 10 1----------------Mobe and Preparation Work-Qty_UM--  1.00-LS-81700.00--+

   A s s i g n m n t s   S c h e d u l e            C o s t  Unit    Total
   --------------------   ------------------         ----------------------------
   Bid Item Line      1 Duration                    Source    Details
   Quote Grp    NONE    Bar Start
   Qry Lbl 1            Bar Finsh                    Lab Base
   Qry Lbl 2            CPM ES                       Lab Brdn
   Qry Lbl 3            CPM EF                       Eqp Own
   Acnt                 CPM LS                       Eqp Op
                        CPM LF                       Rental
   S h i f t s          Float                        Material  20000.00   20000
   ------------------                                Supply
   Work Hrs/Shift   10  P r o d u c t i o n          Subconts  52200.00   52200
   Pay Hrs/Shift    10  -------------------          Other     9500.000    9500
   Shifts/Day      1.0  Days                         Tax RMSO
   Days/Week       5.0  Shifts                       +--------------------------+
                        Hours                        |Total     81700.00   81700|
   R a t e s            Man-Hrs                       +--------------------------+
   ----------           Eqp-Hrs                      > Level 2 81700.00
   Rates Unique         UM/Da
                        UM/Sh                                  >DBE Goal
                        UM/Hr                                  >MBE Goal
                        UM/MH                                  >WBE Goal
                        MH/UM                                  >O-1 Goal
                                                               >O-2 Goal
   +----------------------------------------------------------------------------+



   +33 10 1 1----------------Install Access Roads------Qty_UM--300.00-LF-  84.00--+

   A s s i g n m n t s   S c h e d u l e            C o s t  Unit    Total
   --------------------   ------------------         ----------------------------
   Bid Item Line      1 Duration                    Source    Plug
   Quote Grp    NONE    Bar Start
   Qry Lbl 1            Bar Finsh                    Lab Base
   Qry Lbl 2            CPM ES                       Lab Brdn
   Qry Lbl 3            CPM EF                       Eqp Own
                        CPM LS                       Eqp Op
                        CPM LF                       Rental
                        Float                        Material
                                                     Supply
                                                     Subconts   84.000   25200
                                                     Other
                                                     Tax RMSO
                                                     +--------------------------+
                                                     |Total      84.000   25200|
                                                     +--------------------------+
                                                     > Level 3 25200.00
   +----------------------------------------------------------------------------+
```

(SPMEFALT) spme final Alternate
A C T I V I T Y - D E T A I L   F O R M S

```
+33 10 1 2----------------Install Secondary Contain-Qty_UM--5000.0-SF-    2.40--+
|                                                                              |
|Assignmnts   Schedule              C o s t  Unit    Total                     |
|-------------------  -----------------    ---------------------------         |
|Bid Iten Line      1 Duration             Source    Plug                      |
|Quote Grp    NONE    Bar Start                                                |
|Qry Lbl 1            Bar Finsh            Lab Base                             |
|Qry Lbl 2            CPM ES               Lab Brdn                            |
|Qry Lbl 3            CPM EF               Eqp Own                             |
|                     CPM LS               Eqp Op                             |
|                     CPM LF               Rental                             |
|                     Float                Material                           |
|                                          Supply                             |
|                                          Subconts    2.400      12000        |
|                                          Other                              |
|                                          Tax RMSO                           |
|                                          +--------------------------        |
|                                          |Total      2.400      12000        |
|                                          +--------------------------        |
|                                          > Level 3 12000.00                  |
|                                                                              |
|                                                                              |
|                                                                              |
+------------------------------------------------------------------------------+
```

```
+33 10 1 3----------------Pre-fab Buildings---------Qty_UM--  2.00-EA-10000.00--+
|                                                                              |
|Assignmnts   Schedule              C o s t  Unit    Total                     |
|-------------------  -----------------    ---------------------------         |
|Bid Iten Line      1 Duration             Source    Plug                      |
|Quote Grp    NONE    Bar Start                                                |
|Qry Lbl 1            Bar Finsh            Lab Base                            |
|Qry Lbl 2            CPM ES               Lab Brdn                            |
|Qry Lbl 3            CPM EF               Eqp Own                             |
|                     CPM LS               Eqp Op                             |
|                     CPM LF               Rental                             |
|                     Float                Material  10000.00     20000        |
|                                          Supply                             |
|                                          Subconts                           |
|                                          Other                              |
|                                          Tax RMSO                           |
|                                          +--------------------------        |
|                                          |Total    10000.00     20000        |
|                                          +--------------------------        |
|                                          > Level 3 20000.00                  |
|                                                                              |
|                                                                              |
|                                                                              |
+------------------------------------------------------------------------------+
```

```
                        (SPMEFALT) spme final Alternate
                      A C T I V I T Y - D E T A I L   F O R M S


  +33 10 1 4----------------Site Utilities-----------Qty_UM--  1.00-LS-15000.00--+

   A s s i g n m n t s    S c h e d u l e          C o s t  Unit     Total
   --------------------   -----------------         ---------------------------
   Bid Iten Line      1 Duration                    Source    Plug
   Quote Grp    NONE    Bar Start
   Qry Lbl 1            Bar Finsh                    Lab Base
   Qry Lbl 2            CPM ES                       Lab Brdn
   Qry Lbl 3            CPM EF                       Eqp Own
                        CPM LS                       Eqp Op
                        CPM LF                       Rental
                        Float                        Material
                                                     Supply
                                                     Subconts  15000.00     15000
                                                     Other
                                                     Tax RMSO
                                                     +---------------------------
                                                     |Total     15000.00     15000
                                                     +---------------------------
                                                     > Level 3 15000.00



  +--------------------------------------------------------------------------+


  +33 10 1 5----------------Site Temporary Facilities-Qty_UM--  1.00-MO- 9500.00--+

   A s s i g n m n t s    S c h e d u l e          C o s t  Unit     Total
   --------------------   -----------------         ---------------------------
   Bid Iten Line      1 Duration                    Source    Plug
   Quote Grp    NONE    Bar Start
   Qry Lbl 1            Bar Finsh                    Lab Base
   Qry Lbl 2            CPM ES                       Lab Brdn
   Qry Lbl 3            CPM EF                       Eqp Own
                        CPM LS                       Eqp Op
                        CPM LF                       Rental
                        Float                        Material
                                                     Supply
                                                     Subconts
                                                     Other     9500.000      9500
                                                     Tax RMSO
                                                     +---------------------------
                                                     |Total     9500.000      9500
                                                     +---------------------------
                                                     > Level 3 9500.000



  +--------------------------------------------------------------------------+
```

```
                    (SPMEFALT) spme final Alternate
                  A C T I V I T Y - D E T A I L   F O R M S


  +33 10 2------------------Monitoring/Testing/Analys-Qty_UM--  1.00-LS-         --+
  |
  | A s s i g n m n t s    S c h e d u l e        C o s t  Unit    Total
  | ------------------      -----------------     -----------------------------
  | Bid Item Line     1 Duration                  Source   Details
  | Quote Grp    NONE    Bar Start
  | Qry Lbl 1            Bar Finsh                 Lab Base
  | Qry Lbl 2            CPM ES                     Lab Brdn
  | Qry Lbl 3            CPM EF                     Eqp Own
  | Acnt                 CPM LS                     Eqp Op
  |                      CPM LF                     Rental
  | S h i f t s          Float                      Material
  | ------------------                             Supply
  | Work Hrs/Shift  10   P r o d u c t i o n       Subconts
  | Pay  Hrs/Shift  10   -------------------        Other
  | Shifts/Day     1.0   Days                       Tax RMSO
  | Days/Week      5.0   Shifts                     +------------------------+
  |                      Hours                      |Total                   |
  | R a t e s            Man-Hrs                    +------------------------+
  | ---------            Eqp-Hrs                    > Level 2
  | Rates Unique         UM/Da
  |                      UM/Sh                           >DBE Goal
  |                      UM/Hr                           >MBE Goal
  |                      UM/MH                           >WBE Goal
  |                      MH/UM                           >O-1 Goal
  |                                                      >O-2 Goal
  +----------------------------------------------------------------------------+


  +33 10 3------------------Site Work----------------Qty_UM--  1.00-LS-22000.00--+
  |
  | A s s i g n m n t s    S c h e d u l e        C o s t  Unit    Total
  | ------------------      -----------------     -----------------------------
  | Bid Iten Line     1 Duration                  Source   Plug
  | Quote Grp    NONE    Bar Start
  | Qry Lbl 1            Bar Finsh                 Lab Base
  | Qry Lbl 2            CPM ES                     Lab Brdn
  | Qry Lbl 3            CPM EF                     Eqp Own
  |                      CPM LS                     Eqp Op
  |                      CPM LF                     Rental
  |                      Float                      Material  10000.00     10000
  |                                                Supply
  |                                                Subconts  12000.00     12000
  |                                                Other
  |                                                Tax RMSO
  |                                                +------------------------+
  |                                                |Total     22000.00     22000
  |                                                +------------------------+
  |                                                > Level 2 22000.00
  |
  |
  |
  +----------------------------------------------------------------------------+
```

```
                    (SPMEFALT) spme final Alternate
                    A C T I V I T Y - D E T A I L   F O R M S


  +33 10 4-----------------G'water Collection/Contrl-Qty_UM--  1.00-LS- 1909000--+

  |A s s i g n m n t s   S c h e d u l e        C o s t  Unit    Total
  |--------------------   -----------------     --------------------------
  |Bid Item Line      1 Duration                Source   Details
  |Quote Grp    NONE    Bar Start
  |Qry Lbl 1            Bar Finsh               Lab Base
  |Qry Lbl 2            CPM ES                   Lab Brdn
  |Qry Lbl 3            CPM EF                   Eqp Own
  |Acnt                 CPM LS                   Eqp Op
  |                     CPM LF                   Rental
  |S h i f t s          Float                    Material
  |------------------                            Supply
  |Work Hrs/Shift  10                            Subconts   1909000  1909000
  |Pay  Hrs/Shift  10   P r o d u c t i o n      Other
  |Shifts/Day     1.0   -------------------      Tax RMSO
  |Days/Week      5.0   Days                    +-----------------------------+
  |                     Shifts                   |Total      1909000  1909000|
  |                     Hours                   +-----------------------------+
  |R a t e s            Man-Hrs                  > Level 2  1909000
  |---------            Eqp-Hrs
  |Rates Unique         UM/Da
  |                     UM/Sh                           >DBE Goal
  |                     UM/Hr                           >MBE Goal
  |                     UM/MH                           >WBE Goal
  |                     MH/UM                           >O-1 Goal
  |                                                     >O-2 Goal
  +----------------------------------------------------------------------+


  +33 10 4 1----------------Monitoring Wells----------Qty_UM-- 40.00-EA- 2012.50--+

  |A s s i g n m n t s   S c h e d u l e        C o s t  Unit    Total
  |--------------------   -----------------     --------------------------
  |Bid Item Line      1 Duration                Source   Quote
  |Query Grp    NONE    Bar Start
  |Qry Lbl 1            Bar Finsh               Lab Base
  |Qry Lbl 2            CPM ES                   Lab Brdn
  |Qry Lbl 3            CPM EF                   Eqp Own
  |                     CPM LS                   Eqp Op
  |                     CPM LF                   Rental
  |                     Float                    Material
  |                                              Supply
  |                                              Subconts   2012.500   80500
  |                                              Other
  |                                              Tax RMSO
  |                                             +-----------------------------+
  |                                              |Total     2012.500   80500|
  |                                             +-----------------------------+
  |                                              > Level 3 80500.00
  |Awardee
  |         WELLS                                      >DBE Goal
  |                                                    >MBE Goal
  |                                                    >WBE Goal
  |DBE Allow      1.00%                                >O-1 Goal
  |                                                    >O-2 Goal
  +----------------------------------------------------------------------+
```

```
                    (SPMEFALT) spme final Alternate
                  A C T I V I T Y - D E T A I L   F O R M S


+33 10 4 2----------------Recovery Line System------Qty_UM--5800.0-LF-  161.00--+
|                                                                               |
| A s s i g n m n t s   S c h e d u l e        C o s t  Unit     Total          |
| -------------------   ----------------       ---------------------------      |
| Bid Item Line       1 Duration               Source   Quote                   |
| Query Grp    NONE     Bar Start                                               |
| Qry Lbl 1             Bar Finsh              Lab Base                          |
| Qry Lbl 2             CPM ES                 Lab Brdn                          |
| Qry Lbl 3             CPM EF                 Eqp Own                           |
|                       CPM LS                 Eqp Op                            |
|                       CPM LF                 Rental                           |
|                       Float                  Material                          |
|                                              Supply                            |
|                                              Subconts  161.000    933800       |
|                                              Other                             |
|                                              Tax RMSO                          |
|                                              +--------------------------+      |
|                                              |Total    161.000    933800       |
|                                              +--------------------------+      |
|                                               > Level 3 933800.0               |
| Awardee                                                                        |
|       UGMECH                                             >DBE Goal             |
|                                                          >MBE Goal             |
| DBE Allow     1.00%                                      >WBE Goal             |
|                                                          >O-1 Goal             |
|                                                          >O-2 Goal             |
+-------------------------------------------------------------------------------+



+33 10 4 3----------------Injection Line System-----Qty_UM--5200.0-LF-  161.00--+
|                                                                               |
| A s s i g n m n t s   S c h e d u l e        C o s t  Unit     Total          |
| -------------------   ----------------       ---------------------------      |
| Bid Item Line       1 Duration               Source   Quote                   |
| Query Grp    NONE     Bar Start                                               |
| Qry Lbl 1             Bar Finsh              Lab Base                          |
| Qry Lbl 2             CPM ES                 Lab Brdn                          |
| Qry Lbl 3             CPM EF                 Eqp Own                           |
|                       CPM LS                 Eqp Op                            |
|                       CPM LF                 Rental                           |
|                       Float                  Material                          |
|                                              Supply                            |
|                                              Subconts  161.000    837200       |
|                                              Other                             |
|                                              Tax RMSO                          |
|                                              +--------------------------+      |
|                                              |Total    161.000    837200       |
|                                              +--------------------------+      |
|                                               > Level 3 837200.0               |
| Awardee                                                                        |
|       UGMECH                                             >DBE Goal             |
|                                                          >MBE Goal             |
| DBE Allow     1.00%                                      >WBE Goal             |
|                                                          >O-1 Goal             |
|                                                          >O-2 Goal             |
+-------------------------------------------------------------------------------+
```

CH2M HILL
(SPMEFALT) spme final Alternate
A C T I V I T Y - D E T A I L   F O R M S

```
+33 10 4 4--------------Soil Borings-------------Qty_UM-- 75.00-EA-  600.00--+
|                                                                           |
| A s s i g n m n t s   S c h e d u l e         C o s t  Unit    Total       |
| -------------------   -----------------       -------------------------   |
| Bid Iten Line      1 Duration                 Source    Plug              |
| Quote Grp   NONE     Bar Start                                            |
| Qry Lbl 1            Bar Finsh                 Lab Base                    |
| Qry Lbl 2            CPM ES                     Lab Brdn                   |
| Qry Lbl 3            CPM EF                     Eqp Own                    |
|                      CPM LS                     Eqp Op                     |
|                      CPM LF                     Rental                     |
|                      Float                      Material                   |
|                                                 Supply                     |
|                                                 Subconts   600.000   45000 |
|                                                 Other                      |
|                                                 Tax RMSO                   |
|                                                +------------------------   |
|                                                |Total      600.000   45000 |
|                                                +------------------------   |
|                                                 > Level 3 45000.00         |
|                                                                           |
+---------------------------------------------------------------------------+
```

```
+33 10 4 5----------------Soil Samples-------------Qty_UM-- 25.00-EA-  500.00--+
|                                                                           |
| A s s i g n m n t s   S c h e d u l e         C o s t  Unit    Total       |
| -------------------   -----------------       -------------------------   |
| Bid Iten Line      1 Duration                 Source    Plug              |
| Quote Grp   NONE     Bar Start                                            |
| Qry Lbl 1            Bar Finsh                 Lab Base                    |
| Qry Lbl 2            CPM ES                     Lab Brdn                   |
| Qry Lbl 3            CPM EF                     Eqp Own                    |
|                      CPM LS                     Eqp Op                     |
|                      CPM LF                     Rental                     |
|                      Float                      Material                   |
|                                                 Supply                     |
|                                                 Subconts   500.000   12500 |
|                                                 Other                      |
|                                                 Tax RMSO                   |
|                                                +------------------------   |
|                                                |Total      500.000   12500 |
|                                                +------------------------   |
|                                                 > Level 3 12500.00         |
|                                                                           |
+---------------------------------------------------------------------------+
```

```
                         (SPMEFALT) spme final Alternate
                    A C T I V I T Y - D E T A I L   F O R M S


    +33 10 5-----------------Physical Treatment--------Qty_UM--  1.00-LS- 2068950--+
    |
    |A s s i g n m n t s   S c h e d u l e          C o s t  Unit     Total
    |-------------------   ----------------          --------------------------
    |Bid Item Line      1 Duration                  Source     Details  LC
    |Quote Grp    NONE    Bar Start
    |Qry Lbl 1            Bar Finsh                  Lab Base  29752.50    29753
    |Qry Lbl 2            CPM ES                     Lab Brdn
    |Qry Lbl 3            CPM EF                      Eqp Own  25500.00    25500
    |Acnt                 CPM LS                      Eqp Op   4250.000     4250
    |                     CPM LF                     Rental
    |S h i f t s          Float                      Material  1431464  1431464
    |------------------                              Supply   2500.000     2500
    |Work Hrs/Shift   10  P r o d u c t i o n        Subconts 503785.4   503785
    |Pay Hrs/Shift    10  -------------------        Other
    |Shifts/Day      1.0  Days          25.0000      Tax RMSO 71698.20    71698
    |Days/Week       5.0  Shifts        25.0000      +----------------------------+
    |                     Hours        250.0000      |Total      2068950  2068950|
    |R a t e s            Man-Hrs     1250.000       +----------------------------+
    |---------            Eqp-Hrs     1250.000        > Level 2  2068950
    |Rates Unique         UM/Da         0.0400
    |                     UM/Sh         0.0400           >DBE Goal
    |                     UM/Hr         0.0040           >MBE Goal
    |                     UM/MH         0.0008           >WBE Goal
    |                     MH/UM       1250.000           >O-1 Goal
    |                                                    >O-2 Goal
    +-------------------------------------------------------------------------+


    +-33 10 5--------------------Physical Treatment----------------------------------+
    |RESOURCE LIST Cost Detail                       Source Worksheet:              |
    +================================================================================+
    |                                                +--------------------+
    |  Man-Hrs              Eqp-Hrs                   |>Unit    1358e3    |
    |  UM/Man-Hr            UM/Eqp-Hr                 |>TOTAL   1358e3    |
    |  Man-Hr/UM            Eqp-Hr/UM                 +--------------------+
    |
    |
    |
    |Resource                      Quantity UM Count  Unit Cost        TOTAL
    |-------------------------------------------------------------------
    |                                                     Total:   1358e3
    |                                                     Unit:    1358e3
    |MDECANT1 DECANTER (8GPM)           1.00 EA 1.00      21000    21000.00
    |MFILTER1 BAG FILTER (55GPM)        2.00 EA 1.00      1155.0    2310.00
    |MFILTER4 ULTRAFILTRATION UNIT(50G)      EA 1.00
    |MFILTER3 BAG FILTER (44 GPM)            EA 1.00
    |MGENR'R1 STEAM GENERATOR (150HP)   1.00 EA 1.00      48300    48300.00
    |MMIXER1  CHEM STORAGE TANK MIXER   2.00 EA 1.00      5250.0   10500.00
    |MMIXER2  CHEMICAL MIX TANKS MIXER  3.00 EA 1.00      5250.0   15750.00
    |MPUMP1   CHEMICAL FEED PUMP (25G)  2.00 EA 1.00      5250.0   10500.00
    |MPUMP2   CHEMICAL FEED PUMP (55G)  2.00 EA 1.00      8925.0   17850.00
    |MPUMP3   CHEM. FEED INJ.PUMP (15G) 9.00 EA 1.00      7875.0   70875.00
    |MPUMP4   CHEM. FEED INJ. PUMP (5G)13.00 EA 1.00      6825.0   88725.00
    |MPUMP5   VAR.SPEED RECOV.PMP (15G) 9.00 EA 1.00      4305.0   38745.00
    |MPUMP6   VAR.SPEED RECOV.PMP (5G) 16.00 EA 1.00      4305.0   68880.00
    |MPUMP8   TREAT.PLNT.FEED PUMP (65) 4.00 EA 1.00      4200.0   16800.00
    +-------------------------------------------------------------------------+
```

```
                        (SPMEFALT) spme final Alternate
                       A C T I V I T Y - D E T A I L   F O R M S


+-33 10 5--------------------Physical Treatment------------------------------+
|RESOURCE LIST Cost Detail                      Source Worksheet:            |
+=================================================================================+
                                      +--------------------+
  Man-Hrs              Eqp-Hrs        |>Unit    1358e3     |
  UM/Man-Hr            UM/Eqp-Hr      |>TOTAL   1358e3     |
  Man-Hr/UM            Eqp-Hr/UM      +--------------------+



  Resource                     Quantity UM Count  Unit Cost      TOTAL
  -----------------------------------------------------------------------
                                                  Total:  1358e3
                                                  Unit:   1358e3
  MSTRIP2  STEAM STRIP. COLM. (42G)    1.00 EA 1.00   446250   446250.00
  MSUMP1   GALV. CMP W/ COVER         51.00 EA 1.00   525.00    26775.00
  MTANK1   CHEM. STORAGE TANK (10K)    2.00 EA 1.00    18375    36750.00
  MTANK2   CHEM. MIX TANK (20K)        3.00 EA 1.00    36750   110250.00
  MTANK3   EQUALIZATION TANK (20K)     4.00 EA 1.00    36750   147000.00
  MTANK4   CONCENTRATE STOR.TNK.(20K   1.00 EA 1.00    36750    36750.00
  MVPORIZR PROPANE VAPORIZER           1.00 EA 1.00    68250    68250.00
  MCONDENS CONDENSOR (10GPM)           1.00 EA 1.00    36750    36750.00
  MTANK5   SBR TANKS (300K)                 EA 1.00
  MBLOWER1 CENTRIFUGAL BLOWER(15PSI)        EA 1.00
  MFLTRPRS FILTER PRESS (SMALL)             EA 1.00
  MFEEDSYS CHEMICAL FEED SYSTEM        2.00 EA 1.00   2100.0     4200.00
  SMISC100 MISC. SUPPLIES             25.00 EA 1.00   105.00     2625.00
  USUBPRDM SUBCONTRACTOR PER DIEM    175.00 MD 1.00    65.00    11375.00
+---------------------------------------------------------------------------+

+-33 10 5--------------------Physical Treatment------------------------------+
|RESOURCE LIST Cost Detail                      Source Worksheet:            |
+=================================================================================+
                                      +--------------------+
  Man-Hrs              Eqp-Hrs        |>Unit    1358e3     |
  UM/Man-Hr            UM/Eqp-Hr      |>TOTAL   1358e3     |
  Man-Hr/UM            Eqp-Hr/UM      +--------------------+



  Resource                     Quantity UM Count  Unit Cost      TOTAL
  -----------------------------------------------------------------------
                                                  Total:  1358e3
                                                  Unit:   1358e3
  MOWSEPER Oil Water Seperator         1.00 EA 1.00    21000    21000.00
+---------------------------------------------------------------------------+
```

```
                          (SPMEFALT) spme final Alternate
                        A C T I V I T Y - D E T A I L   F O R M S


+-33 10 5--------------------Physical Treatment--------------------------------+
|CREW CONFIGURATION Cost Detail                    Source Worksheet:           |
+==============================================================================+
|                                                                              |
|Days        25.000  UM/Da  0.0400   Da/UM  25.000  +---------------------+    |
|Shifts      25.000  UM/Sh  0.0400   Sh/UM  25.000  |>Unit      59503     |    |
|Hours       250.00  UM/Hr  0.0040   Hr/UM  250.00  |>TOTAL     59503     |    |
|Man-hours   1250.0  UM/MH  0.0008   MH/UM  1250.0  +--------- ------------+    |
|                                                                              |
|                                                                              |
|                                                                              |
|Resource    Qty                                  Unit Cost   UM      TOTAL    |
|-----------------------------------------------------------------------       |
|Man-Count:   5.00                                         Total:    59503     |
|Eqp-Count:   5.00                                         Unit:     59503     |
|LTCCI5      1.00   CCI TECHNICIAN 5                27.60    HR     6900.00     |
|LCM5        1.00   CONSTRUCTION MGR. 5             34.62    HR     8655.00     |
|LTCCI3      3.00   CCI TECHNICIAN 3                18.93    HR    14197.50     |
|ECAT960     1.00   CAT 960 WHEEL LOADER            34.00    HR     8500.00     |
|ECRN60      1.00   TRUCK CRANE - 60 TN             35.00    HR     8750.00     |
|EPUTRK      2.00   PICKUP TRUCK                     6.00    HR     3000.00     |
|ECAT215     1.00   CAT 215 EXCAVATOR               38.00    HR     9500.00     |
+------------------------------------------------------------------------------+


+==============================================================================+
+33 10 5 1----------------Electrical Package--------Qty_UM-- 1.00-LS-204820.4--+
|                                                                              |
|A s s i g n m n t s   S c h e d u l e         C o s t  Unit     Total         |
|--------------------   ----------------       ---------------------------     |
|Bid Item Line     1 Duration                  Source    Details  L            |
|Quote Grp    NONE   Bar Start                                                 |
|Qry Lbl 1           Bar Finsh                 Lab Base                        |
|Qry Lbl 2           CPM ES                    Lab Brdn                        |
|Qry Lbl 3           CPM EF                     Eqp Own                        |
|Acnt                CPM LS                     Eqp Op                         |
|                    CPM LF                    Rental                          |
|S h i f t s         Float                     Material                        |
|-------------------                           Supply                          |
|Work Hrs/Shift   10 P r o d u c t i o n       Subconts   204820.4   204820    |
|Pay  Hrs/Shift   10 -------------------       Other                           |
|Shifts/Day      1.0 Days                      Tax RMSO                        |
|Days/Week       5.0 Shifts                    +----------------------------+  |
|                    Hours                     |Total      204820.4   204820|  |
|R a t e s           Man-Hrs                   +----------------------------+  |
|---------           Eqp-Hrs                   > Level 3 204820.4              |
|Rates Unique        UM/Da                                                     |
|                    UM/Sh                              >DBE Goal              |
|                    UM/Hr                              >MBE Goal              |
|                    UM/MH                              >WBE Goal              |
|                    MH/UM                              >O-1 Goal              |
|                                                       >O-2 Goal              |
+------------------------------------------------------------------------------+


ALL PRICES ARE FURNISH AND INSTALLED.
```

```
                    (SPMEFALT) spme final Alternate
                  A C T I V I T Y - D E T A I L   F O R M S


+--33 10 5 1-------------------Electrical Package----------------------------+
|RESOURCE LIST Cost Detail                     Source Worksheet:             |
+============================================================================+
|                                        +-----------------------+          |
|   Man-Hrs             Eqp-Hrs          |>Unit      132325      |          |
|   UM/Man-Hr           UM/Eqp-Hr        |>TOTAL     132325      |          |
|   Man-Hr/UM           Eqp-Hr/UM        +-----------------------+          |
|                                                                           |
|                                                                           |
| Resource                      Quantity UM Count  Unit Cost        TOTAL   |
| ------------------------------------------------------------------------- |
|                                                      Total:    132325     |
|                                                      Unit:     132325     |
|                                                                           |
|U2C4WIRE 2" CONDUIT W/ 3-#1 WIRE    262.00 LF 1.00    20.20       5292.40   |
|U480VPNL 480 VOLT PANEL BOARD         1.00 EA 1.00    10200      10200.00   |
|UMPC     MINI POWER CENTER            8.00 EA 1.00    2100.0     16800.00   |
|U1C3WIRE 1" CONDUIT W/ 2-#8 WIRE    200.00 LF 1.00    11.40       2280.00   |
|UMISCW&E MISC. WIRING AND ELECTRIC    1.00 LS 1.00    32830      32830.00   |
|U2C4WIRE 2" CONDUIT W/ 3-#1 WIRE   3214.00 LF 1.00    20.20      64922.80   |
+---------------------------------------------------------------------------+


+============================================================================+
+33 10 5 2----------------Plumbing Package----------Qty_UM--  1.00-LS-446417.2--+
|                                                                           |
| A s s i g n m n t s   S c h e d u l e        C o s t  Unit    Total        |
| -------------------   ----------------        -------------------          |
|Bid Item Line      1 Duration                 Source   Details  L          |
|Quote Grp    NONE    Bar Start                                             |
|Qry Lbl 1            Bar Finsh                Lab Base                      |
|Qry Lbl 2            CPM ES                   Lab Brdn                      |
|Qry Lbl 3            CPM EF                    Eqp Own                      |
|Acnt                 CPM LS                    Eqp Op                       |
|                     CPM LF                    Rental                       |
| S h i f t s         Float                     Material 151264.0   151264   |
| ------------------                            Supply                       |
|Work Hrs/Shift  10   P r o d u c t i o n       Subconts 287590.0   287590   |
|Pay  Hrs/Shift  10   -------------------        Other                       |
|Shifts/Day     1.0   Days                      Tax RMSO 7563.200     7563   |
|Days/Week      5.0   Shifts                   +---------------------------+ |
|                     Hours                    |Total    446417.2   446417| |
| R a t e s           Man-Hrs                  +---------------------------+ |
| ---------           Eqp-Hrs                   > Level 3 446417.2           |
|Rates Unique         UM/Da                                                  |
|                     UM/Sh                                 >DBE Goal        |
|                     UM/Hr                                 >MBE Goal        |
|                     UM/MH                                 >WBE Goal        |
|                     MH/UM                                 >O-1 Goal        |
|                                                           >O-2 Goal        |
+---------------------------------------------------------------------------+
```

(SPMEFALT) spme final Alternate
A C T I V I T Y - D E T A I L   F O R M S

```
+-33 10 8--------------------Demobilization--------------------------------+
|RESOURCE LIST Cost Detail                      Source Worksheet:          |
+=========================================================================+
|                                          +--------------------+          |
|   Man-Hrs           Eqp-Hrs              |>Unit    2100.0     |          |
|   UM/Man-Hr         UM/Eqp-Hr            |>TOTAL   2100.0     |          |
|   Man-Hr/UM         Eqp-Hr/UM            +--------------------+          |
|                                                                         |
|                                                                         |
|                                                                         |
| Resource                       Quantity UM Count  Unit Cost      TOTAL  |
| ----------------------------------------------------------------------- |
|                                                     Total:     2100.0   |
|                                                     Unit:      2100.0   |
|SMISC200 MISC. SUPPLIES           10.00 EA 1.00      210.00     2100.00  |
+-------------------------------------------------------------------------+
```

```
+-33 10 8--------------------Demobilization--------------------------------+
|CREW CONFIGURATION Cost Detail                 Source Worksheet:          |
+=========================================================================+
|                                          +--------------------+          |
|Days      5.0000  UM/Da 0.2000  Da/UM 5.0000 |>Unit    11068    |        |
|Shifts    5.0000  UM/Sh 0.2000  Sh/UM 5.0000 |>TOTAL   11068    |        |
|Hours    50.000   UM/Hr 0.0200  Hr/UM 50.000 +--------------------+      |
|Man-hours 200.00  UM/MH 0.0050  MH/UM 200.00                            |
|                                                                         |
|                                                                         |
| Resource   Qty                           Unit Cost  UM      TOTAL       |
| ----------------------------------------------------------------------- |
|Man-Count:   4.00                             Total:      11068          |
|Eqp-Count:   4.00                             Unit:       11068          |
|ONSITE     1.00  ONSITE MGMT. CREW           221.37    HR  11068.44      |
+-------------------------------------------------------------------------+
```

```
+=========================================================================+
+-33 10 11----------------Construction Management---Qty_UM--104.00-DA- 3333.40--+
```

A s s i g n m n t s		S c h e d u l e		C o s t Unit		Total
Bid Item Line	1	Duration		Source	Details	
Quote Grp	NONE	Bar Start				
Qry Lbl 1		Bar Finsh		Lab Base	1177.618	122472
Qry Lbl 2		CPM ES		Lab Brdn	977.488	101659
Qry Lbl 3		CPM EF		Eqp Own		
Acnt		CPM LS		Eqp Op		
		CPM LF		Rental	240.096	24970
S h i f t s		Float		Material		
				Supply	20.192	2100
Work Hrs/Shift	10	P r o d u c t i o n		Subconts		
Pay Hrs/Shift	10			Other	869.567	90435
Shifts/Day	1.0	Days	145.6000	Tax RMSO	48.434	5037
Days/Week	5.0	Shifts	145.6000			
		Hours	1456.000	Total	3333.395	346673
R a t e s		Man-Hrs	4777.000			
		Eqp-Hrs	4160.000	> Level 2 346673.1		
Rates Unique		UM/Da	0.7143			
		UM/Sh	0.7143		>DBE Goal	
		UM/Hr	0.0714		>MBE Goal	
		UM/MH	0.0218		>WBE Goal	
		MH/UM	45.9327		>O-1 Goal	
					>O-2 Goal	

```
                         (SPMEFALT) spme final Alternate
                      A C T I V I T Y - D E T A I L   F O R M S


  +-33 10 5 2-------------------Plumbing Package----------------------------------+
  |RESOURCE LIST Cost Detail                        Source Worksheet:             |
  +=====================================================================================+
                                          +---------------------+
     Man-Hrs            Eqp-Hrs           |>Unit    287590      |
     UM/Man-Hr          UM/Eqp-Hr         |>TOTAL   287590      |
     Man-Hr/UM          Eqp-Hr/UM         +---------------------+



   Resource                     Quantity UM Count  Unit Cost       TOTAL
   ---------------------------------------------------------------------------------
                                                      Total:   287590
                                                      Unit:    287590
  UPIPE2   2" BLACK PIPE            3936.00 LF 1.00      16.00   62976.00
  UPIPE2IN 2" BLACK PIPE (INJECTION) 3542.00 LF 1.00    16.00   56672.00
  UPIP2DB1 2" BLACK PIPE (DBL. WALL) 3838.00 LF 1.00    19.00   72922.00
  UPIP2DB2 2" BLACK PIPE (IWTP)      524.00 LF 1.00     19.00    9956.00
  UMISCP&F MISC. PIPE & FITTINGS       1.00 LS 1.00     75000    75000.00
  UPIPE4   4" BLACK PIPE              32.00 LF 1.00      31.00     992.00
  UPIPE15S 1.5" BLACK PIPE (STRIPPER  100.00 LF 1.00    12.00    1200.00
  UPIPE15T 1.5" BLACK PIPE (TREAT.)  656.00 LF 1.00     12.00    7872.00


  +=====================================================================================+
  +33 10 8-------------------Demobilization------------Qty_UM--  1.00-LS-13168.44--+

  A s s i g n m n t s   S c h e d u l e          C o s t  Unit    Total
  -------------------   ----------------         ----------------------------
  Bid Item Line      1 Duration                  Source    Details  LC
  Quote Grp   NONE     Bar Start
  Qry Lbl 1            Bar Finsh                  Lab Base  5008.000     5008
  Qry Lbl 2            CPM ES                     Lab Brdn  4887.440     4887
  Qry Lbl 3            CPM EF                     Eqp Own
  Acnt                 CPM LS                     Eqp Op
                       CPM LF                     Rental    1150.000     1150
  S h i f t s          Float                      Material
  -----------------                              Supply    2000.000     2000
  Work Hrs/Shift    10 P r o d u c t i o n       Subconts
  Pay  Hrs/Shift    10 -------------------        Other
  Shifts/Day       1.0 Days        5.0000        Tax RMSO   123.000      123
  Days/Week        5.0 Shifts      5.0000        +---------------------------+
                       Hours       50.0000       |Total      13168.44    13168|
  R a t e s            Man-Hrs     200.0000      +---------------------------+
  ---------            Eqp-Hrs     200.0000       > Level 2 13168.44
  Rates Unique         UM/Da       0.2000
                       UM/Sh       0.2000                  >DBE Goal
                       UM/Hr       0.0200                  >MBE Goal
                       UM/MH       0.0050                  >WBE Goal
                       MH/UM       200.0000                >O-1 Goal
                                                           >O-2 Goal

  +---------------------------------------------------------------------------------+
```

(SPMEFALT) spme final Alternate
A C T I V I T Y - D E T A I L   F O R M S

```
+-33 10 8--------------------Demobilization------------------------------+
|RESOURCE LIST Cost Detail                    Source Worksheet:          |
+========================================================================+
|                                        +------------------------+      |
|   Man-Hrs            Eqp-Hrs           |>Unit    2100.0          |      |
|   UM/Man-Hr          UM/Eqp-Hr         |>TOTAL   2100.0          |      |
|   Man-Hr/UM          Eqp-Hr/UM         +------------------------+      |
|                                                                        |
|                                                                        |
|                                                                        |
| Resource                     Quantity UM Count  Unit Cost     TOTAL    |
| ---------------------------------------------------------------------- |
|                                                     Total:   2100.0    |
|                                                      Unit:   2100.0    |
|SMISC200 MISC. SUPPLIES          10.00 EA 1.00     210.00     2100.00   |
+------------------------------------------------------------------------+

+-33 10 8--------------------Demobilization------------------------------+
|CREW CONFIGURATION Cost Detail               Source Worksheet:          |
+========================================================================+
|                                        +------------------------+      |
|Days        5.0000  UM/Da 0.2000  Da/UM 5.0000 |>Unit    11068   |      |
|Shifts      5.0000  UM/Sh 0.2000  Sh/UM 5.0000 |>TOTAL   11068   |      |
|Hours       50.000  UM/Hr 0.0200  Hr/UM 50.000 +----------------+      |
|Man-hours   200.00  UM/MH 0.0050  MH/UM 200.00                          |
|                                                                        |
|                                                                        |
| Resource   Qty                            Unit Cost   UM      TOTAL    |
| ---------------------------------------------------------------------- |
|Man-Count:  4.00                               Total:        11068      |
|Eqp-Count:  4.00                                Unit:        11068      |
|ONSITE   1.00  ONSITE MGMT. CREW             221.37    HR  11068.44     |
+------------------------------------------------------------------------+

+========================================================================+
+33 10 11-----------------Construction Management---Qty_UM--104.00-DA- 3333.40--+
|                                                                        |
| A s s i g n m n t s   S c h e d u l e      C o s t   Unit    Total     |
| --------------------   --------------      ------------------------    |
| Bid Item Line      1 Duration              Source    Details          |
| Quote Grp    NONE    Bar Start                                         |
| Qry Lbl 1            Bar Finsh             Lab Base  1177.618   122472 |
| Qry Lbl 2            CPM ES                Lab Brdn   977.488   101659 |
| Qry Lbl 3            CPM EF                Eqp Own                     |
| Acnt                 CPM LS                Eqp Op                      |
|                      CPM LF                Rental    240.096    24970 |
| S h i f t s          Float                 Material                    |
| -----------------                          Supply     20.192     2100 |
| Work Hrs/Shift   10  P r o d u c t i o n   Subconts                    |
| Pay  Hrs/Shift   10  -----------------     Other     869.567    90435 |
| Shifts/Day      1.0  Days      145.6000    Tax RMSO   48.434     5037 |
| Days/Week       5.0  Shifts    145.6000    +------------------------+  |
|                      Hours     1456.000   |Total     3333.395   346673| |
| R a t e s            Man-Hrs   4777.000    +------------------------+  |
| ---------            Eqp-Hrs   4160.000    > Level 2 346673.1          |
| Rates Unique         UM/Da       0.7143                                |
|                      UM/Sh       0.7143         >DBE Goal              |
|                      UM/Hr       0.0714         >MBE Goal              |
|                      UM/MH       0.0218         >WBE Goal              |
|                      MH/UM      45.9327         >O-1 Goal              |
|                                                 >O-2 Goal              |
+------------------------------------------------------------------------+
```

(SPMEFALT) spme final Alternate
A C T I V I T Y - D E T A I L   F O R M S

```
+33 10 11 1-------------Mobe/Demobe--------------Qty_UM-- 30.00-DA- 3324.40--+

A s s i g n m n t s    S c h e d u l e           C o s t  Unit    Total
--------------------   -----------------         --------------------------
Bid Item Line       1 Duration                   Source    Details  LC
Quote Grp    NONE     Bar Start
Qry Lbl 1             Bar Finsh                   Lab Base  1177.553   35327
Qry Lbl 2             CPM ES                       Lab Brdn   977.488   29325
Qry Lbl 3             CPM EF                       Eqp Own
Acnt                  CPM LS                       Eqp Op
                      CPM LF                       Rental     240.000    7200
S h i f t s           Float                        Material
-----------------                                  Supply      20.000     600
Work Hrs/Shift   10   P r o d u c t i o n          Subconts
Pay Hrs/Shift    10   -------------------          Other      861.333   25840
Shifts/Day      1.0   Days        42.0000          Tax RMSO    48.025    1441
Days/Week       5.0   Shifts      42.0000         +--------------------------+
                      Hours      420.0000         |Total     3324.400   99732|
R a t e s             Man-Hrs   1378.000          +--------------------------+
---------             Eqp-Hrs   1200.000          > Level 3  958.961
Rates Unique          UM/Da        0.7143
                      UM/Sh        0.7143                   >DBE Goal
                      UM/Hr        0.0714                   >MBE Goal
                      UM/MH        0.0218                   >WBE Goal
                      MH/UM       45.9333                   >O-1 Goal
                                                           >O-2 Goal
+---------------------------------------------------------------------------+


+-33 10 11 1------------------Mobe/Demobe-----------------------------------+
|RESOURCE LIST Cost Detail                    Source Worksheet:             |
+===========================================================================+

    Man-Hrs    178.00    Eqp-Hrs              +----------------------+
    UM/Man-Hr  0.1685    UM/Eqp-Hr            |>Unit      764.21     |
    Man-Hr/UM  5.9333    Eqp-Hr/UM            |>TOTAL    22926       |
                                              +----------------------+




Resource                      Quantity UM Count  Unit Cost      TOTAL
--------------------------------------------------------------------------
                                                      Total:    22926
                                                      Unit:    764.21
LCM5      CONSTRUCTION MGR. 5    120.00 HR 1.00      34.62     4154.40
LCCIOFF   CCI OFFICE/ADMIN.       48.00 HR 1.00      15.40      739.20
LCM6      CONSTRUCTION MGR. 6     10.00 HR 1.00      38.50      385.00
SMISC100  MISC. SUPPLIES           6.00 EA 1.00     105.00      630.00
UAIRFAR1  Not Found                         1.00
UCAR1     RENTAL CAR               6.00 DA 1.00      50.00      300.00
UPRDIEM1  PER DIEM W/ HOTEL        3.00 DA 1.00      89.25      267.75
UPRDIEM2  PER DIEM W/OUT HOTEL     3.00 DA 1.00      25.00       75.00
UAIRFAR1  Not Found                         1.00
UPRDIEM1  PER DIEM W/ HOTEL      180.00 DA 1.00      89.25    16065.00
UMILEAGE  POV MILEAGE           1000.00 MI 1.00       0.31      310.00
+--------------------------------------------------------------------------+
```

```
                         (SPMEFALT) spme final Alternate
                       A C T I V I T Y - D E T A I L   F O R M S


+-33 10 11 1------------------Mobe/Demobe----------------------------------+
|CREW CONFIGURATION Cost Detail                    Source Worksheet:       |
+=========================================================================+
|                                               +----------------------+  |
|Days          30.000  UM/Da  1.0000    Da/UM  1.0000  |>Unit    2213.7 |  |
|Shifts        30.000  UM/Sh  1.0000    Sh/UM  1.0000  |>TOTAL   66411  |  |
|Hours         300.00  UM/Hr  0.1000    Hr/UM  10.000  +----------------+  |
|Man-hours     1200.0  UM/MH  0.0250    MH/UM  40.000                      |
|                                                                         |
|                                                                         |
|                                                                         |
|Resource    Qty                                   Unit Cost  UM    TOTAL  |
|--------------------------------------------------------------------------
|Man-Count:    4.00                                     Total:    66411    |
|Eqp-Count:    4.00                                     Unit:     2213.7   |
|ONSITE      1.00  ONSITE MGMT. CREW                  221.37   HR  66410.64 |
+-------------------------------------------------------------------------+


+=========================================================================+
+33 10 11 2----------------Operations----------------Qty_UM-- 74.00-DA- 3337.04--+
|                                                                         |
|A s s i g n m n t s    S c h e d u l e        C o s t  Unit    Total     |
|--------------------   ------------------     ---------------------------|
|Bid Item Line       1 Duration                Source    Details  LC      |
|Quote Grp     NONE    Bar Start                                          |
|Qry Lbl 1             Bar Finsh               Lab Base  1177.644   87146  |
|Qry Lbl 2             CPM ES                   Lab Brdn   977.488   72334  |
|Qry Lbl 3             CPM EF                   Eqp Own                    |
|Acnt                  CPM LS                   Eqp Op                     |
|                      CPM LF                   Rental    240.135   17770  |
|S h i f t s           Float                    Material                   |
|-----------------                             Supply     20.270    1500  |
|Work Hrs/Shift   10   P r o d u c t i o n     Subconts                   |
|Pay Hrs/Shift    10   -------------------     Other     872.905   64595  |
|Shifts/Day      1.0   Days        103.6000    Tax RMSO   48.600    3596  |
|Days/Week       5.0   Shifts      103.6000    +--------------------------+
|                      Hours       1036.000    |Total    3337.042  246941 |
|R a t e s             Man-Hrs     3399.000    +--------------------------+
|---------             Eqp-Hrs     2960.000    > Level 3 2374.434         |
|Rates Unique          UM/Da          0.7143                              |
|                      UM/Sh          0.7143           >DBE Goal          |
|                      UM/Hr          0.0714           >MBE Goal          |
|                      UM/MH          0.0218           >WBE Goal          |
|                      MH/UM         45.9324           >O-1 Goal          |
|                                                      >O-2 Goal          |
|                                                                         |
+-------------------------------------------------------------------------+
```

```
                        (SPMEFALT) spme final Alternate
                    A C T I V I T Y - D E T A I L   F O R M S


    +-33 10 11 2------------------Operations-------------------------------+
    |RESOURCE LIST Cost Detail                      Source Worksheet:      |
    +========================================================================+
                                               +--------------------+
         Man-Hrs   439.00    Eqp-Hrs           |>Unit    765.79     |
         UM/Man-Hr 0.1686    UM/Eqp-Hr         |>TOTAL   56668      |
         Man-Hr/UM 5.9324    Eqp-Hr/UM         +--------------------+


         Resource                      Quantity UM Count  Unit Cost    TOTAL
         --------------------------------------------------------------------
                                                        Total:    56668
                                                        Unit:     765.79
         LCM5      CONSTRUCTION MGR. 5   296.00 HR 1.00     34.62    10247.52
         LCCIOFF   CCI OFFICE/ADMIN.     118.00 HR 1.00     15.40     1817.20
         LCM6      CONSTRUCTION MGR. 6    25.00 HR 1.00     38.50      962.50
         SMISC100  MISC. SUPPLIES         15.00 EA 1.00    105.00     1575.00
         UAIRFAR1  Not Found                       1.00
         UCAR1     RENTAL CAR             15.00 DA 1.00     50.00      750.00
         UPRDIEM1  PER DIEM W/ HOTEL       8.00 DA 1.00     89.25      714.00
         UPRDIEM2  PER DIEM W/OUT HOTEL    8.00 DA 1.00     25.00      200.00
         UAIRFAR1  Not Found                       1.00
         UPRDIEM1  PER DIEM W/ HOTEL     444.00 DA 1.00     89.25    39627.00
         UMILEAGE  POV MILEAGE          2500.00 MI 1.00      0.31      775.00
    +------------------------------------------------------------------------+

    +-33 10 11 2------------------Operations-------------------------------+
    |CREW CONFIGURATION Cost Detail                 Source Worksheet:      |
    +========================================================================+
                                                   +--------------------+
    |Days       74.000  UM/Da 1.0000  Da/UM 1.0000 |>Unit    2213.7     |
    |Shifts     74.000  UM/Sh 1.0000  Sh/UM 1.0000 |>TOTAL   163813     |
    |Hours      740.00  UM/Hr 0.1000  Hr/UM 10.000 +--------------------+
    |Man-hours  2960.0  UM/MH 0.0250  MH/UM 40.000


    |Resource   Qty                               Unit Cost  UM     TOTAL
    |--------------------------------------------------------------------
    |Man-Count:  4.00                                    Total:   163813
    |Eqp-Count:  4.00                                    Unit:    2213.7
    |ONSITE      1.00  ONSITE MGMT. CREW              221.37   HR 163812.91
    +------------------------------------------------------------------------+
```

Printed and bound by CPI Group (UK) Ltd, Croydon, CR0 4YY

28/10/2024

01780266-0001